Australian Water Bugs

Their Biology and Identification

(Hemiptera-Heteroptera, Gerromorpha & Nepomorpha)

Nils M. Andersen & Tom A. Weir

Australian Water Bugs

Their Biology and Identification

(Hemiptera-Heteroptera,
Gerromorpha & Nepomorpha)

Entomonograph Volume 14
A series facing global biodiversity in insects

Apollo Books, Denmark

CSIRO PUBLISHING, Australia

2004

Printed by: Vinderup Bogtrykkeri, Vinderup.

Bound by: J. P. Møller Bogbinderi A/S, Haderslev.

Published by:

Apollo Books
Kirkeby Sand 19
DK-5771 Stenstrup
Denmark

Telephone + 45 62 26 37 37
Fax + 45 62 26 37 80
apollobooks@vip.cybercity.dk
www.apollobooks.com

ISBN 87-88757-78-1
ISSN 0106-2808

Published exclusively in Australia and New Zealand by CSIRO Publishing.

National Library of Australia Cataloguing-in-Publication entry

Andersen, Nils M.

Australian water bugs. Their biology and identification (Hemiptera-Heteroptera, Gerromorpha & Nepomorpha).

Bibliography.

Includes index.

ISBN 0 643 09051 7.

1. Hemiptera – Australia. 2. Aquatic insects – Australia – Identification. 3. Semiaquatic bugs – Australia – Identification. I. Weir, Tom A. II. CSIRO. III. Title. (Series: Entomonograph ; v. 14).
595.7540994

Available from:

CSIRO PUBLISHING
150 Oxford Street (PO Box 1139)
Collingwood VIC 3066
Australia
Email: publishing.sales@csiro.au

Telephone: +61 3 9662 7666
Local call: 1300 788 000 (Australia only)
Fax: +61 3 9662 7555

Web site: www.publish.csiro.au

Authors' addresses:

Nils M. Andersen
Professor (Docent) & Senior Curator
Zoological Museum
University of Copenhagen
Universitetsparken 15
DK-2100 Copenhagen Ø.
Denmark
E-mail: nmandersen@zmuc.ku.dk

Tom A. Weir
Senior Curator
Australian National Insect Collection
CSIRO Entomology
GPO Box 1700
Canberra, ACT 2601
Australia
E-mail: tom.weir@csiro.au

Front cover, left: Water strider, Tenagogerris euphrosyne (Gerridae) (photograph by Paul Zborowski); right: Giant water bug, Lethocerus distinctifemur (Belostomatidae) (photograph by David McClenaghan).

Published with support from Carlsbergfondet

Contents

Abstract

This book is an introduction to the fascinating world of water bugs with emphasis on the Australian fauna. Water bugs belong to the order Hemiptera, the largest insect order with incomplete metamorphosis, and include familiar aquatic insects like water striders, water scorpions, water boatmen, and backswimmers. There are basically two kinds: (1) the semiaquatic bugs (Gerromorpha) which live upon the water surface and (2) the true water bugs (Nepomorpha) which live beneath the water surface. Water bugs are found in a wide variety of natural habitats from small, temporary pools to larger ponds and lakes, from small streams to rivers, and from inland freshwater bodies to coastal mangroves, tidal pools on coral reefs, and the surface of the ocean. Water bugs are chiefly predators or scavengers, feeding on any prey they can master, from tiny crustaceans and insects to tadpoles and small fish. They play a major role in aquatic ecosystems and may serve as indicators of the biological quality of aquatic habitats. They are chiefly beneficial to man since many species prey on mosquitoes and are themselves preyed upon by fish.

The fauna of water bugs of Australia is large and diverse. Australia is currently known to have representatives of 6 families, 7 subfamilies, 30 genera, and 129 described species of semiaquatic bugs (Gerromorpha), and 9 families, 10 subfamilies, 24 genera, and 132 described species of true water bugs (Nepomorpha). The water bug fauna of Australia is also unique in several respects. First, a high percentage of Australian semiaquatic species are marine (c. 24% compared to c. 10% world-wide). Second, Australia houses several forms a great importance in understanding the evolutionary history of water bugs. Third and finally, the Australian fauna includes many groups that probably originated and diversified within the continent itself rather than invaded the continent from adjacent regions.

This handbook is the first comprehensive guide facilitating the identification of Australian water bugs. It provides an overview of all 15 families, 17 subfamilies, and 54 genera known to occur on mainland Australia, Tasmania and nearby islands. Illustrated keys, featuring a minimum of technical language, are offered to assist with the identification of adult water bugs. For each genus, the handbook includes a description of the characters used to identify the genus and to separate the genus from similar genera, usually an illustration to show overall appearance ("habitus") of a representative species, an illustrated key to species recorded from Australia, an overview of the biology of the genus, a summary of the more important publications dealing with the genus, and a map showing the locations where the genus has been found. In addition, introductory chapters give overviews of the distribution, ecology, biology, classification, phylogeny of water bugs, methods for collecting and preserving water bugs, and explain the terms used in the keys and generic discussions. A check list of Australian water bugs with their distribution is appended.

Acknowledgements

The work leading to this book was initiated by a grant from the Australian Biological Resources Study (ABRS), Canberra, in 1992-1994, and completed within the Australian National Insect Collection, CSIRO Entomology (ANIC), Canberra, and the Zoological Museum, University of Copenhagen, Denmark (ZMUC). The authors are grateful to CSIRO Entomology, Canberra, for their support over the years with this project and in particular for their hospitability enjoyed by the first authors during several visits. We thank Jean Just, former director at Australian Biological Resources Study (ABRS), Canberra, for suggesting the project 'Semiaquatic bugs of Australia' to us. The late Tom Woodward, formerly of University of Queensland, first suggested to the second author that Veliidae would be a good family to work on and acted as a mentor in the late 1960's. The late Tim Angeles provided logistical support for collecting in the Northern Territory while the second author was stationed in Darwin from 1971 to 1974. We are further indebted to Paul Zborowski and Alison Roach for assistance with field work in 1992 and 1994, to Ben Boyd and Eric Hines, CSIRO Entomology, Canberra, for assistance with the SEM work, and to Natalie Barnett, Ben Boyd and David McClenaghan, CSIRO Entomology, Canberra, and Gert Brovad, ZMUC, Copenhagen, for photographical assistance. Birgitte Rubaek, ZMUC, assisted with the preparation of some of the habitus figures, and Martin B. Hebsgaard, Zoological Institute, Copenhagen University, prepared other figures for the book. Gerry Cassis, Australian Museum, Sydney, kindly provided the illustrations for the Gelastocoridae and Ivor Lansbury, Hope Museum of Zoology, University of Oxford, U.K. kindly permitted us to use figures from his numerous publications about Australian water bugs. We appreciate the permissions given by Cooperative Research Centre for Freshwater Ecology, Thurgoona, N.S.W., CSIRO Publishing, Melbourne, and the editors of the journal Invertebrate Taxonomy/ Invertebrate Systematics, for permission to reproduce figures from their publications. We also thanks John Gooderham, School of Zoology, University of Tasmania, Hobart, Edward Tsyrlin, Water Studies Centre, Faculty of Science, Monash University, Melbourne, and Paul Zborowski, CLOSE-UP Photo Library, Kuranda, for providing numerous images of Australian water bugs for the colour-plates of this book.

Thanks are due to the following persons for permission to study specimens under their care: R.W. Brooks, Snow Entomological Museum, University of Kansas, Lawrence, Kansas, U.S.A.; G. R. Brown and A. Wells, Northern Territory Museum, Darwin; G. Cassis, Australian Museum, Sydney; G. Cockayne and A. Steward, Queensland Department of Natural Resources and Mines, Rocklea; J. Forrest, South Australian Museum, Adelaide, S.A.; R. C. Froeschner and D. A. Polhemus, National Museum of Natural History, Smithsonian Institution, Washington, D.C., U.S.A.; T. Houston, Western Australian Museum, Perth; I. Lansbury, Hope Museum of Zoology, Oxford, U.K.; G. B. Monteith, Queensland Museum, Brisbane; D. A. Polhemus, Bernice P. Bishop Museum, Honolululu, U.S.A.; J. T. Polhemus, Englewood, Colorado, U.S.A.; T. Postle and K. Richards, Western Australian Department of Agriculture, Perth; M. Scanlon, Western Australian Department of Conservation and Land Management, Woodvale; M. A. Schneider, University of Queensland Insect Collection, Brisbane; K. Walker, Museum of Victoria, Melbourne; M. Webb, The Natural History Museum, London, U.K; and B. Wiklund and the late P. Lindskog, Swedish Museum of Natural History Museum, Stockholm, Sweden.

This work is part of a project supported by grants from the Australian Biological Resources Study, Canberra, and the Danish Natural Science Research Council, Copenhagen (Grants nos. 11-0090, 9502155, and 9801904), to the first author. The cost associated with the SEM work was partly covered by a grant from the ANIC Fund, Canberra. The publication of this book was supported by a grant from the Carlsberg Foundation, Copenhagen.

This work is dedicated to Ebbe S. Nielsen, former director of ANIC, CSIRO Entomology, Canberra. Ebbe was a close friend and although water bugs were not his favourite group of insects (which were butterflies and moths), his enthusiasm and encouragement was a great inspiration for us in continuing our work towards completing this book. His untimely passing in March 2001 was a deep loss and will be felt for a long time to come.

For permissions to reproduce illustrations we acknowledge:

© Nauman, I. D. (Ed.): *The Insects of Australia. A textbook for students and research workers (1991)*, published by CSIRO Entomology & Melbourne University Press, Melbourne. - Figs 13.2, 13.11A, 14.1A, 14.4A, 15.2, 16.1A, 17.1A, 18.1A, 19.1, 20.8A, and 21.1A:

© Edward Tsyrlin & John Gooderham:
The Waterbug Book: A Guide to the Freshwater Macroinvertebrates of Temperate Australia (2002), published by CSIRO PUBLISHING, Melbourne. - Fig. 5.1G; Plate 6, B-E, I, K; Plate 7, F-H; and Plate 8, A-E, G, J:

© Paul Zborowski, CLOSE-UP Photo Library, Kuranda. - Front cover (left), Plate 2, B; Plate 3, A-B; Plate 6, A, F-H, J; Plate 7, A, F, H-I; and Plate 8 F, H-I.

© David McClenaghan, CSIRO Entomology, Canberra. - Front cover (right), Plate 5, A-I and Plate 7, B.

1. Introduction

Water bugs are familiar insects in aquatic habitats throughout the World. They belong to the order Hemiptera, the largest insect order with incomplete metamorphosis. There are basically two kinds: (1) the semiaquatic bugs (Gerromorpha) which live upon the water surface (Fig. 1.1, Plate 6) and (2) the true water bugs (Nepomorpha) which live beneath the water surface (Fig. 1.2, Plates 6-8). Water bugs are found in a wide variety of natural habitats from small, temporary pools to larger ponds and lakes, from small streams to rivers, and from inland freshwater bodies to coastal mangroves and tidal pools of coral reefs (Plates 1-4). Also included are the sea skaters (*Halobates*), the only insects which have successfully colonised the open ocean. Water bugs are chiefly predators or scavengers, feeding on any prey they can master, from tiny crustaceans and insects to tadpoles and small fish. They play a major role in aquatic ecosystems and may serve as indicators of the biological quality of aquatic habitats. They are chiefly beneficial to man since many species prey on mosquitoes and are themselves preyed upon by fish. Because of their diverse lifestyles and because they are easily observed in their natural habitats, water bugs are excellent model organisms in evolutionary biology, ecology, and conservation biology. Adaptations for prey catching, for respiration and for locomotion show such a bewildering diversity in water bugs that their pioneer British student, G. W. Kirkaldy (1873-1910), called them "the most fascinating, morphologically and biologically, of all the Hemiptera - that is to say, of all animals".

The first water bug to be recorded from Australia was a water strider named *"Cimex"* [currently *Limnometra*] *cursitans* by the Danish entomologist Iohan Christian Fabricius in 1775. He studied material from "Nova Hollandia" (an old name for Australia) collected by the British naturalist Joseph Banks during the voyage of Captain James Cook's ship "Endeavour" along the east coast of Australia in 1770. During the 19th century, additional Australian water bugs were described by Erichson (1842), Fieber (1951), Stål (1854-1876), Dufour (1863), Kirkaldy (1887-1899), Skuse (1891-1893), Carpenter (1892), Bergroth (1893), and Montandon (1898-1899). During the early parts of the 20th century, Horvath (1902), Kirkaldy (1902), Hale (1922-1926), Lundblad (1933), Hungerford & Evans (1934), and Jaczewski (1939) added several new species from Australia. More recently Brooks (1951), Todd (1955-1960), Lansbury (1964-1995), Baehr (1989-1990), J. Polhemus & D. Polhemus (1982-1996), Andersen & Weir (1994-2003), and Cassis & Silveira (2001, 2002) have made important contributions to the knowledge of the Australian water bug fauna.

This handbook is the first comprehensive guide facilitating the identification of Australian water bugs. It provides an overview of all 15 families, 17 subfamilies, and 54 genera known to occur on mainland Australia, Tasmania and nearby islands. Illustrated keys, featuring a minimum of technical language, are offered to assist with the identification of adult water bugs. For each genus, the handbook includes a description of the characters used to identify the genus and to separate the genus from similar genera, usually an illustration to show overall appearance ("habitus") of a representative species, an illustrated key to species recorded from Australia, an overview of the biology of the genus, a summary of the more important publications dealing with the genus, and a map showing the locations where the genus has been found. In addition, introductory chapters give overviews of the distribution, ecology, biology, classification, phylogeny of water bugs, methods for collecting and preserving water bugs, and explain the terms used in the keys and generic discussions. A check list of Australian water bugs with their distribution is appended.

Water bugs in Australia

The waterbug fauna of Australia is large and diverse. Australia is currently known to have representatives of 6 families, 7 subfamilies, 30 genera, and 129 described species of semiaquatic bugs (Gerromorpha), and 9 families, 10 subfamilies, 24 genera, and 132 described species of true water bugs (Nepomorpha). World-wide, there are 8 families, 18 subfamilies, 157 genera, and about 1820 described species of Gerromorpha and 10 families, 22 subfamilies, 109 genera, and about 1980 species of Nepomorpha. Thus Australia currently has representatives of about half of the world's families and subfamilies of water bugs, about 18.5% of its genera, and as far as we know about 7% of its species. 14 of the genera found in Australia occur nowhere else (i.e. are *endemic* to Australia), and some additional genera are shared with only our closest neighbours. Most of the species of water bugs are only known from Australia, with a minority occurring in both Australia and neighbouring areas, in particular New Guinea.

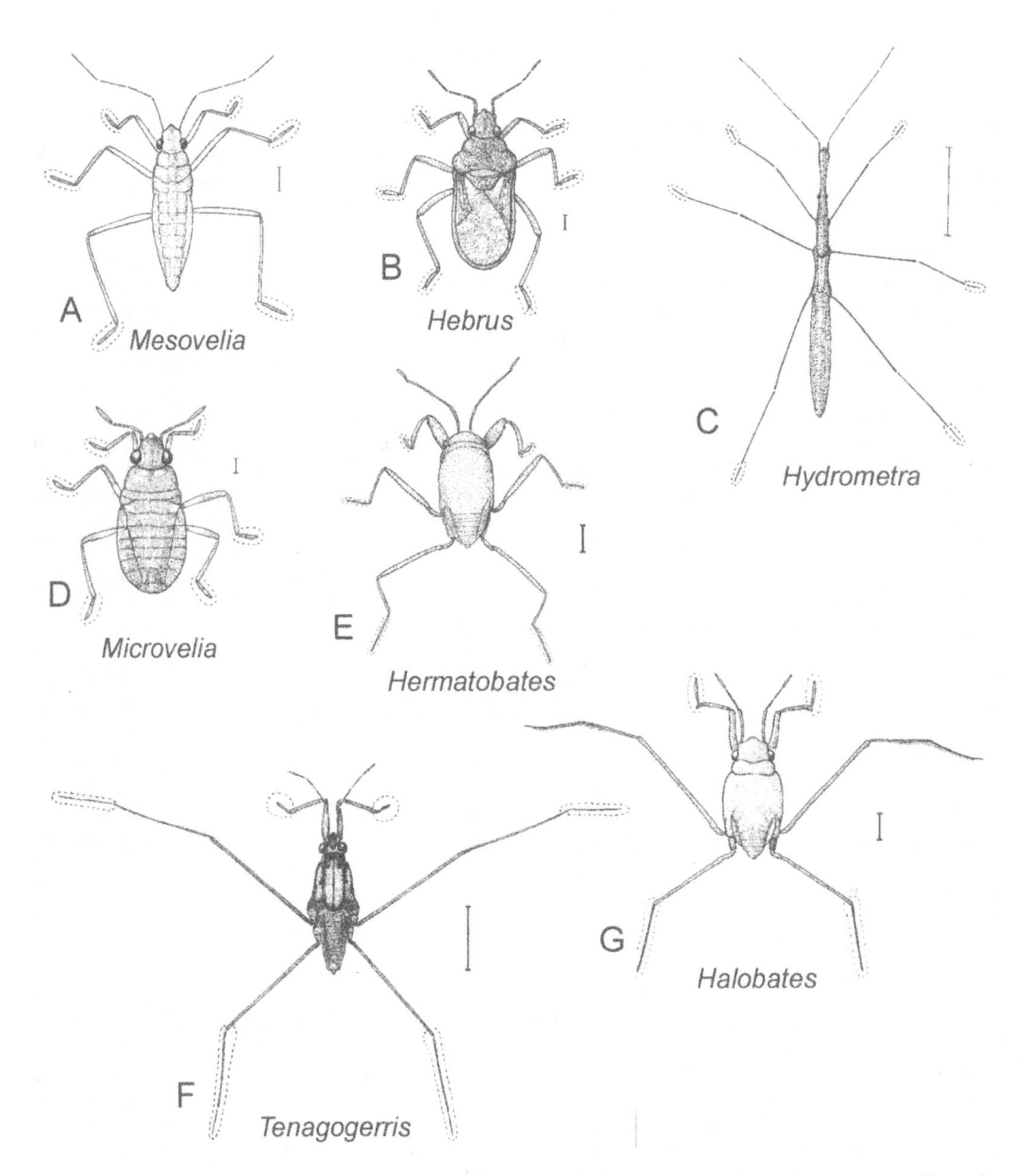

Figure 1.1. A-G, semiaquatic bugs (Gerromorpha) in typical postures resting on a water surface: A, *Mesovelia* (Mesoveliidae); B, *Hebrus* (Hebridae); C, *Hydrometra* (Hydrometridae); D, *Microvelia* (Veliidae). E, *Hermatobates* (Hermatobatidae). F, *Tenagogerris* (Gerridae-Gerrinae). G, *Halobates* (Gerridae-Halobatinae). Scale bars indicate natural size. A-D, reproduced with modification from Andersen (1982); E-F, original.

Australia has overall fewer genera and species of water bugs than any other of the worlds major regions. Australia has about the same number of genera (but fewer species) of semiaquatic bugs (Gerromorpha) as Europe and northern Africa but a higher number of genera (and about the same number of species) than North America. There are about two thirds the number of genera and about half the number of species of true aquatic bugs (Nepomorpha) in Australia as compared to both regions of the northern hemisphere (Henry & Froeschner 1988; Aukema &

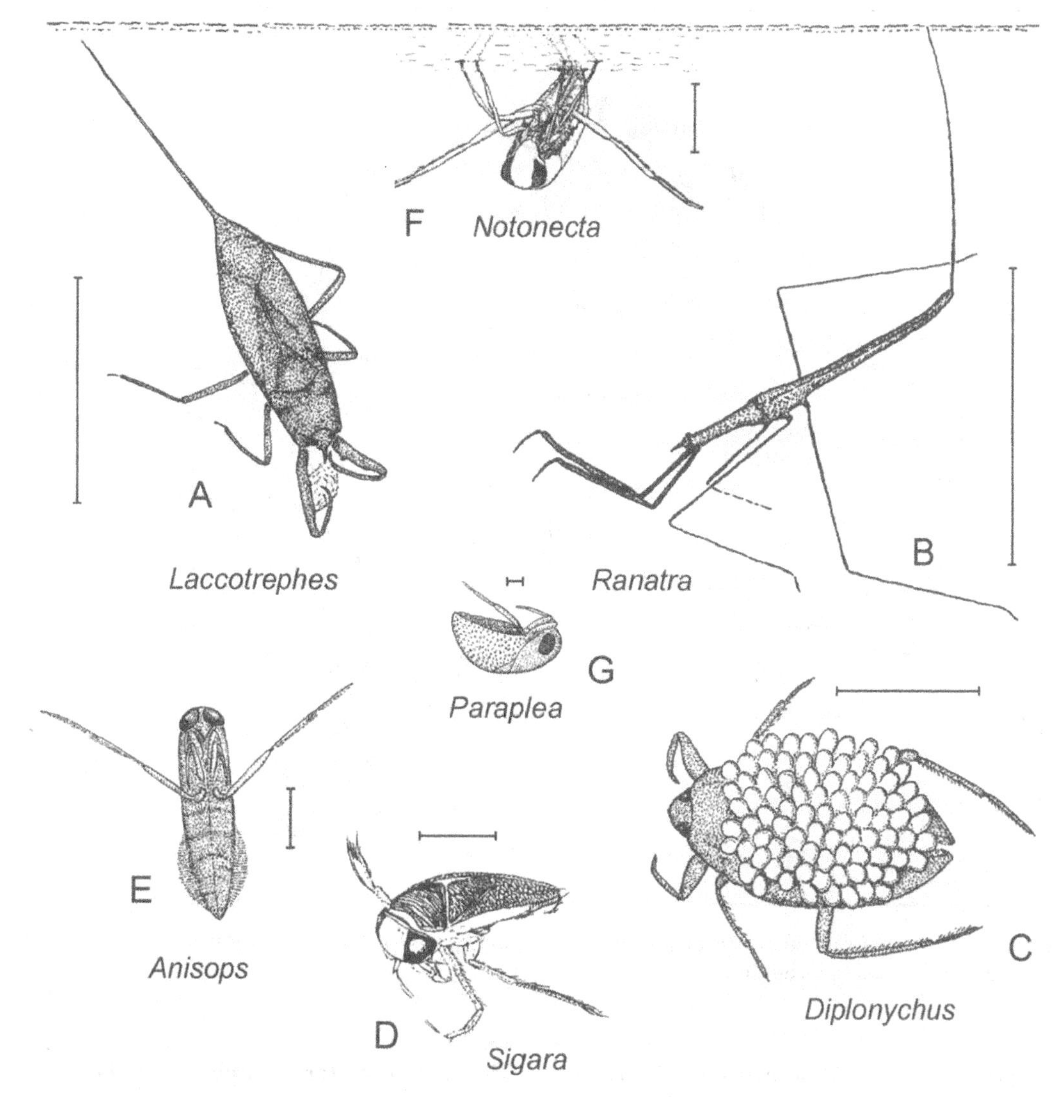

Figure 1.2. A-G, aquatic bugs (Nepomorpha) in typical postures under water: A, *Laccotrephes* (Nepidae-Nepinae). B, *Ranatra* (Nepidae-Ranatrinae). C, *Diplonychus*; male with egg batch on its back (Belostomatidae). D, *Sigara* (Corixidae). E, *Anisops* (Notonectidae-Anisopinae). F, *Notonecta* (Notonectidae-Notonectinae). G, *Paraplea* (Pleidae). Scale bars indicate natural size. D, reproduced from Jansson (1996); A-C, E-G, original.

Rieger 1995). The Oriental region (Asia, from Pakistan to New Guinea) and the Neotropical region (Central and South America) have much higher numbers of genera and species of both semiaquatic and true aquatic bugs than Australia, with the Afrotropical region following next.

The water bug fauna of Australia is unique in several respects. First, a high percentage of Australian semiaquatic species are marine (c. 24% compared to c. 10% world-wide), living in coastal habitats such as mangrove swamps, intertidal coral reef flats, or on the sea surface. Overall, Australia has the most diverse fauna of marine water striders in the World, and studies of the Australian fauna have significantly improved our understanding of the evolution and biogeography of such groups (e.g., Andersen & Weir 1994a, 1994b). Second, Australia houses several forms a great importance in understanding the evolutionary history of water bugs, including the

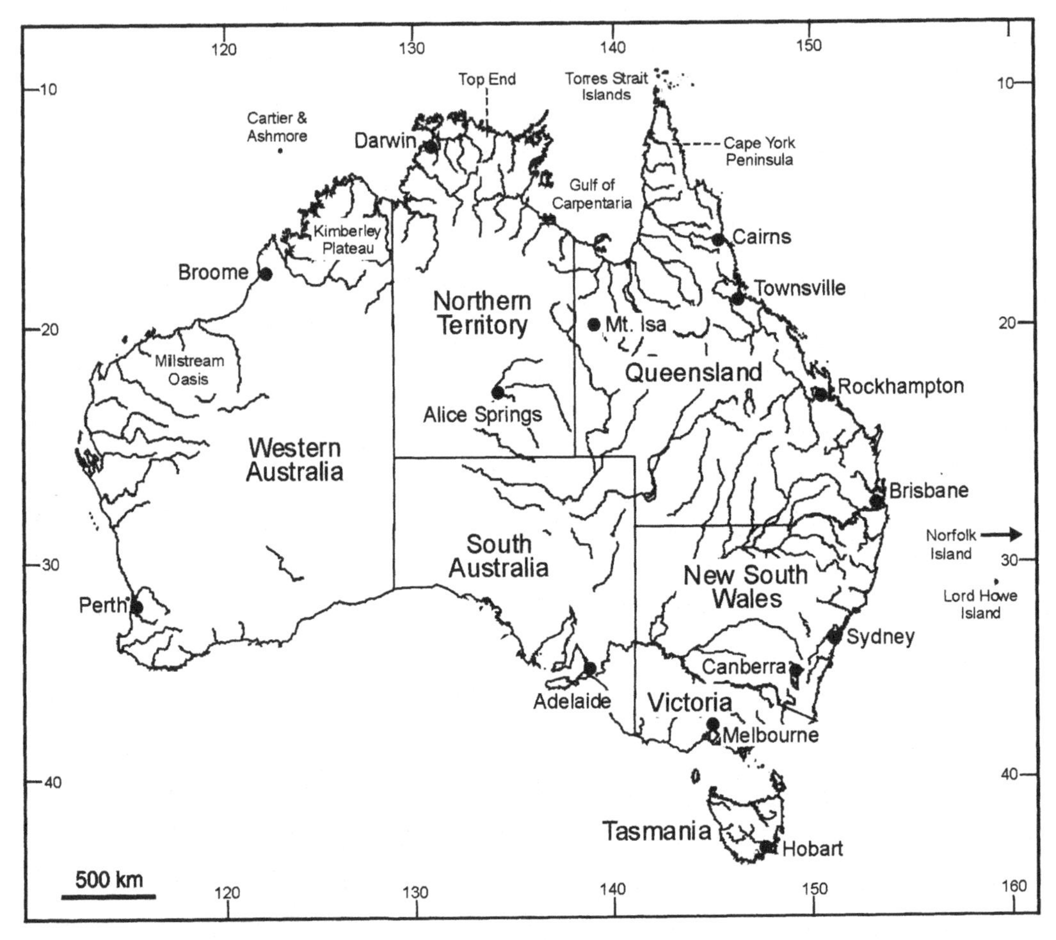

Figure 1.3. Map of Australia with major cities and other geographical places indicated for reference throughout this book. Original.

genera *Austrovelia* (Mesoveliidae), *Hermatobates* (Hermatobatidae), *Drepanovelia, Lacertovelia, Petrovelia, Phoreticovelia* (Veliidae), *Rhagadotarsus, Rheumatometra, Tenagogerris* (Gerridae), *Austronepa, Goondnomdanepa* (Nepidae), *Diaprepocoris* (Corixidae), and the genus *Megochterus* (Ochteridae). Third and finally, the importance of Australia in biogeographical studies recognised for centuries by botanists and vertebrate zoologists may also apply to insects including water bugs (Cranston & Nauman 1991). The Australian fauna includes many groups that probably originated and diversified within the continent itself rather than invaded the continent from adjacent regions, in particular the Oriental region. Backed by well corroborated phylogenetic hypotheses, studies of the distribution of such groups may help us understanding how the present Australian fauna came about.

Distribution patterns within Australia

Australia is the driest of all continents and fresh water is generally rare, particularly in the vast inland areas. This fact is of course reflected in the distribution of aquatic insects including water bugs which are most abundant in areas with a relatively high annual rainfall (Fig. 1.4). Overall, the largest diversity of species is found in the temperate and subtropical, southeastern coastal areas of Queensland, New South Wales, and Victoria, and tropical northeastern Queensland and northern coastal areas of Northern Territory and Western Australia. Much less diverse is the fauna of Tasmania, South Australia, and southwestern Western Australia. Water bugs are very sparsely distributed in the southern part of Northern Territory (except around Alice Springs) and most of South Australia and Western Australia.

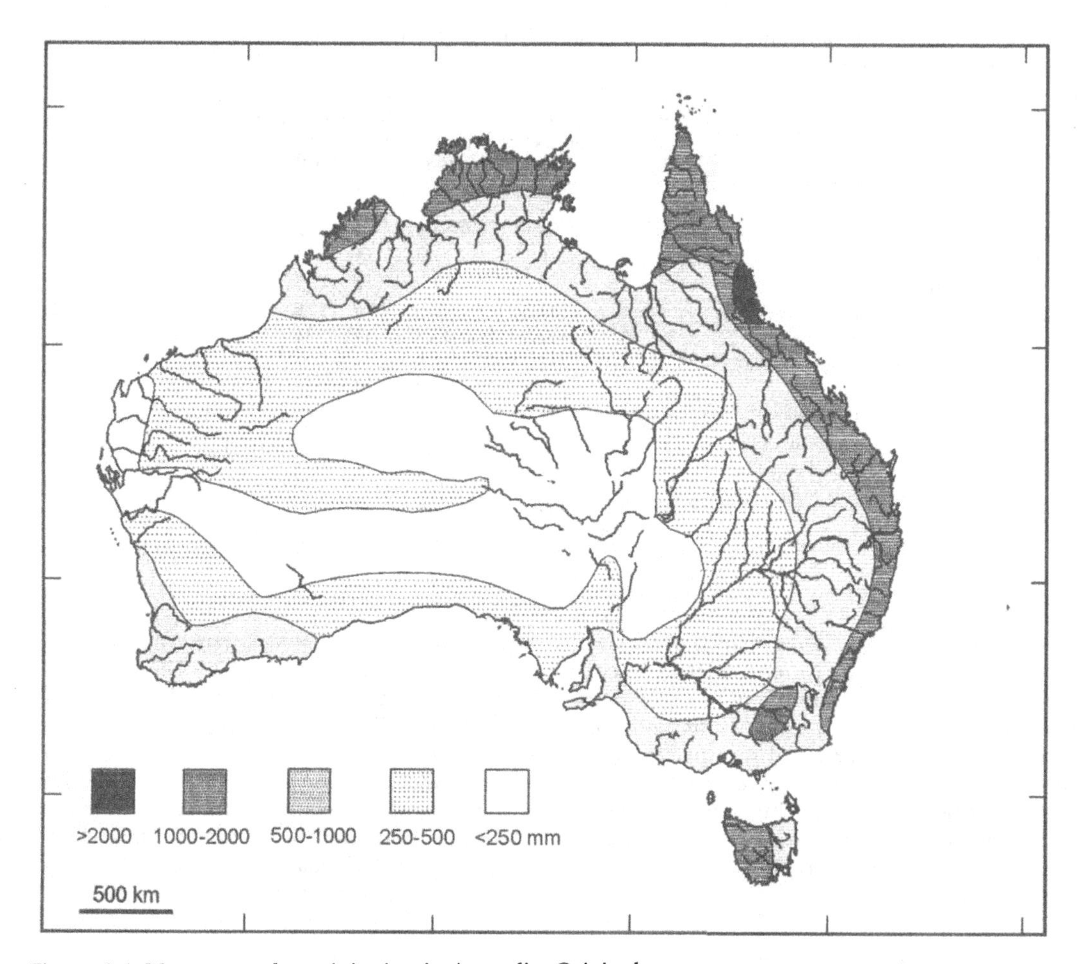

Figure 1.4. Mean annual precipitation in Australia. Original.

Tables 1-2 show the numbers of genera and species of Australian water bugs and their distribution by states. Queensland has by far the richest fauna, followed by Northern Territory, Western Australia, and New South Wales.

The classical biogeographical zones of Australia, the Toressian, Bassian, and Eyrean zones only have limited use in categorising the distribution of water bugs within the continent. Instead, we can group species which have similar ranges of distribution in the following categories:

(A) *Southeastern distribution,* covering the southern, coastal parts of Queensland, New South Wales, Victoria, and (eventually) adjacent parts of South Australia and Tasmania. Gerromorphans with this distribution are: *Mesovelia hackeri* (Mesoveliidae), *Drepanovelia* spp. (except for one species in northern Queensland), *Lacertovelia hirsuta,* the *Microvelia* (*Austromicrovelia*) *fluvialis* group (Veliidae), and *Rheumatometra* spp. (Gerridae). Nepomorphans with this distribution are: *Diaprepocoris barycephala,* several *Agraptocorixa* spp., and *Sigara* (*Tropocorixa*) spp. (Corixidae). Few species, like *Nerthra plauta* (Gelastocoridae), are restricted to South Australia. Almost no gerromorphan bugs but several nepomorphans are endemic to Tasmania, e.g., *Diaprepocoris pedderensis, Sigara* (*Tropocorixa*) *neboissi, S. tasmaniae* (Corixidae), *Nerthra suberosa, N. tasmaniensis* (Gelastocoridae), *Anisops evansi,* and *A. tasmaniaensis* (Notonectidae).

(B) *Northeastern distribution,* covering the northeastern parts of Queensland. Here belong many species and species groups, e.g., belonging to the gerromorphan genera *Halovelia, Xenobates, Microvelopsis,* and *Rhagovelia* (Veliidae), *Tenagogonus, Austrobates,* and *Halobates,* and the nepomorphan genera *Cnethocymatia* (Corixidae),

Table 1. Numbers of genera and species of semiaquatic bugs (Hemiptera, Gerromorpha) in Australia and their distribution on states.

Family	Subfamily	Nos of genera	Nos of species	Distribution in Australia							
				ACT	NSW	NT	QLD	SA	TAS	VIC	WA
MESOVELIIDAE		2	7		2	5	7	1	1	1	4
HEBRIDAE		3	6		3	4	4	1	1	2	4
HYDROMETRIDAE		1	9	1	4	5	9	1	1	1	4
HERMATOBATIDAE		1	2			2	1				1
VELIIDAE	Haloveliinae	2	13			6	8				2
VELIIDAE	Microveliinae	8	53	2	19	19	36	2	4	7	12
VELIIDAE	Rhagoveliinae	1	1				1				
GERRIDAE	Gerrinae	5	13	2	2	9	10	2		2	6
GERRIDAE	Halobatinae	2	15		4	9	12				6
GERRIDAE	Rhagadotarsinae	1	1		1	1	1				1
GERRIDAE	Trepobatinae	4	7	1	2	4	3			2	0
Total		30	127	6	37	64	92	7	7	15	40

Ochterus (Ochteridae), *Nerthra* (Gelastocoridae), *Anisops*, and *Enithares* (Notonectidae). Several species extend their distribution to cover both (A) and (B) like the *Microvelia* (*Austromicrovelia*) *mjobergi* group (Veliidae), and the species *Aquarius antigone* and *Tenagogerris euphrosyne* (Gerridae), *Ochterus eurythorax* (Ochteridae), species of the *Nerthra elongatus* group (Gelastocoridae), and several species of the genera *Anisops* and *Enithares* (Notonectidae).

(C) *Northern distribution*, covering tropical northeastern Queensland, the "Top End" of Northern Territory, and the Kimberley district of Western Australia. Essentially the classical Torresian zone of Australia. Gerromorphan bugs with this type of distribution are, e.g., several species of *Mesovelia* (Mesoveliidae), the *Hydrometra feta* group (Hydrometridae), *Hermatobates* spp. (Hermatobatidae), *Halovelia hilli*, several species of *Microvelia* (*Austromicrovelia*), *M.* (*Picaultia*) (Veli-

Table 2. Numbers of genera and species of aquatic bugs (Hemiptera, Nepomorpha) in Australia and their distribution on states.

Family	Genus	Nos of genera	Nos of species	Distribution in Australia							
				ACT	NSW	NT	QLD	SA	TAS	VIC	WA
BELOSTOMATIDAE	Belostomatinae	1	2		1	2	2	1	1	1	1
BELOSTOMATIDAE	Lethocerinae	1	2		1	2	2				
NEPIDAE	Nepinae	1	1		1	1	1	1	1	1	1
NEPIDAE	Ranatrinae	4	8	1	1	5	4	2	1	2	5
CORIXIDAE	Corixinae	2	13		8	1	6	7	4	6	6
CORIXIDAE	Cymatiainae	1	1				1				
CORIXIDAE	Diaprepocorinae	1	3		1		1	1	2	1	1
CORIXIDAE	Micronectinae	1	18	3	8	4	10	4	3	6	5
APHELOCHEIRIDAE		1	1			1	1				
NAUCORIDAE		1	6	1	1	3	3	1	1	1	2
OCHTERIDAE		2	13	2	3	3	8	4	3	3	5
GELASTOCORIDAE		1	24	2	8	3	9	1	2	5	6
NOTONECTIDAE	Anisopinae	3	35	4	11	15	20	9	7	9	17
NOTONECTIDAE	Notonectinae	3	7	2	2	4	4	2	1	2	3
PLEIDAE		1	3		1	2	1	2	1	1	1
Total		24	137	15	47	46	73	35	27	38	53

idae), all species of the genera *Limnogonus* and *Limnometra*, and *Rhagadotarsus anomalus* (Gerridae). Nepomorphan bugs with this distribution are *Austronepa angusta* (Nepidae), and several species of *Naucoris* (Naucoridae), *Anisops*, *Enithares*, and *Nychia* (Notonectidae). Other species, like *Aquarius fabricii*, *Tenagogerris femoratus*, *Calyptobates minimus* (Gerridae), *Goondnomdanepa* (Nepidae), and several species of *Anisops* (Notonectidae) are confined to the "Top End" of Northern Territory and northwestern part of Western Australia.

(D) *Southwestern distribution*, covering the southwestern parts of Western Australia. Remarkably, no semiaquatic bugs are exclusively found in this area whereas several true water bugs are, e.g., *Diaprepocoris personatus*, *Agraptocorixa gambrei*, *Sigara* (*Tropocorixa*) *mullaka* (Corixidae), *Megochterus occidentalis*, *Ochterus occidentalis* (Ochteridae), *Nerthra adspersa*, *femoralis*, *hirsuta*, *stali*, *tuberculata* (Gelastocoridae), *Anisops baylii*, *Enithares gwini*, *Notonecta* (*Enitharonecta*) *handlirschi*, and *Paranisops endymion* (Notonectidae).

A few species of water bugs are more or less widespread in Australia, found in all areas where there are suitable freshwater bodies, either stagnant or flowing. In the Gerromorpha these species are *Mesovelia hungerfordi* (Mesoveliidae), *Hebrus axillaris* (Hebridae), *Hydrometra strigosa* (Hydrometridae), *Microvelia* (*Austromicrovelia*) *peramoena*, and *M.* (*Pacificovelia*) *oceanica*. In the Nepomorpha the species are *Laccotrephes tristis*, *Ranatra dispar* (Nepidae), *Anisops deanei*, *A. thienemanni*, and *Enithares woodwardi* (Notonectidae).

Which factors are responsible for the actual distribution of Australian water bugs? *Ecological biogeographers* try to explain the distribution of a particular species on the basis of knowledge about the range of optimal conditions (physical and climatic) for that species. Relatively few water bugs are widely ecologically tolerant (*eurytopic*) whereas most species require more or less specific types of habitats and living conditions. This as exemplified by Australian water striders belonging to the subfamily Gerrinae (Fig. 1.5). Andersen & Weir (1997) arranged species in a sequence after the durational stability of their breeding habitats. The most stable habitats (rivers, streams) are occupied by *Aquarius antigone*, *A. fabricii*, *Tenagogonus australiensis*, *Limnometra cursitans*, *L. lipovskyi*, *Tenagogerris femoratus*, and *Limnogonus windi*. The most temporary habitats (shallow, temporary pools and swamps) are occupied by *Limnogonus fossarum gilguy*, *L. luctuosus*, *L. hungerfordi*, *Tenagogerris euphrosyne*, and *T. pallidus*. Likewise, the ecological distribution of marine water striders (Gerridae, Hermatobatidae, and Veliidae), can be placed on an idealised transect of habitats, from limnic (rivers) through estuaries, mangrove swamps, and intertidal coral reef flats, to the sea (Fig. 1.6). The sea skaters, *Halobates*, can be found in almost every type of habitat, whereas the mangrove bugs, *Xenobates*, are confined to mangroves, and the coral bugs,

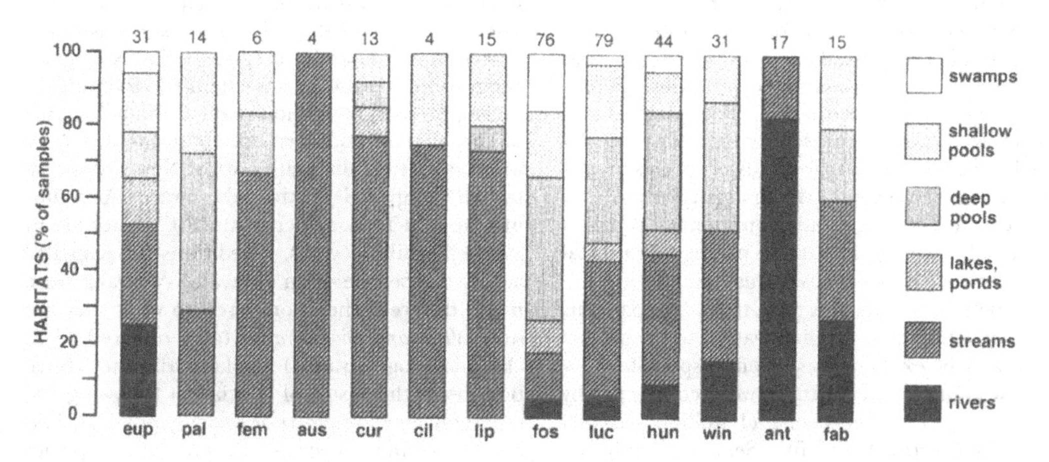

Figure 1.5. Distribution on habitat types of Australian gerrine water striders shown as frequencies of total nos. of samples (listed on top of species columns). Abbreviations: ant, *Aquarius antigone*; aus, *Tenagogonus australiensis*; cil, *Limnometra ciliodes*; cur, *L. cursitans*; eup, *Tenagogerris euphrosyne*; fab, *Aquarius fabricii*; fem, *Tenagogerris femoratus*; fos, *Limnogonus fossarum gilguy*; hun, *L. hungerfordi*; lip, *Limnometra lipovskyi*; luc, *Limnogonus luctuosus*; pal, *Tenagogerris pallidus*; win, *Limnogonus windi*. Reproduced from Andersen & Weir (1997).

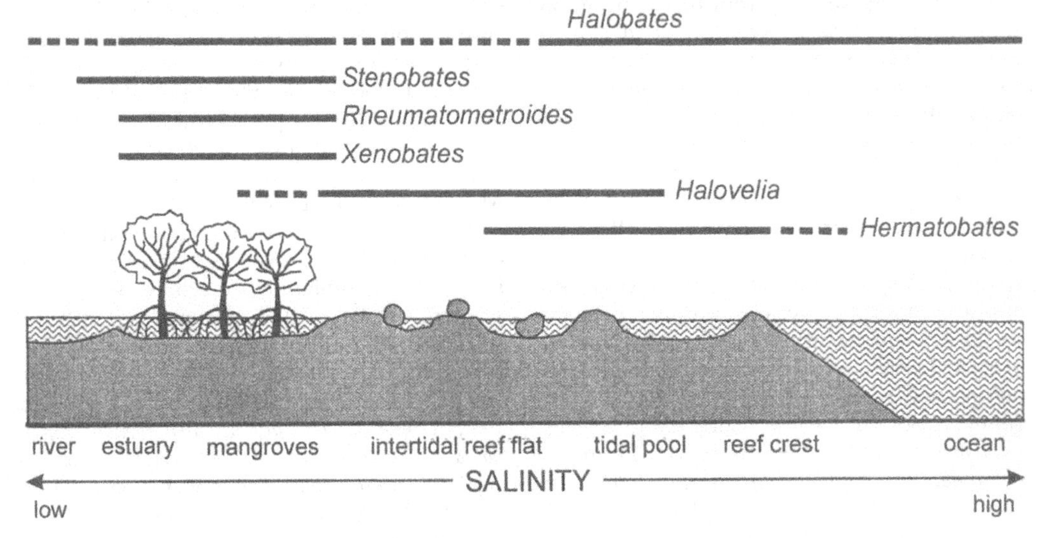

Figure 1.6. Habitats of marine water striders of Australia on an idealised transect from rivers to open ocean. Reproduced with modification from Andersen & Weir (1999).

Halovelia, to intertidal reef flats. However, although environmental factors play an important role in constraining the distribution of water bugs, historical factors are equally important in determining more fundamental patterns (Cranston & Naumann 1991).

For a long time it was a common belief among biogeographers that most of the present day flora and fauna of Australia had originated somewhere else and later colonised the continent. The ideas of *dispersal biogeographers* were prevailing before the general acceptance of the role of plate tectonics in shaping the present geography on the earth. Any disjunct distributions of species or species groups led dispersalists to argue for mobility of those taxa between static continents. Even with present knowledge about the paleogeographical changes that have affected Australia during the past 60 million years, it is true, that some patterns of distribution of Australian water bugs can most profitably be explained as recent dispersal.

New Guinea and Australia were connected by land during the Pleistocene (1 million - 10,000 years ago) across the Arafura Sea, facilitating the exchange of fauna and flora between these areas. There can be little doubt that Australia has received several species of water bugs by dispersal across this land bridge, in particular species that are now widespread in the Indo-Malayan region, such as *Mesovelia horvathi* and *M. vittigera* (Mesoveliidae), *Hydrometra orientalis*, *H. papuana* (Hydrometridae), *Limnogonus fossarum* and *L. hungerfordi* (Gerridae), *Microvelia (Picaultia) douglasi* (Veliidae), *Laccotrephes tristis* (Nepidae), *Micronecta quadristrigata* (Corixidae), *Anisops nodulatus*, *A. philippinensis*, *A. stali*, *A. tahitiensis*, *Nychia sappho* (Notonectidae), and *Paraplea liturata* (Pleidae). Most of these species are now confined to the tropical northern parts of Australia.

It is now generally accepted that Australia's geographical position has changed dramatically during the past 60 million years. From being part of the ancient southern supercontinent, Gondwanaland, Australia (with part of New Guinea attached) migrated northwards towards Asia during the mid-Tertiary, across the 30° of latitude in about 20 million years. Conditions for dispersal across the sea between Asia and Australia were much different then compared to what they are now. *Vicariance biogeography* (also referred to as cladistic biogeography) explains disjunct distributions as the result of speciation following the emergence of some sort of (usually physical) barrier. Following this event, closely related species may have a vicariant pattern of distribution with respect to this barrier. This sort of explanation required well corroborated phylogenetic hypotheses for the species in question and those are still missing for the majority of Australian water bugs. By superimposing the distribution of species upon such a phylogeny (cladogram), area cladograms are produced which can be used to infer relationships between areas of endemism and common patterns of vicariance when two or more groups show congruent distributions (e.g., Andersen 1991a, 1998a).

Botanists have produced area cladograms for plant taxa occurring along the east coast of Australia, finding congruence of vicariance associated with the Macpherson-Macleay overlap (near the eastern coast of northern New South Wales). Comparable studies for insects are rare, but the significance of the Macpherson-Macleay overlap is evident, for example in dragonflies (Odonata) (Cranston & Naumann 1991). However, based on a phylogenetic analysis of the *Nerthra elongata* group (Gelastocoridae), Cassis & Silveira (2002) failed to find any vicariant pattern of distribution across the Macpherson-Macleay overlap. Neither are there any obvious examples in groups of semiaquatic bugs studied by the present authors.

The most common vicariance pattern for Australian water bugs is that across the Gulf of Carpentaria, separating northeastern Queensland and Cape York Peninsula from the northernmost parts of Northern Territory and Western Australia. Examples of this type of vicariance are the sister species *Xenobates major* - *X. angulanus* (Andersen & Weir 1999), *Petrovelia agilis* - *P. katherinae* (Veliidae) (Andersen & Weir 2001), *Aquarius antigone* - *A. fabricii*, and *Tenagogerris euphrosyne* - *T. pallidus* (Gerridae) (Andersen & Weir 1997). Cassis & Silveira (2001) in their cladistic analysis of the *Nerthra alaticollis* group (Gelastocoridae) found a Bassian geographical vicariance between the sister species *N. adspersa* (southern Western Australia) and *N. plauta* (South Australia). Other examples of this type of vicariance are *Megochterus nasutus* - *M. occidentalis* (Gelastocoridae) (Baehr 1990) and *Paranisops inconstans* - *P. endymion* (Notonectidae) (Lansbury 1964).

Additional reading

Andersen, N. M. (1982). The Semiaquatic Bugs (Hemiptera, Gerromorpha). Phylogeny, adaptations, biogeography, and classification. *Entomonograph* 3: 1-455.

Cassis, G. & Gross, G. F. (1995). Hemiptera: Heteroptera (Coleorrhyncha to Cimicomorpha). Pp 1-506 *in* Houston, W. W. K. & Maynard, G. V. (eds): *Zoological Catalogue of Australia. Vol. 27.3A*. CSIRO Australia, Melbourne.

Chen, P., Nieser, N. & Zettel, H. (in press). The aquatic and semiaquatic bugs of Malesia and adjacent areas (Heteroptera: Nepomorpha & Gerromorpha). *Fauna Malesiana.*

Cobben, R. H. (1968). *Evolutionary trends in Heteroptera. Part I. Eggs, architecture of the shell, gross embryology and eclosion.* 475 pp. Centre for Agricultural Publishing and Documentation, Wageningen.

Cobben, R. H. (1978). Evolutionary trends in Heteroptera. Part. II. Mouthpart-structures and feeding strategies. *Mededelingen Landbouwhogeschool Wageningen* 78-5: 1-407.

CSIRO (eds.) (1991). *The Insects of Australia. A textbook for students and research workers, 2nd edition.* Vols 1-2. xvi + 1137 pp. Melbourne University Press, Melbourne.

Davis, J. & Cristidis, F. (1997). *A Guide to Wetland Invertebrates of Southwestern Australia.* vi + 177 pp. Western Australia Museum, Perth.

Gooderham, J. & Tsyrlin, E. (2002). *The Waterbug Book.* 232 pp. CSIRO Publishing, Melbourne.

Hawking, J. H. & Smith, F. J. (1997). *Colour Guide to Invertebrates of Australian Inland Waters.* Cooperative Research Centre for Freshwater Ecology. Identification and Ecology Guide No. 8, 213 pp.

Hungerford, H.B. (1920). The biology and ecology of aquatic and semiaquatic Hemiptera. *Kansas University Science Bulletin* 11 [1919]: 1-328.

Lansbury, I. & Lake, P. S. (2002). *Tasmanian Aquatic & Semi-Aquatic Hemipterans.* Cooperative Research Centre for Freshwater Ecology. Identification and Ecology Guide No. 40, 64 pp.

Lundblad, O. M. (1933). Zur Kenntniss der aquatilen und semiaquatilen Hemipteren von Sumatra, Java und Bali. *Archiv für Hydrobiologie Supplement* 12: 1-195, 263-489.

Mahner, M. (1993). Systema Cryptoceratorum Phylogeneticum (Insecta, Heteroptera). *Zoologica* 143: 1-302.

McCafferty, W. P. (1998). *Aquatic Entomology. The Fisherman's and Ecologist's Illustrated Guide to Insects and Their Relatives.* 480 pp. Jones & Bartlett, Boston.

Menke, A. S. (Ed.) (1979). *The Semiaquatic and Aquatic Hemiptera of California (Heteroptera: Hemiptera).* ix + 166 pp. Bulletin of the Californian Insect Survey 21.

Merritt, R. W. & Cummins, K. W. (eds) (1996). *An Introduction to the Aquatic Insects of North America. 3rd edition.* xii + 441 pp. Kendall/Hunt, Dubuque.

Nilsson, A. (Ed.) (1996). *Aquatic Insects of Northwest Europe, Vol I.* 274 pp. Apollo Books, Stenstrup, Denmark.

Poisson, R. (1957). Hémiptères aquatiques. *Faune de France* 61: 1-263.

Savage, A. A. (1989). Adults of the British Aquatic Hemiptera Heteroptera. A key with ecological notes. *Freshwater Biological Association Scientific Publication* 50: 1-173.

Schuh, R. T. & Slater, J. A. (1995). *True Bugs of the World (Hemiptera: Heteroptera).* xii + 336 pp. Cornell University Press, Ithaca and London.

Southwood, T. R. E. & Leston, D. (1959). *Land and Water Bugs of the British Isles.* The Wayside and Woodland Series. 436 pp. Frederick Warne & Co., London & New York.

Usinger, R. L. (Ed). (1956). *Aquatic Insects of California.* x + 508 pp. University of California Press, Berkeley and Los Angeles.

Wesenberg-Lund, C. (1943). *Biologie der Süsswasserinsekten.* viii + 682 pp. Gyldendal, Copenhagen.

Wichard, W., Arens, W. & Eisenbeis, G. (2002). *Biological Atlas of Aquatic Insects.* 330 pp. Apollo Books, Stenstrup, Denmark.

Williams, D. D. & Feltmate, B. W. (1992). *Aquatic Insects.* xii + 358 pp. C.A.B. International, Wellingford, U.K.

Williams, W. D. (1980). *Australian Freshwater Life. 2nd edition.* 321 pp. Macmillan, Melbourne.

Zborowski, P. & Storey, R. (1995). *A Field Guide to Insects in Australia.* 207 pp. Reed Books, Chatswood, N.S.W.

2. Biology and Ecology

Habitats

Australian water bugs occupy a wide variety of habitats, from temporary pools to lakes, from small streams to rivers and from humid terrestrial habitats to the open ocean (see Plates 1-4). We distinguish between the categories and subcategories of aquatic habitats listed in Table 3. Semiaquatic bugs (Gerromorpha) live in a wide variety of these habitats. Humid, terrestrial habitats (A) are inhabited by species belonging to *Austrovelia* (Mesoveliidae) and *Hebrus* (Hebridae). The most frequent genera in still freshwater (limnic) habitats (B) are *Mesovelia* (Mesoveliidae), *Hydrometra* (Hydrometridae), *Drepanovelia, Microvelia* (Veliidae), *Limnogonus, Rheumatometra,* and *Tenagogerris* (Gerridae). Species of *Rhagovelia* (Veliidae), *Aquarius, Limnometra,* and *Tenagogonus* (Gerridae) are especially found in flowing (lotic) waters (C). Most remarkable for the Australian water bug fauna are the relatively high number (30, or about 24%) of species living in marine habitats (D), belonging to the genera *Halovelia, Xenobates* (Veliidae), *Halobates, Rheumatometroides,* and *Stenobates* (Gerridae). Two species of the exclusively marine genus *Hermatobates* (Hermatobatidae) occur in Australia. Fig. 1.6 shows the distribution of different marine genera on an idealised transect of aquatic habitats, from rivers to the open ocean.

True aquatic bugs (Nepomorpha) are most abundant in lakes, ponds, and quiet parts of

Table 3. Categories and subcategories of aquatic habitats harbouring Australian water bugs. See also Plates 1-4.

(A) *Humid, terrestrial habitats*
Mosses, litter, and debris on moist soil (not necessarily close to free water); seeping rock-faces; mosses and epiphytic plants on trees, etc.

(B) *Still, limnic waters*
- (a) *Temporary pools, swamps, etc.*: Shallow, muddy, still waters, usually vegetated; dry out in summer (Plate 1, A-D).
- (b) *Shallow pools*: small, usually shallow bodies of still water (also in dry river or stream beds), usually vegetated along margins and with floating plants; usually temporary (Plate 1, E, F; Plate 2, A, B).
- (c) *Riverine pools (lagoons, billabongs)*: deep waterholes in river beds (sometimes in streams), still during part or much of the year. Vegetated along margins and surface often covered with floating plants. Some are temporary (Plate 2, C-F).
- (d) *Ponds*: small, often shallow bodies of still water, usually vegetated, but with extensive open water. Many farm ponds are included here. In areas with strong seasonal rainfall (summer in the north, winter in the south), ponds are often temporary (Plate 3, B).
- (e) *Lakes*: large, deep bodies of permanent, still water. Many have extensive growth of aquatic plants, submerged as well as emergent, particularly at their margins (e.g., rushes and bulrushes) (Plate 3, A).

(C) *Flowing, limnic waters*
- (a) *Streams*: small, permanent or semipermanent waters, usually flowing rapidly, at least during the rainy season (Plate 3, C-F).
- (b) *Rivers*: broader, permanent waters, usually flowing slowly throughout the year (Plate 4, B).

(D) *Marine waters*
- (a) *Estuaries and mangrove swamps*: saline rivers, tidal streams and pools, usually with a pronounced salinity gradient from euhaline to limnic as one progress inland (Plate 4, C).
- (b) *Coastal lagoons and lakes*: saline waters usually protected from wind and wave action.
- (c) *Tidal pools on coral reef flats or rocky coasts*: saline waters subject to a distinct tidal cycle (Plate 4, D, E).
- (d) *Sea surface*: saline waters exposed to wind and wave action, from coastal localities to the open ocean (Plate 4, F).

streams and rivers (B and C) where the dominant genera are *Laccotrephes, Ranatra* (Nepidae), *Diplonychus* (Belostomatidae), *Agraptocorixa, Diaprepocoris, Micronecta, Sigara* (Corixidae), *Naucoris* (Naucoridae), *Anisops, Enithares* (Notonectidae), and *Paraplea* (Pleidae). This group also includes littoral, non-aquatic species belonging to the genera *Megochterus, Ochterus* (Ochteridae), and *Nerthra* (Gelastocoridae); the latter genus also with species living in fairly dry situations, far from water.

Although each genus occupies a rather characteristic set of habitats and in most cases exhibits distinctive behaviour patterns, such information cannot be easily organised in a useful way. Hungerford (1920) gave a "habitat key" for North American water bugs which was updated by Usinger (1956) and Menke (1979) for Californian water bugs. We have attempted to produce a similar key to Australian species which, however, can only be used as a rough guide because a single genus may be found in more than one type of habitat or several different genera may inhabit similar habitats. Such problems occur primarily among the semiaquatic bugs.

Key to Australian water bugs based on habitats and habits

1. Semiaquatic bugs, living upon the water surface, on the shore of aquatic habitats or in humid, terrestrial habitats 2
– True aquatic bugs, living beneath the water surface 20
2. Water striders, walking or skating on the water surface 3
– Bugs living in humid, terrestrial habitats, on the shore of aquatic habitats, or on floating mats of vegetation (may walk or run on water surface when disturbed) 14
3. Living in freshwater habitats (still or flowing) 4
– Living in marine habitats (mangroves, coral coasts, sea surface) 9
4. Living on marginal waters of ponds and streams 5
– Living on open water surface 6
5. Slow moving species on long, stilt-like legs in protected places (HYDROMETRIDAE) .... *Hydrometra*
– Rapid moving species on short legs in open places; gregarious, resting on surface film, submerged rocks, logs, or floating vegetation near shore (VELIIDAE) .... *Drepanovelia, Lacertovelia, Microvelia, Microvelopsis, Petrovelia, Phoreticovelia*
6. Living on ponds, lakes, and quiet waters of streams (GERRIDAE) .... *Calyptobates, Limnogonus, Limnometra, Rhagadotarsus, Tenagogerris, Tenagogonus*
– Living on streams and small rivers 7
7. Living on open water, usually of large rivers (GERRIDAE) .... *Aquarius, Halobates* (rare)
– Living on riffles in streams and small rivers 8
8. Middle tarsi with plumose swimming fan. Hind femora thickened (VELIIDAE).. *Rhagovelia*
– Middle tarsi without swimming fan. Hind femora slender. (GERRIDAE) ....*Austrobates, Rheumatometra*
9. Living on sea surface far from shore (GERRIDAE) .... *Halobates*
– Living in coastal marine habitats 10
10. In tidal streams of mangrove swamps and coastal lagoons 11
– In pools on intertidal coral reef flats 12
11. Hind femora short, length less than width of body (VELIIDAE) .... *Xenobates*
– Hind femora long, length greater than width of body (GERRIDAE) .... *Halobates, Rheumatometroides, Stenobates*
12. Hind femora long, length greater than width of body (GERRIDAE) .... *Halobates*
– Hind femora short, length less than width of body 13
13. Head very broad, at least three time as wide as long. Tarsi 3-segmented (HERMATOBATIDAE) .... *Hermatobates*
– Head longer than wide. Tarsi 2-segmented (VELIIDAE) .... *Halovelia*
14. Living in the open 15
– Hidden in cryptic, littoral habitats 17
15. Found on mosses or other shore plants or on floating mats of vegetation 16
– Littoral species found on sand or mud 17
16. Walking on the water surface when disturbed (MESOVELIIDAE) ....*Mesovelia*
– Rarely if ever walking on the water when disturbed (HEBRIDAE) .... *Hebrus, Merragata*
17. Burrowing in mud under rocks and debris, body covered with mud (GELASTOCORIDAE) .... *Nerthra*
– Non-burrowing species 18
18. Fast running species that hide in cryptic situations (OCHTERIDAE).. *Megochterus, Ochterus*
– Slow moving species living in moss or in crevices of loose gravel 19
19. Living on sloping banks of ponds or streams (HEBRIDAE) .... *Hebrus*
– Living on the forest floor in mist forests (MESOVELIIDAE) .... *Austrovelia*
20. Swimming with venter up 21
– Swimming with dorsum up 23

21. Swimming with short, slender hind legs (PLEIDAE) .... *Paraplea*
- Swimming with long oar-like hind legs (NOTONECTIDAE) 22

22. Resting at surface film or on submerged objects, not in equilibrium with water, rising toward surface when not swimming......................... *Enithares, Notonecta, Nychia*
- Resting poised in mid-water, in equilibrium with water, not rising toward surface when resting ....*Anisops, Paranisops, Wallambianisops*

23. Living on the bottom, not depending on access to surface for breathing (APHELOCHEIRIDAE) .... *Aphelocheirus*
- Living in the water or among submerged vegetation (occasionally resting on the bottom), depending on access to surface for breathing 24

24. Air-store replenished by breaking surface film with pronotum; usually feeding on bottom detritus (CORIXIDAE) .... *Agraptocorixa, Cnethocymatia, Diaprepocoris, Micronecta, Sigara*
- Air-store replenished by breaking surface film with tip of abdomen or by apical abdominal appendages; predaceous species 25

25. Air-store replenished by breaking surface film with tip of unspecialised abdomen (NAUCORIDAE) .... *Naucoris*
- Air-store replenished by long, slender siphon or short, retractable straps at end of abdomen 26

26. Strong swimmers, air-store replenished by retractile air-straps (BELOSTOMATIDAE) 7
- Slow swimmers, air-store replenished by long, slender siphon tube (NEPIDAE) 28

27. Eggs laid on backs of males........... *Diplonychus*
- Eggs laid on stems of rushes and other emergent vegetation .... *Lethocerus*

28. Living under rocks and stones in streams .... *Goondnomdanepa*
- Living on the bottom or among submerged vegetation in lakes, ponds or slow moving streams 29

29. Living at the edge of water usually in very shallow muddy places... *Austronepa, Laccotrephes*
- Living at the edge of water among submerged vegetation .... *Cercotmetus, Ranatra*

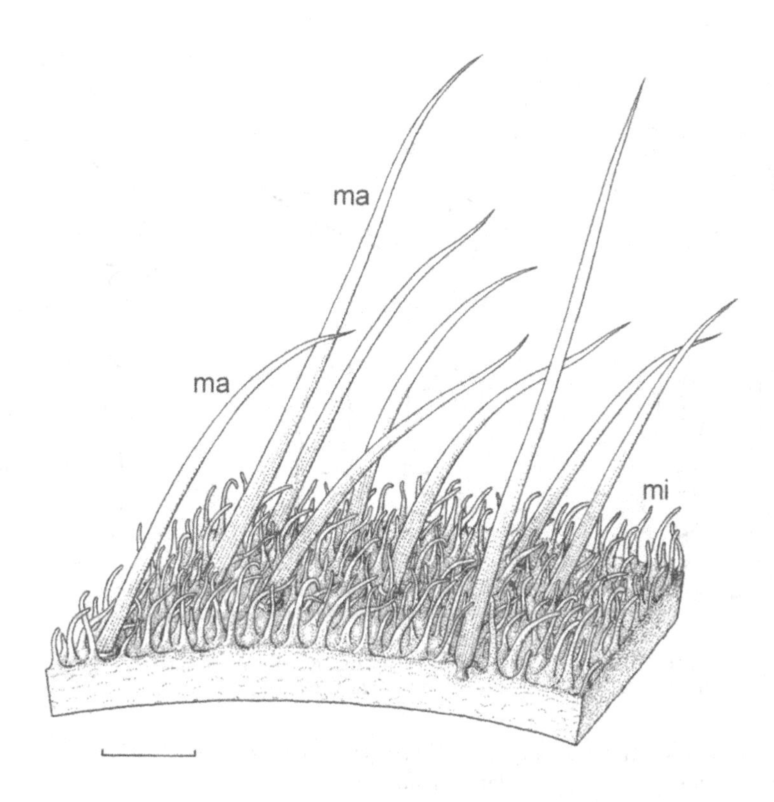

Figure 2.1. Reconstruction of body surface structure of *Gerris* based upon Scanning Electron Micrographs (ma, macro-hair; mi, micro-hair). Reproduced with modification from Andersen (1982).

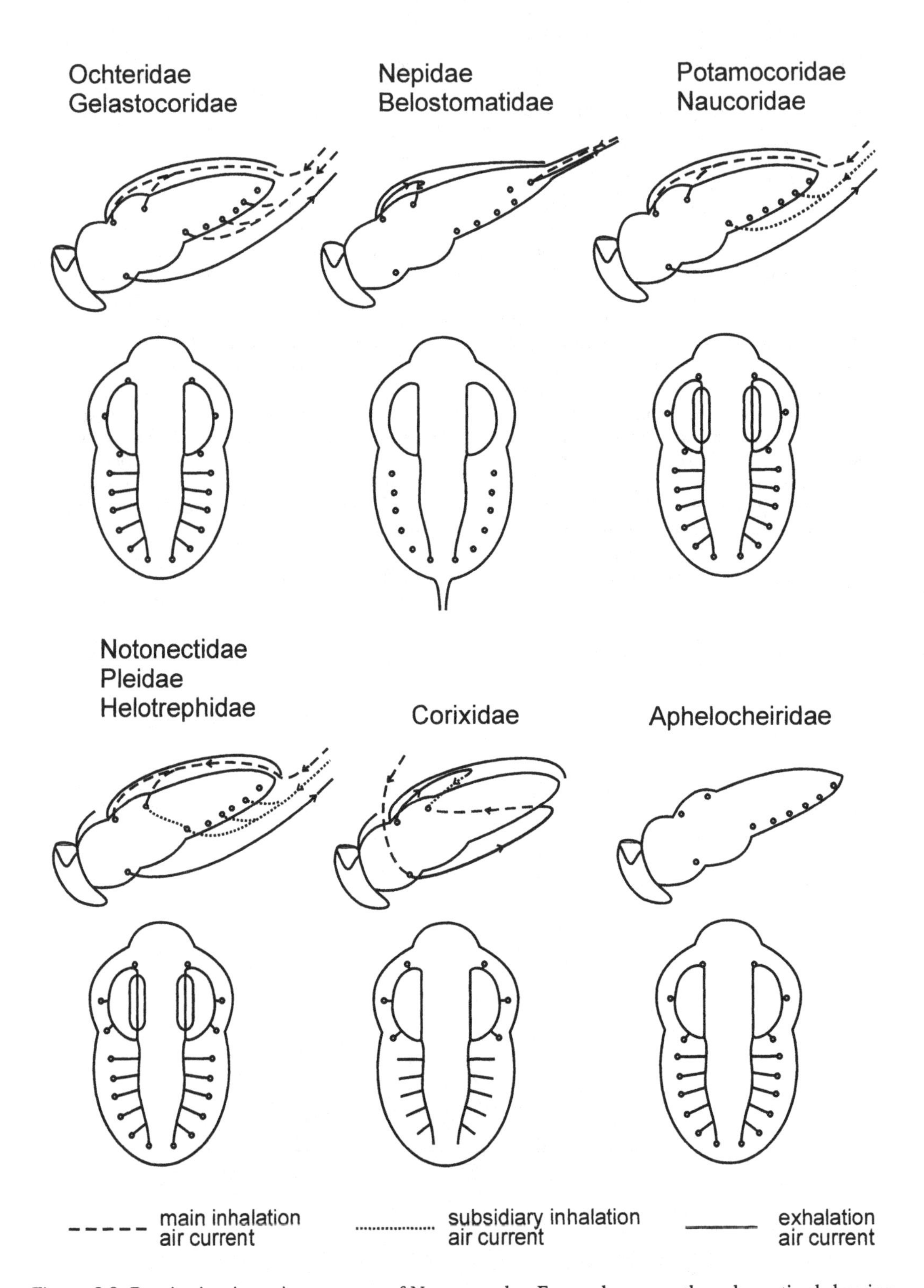

Figure 2.2. Respiration in various groups of Nepomorpha. For each group, the schematised drawing on top shows the air-flow during inhalation (punctured line; dotted line indicate less important routes) and exhalation (unbroken line). The schematised drawing at bottom shows the air-transport system (trachea and air-sacs) wthin the bug. Modified from Popham (1960) by Martin B. Hebsgaard.

Respiration

Like most insects, water bugs are dependent upon access to atmospheric oxygen in order to breathe and survive. Above water, air enters the body of a bug through its spiracles of which there are two pairs on the thorax and 8 pairs on the abdomen. Within the body, oxygen is distributed by way of the tracheal system which effectively reaches all internal organs.

Most semiaquatic bugs (Gerromorpha) are able to live on the water surface without being wet or breaking through the surface film. Their body is covered by an elaborate system of unwettable (hydrofobe) hair-layers, composed by macrohairs or *setae* (Fig. 2.1; ma) and micro-hairs or *microtrichia* (mi). These hair-layers are responsible for the usual dull, velvety appearance of these bugs. Apart from being covered by unwettable hair-layers, gerromorphan bugs do not have any specialised structures for breathing, even if some species occasionally submerge below water, e.g. to lay their eggs. In contrast to this, each of the truly aquatic families (Nepomorpha) has a distinctive type of respiration which will be described in their respective chapters (see also Fig. 2.2). Air layers adhere to these bugs when submerged and are renewed at the water surface through respiratory siphons (Nepidae) or air-straps (Belostomatidae) arising from the abdominal end, the tip of abdomen (Naucoridae, Notonectidae, Pleidae), or through spaces between the head and pronotum (Corixidae). A large part of the air is usually concealed dorsally between the wings and the abdomen, and another part is exposed on the venter where it is held in place by hydrofuge hairs.

The air carried by aquatic bugs has a hydrostatic or buoyancy function as well as serving as a store of oxygen. Even more important, the air-stores on the body surface have the function of a "physical gill" which extracts dissolved oxygen from the water. When replenished at the water surface, the air store has the same composition as the atmosphere (21% O_2, 78% N_2 and a little CO_2). When the oxygen in the air store is used up by the bug, an excess of nitrogen is left behind. One would expect that some of this nitrogen would dissolve in the surrounding water to restore the balance. However, since oxygen diffuses into and out of water more rapidly than nitrogen, more oxygen than nitrogen leaves the surrounding water to replace the oxygen that is used by the bug. Some nitrogen is lost at the same time, of course, but as long as any nitrogen remains the process can go on. When the bug finally rises to the surface it is as much to renew its store of nitrogen as to replace the oxygen.

An air layer held by a simple pile of hairs gradually dissolves, and the bug must come to the water surface in order to renew it. This is normally the case in true aquatic bugs. However, one Australian water bug, *Aphelocheirus australis* Usinger, has solved this problem by developing a permanent physical gill which enables the bug to stay submerged indefinitely. This bug has an air store of the usual type, but it is carried by a pile of fine hydrofuge hairs which have their tips bent over at right angles. The hairs are so small and closely set that there are about two million of them per square millimetre of body surface. It is extremely difficult to displace this so-called "plastron" of air, even by pressures of 4-5 atmospheres. These bugs live permanently submerged and the European species, *A. aestivalis*, has been reported in the Danube up to a depth of 7 metres.

Locomotion

Most semiaquatic bugs (Gerromorpha) are able to move around on the water surface with great ease. Each leg produces a trough-shaped depression (so-called meniscus) in the surface film. The convex shape of this meniscus indicates that the leg sheds the water (is hydrofuge). This seems to be a general property of the hair layers of surface bugs, including the complex hair layers covering most of the body of these insects (Andersen 1976, 1977). Chiefly because of this, the surface tension of water will easily support the weight of surface dwelling bugs. The hydrofuge property is not permanent, however, and grooming of the legs is an important activity in surface bugs which use specialised grooming structures on their tibiae to keep the hair layers tidily arranged for preventing entry of water.

Locomotion on the water surface is achieved in three different ways (Andersen 1976, 1982):

(1) *Walking*, where the three pairs of legs are moved as alternating tripods composed of one fore leg and one hind leg of one side of the body and one middle leg of the opposite side of the body, whereas the other three legs are lifted and swung forward (Fig. 2.3A).
(2) *Rowing* with simultaneous strokes of the middle legs while at least the hind legs slide on the water surface; this rhythm of leg movement allows a more efficient use of muscle power than the walking rhythm (Fig. 2.3B); and
(3) *Skating*, or jump-and-slide movements with simultaneous power strokes of the middle legs which cause the bug to take off from the water surface (Fig. 2.4A); after touchdown, the bug usually slides forward for a long distance.

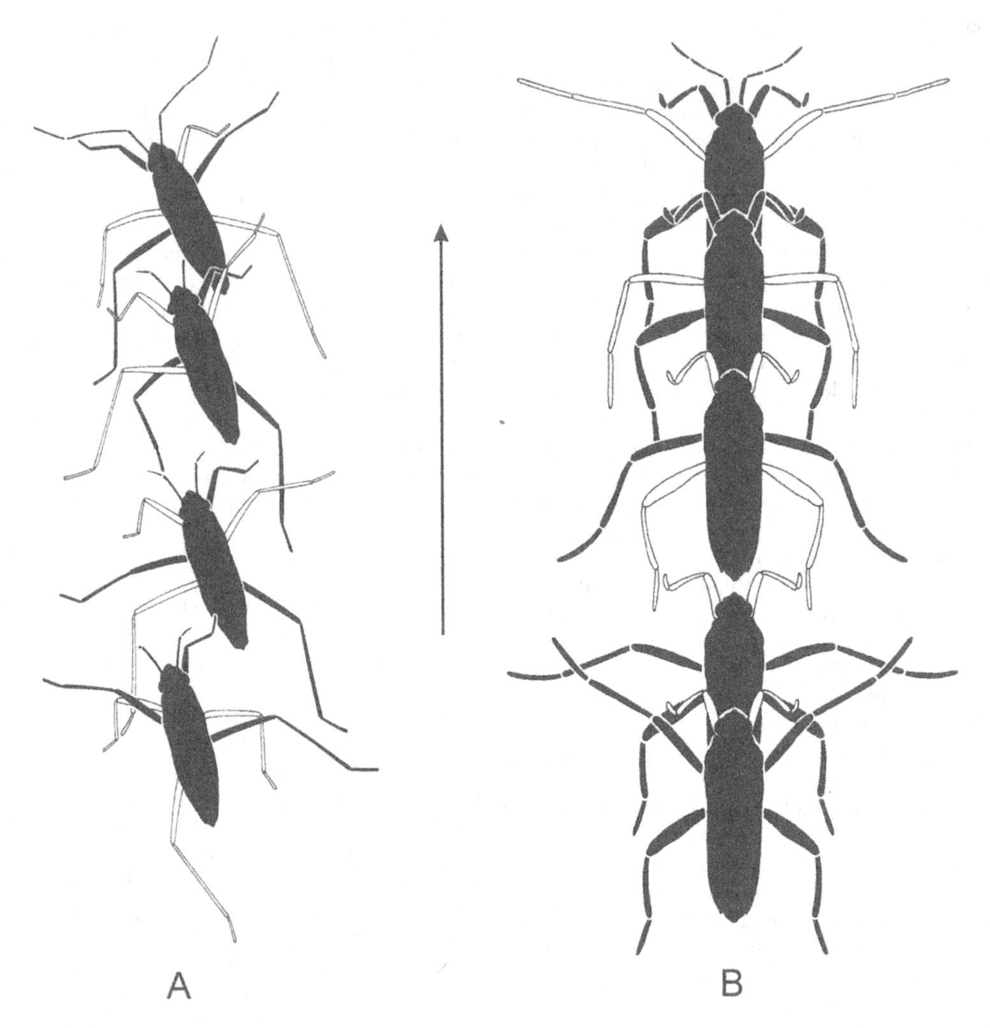

Figure 2.3. Locomotion on the water surface. Supporting legs shown black. Time interval between successive postures 16 mseconds. A, *Mesovelia* (walking). B, *Velia* (rowing). Reproduced with modification from Andersen (1982).

Since the surface film offers so little resistance, the bug has to generate ripples which are then used as "starting blocks" (Fig. 2.4B). Recent studies of the hydrodynamics of water strider locomotion, however, show that the bug also transfers momentum to the underlying water through hemispherical vortices shed by its driving legs (Hu et al. 2003).

Most surface bugs walk on leaves of floating plants or on land, but species of *Mesovelia, Hebrus, Hydrometra,* and *Microvelia* have retained the same rhythm of leg movements when moving around on the water surface. Rowing along the water surface is practiced by some Veliidae, e.g. *Halovelia, Xenobates,* and *Rhagovelia,* in the latter assisted by an elaborate swimming fan arising from the middle tarsi (Fig. 4.4B). All water striders belonging to the family Gerridae are structurally adapted to skate along the water surface, whereas most species move rather clumsily on land.

In nature, the water surface forms a concave meniscus around the edges of floating leaves and emergent vegetation. Small semiaquatic bugs, like *Hebrus* (Hebridae), *Mesovelia* (Mesoveliidae), and *Microvelia* (Veliidae) are not able to climb the steep slope of a concave meniscus. Instead, these small bugs utilise surface tension forces to ascend

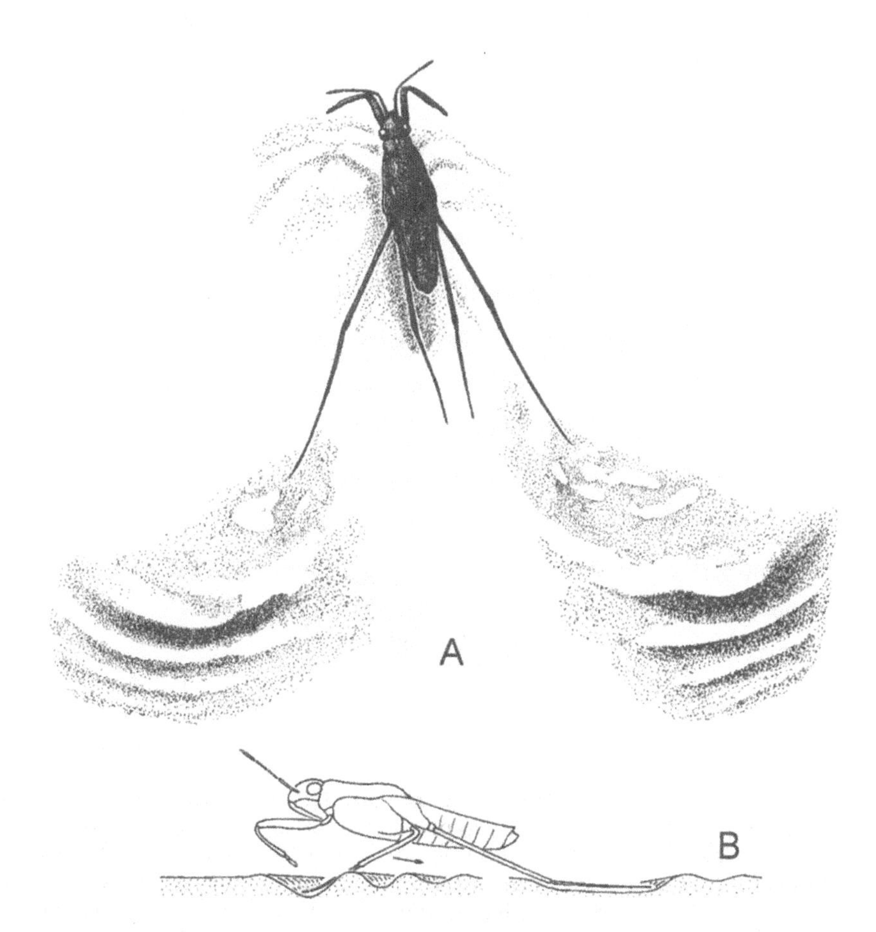

Figure 2.4. A, pondskater, *Gerris*, jumping off the water surface. Drawn from a movie frame showing the shadows thrown by surface waves generated by the bug (dark areas corresponds to troughs, light areas to the crests of surface waves. B, lateral reconstruction of the posture of *Gerris* during the thrust phase showing use of surface waves as starting blocks. Reproduced with modification from: A, Andersen (1982); B, Andersen (1976).

the slope of a meniscus. The pulling up of surface film by the legs (the claws are hydrophilic) creates miniature menisci (Fig. 2.5A). The laws of physics say that menisci of similar shape (concave in this case) attract each other and since the menisci created by the bug are much smaller than the permanent meniscus formed along the edge of a plant, the bug will be drawn up the slope of the larger meniscus. Ascension of a meniscus (also called "hydranapheuxis"; Schuh & Slater 1995) is easily observed by placing one of the small surface bugs in a small dish with water. The bug will run towards the edge and struggle to climb the meniscus formed there. If unsuccessful, the bug will assume one of the postures illustrated (Figs 2.5B-D) and may, seemingly without effort, be drawn up the slope of the meniscus. Even larger surface bugs like *Hydrometra* may use this technique from time to time (Andersen 1976).

Water surfaces behave like a stretched membrane. If a drop of detergent is added, the surface tension will be momentarily lowered and a "hole" formed in the surface film. As a result, any floating object (including a surface bug) will be displaced. Some species of Veliidae (*Microvelia*, *Velia*) use "expansion skating" or "skimming" to escape enemies (Andersen 1976, 1982). The movement is initiated by the spread of a detergent fluid (probably saliva ejected through the rostrum) which lowers the surface tension of water. This is most easily demonstrated if the

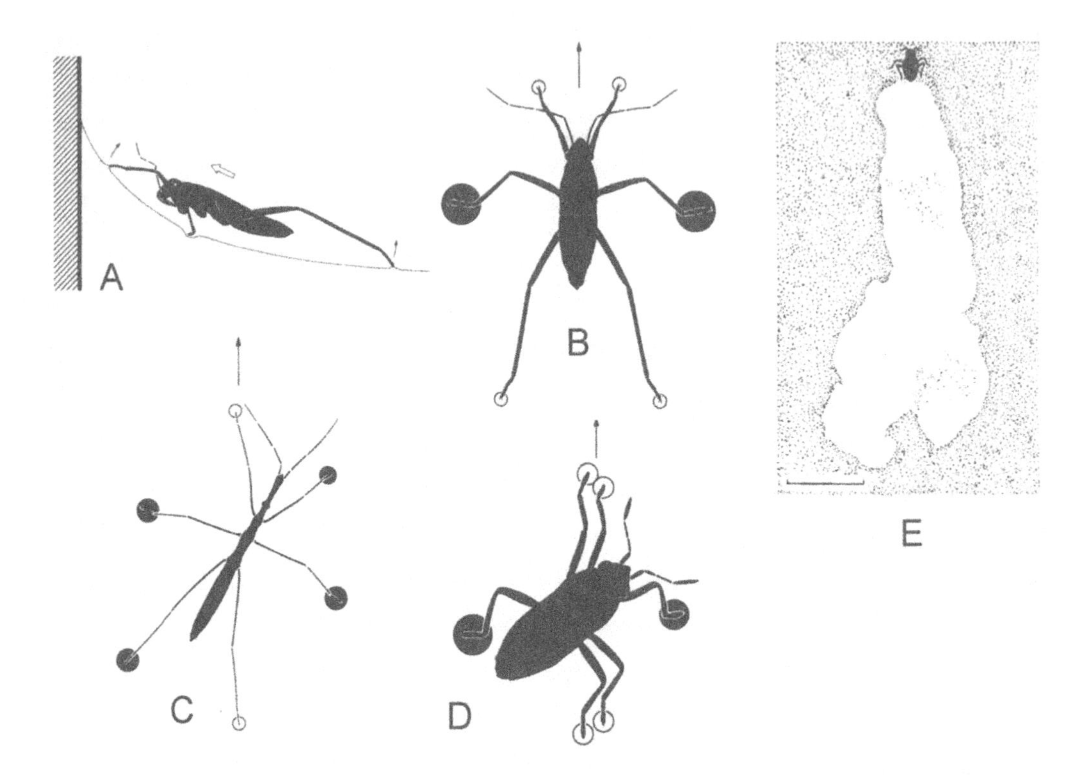

Figure 2.5. A-D, ascension of a concave water meniscus: A, lateral view of *Mesovelia*; the body is supported by the middle legs while the fore and hind legs pull up the surface film (light arrows) and the insect is drawn up the slope (heavy arrow). B-D, dorsal view of body and leg postures during ascension; supporting leg tarsi indicated by black circles, surface-pulling tarsi by white circles. B, *Mesovelia*; C, *Hydrometra*; D, *Microvelia*. E, expandion skating in *Microvelia* traceable in a *Lycopodium* spore layer on the water surface (drawn from a photograph). Scale 5 mm. Reproduced with modification from Andersen (1982).

water surface is covered by dust (alternatively by a layer of *Lycopodium* spores). If the bug is seized by a leg or otherwise pursued, it will sometimes deliver a drop of fluid that will form a "hole" in the dusty surface (Fig. 2.5E). The same individual may produce several such holes at short intervals, but soon becomes exhausted.

True water bugs (Nepomorpha) propel themselves through the water in essentially two different ways:

(1) By synchronous strokes of the middle and hind pair of legs, each pair working alternately, as in the Nepidae and Belostomatidae.
(2) With synchronous, oar-like movements of the hind legs as in the Corixidae, Naucoridae, Notonectidae, and Pleidae.

The water scorpions (Nepidae) are the weakest swimmers and their slender middle and hind legs are not designed for efficient underwater propulsion. The legs of the creeping water bugs (Naucoridae) are also less efficient for swimming and must be moved rapidly to propel the bugs through water. The water boatmen (Corixidae), backswimmers (Notonectidae), and giant water bugs (Belostomatidae) are the most excellent swimmers. Their hind tibia and tarsi are broad, flattened and fringed with long hairs (Fig. 4.4F). These hair-fringes collapse on the forward stroke, thus minimising the drag forces, but spread as a result of water pressure on the backstroke, thus maximising the propulsive forces. The corixids are probably the most agile swimmers, being able to making abrupt changes in the swimming direction. The Notonectidae, which swim upside down, move gracefully with powerful sweeps of their oar-like hind legs. The Belostomatidae, especially the species of *Lethocerus*, which have the broadest swimming legs of all water bugs, are the most powerful swimmers and may stroke synchronously with both posterior leg pairs during periods of vigorous swimming.

Feeding

Most water bugs can be characterised as "predacious fluid-feeders" which feed on the tissue of other animals, chiefly insects and various aquatic invertebrates. Only water boatmen (Nepomorpha: Corixidae) have been reported to feed on plant material (see below). Compared to insects with biting mouthparts (like cockroaches, grasshoppers, beetles, and wasps), the mouthparts of hemipterous insects are modified to form a piercing-sucking beak or rostrum. It is typically composed of the 4-segmented labium which is indented medially to form a groove enclosing two pairs of thread-like stylets (Fig. 4.2E). These stylets correspond to the mandibles and maxillae of insects with biting mouthparts. The maxillary stylets enclose between them one or two canals communicating with the salivary glands and/or the food pump located inside the head capsule. The act of feeding is usually initiated by "harpooning" of the prey with the serrated apices of the mandibular stylets. Only the maxillary stylets are forced deeper into the prey which are rapidly paralysed by the discharge of considerable amounts of toxic, proteolytic saliva. Cobben (1978) reported adult *Hydrometra* (Gerromorpha: Hydrometridae) immobilise fruit flies (*Drosophila*) within a few seconds and common house flies (*Musca domestica*) within 10 seconds (Figs 2.6C-D). In water bugs, the maxillary stylets are provided with barbs and hairs throughout most of their length which lacerate the prey's tissue with their continuous filing action. The stylets have an astonishing flexibility and motility within the prey (easily seen in prey with a transparent cuticle like fruit flies). From the point of insertion, the stylets are able to reach and suck up the contents of a fruit fly, reaching the tip of the abdomen as well as the head.

Most semiaquatic bugs (Gerromorpha) are opportunistic predators and scavengers although small or medium-sized arthropods with a soft integument are preferred as food items in nature. Two basic predator strategies can be recognised (Andersen 1982): the "searching" strategy and the "waiting" strategy. Some overlapping between the two types may occur as well as that the same species may change its strategy from time to time. Pondweed bugs (*Mesovelia*) and water measurers (*Hydrometra*) seem to be chiefly *searching predators* which move slowly around, examining crevices and cavities in the vegetation near the shore or floating on the water surface in search of something edible (Figs 2.6A-B). The prey is probably recognised by the tactile or olfactorial sensory apparatus rather than by vision. The antennae are long and are waved constantly from side to side. The eyes of a searching predator are often small. The rostrum is long and can be extended forward, far in front of the fore legs, which are not raptorial. The approach of the prey is rather slow and careful and food items are usually examined with the antennae before feeding.

Water striders (Gerridae) and some species of Veliidae can, at least in part, be characterised as *waiting predators* which utilise the special properties of the water surface as means of detecting and locating the prey (much in the same way as a spider uses its web). Surface bugs can detect and locate live prey trapped in the surface film of water solely by the surface vibrations generated by the struggling prey. Visual stimuli are probably significant for close-range location of prey, whether moving or not, when the surface bug patrols the water. The large, multifaceted eyes of most gerrids and veliids are probably also necessary during fast locomotion on the water surface. Once the prey has been located, the predator proceeds rapidly and in a more or less straight line to the prey which, depending on its size, usually is grasped with the raptorial fore legs and sucked out through the relatively short rostrum (Fig. 2.6E). Many water striders (Gerridae) and some species of Veliidae aggregate in tight schools on the water surface. Such gregarious behaviour may improve the efficiency of prey location for the population as a whole.

A common predator strategy among true aquatic bugs (Nepomorpha) is that of a waiting or *ambush predator*. Water scorpions (Nepidae) wait, hidden under mud or between water plants, until a potential prey comes in reach of their fore legs. By keeping their long respiratory siphon in contact with the water surface, these bugs are able to replenish their air store while waiting for prey (Figs 1.2A-B). The fore tibia has an apical sense organ, which apparently senses prey vibrations, but visual cues seem to be more important in prey capture. In the water stick insects (*Ranatra*), the orientation of the bug while waiting for prey is usually head down and fore legs stretched forward in an angle of 30° or less from the vertical (Fig. 1.2B) The reach of the raptorial fore legs is considerable due to the lengthening of the fore coxa and femur (Cloarec 1972; Cloarec & Joly 1988). Although giant water bugs (Belostomatidae) are good swimmers, they usually also ambush their prey which includes small vertebrates, such as fish and tadpoles. They may, however, alternate between waiting for prey and actively

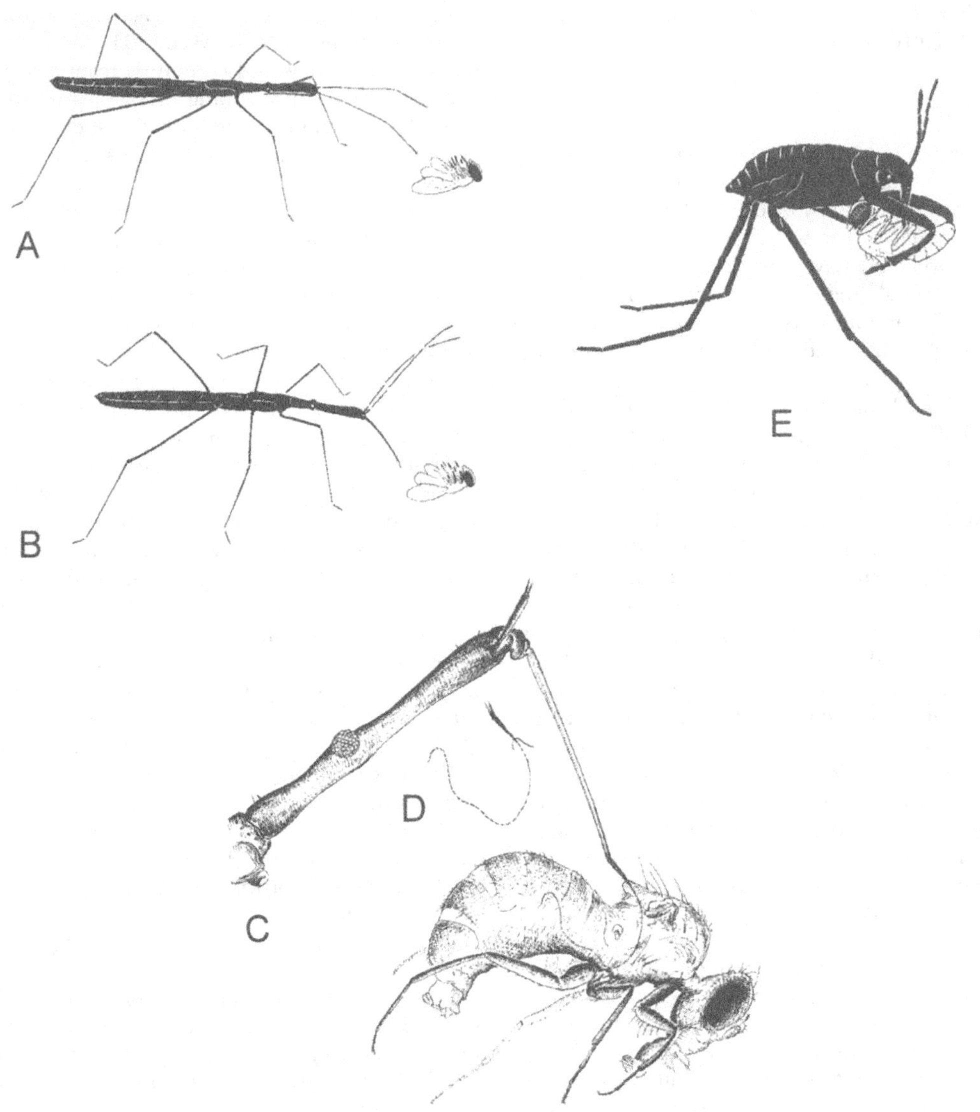

Figure 2.6. A-B, two stages in predation by *Hydrometra* (drawn from photographs): the prey (a fruit fly) is appraoached very carefully and examined with the antennae (A), whereafter the rostrum is lowered, ready to spear and suck out the prey (B). C-D, *Hydrometra* preying on a fruit fly: shortly after hapooning by the mandibular stylets, the maxillary stylets extends far into the prey (C); pathway of stylets shown separately (D). E, a sea skater, *Halobates*, holding a prey with its fore legs and sucking it out through its short rostrum (drawn from a photograph). Reproduced with modification from: A-B, D, Andersen (1982); C, Cobben (1978).

foraging. Backswimmers of the genus *Notonecta* (Notonectidae) wait until a prey comes within sight after which they strike at it. Some species hold to a perch under water, other species float against the surface film and wait for prey falling onto the water surface (Fig. 1.2F). Struggling insects captured in the surface film reveal themselves by emitting surface ripples. Prey lying in front of or at the sides of the backswimmer is detected at up to five or six times the body length of the predator; prey lying behind the bug is observed at a distance up to three to four times the body length of the predator (Markl & Wiese 1969; Giller & McNeil 1981).

The feeding biology of water boatmen (Corixidae) is unique since many species are *detritus feeders*, foraging on the bottom ooze or detritus in their habitats (Fig. 1.2D). The flat, spoon- or scoop-like fore tarsi fringed with long hairs are rotated around each other like twirling thumbs forcing detritus past the tip of the short, triangular rostrum. Detritus particles are ingested through the long stylet groove on the face of the rostrum and broken up by a powerful food pump including tooth-like grinders inside the head capsule. Hungerford (1920, 1948) observed water boatmen feeding on plant material, such as multicellular algae (*Spirogyra*). These observations apparently led to a long hold belief that water boatmen are chiefly herbivores. However, more recent findings have shown that most water boatmen feeding on bottom detritus eat mainly animal food, including living or dead protozoans and various microscopic invertebrates. Some species also hunt for larger animal prey in open water and Hale (1922) kept several Australian species for months on a diet of mosquito larvae only; even newly hatched nymphs were observed to capture tiny mosquito larvae. The results of other rearing experiments show that most water boatmen prefer animal food which in many cases is necessary for reproduction and improves the longevity of the bugs (Jansson 1986). Thus, water boatmen should no longer be regarded as strictly plant feeders.

Reproductive biology

Water bugs are hemimetabolous insects, that is insects with an incomplete metamorphosis. Unlike beetles, butterflies, or ants which pass through distinct larval and pupal stages between the egg and adult insect, water bugs develop more gradually from the egg to the adult insect, through several juvenile stages (called nymphs). Most works on the reproductive biology of water bugs are of the type that describes the "life history" of one particular species, its development from egg to adult insect. Unfortunately, even this type of information is scarce for Australian species and the following account is chiefly compiled from works on European and North American water bugs, in particular Hungerford (1920), Poisson (1924), Ekblom (1926, 1930), Wesenberg-Lund (1943), Southwood & Leston (1959), Cobben (1968), and Menke (1979). Useful information on water bugs from the Indo-Australian region is presented by Chen et al. (in press).

Mating

Prior to egg-laying, females have to mate with one or more males. Mating is usually initiated by the male which has to insert his copulatory organ (phallus) into the genital tract of the female in order to accomplish a transfer of sperm cells. A variety of strategies to conquer the opposite sex are displayed by males of water bugs. The simplest is one where the male lunges on and engages the female very quickly and leaves her almost immediately after the actual transfer of sperm has taken place. Usually the male jumps onto the back of the female and clasps his fore legs around the female's thorax whereas the middle and hind legs rest on some part of the female's body or are hold free in the air or water. In semiaquatic bugs, the delicate nature of the water surface requires that the male positions itself almost symmetrical on the females body in order to minimise the impediment on the female's movements. The duration of the mating act varies greatly, from a few seconds to days or even weeks. Actual copulation may take place at the beginning of the act, but after that the male stays with the female for a shorter or longer period, thereby "guarding" the female from being pursued by other males (and his own sperm from being displaced by that of other males). Among Australian water bugs, species like the water strider *Rheumatometra philarete* (Gerridae) (Fig. 12.41) and, in particular, the coral bug *Halovelia hilli* (Fig. 11.4B) and the Zeus bugs *Phoreticovelia rotunda* and *P. disparata* (Veliidae) (Fig. 11.35A) show prolonged mate-guarding (Andersen & Weir 1998, 1999, 2001). In the latter species, the male feeds on a secretion produced by glands in the female's dorsal thorax (Arnqvist et al. 2003).

Sexual communication

Several species of water striders (Gerridae) are known to communicate by means of surface waves (Wilcox 1979; Spence & Wilcox 1986; Wilcox & Spence 1986). This kind of behaviour has been described for at least one Australian species, *Rhagadotarsus anomalus* (as *R. kraepelini*; Wilcox 1972). Patterns of surface waves are generated by leg movements while the male is free on the surface or while it grasps floating or fixed objects; these objects then become copulation and oviposition sites for the female (Fig. 12.36). Several different signals (waves with different amplitudes and frequency patterns) have been recorded in *R. anomalus*, including aggressive signals towards other males.

Several species of true water bugs are known to produce sound by "stridulation" where one part of the body (the so-called "plectrum") is rubbed against another part ("stridulitrum"). The sound production in water boatmen (Corixidae) has received considerable attention.

In species of the subfamily Corixinae, the mechanism consists of a group of pegs on the base of the fore femur (Fig. 15.1D; functioning as the plectrum) which is rubbed against the margin of the sharp lateral edge of the head (functioning as the stridulitrum). Sound is usually produced by males only, but females may also stridulate (Hungerford 1948; Jansson 1972, 1973, 1976). The sounds produced by corixine water boatmen

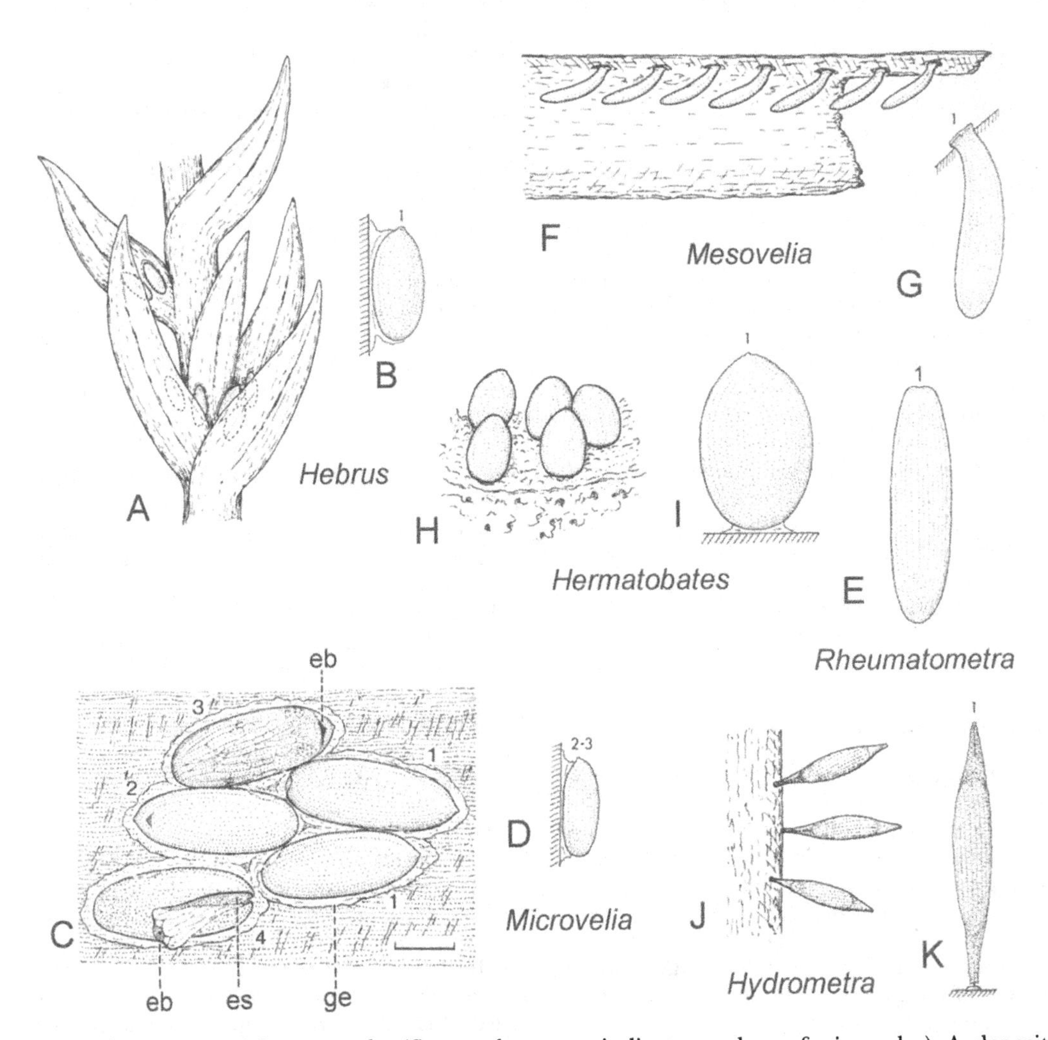

Figure 2.7. A-J, eggs of Gerromorpha (figures above eggs indicate numbers of micropyles). A, deposited eggs of *Hebrus ruficeps* Thomson (Hebridae); eggs deposited between leaves of moss plant. B, egg of *Hebrus* deposited superficially. C, deposited eggs of *Microvelia umbricola* Wroblewski (Veliidae) in various stages of development (1-3) and empty egg shell after eclosion (4) (eb, egg burster (ruptor ovi); es, eclosion split; ge, gelantinous substance). Scale 0.2 mm. D, egg of *Microvelia* deposited superficially. E, egg of *Rheumatometra* (Gerridae). F, eggs of *Mesovelia furcata* Mulsant & Rey (Mesoveliidae) embbed in plat stem (seen from inside). G, egg of *Mesovelia.* H, eggs of *Hermatobates* sp. (Hermatobatidae) deposited in small cavity in dead coral. I, egg of *Hermatobates* deposited upright. J, eggs of *Hydrometra stagnorum* (L.) (Hydrometridae) deposited on erect plant stem. K, egg of *Hydrometra* deposited upright. Reproduced with modification from: A-D, F-K, Andersen (1982); E, Andersen & Weir (1998).

can be perceived by the human ear. More surprisingly, the minute *Micronecta* species also produce audible sounds which are produced by rubbing a ribbed process on one of the parameres (genital claspers) against ridges on the abdominal sternite 8. Males of different species produce species-specific signals which attracts conspecific females (Jansson 1989; King 1999a, 1999c). Acoustic communication is also known in the backswimmer genus *Anisops* (Notonectidae) where males stridulate by rubbing the fore tibial stridulatory comb against the rostral "prong". This type of stridulation used in courtship was described for *A. thienemanni* Lundblad (as *A. hyperion* Kirkaldy) by Hale (1923: 407) who characterises the sound as "a distant grindstone at work". Finally, stridulatory structures have been described in the water stick insect *Ranatra*, composed by a roughened elevated area on the outer surface of the fore coxa which is rubbed against the striate inner surface of the fore acetabula (J. Polhemus 1994a; Schuh & Slater 1995).

Eggs and oviposition

The eggs and habits of oviposition are remarkably diverse in both semiaquatic and aquatic bugs. The egg structure was described and illustrated by Cobben (1968) and Andersen (1982) for most groups of water bugs. The size of the egg increases by the length of the females body, but not in a proportional way. As a rule, if the size of the female increases by a factor 10, the size of the egg only increases with a factor 4.5. Thus, the eggs of small species are relatively bigger than those of large species. This is exemplified by the following figures (length of female followed by size of egg, both in millimeters): *Hebrus* 1.8/ 0.6; *Microvelia* 1.9/ 0.5; *Plea* 2.2/ 1.5; *Hermatobates* 3.9/ 0.9; *Rhagadotarsus* 3.9/ 1.0; *Ochterus* 4.0/ 0.8; *Diaprepocoris* 6.0/ 0.9; *Anisops* 6.1/ 1.0; *Limnogonus* 9.4/ 1.2; *Nerthra* 9.4/ 1.4; *Hydrometra* 11.2/ 1.8; *Notonecta* 13.9/ 2.0; *Aquarius* 15.0/ 1.5; *Limnometra* 17.0/ 2.8 mm; *Nepa* 22.6/ 2.2; and *Lethocerus* 72.0/ 4.0 (data from Cobben 1968; Andersen 1982). This means that, other things being equal, only a few eggs may ripen simultaneously in females of smaller species while females of larger females are able to mature several eggs at the same time.

Most water bugs deposit their eggs in the water or slightly above the water surface. The only exceptions are *Hydrometra*, which places its spindle-shaped eggs well above the water surface (Figs 2.7J-K), and terrestrial species of Ochteridae and Gelastocoridae which lay their eggs on sand or mud, sometimes far from the water margin. The toad bugs, *Nerthra*, lay their eggs in small holes in mud far from shore, the female "guarding" the eggs (Menke 1979). The eggs of water bugs are deposited in three different ways:

(1) *Superficial and lengthwise* on the substrate. Most gerromorphan bugs glue their eggs lengthwise to the substrate, surrounded by a substance that swells in water to a gelatinous mass which may offer the egg protection against desiccation, egg parasites, and predators. Although the eggs are placed superficially, they may be protected in some way. *Hebrus* species tend to deposit their eggs in secluded places, such as the angle between a leaf and a stem of moss plants (Figs 2.7A-B). Species of *Microvelia* and other Veliidae deposit their eggs at or slightly above water (Figs 2.7C-D). Water striders may deposit their eggs in regular rows on floating objects, for example on the underside of a leaf of a water lily. Females may also submerge completely in water to lay their eggs on submerged objects. When suitable objects for oviposition are scarce, several females may oviposit together. Eggs of ocean striders (*Halobates*) are often found in great numbers on any kind of floating object, like bird feathers, shells of squids (*Sepia*), cork, and lumps of tar (Andersen & J. Polhemus 1976).

    Relatively few nepomorphan bugs deposit their eggs superficially and lengthwise to the substrate, but always under water. The eggs of *Aphelocheirus* and *Naucoris* are elongate, slightly bent, with oblique cap-like area at anterior end. Unlike other water boatmen (Corixidae), species of *Diaprepocoris* and *Micronecta* glue their eggs lengthwise to the substrate (Fig. 2.8I). Finally, species of Ochteridae (Fig. 2.8F) and Gelastocoridae glue their eggs lengthwise in small holes on moist sand or mud.

(2) *Superficial and upright* on the substrate. Eggs are deposited on one end, perpendicular to the substrate in two gerromorphan groups, *Hydrometra* (Figs 2.7J-K) and *Hermatobates*. Egg clusters of the latter, exclusively marine group have been found in holes made by boring bivalves in blocks of dead coral (Figs 2.7H-I). Eggs of the nepomorphan subfamilies Corixinae and Cymatiainae (Corixidae) are attached in an upright position to the substrate by a slender stalk (Figs 2.8J-K). The giant water bugs (Belostomatidae) also deposit their eggs in an upright position (Fig. 2.8A). *Lethocerus* females lay their eggs in a mass above water on stems and other objects. The male stays in the water at the base of the

egg batch, defending it against predators and from time to time "water " the eggs. Parental care has been carried a step further in *Diplonychus* and other giant water bugs of the subfamily Belostomatinae where the female lays its egg on the back of the male which "brood" the eggs until they hatch (Smith 1997; further details in Chapter 14).

(3) *Embedded* in the substrate Only two groups of gerromorphan bugs embed their eggs into the substrate, usually plant stems or leaves, *Mesovelia* (Figs 2.7F-G) and *Rhagadotarsus* (Gerridae). In both groups the females have a well developed, serrated ovipositor which is used to drill a hole for the egg. The anterior end of the egg of *Mesovelia* is truncate, with a distinct cap which is pushed open by the pronymph prior to eclosion. Several groups of nepomorphan bugs embed their eggs in to the substrate. The eggs of water scorpions (Nepidae) are unique in having 2 (*Austronepa, Goondnomdanepa, Ranatra*; Fig. 2.8E) or 7-10 (*Laccotrephes*; Fig. 2.8D) threadlike, respiratory filaments at the anterior end. The eggs are inserted in mud and debris at the bottom or in submerged plant tissue. Species of the backswimmer genus *Anisops* (Notonectidae) also insert their eggs in plant tissue (Figs 2.8B-C); the egg has small, cylindrical, bent process at the anterior end, unlike other notonectids (*Enithares, Notonecta*; Fig. 2.8G). Finally, the pygmy backswimmers, *Paraplea*, insert their small eggs in plant tissue (Fig. 2.8H).

Fertilisation of the egg takes place when the egg passes from the ovaries through the genital chamber of the female. During mating, males deposit their sperm in specialised female organs called *spermathecae* where it is stored until fertilisation. The sperm cells are generally very long in water bugs, about as long as the egg. At fertilisation, spermatozoa enter the egg through one or more *micropyles* that penetrate the anterior end of the egg shell. Gerromorphans, except most Veliidae and some Gerridae, have only a single micropyle. So do the nepomorphans *Micronecta* (Corixidae), *Aphelocheirus*, all Notonectidae, *Paraplea* (Pleidae), *Nerthra* (Gelastocoridae), and *Ochterus* (but not *Megochterus* which has two micropyles). Other true water bugs have two or more micropyles with the highest number found in the Belostomatidae (4-14).

When the embryo is fully mature, the pronymph hatches from the egg by pushing its way through the egg shell, usually assisted by an egg burster attached to the cuticle of the pronymph (Fig. 2.7C; eb). Eclosion takes place in several different ways in water bugs (Cobben 1968; Andersen 1982). The simplest way is by a longitudinal split of the egg shell (Fig. 2.9C; es). In *Mesovelia*, Belostomatidae, Nepidae, Corixidae (except *Diaprepocoris* and *Micronecta*), *Ochterus, Aphelocheirus, Naucoris*, and *Anisops* (Notonectidae), the shell splits open along a semicircular or circular line. In a few of these groups (*Mesovelia*, Nepidae, *Aphelocheirus, Naucoris*, and *Anisops*), the anterior end of the egg is modified to form a distinct cap or lid.

Nymphal development.

Water bugs pass through several (usually five) *nymphal* stages (the term *nymph* is preferred to larva to emphasise that the juveniles of true bugs cannot be compared to the larval stage of other insects). Nymphs are similar to the adult bug of the same species, except that they are smaller, usually softer and paler, and lack wings and fully formed genital segments. In addition, nymphs of semiaquatic and aquatic bugs (except Ochteridae) have one-segmented tarsi. This should distinguish them from apterous (wingless) adults found in many species of Gerromorpha which have two or three segments, at least in the middle and hind legs.

Since the body of an insect is surrounded by an exoskeleton that is completely rigid, except for the intersegmental membranes, growth can only take place by *moulting*. Change from one nymphal instar to the next, and from the last nymphal instar to the adult stage, starts with a longitudinal split along the dorsal midline through which the next instar emerges. The new skin is soft and wrinkled and able to expand to the size of the next instar (or adult).

In most cases, the family key to adult water bugs (Chapter 6) also works for nymphs, but 4th and 5th instar nymphs are easiest to key out because some structures are difficult to see in the smaller nymphs.

Phenology

It is relatively straightforward to study the basic life history of a species by rearing individuals taken from the field in the laboratory. Establishing the phenology (annual and seasonal reproductive activity) of natural populations of water bugs is not that easy. It requires that the population is sampled regularly throughout the season, that some quantitative estimates of numbers of different developmental stages (nymphal instars, adults) at a certain time are obtained, and that the reproductive state of adults (in particular

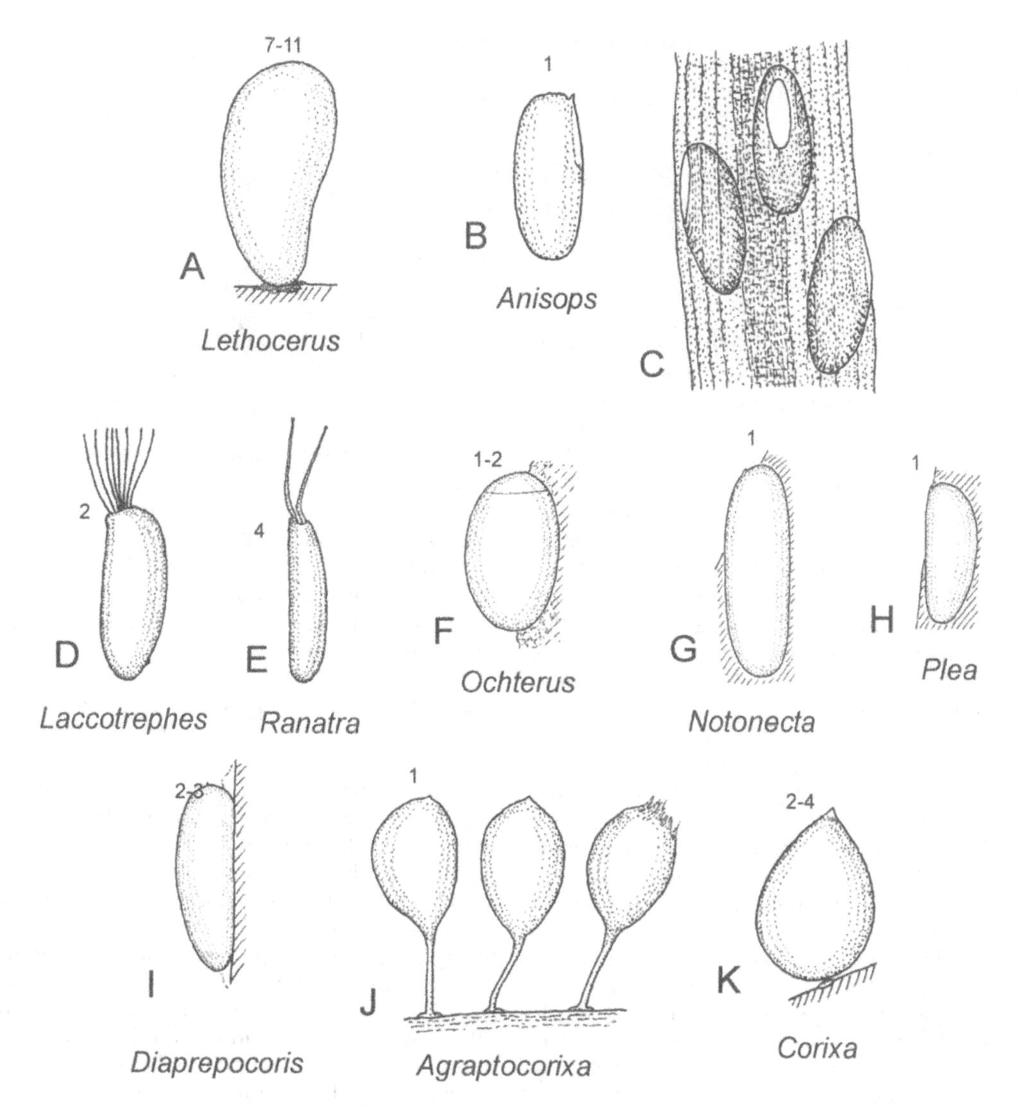

Figure 2.8. A-K, eggs of Nepomorpha (figures above eggs indicate numbers of micropyles). A, egg of *Lethocerus* (Belostomatidae) deposited in upright position. B, egg of *Anisops*. C, eggs of *Anisops* (Notonectidae) deposited in *Potamogeton* stem with anterior end exposed. D, egg of *Laccotrephes* (Nepidae). E, egg of *Ranatra* (Nepidae). F, egg of *Ochterus* (Ochteridae) deposited superficially. G, egg of *Notonecta* (Notonectidae) embedded in substrate. H, egg of *Plea* (relative of *Paraplea*; Pleidae) embedded in substrate. I, egg of *Diaprepocoris* (Corixidae) deposited superficially. J, eggs of *Agraptocorixa* (Corixidae) deposited upright on substrate. K, egg of *Corixa* (relative of *Sigara*; Corixidae) deposited upright. Redrawn from: A, F-I, K, Cobben (1968); B-C, Hale (1923); D-E, Hale (1924); J, Hale (1922).

females) throughout the season is assessed by dissection. For example, Young (1978) compared seasonal cycles of ovarian development in species of Corixidae in Britain (*Corixa, Sigara*) with cycles for species of Corixidae (*Sigara*) and Notonectidae (*Anisops*) in New Zealand. There are many similarities. In both countries, the overwintering adult corixids mature and lay eggs in early spring to give the first adults of the new generation in early summer. These adults immediately begin ovarian development and oviposition to produce a partially second generation. All adults overwinter and lay eggs in the following spring, some for the second time. The main difference between the cycles in the two countries is the much later date of spring ovarian development in Britain, which probably reflects differences in the severity of the winter climates. If the species under study exhibit wing polymorphism, assessment of wing and/or flight muscle development may be added to the above list. An example is given in the next section (Andersen 1973).

Wing polymorphism

Insects owe a great deal of their success to their ability to fly. Nevertheless, secondary reduction and loss of the wings is relatively frequent in many groups of insects, including semiaquatic and aquatic bugs (Menke 1979; Andersen 1982; Schuh & Slater 1995). Unlike the case in butterflies, wasps, and flies, flight does not form part of the every activity of most water bugs, but is probably required for dispersal between aquatic habitats or, in cold climates, between overwintering sites and breeding sites. In winged insects, the development of wings and wing muscles is an energy-demanding process which has to compete with other life processes, especially the production of eggs in females. When flight is no longer necessary for the survival of the species, there is a distinct trend towards flightlessness. This is most clearly demonstrated in marine water striders which live in an extremely stable environment. Adult sea skaters (*Halobates*), for example, always lack wings and wing muscles completely. Most semiaquatic bugs (Gerromorpha) living in freshwater habitats are *wing polymorphic*, with two or more distinct adult *wing morphs*. One is the normal long-winged or *macropterous* morph which presumably can fly (at least at some stage of its life), another is the flightless morph, which may be either wingless (*apterous*) or short-winged (*brachypterous* or *micropterous*). Examples of different wing morphs are illustrated in Fig. 6.1 for *Mesovelia* (B, C) and *Drepanovelia* (H, I) and in the following chapters of this book.

True water bugs (Nepomorpha) are usually long-winged. The only exception is *Aphelocheirus*, which is polymorphic with respect to wing development although most adults are short-winged and flightless (see Chapter 16). The long-winged morph is extremely rare in the European species *A. aestivalis*, but more common in the Australian species, *A. australicus* (Fig. 16.1B). The persistence of well developed wings (at least the fore wings or hemelytra) in nepomorphan bugs is probably associated with the function of the wings in respiration, holding a subelytral air store (see above). However, not all long-winged aquatic bugs are able to fly. Sub-macropterous morphs are frequent among the Corixidae (*Diaprepocoris*, *Micronecta*), Gelastocoridae (*Nerthra*), and Notonectidae (*Anisops*, *Enithares*), distinguished by having slightly shorter hemelytra and narrower thorax than the true macropterous morph. The hind wings are usually reduced in size and the flight muscles absent. But even water bugs with seemingly fully developed wings and thorax do not always fly. The common European water scorpion, *Nepa cinerea*, is usually flightless because its wing muscles are atrophied. Young (1965a, 1965b) found that populations of several British species of Corixidae had a high proportion of individuals with normal wings but undeveloped flight muscles. Scudder & Meredith (1972) showed that such non-flying, long-winged corixids are produced because normal muscle development is arrested in the young (teneral) adult. Experiments showed that flight muscles remained undeveloped if the developing adult was subjected to low temperatures and that growth could be resumed when the same individual was transferred to warmer water.

Andersen (1973) was among the first to demonstrate that wing polymorphism is a seasonal phenomenon in northern European pond skaters (*Gerris* spp.). He sampled populations with two annual generations (that is, bivoltine) at regular intervals during the season, demonstrating that (1) the frequency of short-winged and flightless adults is highest in the overwintering population; (2) that adults of the first generation developing before summer solstice (in late June) were almost 100% flightless, attained reproductive maturity just after their last moult, and produced a second generation of primarily winged offspring; and (3) that adults of the first generation developing after summer solstice was primarily long-winged and entered a reproductive diapause which lasted until next spring. Thus, wing polymorphism is strictly seasonal in these bivoltine pond skaters. This study lead to the hypothesis that wing polymorphism is at least partially determined by environmental cues like temperature and/or day-length (photoperiod). Previous hypotheses that the development of short- and long-winged morphs were mainly determined by genetic factors were thereby questioned. Subsequently, Andersen's (1973) hypothesis has been confirmed and elaborated by several workers in the northern Hemisphere (Spence & Andersen 1994). Factors other than temperature and day-length may be important in regulating the seasonal occurrence of wing morphs. Muraji et. al (1989) found that in the veliid *Microvelia douglasi*, the frequency of long-winged adults decreased significantly in the autumn, and the populations are then almost entirely wingless. The strongest determinant for an increase in the proportion of long-winged forms seems to be the population density, but this factor is modified by day-length.

Much has been written about the ecological and evolutionary significance of wing polymorphism in insects. Darwin (1859) was intrigued by the increased frequency of flightless beetles in some islands and pointed out that if it is a disadvantage for a species to fly (greater risk of being blown to sea), an increasing proportion of populations may become flightless until the entire species has lost this ability. Other workers have associated wing polymorphism with the durational stability of habitats. Species that live in relatively stable habitats are more often flightless than species living in unstable or temporary habitats. In terms of durational stability, water bugs live in habitats ranging from extremely stable (the sea, rivers, and larger lakes), through intermediate (streams, ponds), to extremely unstable or temporary (streams, pools, and marshes that dry up during summer). Following the general explanatory model of wing polymorphism, species that live in stable habitats should be predominantly flightless while species that tolerate temporary or less predictable habitats should be wing dimorphic or long-winged. Andersen & Weir (1997) showed that Australian water striders of the subfamily Gerrinae follow this model quite well. Species which live in rivers and streams (*Tenagogonus australiensis*, *Limnogonus windi*, *Aquarius antigone*, and *A. fabricii*) are predominantly wingless while those species found in habitats that are seasonal or temporary within seasons (*Tenagogerris euphrosyne*, *T. pallidus*, *Limnogonus fossarum gilguy*, *L. luctuosus*, and *L. hungerfordi*) are dimorphic or predominantly long-winged. The most versatile of these are the three *Limnogonus* species, with a wide range of habitats from relatively stable, flowing waters to the most temporary pools, swamps, etc. The flight activity of these species is high which mean that they are able to colonise new and temporary water bodies. These are the most widespread species of water striders in the Indo-West Pacific region where they have colonised even the most remote islands.

Parasites and commensals

Semiaquatic and aquatic bugs are parasitised by immature stages of water mites (Acari: Hydrachnidia). The red, saclike juvenile mites are attached to the body of the host, sometimes in huge numbers (making the host partly or totally reddish). Smith (1989) showed that these mites significantly increased mortality, duration of instars, and variance in age at first moult for parasitised water striders. Mortality and duration of first instars were directly correlated with number of mites per host. The mite *Hydrachna conjecta* Koenike is a parasite of many European species of Corixidae. Adults and larvae feed directly on *Sigara* eggs whereas nymphs attach themselves to the wings of adult corixids and may reduce the fecundity of females (Savage 1989). Ectoparasitic fungi of the order Labaoulbeniales occur on several different groups of water bugs and are usually genus-specific (e.g., Benjamin 1986). Colonial, ciliate protozoans like Peritrichida are common on many true water bugs and Hale (1924) showed that the algae *Cladophora* and *Spirogyra* attached to *Ranatra* and *Laccotrephes*. The last two groups are commensals using the host for transportation without harming it. Water bugs may play an important role in the dispersal of these organisms to new habitats.

Many adult Nepomorpha spread the secretion from the metathoracic scent glands actively over the body. For this purpose the animals crawl out of the water (Naucoridae, Notonectidae, Pleidae) or lie on the water (Corixidae). The secretion has been shown to have an antimicrobial effect and restrains the colonisation of bacteria on the hair pile which holds the air store. The habit of secretion grooming has primarily been found in species with a physical gill. The Nepidae preen only the eyes and the respiratory siphon while *Aphelocheirus* and *Anisops* do not show this behaviour (Kovac & Maschwitz 1989, 1991).

A great variety of endoparasites attack water bugs, including protozoans and nematodes living in the gut of their host (Poisson 1957; Southwood & Leston 1959; Menke 1979). Arnqvist & Mäki (1990) found that trypanosomatid parasites cause significant reductions in vigor of their gerrid hosts. Egg parasitoids of the wasp families Scelionidae, Mymaridae, and Trichogrammatidae also have major impacts on populations of water striders and may restrict some species to temporary habitats unsuitable for overwintering wasps (Spence 1986).

Economic importance

Most water bugs are harmless but biologically fascinating creatures without any particular significance to man apart from being an integral and essential part of aquatic ecosystems. Only a few possible direct links to economic importance have been suggested for semiaquatic bugs (Schaefer & Panizzi 2000). Species of the pond skater genus *Limnogonus* have been reported to be significant predators of the brown plant hopper, *Nilaparvata lugens* (Stål), one of the world's most important rice pests. A detailed assessment of the predatory impact of the veliid *Microvelia douglasi atrolineata* (Bergroth) on *N. lugens* has been presented by Nakasuji & Dyck (1984) who conclude that this species is "one of the most important natural enemies of the brown planthopper". Since surface bugs are known to be highly susceptible to organic insecticides, the use of broad-spectrum insecticides likely diminishes the utility of these bugs as biological control agents in highly managed paddies.

Several species of gerromorphans have been observed consuming mosquito eggs and larvae, but their actual significance as natural enemies of biting flies has not been established. Nummelin (1988) studied the potential of two common East African gerrids, *Gerris swakopenis* (Stål) and *Limnogonus cereiventris* (Signoret), on newly emerged adults of *Aedes aegypti* L., the principal vector of yellow fever. In his experiments, adult gerrids of both species, without access to other food, killed 5-20% of teneral mosquitoes before they could take flight in simple laboratory experiments run under rather high densities of water striders. Gerrid nymphs, however, were not particularly effective at catching the hatchling mosquitoes, which escaped easily by walking across the surface, even before they could take flight. Nummelin concluded that the potential of these two gerrids was "too low for use of water striders as a controlling agent of mosquitoes". Thus, it appears that although water striders provide some background mortality of mosquitoes, their potential as biological control agents for these pests is actually rather low.

Some species of true water bugs (Nepomorpha) may be of economic importance, both in a negative and positive way. Giant water bugs (Belostomatidae) and water scorpions (Nepidae) are powerful predators which include small fish, fish larvae, and fish eggs in their diet. They may therefore be of economic importance in fish hatcheries. On the other hand, some species feed on freshwater snails and may play a role in the control of human and veterinary diseases caused by parasitic worms which have snails as intermediate hosts. As dedicated predators, water bugs bite (or rather "sting") a prey by pushing their stylets into the prey and ejecting a portion of salivary fluid. This fluid is proteolytic and dissolves the tissue of the prey. Several groups of true water bugs are known to inflict bites that are painful to man. In the worst case, however, a bite by a giant water bug or water scorpion will cause an immediate burning sensation which may last for several hours. No permanent damage or side-effects have been reported although Wesenberg-Lund (1943) characterised the bite by the creeping water bug *Ilyocoris cimicoides* (a relative of *Naucoris*) as "worse than a bee sting". Giant water bugs and backswimmers may occasionally fly into swimming pools where they can be a nuisance.

Being so abundant in most freshwater habitats, water bugs play a great role as food for fish such as trout exploited for commercial or recreational purposes. In some cultures water bugs even serve as food for humans. For example, giant water bugs, *Lethocerus*, may be obtained in the food markets in southern China and in Mexico all stages of certain corixids are used as human food (Lauck 1979). Most water bugs feed on larvae of aquatic insects including biting flies and mosquitoes. There are mixed reports about the actual importance of water bugs in controlling these insect pests (Schaefer & Panizzi 2000), but where they are abundant, water bugs undoubtedly play a role which should be taken into consideration before pesticides are used in mosquito control in natural habitats. Using a species of water scorpion, *Laccotrephes*, Hoffmann (1927) reported success for mosquito control in small garden ponds and pots with marsh plants in Guangdong, China. Another nepid, *Cercotmetus asiaticus*, was reported to feed almost exclusively on mosquito larvae in West Malaysia and Laird (1956) suggested introducing it into permanent water bodies on some Pacific Islands in order to help in mosquito control. Care should be taken, however, to ensure that such introduction does not affect non-target organisms important in maintaining the balance of the ecosystem.

Water bugs and environmental monitoring

Water bugs are conspicuous components of undisturbed aquatic and wetland habitats worldwide. Although these bugs are perhaps not critical functional elements of such ecosystems, their presence and local diversity in anthropogenic landscapes, may provide good indications of how human activities affect these systems in comparison to areas less dominated by human activity. For example, in western and northern Europe the large-bodied *Aquarius najas* (DeGeer) appears to be threatened due to an overall reduction of suitable habitats (Nieser & Wasscher 1986; Damgaard & Andersen 1996). This species depends on small streams running through forested areas and which are not polluted by high quantities of organic waste associated with animal agriculture. Declines in *A. najas* populations could indicate that conservation attention is required for the larger suite of species associated with such habitats that were once more common in western and northern Europe. Populations may also increase in response to anthropogenic effects. For example, Nieser & Wasscher (1986) attribute a conspicuous increase in the records of *Aquarius paludum* (Fabricius) to an increase in numbers of "recreative" ponds built and stocked for angling in the southern part of the Netherlands. With sufficient study of metapopulation dynamics, such dispersive species of water bugs might have value as indicators of landscape level changes in wetland habitats.

Savage (1982, 1989) used water boatmen (Corixidae) in the classification of lakes. The species succession in relation to the presence of vegetation and accumulation of organic matter was first demonstrated by Macan (1938) in essentially oligotrophic lakes. Completely different species successions are observed in eutrophic lakes whereas intermediate characteristics occur in more mesotrophic lakes. The distribution and abundance of various *Micronecta* species (Corixidae) may also be used to monitor water quality in lakes. In Lake Vasijärvi in southern Finland, *M. griseola* Horvath was found in eutrophic or eutrophicated waters, *M. poweri* (Douglas & Scott), favoured oligotrophic waters, and *M. minutissima* (L.) was intermediate between the other two species. None of the species could survive in heavily polluted waters, but when the situation improved, *M. minutissima* was the species best tolerating pollution (Jansson 1977a, 1977b, 1987). In southwestern Australia, Davis & Crisidis (1997) found that *Micronecta robusta* Hale and *Agraptacorixa hirtifrons* (Hale) respond positively while *Sigara truncatipala* (Hale) responds negatively to eutrophication. Acidification of many lakes in industrialized countries, owing partly to atmospheric pollution, has also had some surprising consequences. For example, the predominantly carnivorous water boatman, *Glaenocorisa propingua* (Fieber), greatly increased its range in acidified lakes in Sweden where fish populations had become depleted or extinct (Henrikson & Oscarson, 1978, 1981, 1985). Likewise, Bendell & McNicol (1987) reported an increase in abundance of water striders (Gerridae) with increasing acidity in Ontario, Canada, and attributed this change to the drastic reduction in fish populations that accompanies acidification.

In general, water bugs, along with other macroinvertebrates, may be important indicators of water quality in natural water bodies, in particular streams and rivers. By scoring the presence of various species or higher groups of water bugs, a relative number can be calculated using the SIGNAL method ("Stream Invertebrate Grade Number - Average Level") which assigns a certain grade (1 - 10) to various groups of water bugs (Gooderham & Tsyrlin 2002). The higher the grade the more sensitive to pollution is the group. In this way an objective opinion about the "health" of a certain stream or river may be formed. It is now generally recognised that freshwater habitats are among the most threatened biotas throughout the World and those most in need of conservation efforts (Polhemus 1993). Water bugs may not only be useful for environmental monitoring in freshwater localities. The Australian water bug fauna is unique in its relatively high number of species confined to coastal, marine habitats, such as mangrove swamps and intertidal coral reef flats, which are seriously threatened by human activities.

3. Classification and Phylogeny

Classification

The pre-evolutionary entomologist Léon Dufour (1833) classified the true bugs (Hemiptera-Heteroptera) into three "families" after their preferred habitats: (1) *Hydrocorises* (aquatic bugs), (2) *Amphibicorises* (semiaquatic bugs), and (3) *Geocorises* (terrestrial bugs). Although this classification is strictly "typological" (based on only one set of characters), Dufour's groups have been in use up to quite recently as the "Divisions" or "Series" Hydrocorisae, Amphibiocorisae, and Geocorisae. What Cobben (1968) called the "-morpha" era in the classification of Heteroptera began with a short note by Leston, Pendergrast & Southwood (1954) in which one of Dufour's "families", Geocorisae, was subdivided into two groups, the Cimicomorpha (those resembling bed bugs) and Pentatomomorpha (those resembling stinkbugs). This action questioned the long-assumed "naturalness" (or monophyly in modern terms) of the terrestrial bugs. The Russian paleontologist Y. Popov was the first to introduce the category "infraorder" into the classification of true bugs with a consistent use of a compound name formed by the name of the type-genus ending with "-morpha". Popov (1971a, 1971b) recognised five extant infraorders: Nepomorpha (aquatic bugs), Leptopodomorpha (shore bugs and their allies, semiaquatic bugs), Enicocephalomorpha (a small terrestrial group including Dipsocoridae and their allies as well), Cimicomorpha, and Pentatomomorpha. This classification was emended by Stys & Kerzhner (1975) with the addition of the infraorders Dipsocoromorpha and Gerromorpha (semiaquatic bugs excluding shore bugs) to produce the classification of the Heteroptera currently in use (e.g., CSIRO 1991; Cassis & Gross 1995; Schuh & Slater 1995).

The currently accepted classification of the infraorder Gerromorpha stems from the work by Andersen (1982). He recognised the following 4 superfamilies and 8 families [taxa marked with an asterisk (*) not represented in Australia]:

Superfamily Mesoveloidea
  Family Mesoveliidae
Superfamily Hebroidea
  Family Hebridae
Superfamily Hydrometroidea
  Family Paraphrynoveliidae*
  Family Macroveliidae*
  Family Hydrometridae
Superfamily Gerroidea
  Family Hermatobatidae
  Family Veliidae
  Family Gerridae

The most recent classification of the infraorder Nepomorpha is that of Stys & Jansson (1988). These authors recognise 6 superfamilies and 11 families [taxa marked with an asterisk (*) not in Australia]:

Superfamily Nepoidea
  Family Nepidae
  Family Belostomatidae
Superfamily Corixoidea
  Family Corixidae
Superfamily Naucoroidea
  Family Aphelocheiridae
  Family Potamocoridae*
  Family Naucoridae
Superfamily Ochteroidea
  Family Ochteridae
  Family Gelastocoridae
Superfamily Notonectoidea
  Family Notonectidae
Superfamily Pleoidea
  Family Pleidae
  Family Helotrephidae*

[Note: some authors, e.g., Chen et al. (in press), treat the corixid subfamilies Diaprepocorinae and Micronectinae as separate families].

Phylogeny

It is generally agreed that classifications in order to be "natural" should only include groups that are monophyletic in the strictest sense, that is containing all and only the descendants of a single ancestral species. A system following this rule is built using the principles and methods of *phylogenetic systematics* (sometimes also referred to as cladistics), first laid out by the German entomologist Willi Hennig (e.g., Hennig 1966; for an overview, see Andersen 2001b). A phylogenetic classification is usually presented as a phylogeny (or cladogram), a branching diagram picturing the relationships between *terminal taxa*, which may be species, genera, or groups of higher rank (subfamilies, families, superfamilies, infraorders).

Schuh (1979) presented the first documented higher-level phylogeny for the seven infraorders of Heteroptera (see above), chiefly based upon morphological data assembled by Cobben (1968, 1978). Schuh's phylogeny placed the infraorders

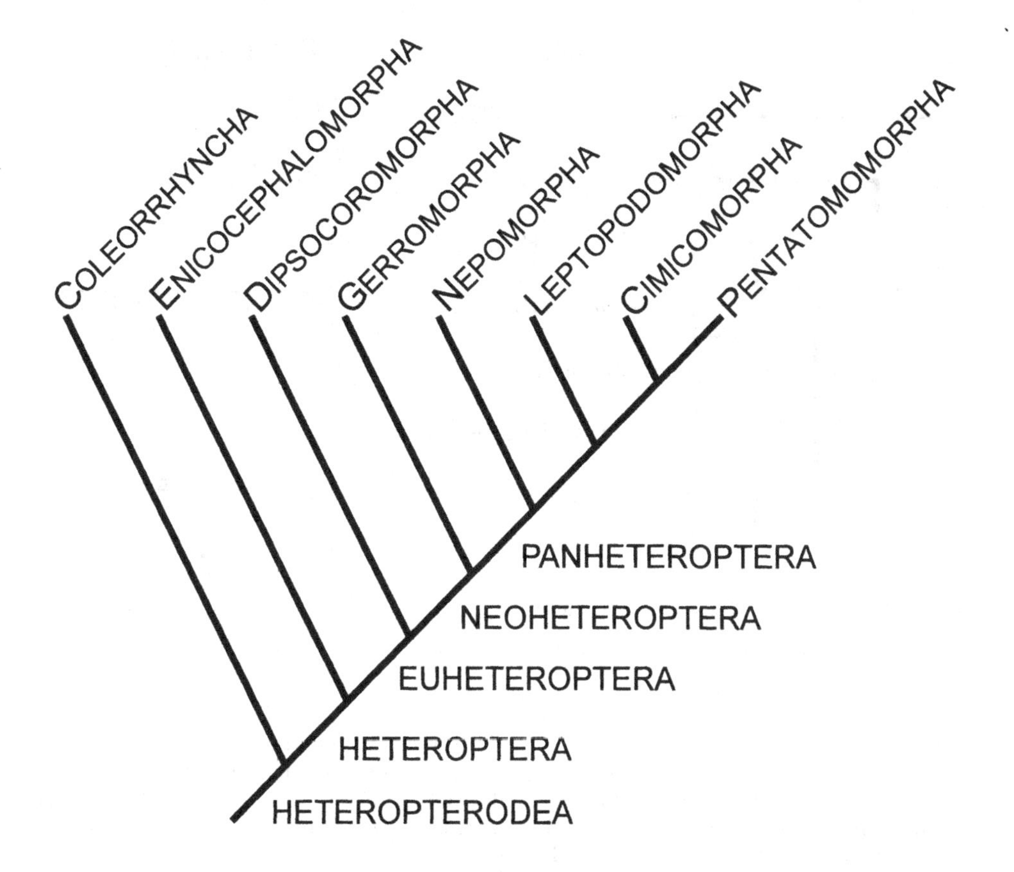

Figure 3.1. Phylogeny of the infraorders of Hemiptera-Heteroptera.

Enicocephalomorpha, Dipsocoromorpha, and Gerromorpha as the most basal groups, whereas the infraorders Cimicomorpha, Pentatomomorpha, Leptopodomorpha, and Nepomorpha (the last two as close relatives) were placed as increasingly more derived. Little new evidence for higher-level relationships within the Heteroptera was added until Wheeler et al. (1993) published molecular data (18S rDNA sequences) for 29 Hemiptera taxa, representing all infraorders and six "outgroup taxa" (non-heteropteran species added to "root" the phylogeny). Their phylogeny (Fig. 3.1) indicates substantial agreement (congruence) between the molecular data and the morphological data used by Schuh (1979). The infraorders Enicocephalomorpha, Dipsocoromorpha, and Gerromorpha branch off in that order, followed by the Nepomorpha, Leptopodomorpha, Cimicomorpha, and Pentatomomorpha. The Coleorrhyncha (Peloridiidae) - a small group of hemipterans inhabiting South America, New Zealand, New Caledonia, and Australia (incl. Lord Howe Island) - is found to be the closest relatives of the true bugs, Heteroptera. Subsequently, Stys (1985) provided a set of names for the basal inclusive groups on the phylogeny (Fig. 3.1).

The phylogenetic hypothesis (Fig. 3.1) is at variance with traditional ideas about the evolution of the Heteroptera (Schuh & Slater 1995). First, true bugs are most likely primitively predaceous, contrary to the beliefs of many previous authors. Second, the Enicocephalomorpha is the basal heteropteran lineage, rather than the Nepomorpha (as proposed by China 1933) or Gerromorpha (as proposed by Cobben 1978). Third, the "hemelytron" (fore wing composed of a coriaceous basal part and membranous distal part) is not a "ground plan" character (found in the common ancestor) for the Heteroptera, but rather a synapomorphy (derived character decisive for monophyly) for the "Panheteroptera" (Fig. 3.1).

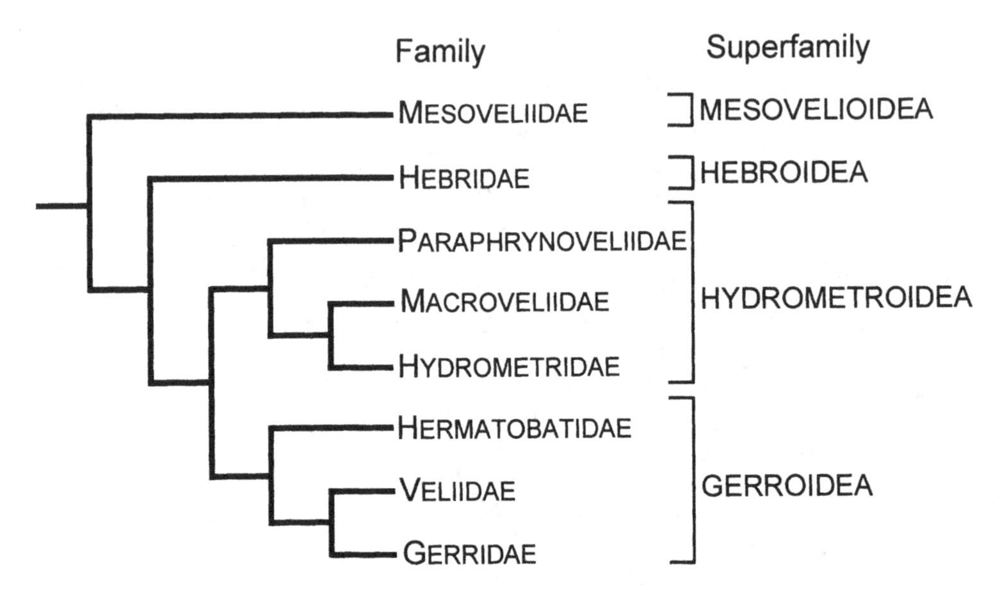

Figure 3.2. Phylogeny of the families of Gerromorpha.

Gerromorpha

The infraorder Gerromorpha (semiaquatic bugs) is the sister group of the Panheteroptera (Fig. 3.1). The monophyly of the Gerromorpha is supported by the following apomorphic (derived) characters: (1) head with three pairs of trichobothria (long slender setae) inserted in cuticular pits; (2) mandibular levers (internal rod associated with the base of the mandibular stylets) quadrangular; (3) pretarsus with two arolia, one dorsal and one ventral; (4) female genital tract with gynatrial complex, including a long tubular spermatheca and a secondary fecundation canal. The phylogenetic relationships between its 8 families (see classification above) was reconstructed by Andersen (1982) based upon a manual cladistic analysis of a data set of 50 characters covering most of the morphological variation within the infraorder. Each of the eight gerromorphan families was scored for their ground plan characters, i.e., the character states inferred to be ancestral to the respective families. A quantitative cladistic reanalysis of these data by Andersen & Weir (in press) supports Andersen's (1982) initial hypothesis (Fig. 3.2).

The family Mesoveliidae includes surface bugs that are more primitive than other gerromorphans in a more generalised thorax structure, metathoracic scent apparatus, male genital tract, etc. The Mesoveliidae therefore assume a basal position in the cladogram of relationships between the families of Gerromorpha (Fig. 3.2). The family Hebridae comprises a very distinct group and its monophyly is supported by several derived characters, e.g., presence of plate-shaped bucculae, two-segmented tarsi, fore wing venation reduced, and genital segments inserted before apex of abdomen. Hebrids are more apomorphic than members of the family Mesoveliidae in having a reduced meso-scutellum, metathoracic spiracles located laterally on thorax, coxae laterally displaced on thorax, separated by a rostral groove, and a male genital tract without mesadene (accessory) glands.

Macropterous (winged) adults of the Macroveliidae, Hydrometridae, Veliidae, and Gerridae all have the notal part of pronotum produced backwards, forming a so-called pronotal lobe which covers the mesonotum and median part of metanotum. The meso-scutellum is rudimentary or absent. These character states, however, cannot be scored for the families Paraphrynoveliidae and Hermatobatidae where only the apterous (wingless) adult form is known. Gerromorphan families except the Mesoveliidae and Hebridae share a derived metasternal scent apparatus, in particular the glandular reservoir, and presence of sclerotisations associated with the accessory scent gland. These characters support (as synapomorphies) the relationship between these six gerromorphan families (Fig. 3.2).

The close relationship between the Hydrometridae and the small non-Australian families Macroveliidae (North and South America) and Paraphrynoveliidae (South Africa) was suggested by Andersen (1978, 1982) based on the anteriorly produced bucculae (ventral lobes of head) and presence of a metepisternal process. The two known species of *Paraphrynovelia* are only known in the apterous (wingless) adult form and the missing observations for structures associated with wings complicate a correct placement of this family. The sister group relationship between the Macroveliidae and Hydrometridae is supported by the elongate shape of the head with eyes distinctly removed from the anterior margin of prothorax, the dorsal location of the metathoracic spiracle, and presence of paired ridges on the basal abdominal terga. The monophyly of the family Hydrometridae is strongly supported by the prolonged head capsule, tuberculate posterior pair of cephalic trichobothria, and apically modified antennae.

Previous authors (e.g. China 1957) treated the small, exclusively marine family Hermatobatidae as a member of the Gerridae, chiefly based on its superficial resemblance to the likewise marine gerrid genus *Halobates*. The hermatobatids are, however, more plesiomorphic than both the Veliidae and Gerridae in the characters used to support the sister group relationship between these two families. Species belonging to the Hermatobatidae share the clypeal embryonic egg burster with the veliids, but since the frontal egg burster of the Gerridae is assumed to have evolved from the clypeal type (Cobben 1968), this character may be a synapomorphy for the three families. The presence of intercalary sclerites between the third and fourth labial segments, a female gynatrial complex with a glandular gynatrial sac, and only four ovarioles in each ovarium, are other synapomorphies supporting the close relationship of the Hermatobatidae with the veliid-gerrid clade.

Arguments in favor of the monophyly of the Veliidae used by Andersen (1982) included the anteriorly deflected head capsule, the presence of a specialised grasping comb on the male fore tibiae, and a row of trichobothria-like setae on the middle tibiae. The South African genus *Ocellovelia*, however, lacks the protibial grasping comb and mesotibial setae and is the only veliid genus which includes species with ocelli. On the other hand, the protibial grasping comb and mesotibial setae have been lost in some veliids, e.g. the Haloveliinae (except *Halovelia*) and the absence of these structures in *Ocellovelia* may be secondary. Supporting characters for the monophyly of the Veliidae are the deflected head, a median, longitudinal impressed line on the dorsal head surface, metathorax with lateral scent channels (shared with some Gerridae), and the egg with more than one micropyle.

The Gerridae is a species-rich but structurally quite homogenous and undoubtedly monophyletic group with a number of apomorphic characters, such as the loss of ocelli, increase of cephalic trichobothria from three to four, prolongation of the mesothorax, rotation of the middle and hind coxal axes from oblique to horizontal orientation, reduction in number of tarsal segments from three to two, and change of the embryonic egg burster from clypeal to frontal. The sister group relationship between the Gerridae and Veliidae (Fig. 3.2) is strongly supported by several synapomorphies, including the presence of cuticular thorns on the body ("grooved setae" of Cobben 1978; found in all veliids and in the gerrid genus *Rhagadotarsus*), salivary pump being laterally inflected from behind, preapical insertion of claws, asymmetrical or reduced parempodia, and loss of abdominal scent apparatus.

The phylogeny of the Gerromorpha outlined above has not been seriously challenged since its original presentation by Andersen (1982). However, Muraji & Tachikawa (2000) recently presented the results of an analysis of 32 species of Gerromorpha based on partial sequences of the mitochondrial 16S rDNA (c. 400 base-pairs) and nuclear 28S rDNA (c. 500 bp) genes. They sequenced 24 species and subspecies of Gerridae representing three subfamilies, one species of Hermatobatidae, and five species of Veliidae representing three subfamilies. They used one species of *Mesovelia* (Mesoveliidae) and one species of *Hydrometra* (Hydrometridae) as outgroups. Muraji & Tachikawa (2000) confirmed the monophyly of the superfamily Gerroidea (Gerridae + Hermatobatidae + Veliidae), but found difficulties in distinguishing between taxa belonging to the families Gerridae and Veliidae, in particular *Halovelia* (Haloveliinae). The authors concluded that ribosomal DNA can be used for studying the relationships among higher taxa (subfamilies, families), but recommended that such studies are supplemented with analyses of several other DNA regions. We may add that phylogenetic analyses of combined molecular and morphological data sets also seem relevant and desirable. Such work is ongoing (Damgaard et al. unpublished).

Previous ideas about the ecological evolution of semiaquatic bugs suggested an early colonisation of the water surface by "littoral" bugs (China 1955; the Saldidae or shore bugs were thought to be the sister group of the Gerromorpha). These bugs included the ancestor of all families except the Mesoveliidae and Hebridae. Terrestrial or semi-terrestrial forms of the families Macroveliidae and Hydrometridae were seen as examples of secondary return to land. Based upon reconstructed phylogenies of higher taxa, several elements of China's (1955) scenario have been falsified (Andersen 1979, 1982). Semiterrestrial habitats (the "hygropetric adaptive zone") were inferred to be ancestral to all gerromorphan lineages except the Gerridae, and the adaptive transitions between such habitats and the water surface (the "pleustonic adaptive zone") were visualized as being a gradual process where different lineages, independently of each other, passed through a transitional adaptive zone (the "intersection zone") and reached the pleustonic zone where extensive adaptive radiation took place. However, a cladistic test supports only some parts of this scenario (Andersen 1995c). The hypothesis that the hygropetric zone is the ancestral one is confirmed for all lineages except the Gerroidea. There is only weak support for the hypothesis that the intersection zone was a sort of transitional zone during the ecological evolution of pleustonic bugs. On the contrary, most of the unique structural and behavioural adaptations of water striders seem to have evolved after the inferred, direct transition between the hygropetric and pleustonic zones. Such cladistic tests have also been successfully applied to other aspects of water strider biology as exemplified by studies of wing polymorphism, sexual size dimorphism, and mating strategies (Andersen 1993a, 1994, 1997, 2000b; Spence & Andersen 1994).

Nepomorpha

The infraorder Nepomorpha (aquatic bugs) is the most basal lineage of the Panheteroptera and sister group of the major groups of terrestrial bugs (Fig. 3.1). The monophyly of the Nepomorpha is supported by the following apomorphic characters: (1) antennae short, usually concealed in a groove or concavity beneath the eyes; (2) fore wings with well developed wing-to-body coupling mechanism; (3) fore wings in the form of hemelytra, with coriaceous basal part and membranous distal part (shared with other Panheteroptera).

China (1955) proposed a scheme of relationships among the families of aquatic bugs, treating the Ochteridae as the most primitive family, on

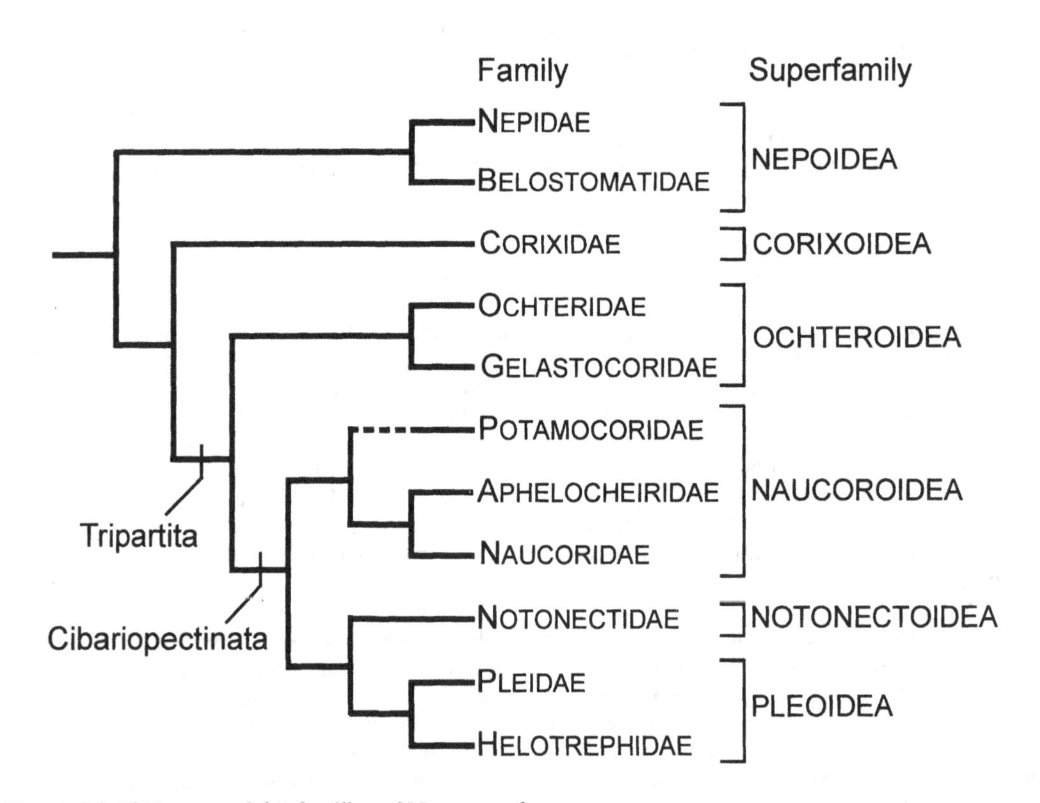

Figure 3.3. Phylogeny of the families of Nepomorpha.

the basis of the presence ocelli and a respiratory system typical of terrestrial bugs. More recent authors (Popov 1971a; Rieger 1976; Mahner 1993), have placed the Nepoidea (Nepidae + Belostomatidae) as the sister group of the remaining Nepomorpha. The character analyses of Rieger (1976) and Mahner (1993) show a substantial increase in breath of coverage and methodological sophistication over previous efforts. Their conclusions, however, contradict one another mainly in the placement of the Corixidae and relationships among those families placed in the Naucoroidea, the Aphelocheiridae, Naucoridae, and Potamocoridae (a small group of naucorid-like bugs confined to the Neotropics). Fig. 3.3 depicts a phylogeny of the Nepomorpha presented by Andersen (1995d) based upon a critical review of Mahner's (1993) work. Ongoing analyses involving both morphological and molecular data support crucial elements of this hypothesis (Hebsgaard et al. in press).

Mahner (1993) - using the old name Crytocerata for aquatic bugs (from the concealed antenna) - presented the most comprehensive morphology-based phylogenetic analysis of the Nepomorpha to date. Unfortunately, in his classification, Mahner adheres to a strict subordination of taxa and name these without regard to conventional rank and nomenclature. For example, the superfamily Notonectoidea sensu Mahner includes both a family, Notonectidae, and another superfamily, Pleoidea (Pleidae + Helotrephidae). In his phylogenetic analysis, Mahner (1993) emphasises the importance of the food pump and associated structures and for the first time describes these for the Pleidae, Helotrephidae, and Potamocoridae. Along with other traits, a tripartite food pump is used to support the monophyly of a group composed by the Ochteroidea (Ochteridae + Gelastocoridae), Naucoroidea (Naucoridae + Aphelocheiridae), Notonectidae, and Pleoidea (Pleidae + Helotrephidae) for which Mahner proposes the name Tripartita. The food pump of these groups except the Ochteroidea also possess so-called cibariopectines and Mahner (1993) proposes the name Cibariopectinata for a monophylum composed of these groups.

The sister group relationship between the Nepidae and Belostomatidae is above all supported by the presence in both groups of paired respiratory processes on the apex of abdomen. In the Nepidae, these processes form a respiratory siphon of varying length whereas they form so-called air-straps in the Belostomatidae. The monophyly of the Nepidae is further supported by the presence of static sense organs on the abdominal venter and the presence of respiratory horns on the eggs. The monophyly of the Belostomatidae is supported by the conical hind coxae which are firmly associated with metapleuron and the hind tibia modified as swimming legs (flattened, fringed with hairs).

Previous studies on the phylogeny of the Nepomorpha have not been able to find a consensus for the relative position of the Corixidae and Ochteroidea (Ochteridae + Gelastocoridae). Rieger (1976) placed the Corixidae as sister group to the Naucoroidea + Notonectidae + Pleoidea based on the aquatic way of life and presence of air-stores (as opposed to the Ochteroidea). These traits, however, are most likely subject to parallel evolution. Mahner (1993) united the Ochteroidea with the three groups mentioned above to form a monophylum which he called Tripartita on account of their tripartite food pump (Fig. 3.3). The monophyly of the Corixidae is unquestionable because of the unique structure of the rostrum (short, triangular with indistinct segmentation) and fore tarsus (forming a pala). the presence of ocelli in one subfamily, Diaprepocorinae, suggests that ocelli may have been present in the common ancestor of corixids and subsequently lost in most subfamilies.

The sister group relationship between the Ochteridae and Gelastocoridae is based on several characters such as the presence of a cross-fold on the clypeus, rotation of the male genital capsule, and the one or two-segmented fore tarsi. Whereas the Ochteridae has relatively few derived characters (e.g., the ridged clypeus), the Gelastocoridae has several, e.g., raptorial fore legs, short and robust rostrum, tarsal formula 1 : 2 : 3, and generally "warty" appearance.

Based on the modified middle part of the food pump (presence of cibariopectines), Mahner (1993) proposed a monophylum called Cibariopectinata for the Naucoroidea + Notonectidae + Pleoidea (Fig. 3.3). The relationships of the three families Aphelocheiridae, Naucoridae, and Potamocoridae (together forming the Naucoroidea) have been much discussed. The family status of the latter was proposed by Stys & Jansson (1988) following a suggestion by Cobben (1978). Rieger (1976) considered the Aphelocheiridae to be closer to the Notonectidae + Pleoidea than to the other Naucoroidea. Finally, Mahner (1993) maintained a Naucoroidea composed of Naucoridae + Aphelocheiridae, but left the Potamocoridae unplaced within his Cibariopectinata (Fig. 3.3). Recent evidence from molecular data (Hebsgaard et al. unpublished) strongly suggests a sister group relationship between the Aphelocheiridae and Potamocoridae. These data, however, also question the monophyly of the family Naucoridae.

The strongly convex dorsum and unique way of swimming upside-down supports the sister group relationship between the Notonectidae and Pleoidea (Pleidae + Helotrephidae). The monophyly of the Notonectidae is supported by the presence of respiratory grooves on abdominal venter and oar-like hind legs fringed with swimming hairs. The firm association between the head and prothorax supports the sister group relationship between the families Pleidae and Helotrephidae (head and pronotum forms a cephalonotum in the latter family).

Mahner (1993) discussed the ecological evolution of true water bugs based (at least in part) on his reconstructed phylogeny. All nepomorphan families, except the Ochteridae and Gelastocoridae, are aquatic in their way of life. The similarities in spiracle structure, distribution of air stores, and the shortened (cryptocerate) antennae lead Mahner to believe that ancestral nepomorphans were aquatic rather than terrestrial insects (contrary to the belief of China 1955, who saw the terrestrial Ochteridae as the most ancestral group). Mahner (1993) hypothesized that the ancestral mode of air intake was naucorid-like, i.e., at the abdominal end when the body is placed in normal, dorso-ventral position. From this state, the evolution of respiratory behaviour took different courses: air intake by way of a siphon (Nepoidea), through channels on the abdominal venter when the body is placed upside-down (as in the backswimmers, Notonectidae), and by way of the shield-like extended head and pronotum (Corixidae). In any case, the most parsimonious sequence of transitions between the aquatic and terrestrial environment is that ancestral nepomorphans were aquatic and that the terrestrial way of life in ochterids and gelastocorids has evolved secondarily. This hypothesis is supported by the presence of thoracic air stores in these families which are not found in other terrestrial bugs. The shortened antennae (especially in the Gelastocoridae) also support this scenario. Swimming legs (where the segments are flattened and provided with hair fringes) probably evolved independently several times in the course of the evolution of water bugs.

Fossils

The fossil record of water bugs is better than for any other group of Heteroptera. That record has been reviewed by Popov (1971a), Carpenter (1992), Nel & Paicheler (1992-1993), Andersen (1998b), and in Rasnitsyn & Quicke (2002). Most fossils are preserved as impressions in rocks but,

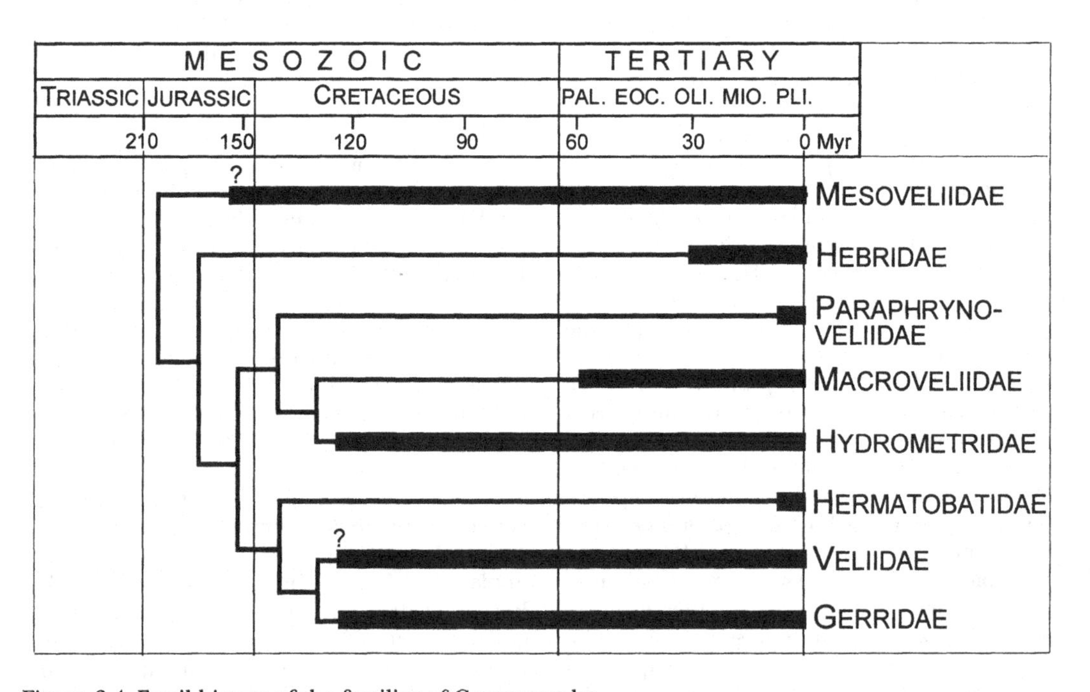

Figure 3.4. Fossil history of the families of Gerromorpha.

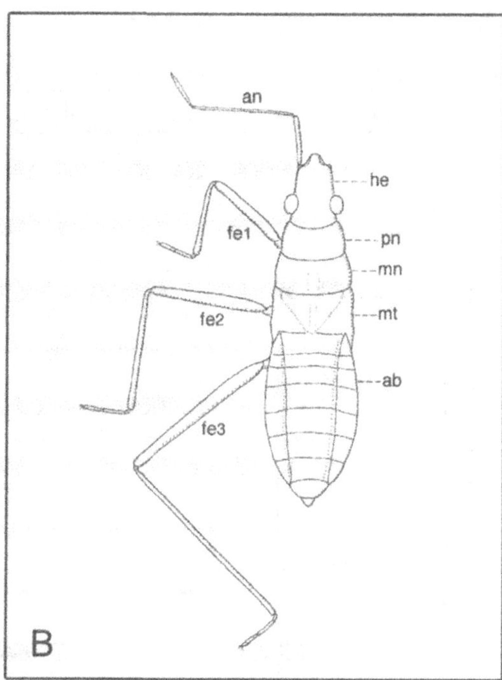

Figure 3.5. *Duncanovelia extensa* (Mesoveliidae), from the Lower Cretaceous of Australia. A, holotype, Koonwarra formation, Victoria. Photograph by David Grimaldi. B, reconstruction (ab, abdomen; an, antenna; he, head; fe1, fe2, fe3, fore, middle, and hind femur; mn, mesonotum; mt, metanotum; pn, pronotum). Reproduced from Andersen (1998b).

surprisingly, also found as inclusions in fossiliferous amber. Fossil water bugs are known from geological formations in most continents, including Australia. Semiaquatic bugs (Gerromorpha) have a fossil record spanning more than 120 million years of geological history (Fig. 3.4) with fossil forms representing six out of eight families. Most fossils are of Cenozoic (Tertiary) age. The Mo-Clay (Fur and Ølst Formations) of northern Denmark (Paleocene-Eocene transition) contains an unusually rich fauna of insects, including seven species classified in four genera and three families, Macroveliidae, Hydrometridae, and Gerridae (Andersen 1998b). A species of the marine genus *Halobates* Eschscholtz (Gerridae) has been described from marine deposits of northern Italy (Andersen et. al 1994). The Eocene Baltic amber is famous for its inclusions of primarily terrestrial insects whereas aquatic insects are quite rare (Larsson 1978; Poinar 1992; Wichard & Weitschat 1996; Weitschat & Wichard 1998). Nevertheless, Andersen (2000c) described fossils from Baltic amber representing the families Hydrometridae, Veliidae, and Gerridae. Water striders have also been found in the Oligocene/Miocene Dominican amber, including an extinct subfamily of Gerridae, several species of *Microvelia* Westwood, and a species of the marine genus *Halovelia* Bergroth (Andersen & Poinar 1992, 1998; Andersen 2001a).

An extinct genus of Hydrometridae, *Carinametra* Andersen & Grimaldi (2001) was described from the Middle Cretaceous Burmese amber, but the oldest members of the Gerromorpha are of Lower Cretaceous (Aptian) age and include species of Hydrometridae, Mesoveliidae, and Gerridae (Andersen 1998b; Perrichot et al. in press). Some of these fossils are from Australia. Jell & Duncan (1984) described *Duncanovelia extensa* (Mesoveliidae) and recorded an undescribed veliid from the Koonwarra formation, Victoria (Fig. 3.4). The infraorder Gerromorpha probably evolved in the Lower Mesozoic and putative gerromorphans have been recorded from the Upper Jurassic (Rasnitsyn & Quicke 2002).

The fossil record of Nepomorpha is well documented through the works by Popov (1971a), Nel & Paicheler (1992-1993), and Rasnitsyn & Quicke

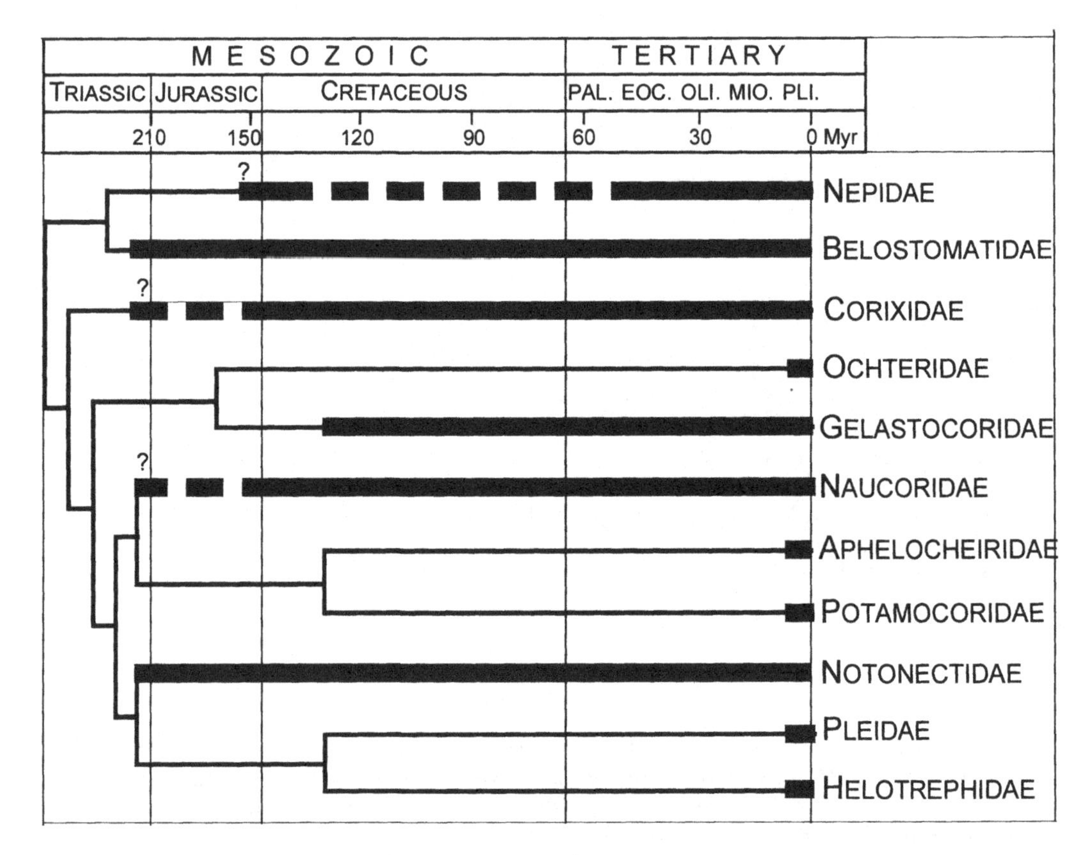

Figure 3.6. Fossil history of the families of Nepomorpha.

(2002). Most fossil forms are preserved as impressions in rocks which usually do not reveal the details crucial for safe identification. Aquatic bugs have a fossil record spanning more than 200 million years of geological history (Fig. 3.6) with fossil forms representing six out of eleven families. There are many fossil nepomorphans in Tertiary, lacustrine deposits, including species belonging to extant genera of Belostomatidae, Corixidae and Notonectidae. There are only few Baltic amber inclusions of "real" aquatic insects (where juveniles or adults normally live in the water), but these include nymphs of water boatmen (Corixidae). The oldest Nepomorpha are found in Upper Triassic deposits of Virginia, U. S. A. and belong to the families Belostomatidae, Notonectidae, and (possibly) Naucoridae (Fraser et al. 1996). Upper Jurassic belostomatids from the famous lithographic shale of Solenhofen, Germany, include typical forms like *Mesobelostomum* Haase and *Mesonepa* Handlirsch, but also the peculiar *Stygeonepai* Popov (with very broad, oar-shaped hind legs). The Lower Cretaceous genus *Clypostemma* Popov was referred to the extant genus *Notonecta* by Nel & Paicheler (1992-1993). The family Gelastocoridae is represented by nymphs found in the Lower Cretaceous Koonwarra formation, Victoria (Jell & Duncan 1984). Judging from the fossil record, the infraorder Nepomorpha undoubtedly evolved at the transition between the Upper Palaeozoic (Permian) and Lower Mesozoic (Triassic) and was already quite diverse in the Jurassic and Early Cretaceous (Rasnitsyn & Quicke 2002).

4. Identification

Terms used in the keys and descriptions

Major parts of the body

The body of a water bug is composed of (Figs 4.1A-C): (1) the *head* (he) with its appendages, the two antennae (an) and the rostrum (ro); (2) the *thorax* (th) with its appendages, the three pairs of legs and the two pairs of wings (when present); and (3) the *abdomen* (ab) with the genital segments (gs).

Head

The head is very variable in shape, ranging from extremely short and wide (Hermatobatidae, Corixidae, Naucoridae) to extremely long and slender (Hydrometridae). It is typically elongate, with prominent, globular *eyes* (Fig. 4.1A; ey). The minimum distance between the eyes is commonly referred to as the *synthlipsis* (e.g., Fig. 20.1A, sy). Two *ocelli* (Figs 4.2A, B; oc) are found in a few families (Hebridae, Mesoveliidae, Gelastocoridae, Ochteridae, Corixidae-Diaprepocorinae), located near the inner margins of the eyes. The frons is not delimited from the clypeal region. Anteriorly there are a pair of lateral grooves which divide the clypeal region into an *anteclypeus* (Fig. 4.2A; ac) and a *postclypeus* (pc). The lateral grooves continue antero-laterally on either side into the deep clypeal clefts separating the anteclypeus from the lateral *maxillary plates* (mx). The anterior margin of the anteclypeus is joined by a membrane to the broad base of the triangular *labrum* (lr). The part of the head lying immediately dorsal to the maxillary plate is commonly identified as the *mandibular plate.* The lowermost side of the maxillary plate borders the ventral plate or *buccula* (Fig. 4.2B; bu) which typically is quite distinctly separated from the maxillary plate by a groove. The bucculae overlap the first labial segment (la1) and often continue posteriorly along the ventral head surface (gula) as a pair of ridges.

The *antennae* (Fig. 4.1A; an) are inserted in front of the eyes on the antennal tubercles or *antennal sockets* (Figs 4.2A, B; at). In the Gerromorpha the antennae are long and slender, with 4 segments (the apparent 5-segmented antenna of some Hebridae is formed by a non-segmental subdivision of the last segment). The bases of the last two segments are usually encircled by narrow, membranous bands separating short *internodial pieces.* In the Nepomorpha, the antennae are short, at most as long as head, often thickened, provided with processes, or with segments fused, situated postero-ventrally on head, often hidden in a groove or concavity.

The *rostrum* (Fig. 4.1B, C; ro) typical of hemipterous insects is composed of a 4-segmented *labium* (Fig. 4.2B; la1-la2) enclosing two pairs of *stylets,* corresponding to the mandibulae and maxillae of insects with biting mouthparts. It is typically long and slender in the Gerromorpha, reaching backward beneath the body when not in use. In the Nepomorpha the rostrum is usually short and stout (except Ochteridae and Aphelocheiridae), often with 3 apparent segments since segment 1 is greatly reduced. The broadly triangular, non-segmented labium of the Corixidae is unique (Fig. 4.2F). The dorsal surface of the extended labium is indented medially to form a groove which encloses the *mandibular* (Fig. 4.2E; md) and *maxillary stylets* (ms). The first and second segments are usually the shortest of the four labial segments whereas the third segment is by far the longest. The last two labial segments may be connected by a pair of *intercalary sclerites.* The maxillary stylets are very motile and flexible, enclosing between them one or two canals communicating with the salivary glands (Fig. 4.2E; sa) and/or the food pump (fo).

Gerromorpha have three pairs of *cephalic trichobothria* (Figs 4.2A, B; ct) inserted on the dorsal head surface. The posterior pair is inserted at the base of the head and usually close to the hind corners of the eyes. The two anterior pairs are inserted on the frontal surface of the head, usually with one pair just behind the other. The cephalic trichobothria are long and slender setae (the "trich"), each arising from a deep cuticular pit. The socket of the trich is there located on a domelike elevation ("bothrium"). This structure is unique to adult gerromorphans whereas in nymphs the trich arises from a domelike elevation projecting from the cuticular surface.

Thorax

Most of the externally visible parts of the *prothorax* (Figs 4.1A-C; pn) are composed of marginal evaginations or lobes. A short cervical constriction surrounds the postocciput of the head an sometimes forms a narrow anterior collar. The pronotum is a broad plate with its lateral and, in particular its posterior margin expanded into a large *pronotal lobe* (Figs 4.1A, C; pn). The lobe typically overlaps and covers most of the mesonotum (ms) and the wing bases where it is elevated as *humeral* projections. The dorsum of the *mesothorax* or mesonotum is the largest part of the

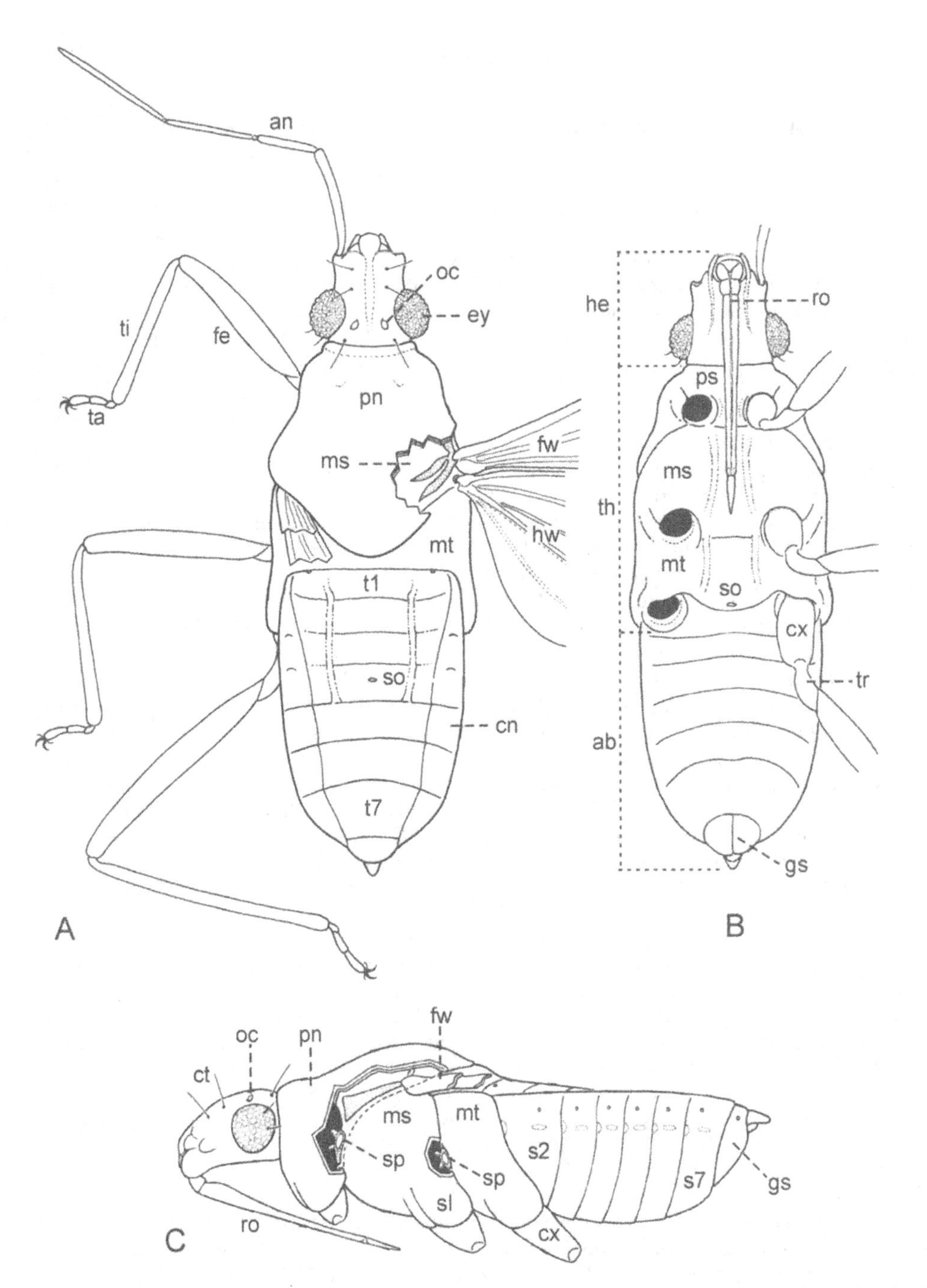

Figure 4.1. A-C. Generalised water bug: A, dorsal view; antennae and legs of right side omitted; wing cut of at base (an, antenna; cn, connexivum; ey, eye; fe, femur; fw, fore wing; hw, hind wing; ms, mesonotum; mt, metanotum; oc, ocellus; pn, pronotum; so, abdominal scent orifice; t1, t7, abdominal mediotergites 1 and 7; ta, tarsus; ti, tibia). B, ventral view; antenna and legs of right side omitted; antenna and legs of left side cut of at base (ab, abdomen; cx, coxa; he, head; ms, mesosternum; mt, metasternum; ps, prosternum; ro, rostrum; so, metasternal scent orifice; th, thorax; tr, trochanter). C, lateral view; antenna and legs except coxa omitted (ct, cephalic trichobothrium; cx, coxa; fw, fore wing; gs, genital segments; ms, mesopleuron; mt, metapleuron; oc, ocellus; pn, pronotum; s2, s7, abdominal sternum 2 and 7; sl, supracoxal lobe; sp, spiracle). Reproduced with modification from Andersen (1982).

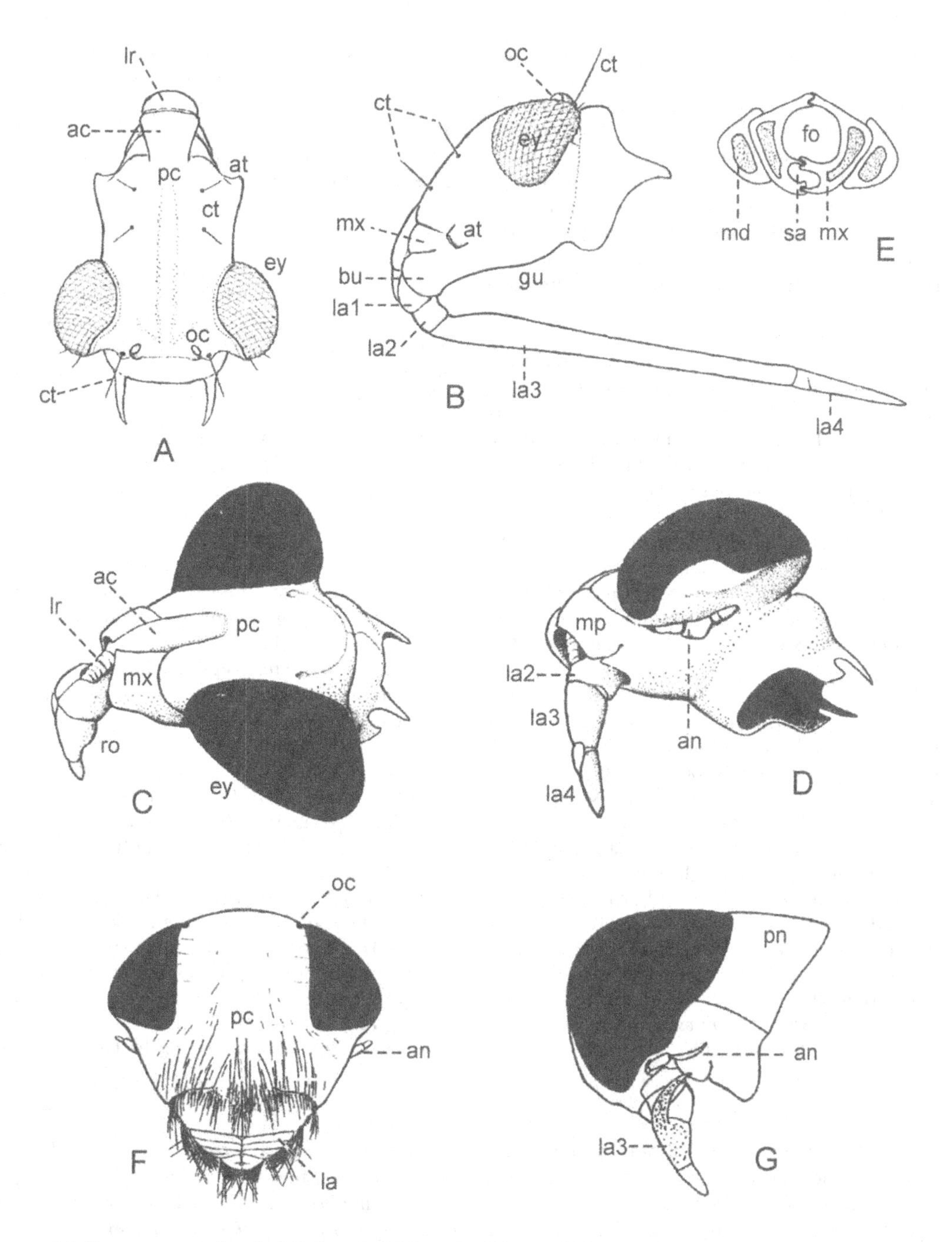

Figure 4.2. Structures on head. A-B, head of *Mesovelia* (Mesoveliidae) in dorsal (A) and lateral view (B); antenna omitted (ac, anteclypeus; at, antennal tubercle; bu, bucculum; ct, cephalic trichobothrium; ey, eye; gu, gula; la1-4, labial segments 1-4; lr, labrum; mx, maxillary plate; oc, ocellus; pc, postclypeus). C-D, head of *Lethocerus* (Belostomatidae) in oblique dorsal (C) and oblique ventral view (D) (ac, anteclypeus; an, antenna; ey, eye; la2-4, labial segments 2-4; lr, labrum; pc, postclypeus; ro, rostrum). E, cross section through rostrum of *Notonecta* (Notonectidae) (fo, food channel; md, mandibular stylet; mx, maxillary stylet; sa, salivary channel). F, frontal view of head of *Diaprepocoris* (Corixidae) (an, antenna; ls, labium (rostrum); oc, ocellus; pc, postclypeus). G, lateral view of head and prothorax of *Anisops* (Notonectidae) (an, antenna; la3, labial segment 3; pn, pronotum). Reproduced with modification from: A-B, Andersen (1982); C-D, Parsons (1968); E-F, Martin (1969); G, Brooks (1951).

pterothorax (the wing-bearing thoracic segments). It consists of a broad anterior scutum and a posterior *scutellum* which, primitively, carries a large triangular scutellar lobe (usually referred to as the scutellum). Among gerromorphan bugs, a typical scutellum is only found in the mesoveliid subfamily Madeoveliinae (not represented in Australia). In all other gerrromorphans the scutellum is reduced (*Mesovelia*; Fig. 4.3B; sc), rudimentary (Hebridae) or absent (all other families). Nepomorphan bugs typically have a large triangular scutellum (Figs 4.3E, G; sc).

The *pleuron* or lateral part of each thoracic segment bear a pair of supracoxal lobes (Fig. 4.1C; sl) produced by the epimeron and episternum, respectively. These lobes form together the *acetabulum* into which the coxa (cx) of the leg fits. The coxal cleft separating these lobes extends anterodorsally to the point where the internal coxal process arises and the coxa articulates with the pleuron. There are two pairs of spiracles on the thorax, each pair located dorsally on the pleuron, in the intersegmental membranes between pro- and mesothorax and meso- and metathorax (Fig. 4.1C; sp).

The *sternum* or ventral part of each thoracic segment, is relatively broad in most Gerromorpha (Fig. 4.1B), usually obscured by closely spaced coxae in the Nepomorpha (Fig. 4.3F). Most adult Heteroptera have active scent glands located ventrally in the metathorax. The metathoracic *scent gland apparatus* typically consists of lateral tubular glands which open into a median reservoir which may by spherical, oval, or bilobed. Accessory glands may be present in the dorsal wall of the reservoir. The secretion from the glands is discharged into the reservoir and released to the external world through a midventral *scent orifice* or *scent ostiole* on metasternum (Figs 4.1B, 4.3C; so). The secretion probably chiefly has a defensive function, but may also serve sexual, alarm, and aggregation functions. The scent glands and associated structures are secondarily lost in some groups (Gerromorpha: *Hydrometra*, some Gerridae; Nepomorpha: Nepidae, Aphelocheiridae).

Legs

The legs have the typical segmentation of insects, with a coxa, trochanter, femur, tibia, and tarsus (Fig. 4.1A). The structure of the legs shows substantial variation within and among groups of Australian water bugs (Figs 4.4A-F). Much of this diversity is correlated with life habits.

(1) *Cursorial legs* (Fig. 4.4A) The legs of some gerromorphan bugs (Hebridae, Mesoveliidae, some Veliidae) and nepomorphan bugs (Ochteridae, Gelastocoridae) are of the cursorial type, with femora and tibiae elongate and relatively slender, increasing in length from fore leg to hind leg. Primarily, they serve the function of walking and running.
(2) *Raptorial legs* (Fig.4.4C). This term has been applied to the fore legs of some obvious predators among the Nepomorpha, such as the Belostomatidae, Naucoridae, and Nepidae. The fore femur is moderately to strongly enlarged and often armed with spines; the fore tibia is also often modified so as to be adpressed against the fore femur in repose. The claws of the fore leg are either reduced to only one or are apposed in such a way as to function as a single claw. The spoon-shaped fore tarsi of water boatmen (Corixidae; Fig. 4.4D) are used to scoop up detritus. In males, specialised fore legs may serve the function of grasping and holding the female during mating (e.g., *Rheumatometra*; Fig. 4.4E)
(3) *Rowing legs.* All members of the Gerromorpha

Figure 4.3. Structures on thorax and abdomen. A, dorsal view of apterous female of *Mesovelia* (Mesoveliidae); antennae and legs omitted (cn, connexivum; ms, mesonotum; mt, metanotum; pn, pronotum; so, abdominal scent orifice; t1, t7, abdominal mediotergites 1 and 7). B-C, thorax of macropterous form of *Mesovelia* (Mesoveliidae) in dorsal (B) and ventral view (C); legs except coxa of left side omitted (cx1, cx2, cx3, fore, middle and hind coxa; fw, fore wing; lt2, abdominal laterotergite 2; mn, metanotal elevation; ms, mesosternum; mt, metanotum (B) or metasternum (C); pn, pronotum; ps, prosternum; s2, abdominal sternum 2; sc, meso-scutellum; so, metasternal scent orifice; t1, t2, abdominal mediotergites 1 and 2). D, ventral view of female abdomen of *Mesovelia* (Mesoveliidae) (go, gonocoxa; gp, gonoplac; s7, abdominal sternum 7). E-G, body of *Diaprepocoris* (Corixidae) in dorsal (E), ventral (F), and lateral view (G); head (F) and legs except coxae on left side omitted (cl, clavus; co, corium; cx3, hind coxa; em, embolium; la, labium (rostrum); ls, abdominal laterosternite; oc, ocellus; pn, pronotum; s2, s6, abdominal sternum 2 and 6; sc, meso-scutellum; sg, subgenital plate (operculum); xi, metaxiphus). Reproduced with modification from: A-D, Andersen (1982); E-G, Parsons (1976).

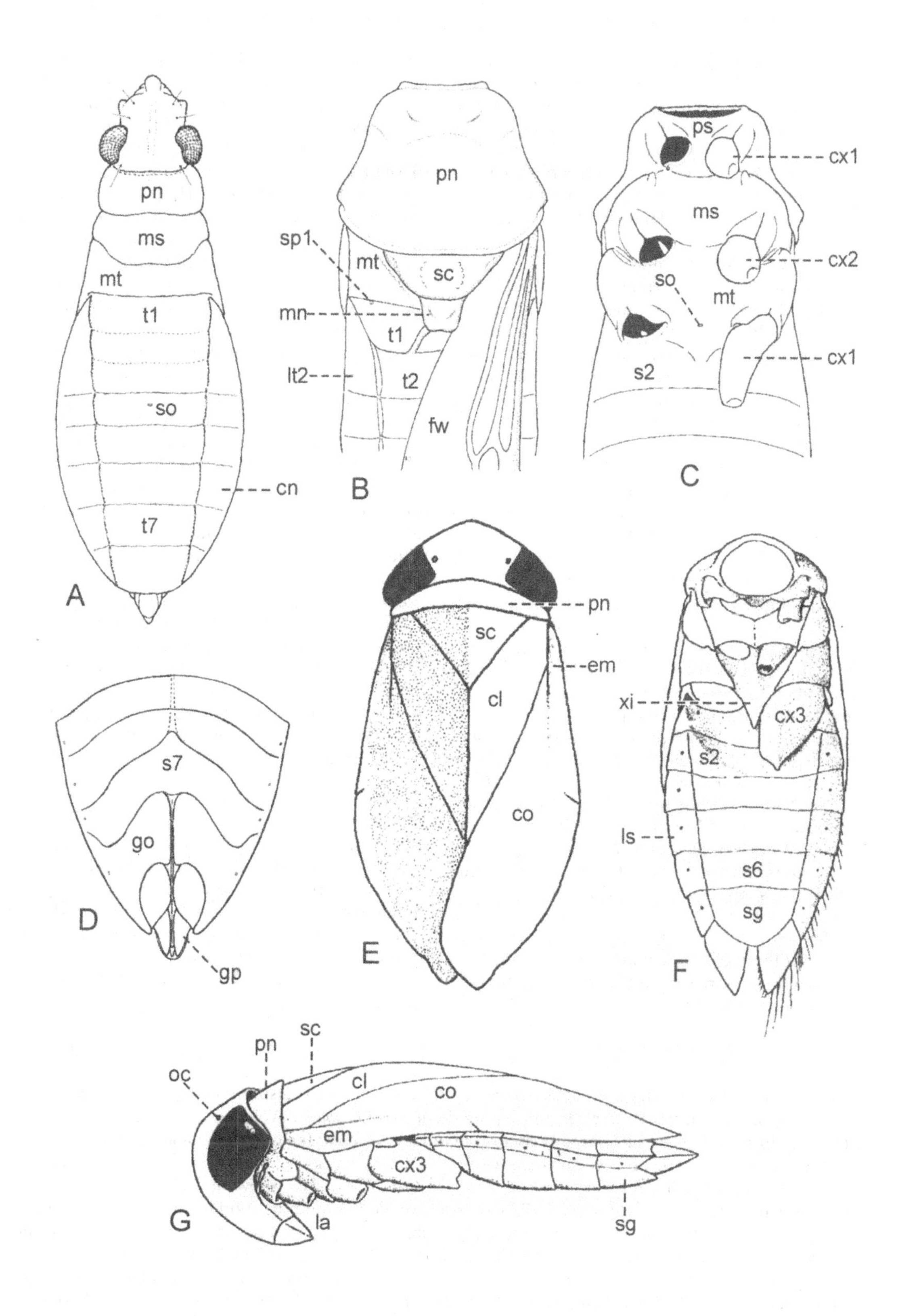
pn
ms
mt
t1
so
cn
t7
A
pn
sp1
mt
sc
mn
t1
lt2
t2
fw
B
ps
cx1
ms
cx2
so
mt
s2
cx1
C
s7
go
D
gp
pn
sc
em
cl
co
E
xi
cx3
s2
ls
s6
sg
F
oc
pn
sc
cl
co
em
cx3
G
la
sg

are capable of walking on the water surface film (see above), but rowing legs are restricted to some species of the Veliidae (Haloveliinae, some Microveliinae, Rhagoveliinae; Fig. 4.4B) and all Gerridae. The middle legs are longer than the hind legs and serve the function of propulsion on the water surface; also, the claws are usually inserted before the apex of the tibia.

(4) *Natatorial legs.* Modifications for swimming among the Nepomorpha are of several kinds. In the Belostomatidae and most Naucoridae, the middle and hind femora and tibiae are usually flattened and fringed with hairs, and both leg pairs participate in the swimming function. The legs of the Nepidae and Pleidae show little modification for true swimming, but appear to be better suited to assist the bugs in "crawling" through the water. In the Corixidae and Notonectidae, the hind legs are flattened and fringed with hairs and function like oars (Fig. 4.4F).

The *coxae* are typically inserted close to the ventral midline of the body, pointing towards each other (Fig. 4.3C, F; cx). In most Gerromorpha, however, the coxae of the middle and hind legs are displaced laterally on the body and in the Gerridae furthermore moved backwards and rotated into the horizontal plane. The *trochanters* are usually simple and relatively short. The *femora* are typically moderately thickened but may be swollen and fitted with spines, teeth, etc. (e.g., fore femora of Gerromorpha: *Hermatobates*; Nepomorpha: Belostomatidae, Gelastocoridae, Naucoridae, Nepidae). The *tibiae* are usually simple and straight, but may be flattened and fringed with hairs to assist in swimming in some Nepomorpha.

The *tarsi* are typically composed of 3 segments in the adult stage. The number of tarsal segments is usually the same for all legs, although some taxa (e.g., Gerromorpha: Veliidae) show variation in this regard. First instar nymphs of the Gerromorpha and many Nepomorpha have only one tarsal segment and some groups add segments during nymphal development. The *pretarsi* have two basic elements present in all groups: the *unguitractor plate* (Fig. 4.4G,H; ut) which is located internally at the apex of the tarsus, and the *claws* (ungues, un) which are attached to the unguitractor plate and the distal end of the tarsus. In addition, one or both of the following structures may be present. The *parempodia* (pe) are usually a symmetrical pair of setiform structures arising from the unguitractor plate between the claws. They are found in all Gerromorpha, but are asymmetrically reduced in the Veliidae and Gerridae. Some Nepomorpha have either 3 parempodia or 2 pairs of parempodia. The *aroliae* are unpaired structures arising between the bases of the claws, dorsal to the unguitractor plate. The *dorsal arolium* (da) is found both nymphs and adults of Gerromorpha and some Nepomorpha. The *ventral arolium* (va) is present in both nymphs and adults of Gerromorpha and in nymphs of all Nepomorpha (at least on some legs). Whereas the pretarsi typically are inserted on the tip of the last tarsal segment (Fig. 4.4G), the claws and associated structures of many Gerromorpha (Veliidae, Gerridae) are seemingly inserted preapically (Fig. 4.4H), in a cleft produced by the prolongation of the anterior (or ventral) part of the last tarsal segment.

Wings

The wings are usually folded flat over the abdomen. The wing venation is relatively simple, but there are no absolute agreement about the homologies of all veins (Schuh & Slater 1995). The *fore wings* of many Gerromorpha, particularly the Gerridae and Veliidae, have a fairly distinct venation (Fig. 4.5A). The anterior margin of the

Figure 4.4. Structure of legs. A, cursorial middle leg of *Mesovelia* (Mesoveliidae) (cx, coxa; fe, femur; ta, tarsus; ti, tibia; tr, trochanter). B, rowing middle leg of *Rhagovelia* (Veliidae) showing swimming fan (fa). C, raptorial fore leg of *Lethocerus* (Belostomatidae) (cl, claw; fe, femur; ta, tarsus; ti, tibia). D, specialised male fore leg of *Agraptocorixa* (Corixidae) (cx, coxa; fe, femur; ta, tarsus; ti, tibia, tr, trochanter). E, grasping fore leg of male *Rheumatometra* (Gerridae) showing apex of tibia at higher magnification (fe, femur; ta, tarsus; ti, tibia; tr, trochanter). F, swimming hind leg of *Sigara* (Corixidae) (fe, femur; ta1, ta2, tarsal segments 1 and 2; ti, tibia). G, apex of last tarsal segment of *Mesovelia* (Mesoveliidae) with apically inserted claws (da, dorsal arolium; pe, parempodia; un, claws (ungues); ut, unguitractor plate; va, ventral arolium). H, apex of last tarsal segment of *Velia* (Veliidae) with preapically inserted claws (abbreviations as in G). Reproduced with modification from: A, G, H, Andersen (1982); D, Zimmermann (1986); E, Andersen & Weir (1998); F, Lansbury & Lake (2002). C, original.

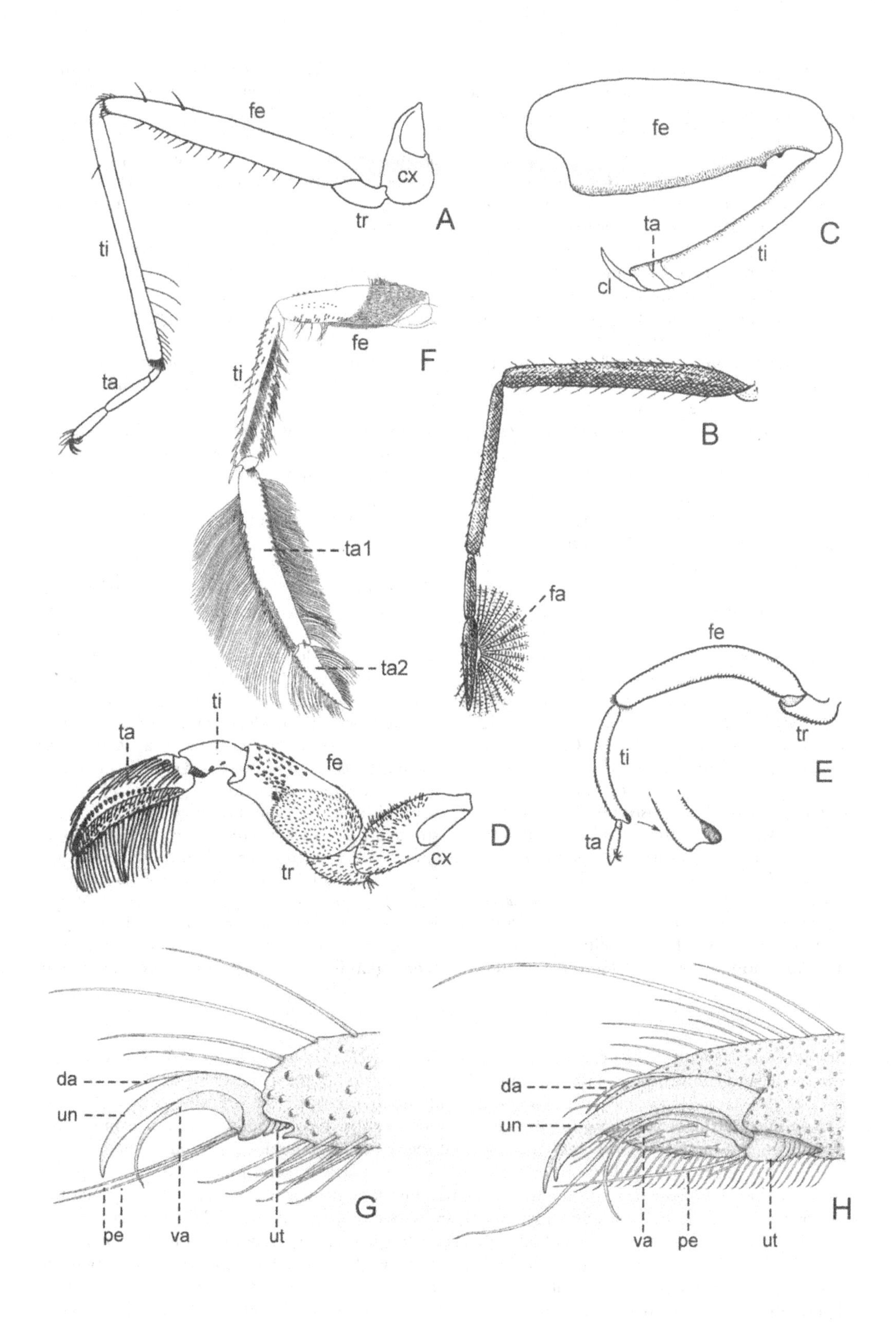

fe
cx
tr
A
ti
ta
fe
ta
C
ti
cl
ti
fe
F
ta1
ta2
B
fa
fe
tr
E
ti
ta
ti
ta
fe
D
cx
tr
da
un
pe
va
ut
G
da
un
va
pe
ut
H

wing, which lies laterally in repose, is thickened for about three fourths of its length; it is without doubt a complex of veins, probably the *subcosta* (Sc) plus the *radius* (Sc + R). A vein arising from the wing base next to the thickened anterior region is probably the *cubitus* (Cu) or the *media* plus the cubitus (M + Cu) since it is associated with the 1st and 2nd axillary sclerites via a system of sclerotised plates. The most posterior of the basal veins is fused to a thickened plate which is again connected to the 3rd axillary sclerite. This vein must be the 1st anal vein (1A). A very short residual vein is perhaps the 2nd anal vein (2A). A *claval furrow* in front of the anal veins separates the anterior *remigium* from the posterior *clavus*. The distal veins of the fore wing form 2-4 closed cells. Some of these veins are undoubtedly cross-veins, others are probably medial and cubital branches. A very characteristic feature of the fore wing of Gerromorpha is the two free veins extending toward the apex of the wing.

The fore wings of the Nepomorpha are in the form of *hemelytra*, that is, divided into a distinct coriaceous anterior part and a membranous posterior part as in most terrestrial Heteroptera (Figs 4.5B, C). The fore wing venation is largely obscured in the anterior part, but the following regions may be identified. The anterior margin of the wing is often reflexed and thickened ventrally into a *hypocostal ridge* or *lamina*. A fracture line known as the *medial furrow* (Fig. 4.5C; mf) is continuous with the *costal fracture*, a break in anterior margin, distal to which a distinct triangular *cuneus* may be formed. A posterior *clavus* lies adjacent to the scutellum, interlocked with the latter and usually to the opposite fore wing along the *claval commissure* by the *frenum*. The posterior clavus is separated from the anterior *corium* (remigium) by a line of weakness, the *claval furrow* (Figs 4.5B, C; cf), which run obliquely from the basal articulation of the wing toward the postero-distal margin. The outer portion of the corium is often expanded and reflexed and referred to as the *embolium* (Fig. 4.5C; em), delimited posteriorly by the *nodal furrow* (Fig. 4.5C; nf). The venation of the *membrane* (Fig. 4.5B-C; mb) may be totally lacking, composed by a few veins and closed cells, or in some Nepomorpha composed by numerous anastomosing veins (Fig. 4.5B).

The venation of the *hind wing* is similar to that of the fore wing, but often shows even greater reduction, and the wing is membranous except for the veins. Wing-to-body coupling mechanisms are not elaborately developed in the Gerromorpha. The fore wings of the Nepomorpha are coupled to the body antero-laterally by the "Druckknopf"-system and posteriorly (mesially) by the above-mentioned frenum. Wing-to-wing coupling mechanisms consist of a holding structure on the posterior margin of the fore wing, which provides a sliding attachment of the recurved anterior margin of the hind wing.

Abdomen

The abdomen (Fig. 4.1; ab) consists of 10 segments of which the segments 1-7 in females and the segments 1-8 in males constitute the *pregenital abdomen*. The dorsal part (tergum) of each of these segments is divided into a *mediotergite* and a pair of *dorsal laterotergites* and sometimes *ventral laterotergites*, collectively forming the *connexivum* (cn). There are only 6 visible ventral *sterna*, since sternum 1 is fused with metasternum (secondarily retained in the gerrid *Rhagadotarsus*). Primitively there are 8 abdominal spiracles of which spiracle 1 is always located on tergum 1, whereas spiracles 2-8 usually located ventrally, below the upper edge of the connexivum. The abdomen of the fully aquatic Nepomorpha is usually with various modifications for respiration under water.

The nymphs of nearly all Heteroptera have scent glands located on the dorsal abdomen.

Figure 4.5. A-C, structure of fore wings. A, left fore wing of *Microvelia* (Veliidae) (1A, first anal vein; M+Cu, media and cubitus; Sc+R, subcosta and radius). B, left fore wing (hemelytron) of *Goodnomdanepa* (Nepidae) (cf, claval furrow; cl, clavus; co, corium; mb, membrane). C, left fore wing (hemelytron) of *Agraptocorixa* (Corixidae) (cf, claval furrow; cl, clavus; co, corium; em, embolium; mb, membrane; mf, medial furrow; nf, nodal furrow). D-E, male genital segments of generalised gerromorphan bug: D, pygophore (py; segment 9) pulled out of segment 8 (s8). E, partly inflated phallus (ba, basal apparatus of phallus; en, endosoma; cj, conjunctivum; pa, paramere; pr, proctiger (segment 10); pt, phallotheca; py, pygophore or genital capsule (segment 9); s8, segment 8; sp8, spiracle). F, female genital segments of *Gerris* (Gerridae); gonocoxa of left side removed (gp1, gp2, gonapophyses 1 and 2; gx, gonocoxa 1 (segment 8); pr, proctiger (segment 10); sp8, spiracle). G, gynatrial complex of generalised gerromorphan bug (fc, fecundation canal; lo, lateral oviducts; gs, gynatrial sac with ring-gland; sp, tubular spermatheca). Reproduced with modification from: A, D, E, G, Andersen (1982); B, Lansbury (1974a); C, Zimmermann (1986); F, Andersen (1993).

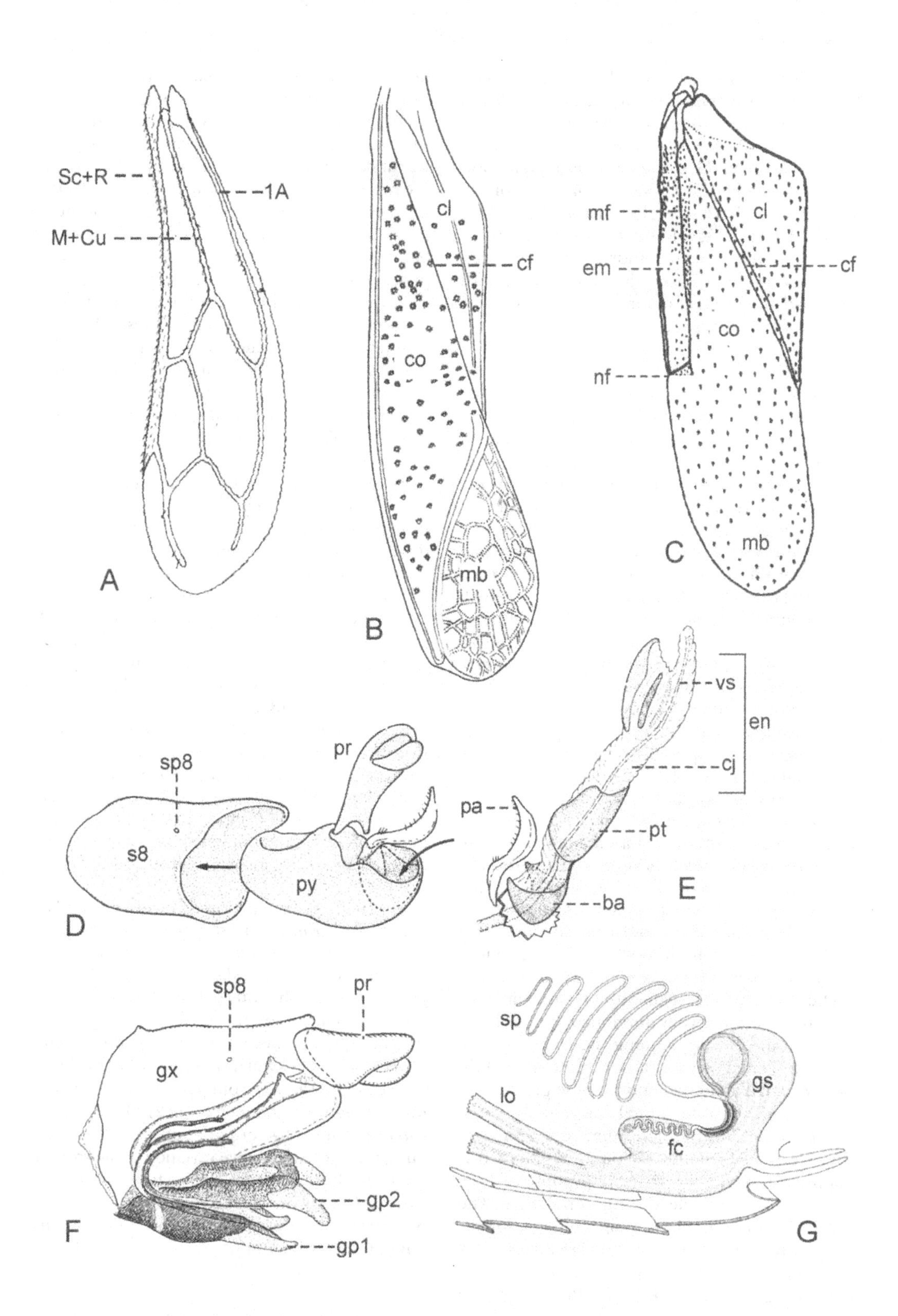
Sc+R
1A
M+Cu
A
cl
cf
co
mb
B
mf
em
nf
cl
cf
co
mb
C
vs
en
cj
pa
pt
ba
E
sp8
pr
s8
py
D
sp8
pr
gx
gp2
gp1
F
sp
gs
lo
fc
G

Typically there are 3 functional glands located on the anterior margins of abdominal terga 4, 5 and 6. This condition is found in nymphal Corixidae (Nepomorpha). A few Gerromorpha (Hebridae, Hermatobatidae, Mesoveliidae) and Nepomorpha (Aphelocheiridae, Naucoridae, Notonectidae, Pleidae) have a single functional gland on tergum 4 (Fig. 4.3A; so), but in most groups there are no abdominal scent glands at all. When present in the nymphal stage, the abdominal glands are usually non-functional in the adult stage and become smaller over time.

Genital segments

The *male genital segments* are composed of the abdominal segments 9 and 10 (Fig. 4.5D). The sternal part of segment 9 forms a boat-shaped *genital capsule* or *pygophore* (py) in which the *phallus* rests. Tergum 9, on the other hand, is a narrow sclerotised bridge which may be small and hardly recognisable. The opening of the pygophore points either caudally or dorsally. Attached to the margins of the pygophore is a pair of *parameres* ("claspers" of Andersen 1982; pa) which vary greatly in size and shape which make them taxonomically useful. A plate-shaped *proctiger* (segment 10; pr) more or less covers the posterodorsal opening of the pygophore; it carries ventrally, before its apex, a semicircular plate which flanks the lower edge of the anal opening. The *phallic organ* is composed of a proximal, sclerotised *phallotheca* (Fig. 4.5E; pt) which articulates with the pygophore, and a distal, membranous and more or less inflatable *endosoma* (en). The latter is divided into the *conjunctivum* (cj) and the *vesica* (vs), which usually have an armature of sclerotised pieces of taxonomic importance. In the Gerromorpha, the last pregenital segment (segment 8; Fig. 4.5D; s8) is always modified to form a cylindrical structure into which the pygophore and phallus are withdrawn when not in function.

The male genital segments of Gerromorpha are usually symmetrically developed and easily visible from the outside. Among the Nepomorpha, the male genital segments are retracted within the pregenital abdomen and in some groups covered ventrally by a subgenital (or genital) *operculum* (sternum 7 in the Belostomatidae and Nepidae; sternum 8 in the Pleidae). Segment 8 is modified in such a way that the pygophore (segment 9) can protrude during copulation. The opening of the pygophore points either caudally or dorsally. In the latter case the lateral and caudal margins are bent dorsad. The male genital segments are symmetrical in the Belostomatidae and Nepidae, asymmetrical in most other groups except some Naucoridae, and many Notonectidae; the asymmetrical nature often includes some pregenital abdominal segments as well.

The *female genital segments* (Fig. 4.5F) are somewhat more uniform in structure, retracted within the abdomen in the Nepomorpha whereas they are more or less plainly visible from the outside in the Gerromorpha (Fig. 4.3D). The ovipositor is formed by the abdominal segments 8 and 9. Ventro-laterally on segment 8 there are two plates, the *first gonocoxae* (first valvifer; Fig. 4.5F; gx), which covers the remaining genital segments in most Gerromorpha. Each of these plates carries ventrally and caudally the first *gonapophysis* (first valvula; gp1). Segment 9 carries a pair of *second gonocoxae* (second valvifer) to which are attached a pair of *second gonapophyses* (second valvulae; gp2). Segment 9 may also bear a pair of gonoplacs (Mesoveliidae; Fig. 4.3D; gp). The two pairs of gonapophyses form the ovipositor proper. The proctiger (segment 10; Fig. 4.5F; pr) is usually cone-shaped, but may be widened and deflected to form an anal lid in some Veliidae (Gerromorpha). The genital segments of some Nepomorpha are covered ventrally by a subgenital (or genital) plate or *operculum* (Figs 4.3F, G; sg) either in both sexes (sternum 7 in the Belostomatidae and Nepidae; sternum 8 in the Pleidae) or in females only (sternum 7 in the Aphelocheiridae and Naucoridae). The development of the ovipositor is correlated with the method of oviposition. In some groups the gonapophyses are elongate and serrate to form a *laciniate ovipositor* (e.g., Mesoveliidae, Gerridae-Rhagadotarsinae) suited for depositing eggs in plant tissue, whereas in others the gonapophyses are more or less reduced and the eggs are laid free or glued to the substrate.

Internally, the female reproductive organs consist of the *ovaries* (of the teletrophic type in Heteroptera), the *lateral oviducts* which are united to form the *common oviduct* which opens into the *genital chamber*. The internal structures associated with copulation, sperm reception, and sperm storage is sometimes referred to as the *gynatrial complex* (Fig. 4.5G which has a very characteristic structure in the Gerromorpha (for a detailed presentation, see Andersen 1982). The *spermatheca* or sperm-storage organ is found in most water bugs, but show distinctive variation from group to group (Schuh & Slater 1995). It is tubular and extremely long in gerromorphan bugs, tubular but shorter in nepomorphans bugs, usually without a distal bulb and pump.

Eggs

The eggs of water bugs are extremely variable in size, structure, and in the way they are deposited as described and illustrated in Chapter 2. The simplest egg type is ovoid, slightly flattened on one side (which is attached to the substrate). The anterior end more blunt than the other and the shell is traversed by one or more *micropyles.* This end of the egg may carry small processes or respiratory projections (Nepidae). In some groups (*Hydrometra,* Corixidae-Corixinae), the posterior end is provided with a *pedicel* or stalk by which the egg is attached to the substrate. *Eclosion* usually takes place by a simple split or rupture (longitudinal, circular or semicircular) of the egg shell. But in some groups (*Mesovelia,* Nepidae, *Aphelocheirus, Naucoris,* Notonectidae-Anisopinae), the anterior end of the egg has a more or less distinct *operculum* or egg cap which is lifted off during eclosion.

Nymphs

The juvenile stages of water bugs (and other insects with incomplete metamorphosis) are called *nymphs* in order to distinguish them from the larvae of, for instance, beetles, butterflies, and flies. As a rule, there are five nymphal instars between the egg and adult stage. In most cases nymphs are quite similar to the adults, just smaller. Nymphs are similar to adults in external morphology but may be paler, more soft-bodied, or have less elaborate colour patterns. In addition the tarsal segmentation is absent or indistinct, wings are absent or not fully formed, and the genital segments are not developed. However, wingless (apterous) adults of several species of Gerromorpha can easily be confused with nymphs. Sexes are not generally distinguishable until the final (usually the 5th) nymphal instar. The different nymphal instars may be distinguished by size differences; the length of leg segments should be used instead of body length. The abdomen of nymphs grows in between moults, depending on the feeding activity of the individual nymph. For long-winged (macropterous) forms, the older nymphs may be distinguished by the following criteria:

Instar III: Anterior wing-pads just visible as dark, sickle-shaped lateral outgrowths on mesonotum.

Instar IV: Both pairs of wing-pads distinct, semicircular, anterior pair not covering posterior pair.

Instar V: Both pairs of wing-pads well developed (Fig. 12.4B), anterior pair as long as and covering posterior pair. Abdominal end sexually differentiated (most easily seen in ventral view; Figs 12.4C, D).

5 Collecting Methods and Specimen Preparation

Collecting water bugs

Water bugs are most easily collected using a net (Fig. 5.1G). A butterfly net is not strong enough. A good aquatic net has a not too fine-meshed bag (about 1 mm, otherwise it easily gets clogged with mud), that is not too deep. It has to have a diamond-shaped or semicircular (not circular) shape so you can get close to the bottom-dwelling bugs. The handle should be long and strong, able to be disassembled in pieces for easy transport. An ordinary kitchen sieve (or soup strainer) attached to a broom handle may also be used. Smaller nets or sieves may be useful when working in confined spaced, e.g., in shallow streams or springs (Plate 5, A-B). When using a net in fast-flowing water with a rocky or stony bed, the net should be held firmly in a vertical position against the bed with its mouth facing upstream. If you lift or stir stones or bottom material just upstream of the net, a cloud of debris and bugs will be washed into the net. Surface-dwelling bugs can be caught by stirring up the vegetation along the banks. In slow or stagnant water the net can be used to stir the bottom and scoop up the debris or used to sweep through submerged, emergent, or floating vegetation, as well as the free water surface.

When the net is full of plant material and debris, the most convenient way of finding the water bugs is to empty the net into an ordinary, white photographic tray or other type of light coloured plastic tray. The tray should be about one third filled with water. The bugs are then picked up from the tray using small forceps, tweezers or large-mouthed pipettes and transferred to a glass or plastic tubes with alcohol (Plate 5, C-D). If you want to keep the bugs alive, they should be placed in a larger container with moist paper towels or water plants (not free water). In order to see the smallest specimens you will need a good magnifying glass as well. This method works well for most water bugs, but water striders may be able to jump or crawl out of the tray. Some bugs may also be able to fly and thus escape from the tray. A general account of collecting methods for aquatic invertebrates is given by Gooderham & Tsyrlin (2002). Many water bugs fly regularly and may be attracted to standard light-traps for collecting nocturnal insects. During day-time, flying bugs may also be caught in window-traps. Semiterrestrial species living in forest litter, moss carpets, etc., may be disclosed using pyrethrum knock-down or a Berlese-funnel.

In tropical and subtropical areas, most water bugs can be collected throughout the year, but in temperate areas most species are only found in their natural habitats from spring to autumn. In establishing the annual life cycle (phenology) of species, the composition of adult and nymphal populations in early summer is the most crucial. Useful information can be gathered by taking semi-quantitative samples (e.g., collecting for a fixed time period) of nymphal and adults populations through most of the summer season. For each sample, the number of nymphs belonging to the last 2-3 instars, the number of newly moulted adults (distinguished by their soft integument) should be recorded and frequencies calculated. The reproductive state of females (e.g., size of ovaries) may be assessed by dissection.

Rearing

Most water bugs can be reared successfully in the laboratory in smaller or larger containers with water, provided with pieces of styrofoam or cork for the surface bugs to rest on when cleaning themselves, for egg-laying, etc., and water plants for the true aquatic bugs. Keeping the water well aerated is important to prevent the formation of a bacterial film and the accumulation of dust on the water surface which, eventually, affects the ability of the semiaquatic bugs to stay on top of the surface film. Nymphs and adults of most water bugs will feed upon all kinds of soft-bodied insects. Dead fruit-, house- or blowflies are most suitable for surface bugs, whereas microcrustaceans (water fleas, cladocerans, copepods), water slaters or sow bugs (isopods), mayfly nymphs (ephemeropterans), or larvae of various dipterous insects such as mosquitoes (culicids) and non-biting midges (chironomids) are excellent food for true aquatic bugs. The larger individuals will also feed on other water bugs and cannibalism will occur if the cultures are crowded or the food-supply limited.

Specimen preparation

Semiaquatic and aquatic bugs are most suitably killed and preserved in 70% to 80% ethanol alcohol (95% to 96% alcohol is needed if the specimens are intended to be used for DNA extraction). If ethanol is not available (or too expensive), a mixture of 75% methylated spirits with 25% water works almost as well (beware of its toxic effects!). The use of formaldehyde should be avoided. The glass or tube containing speci-

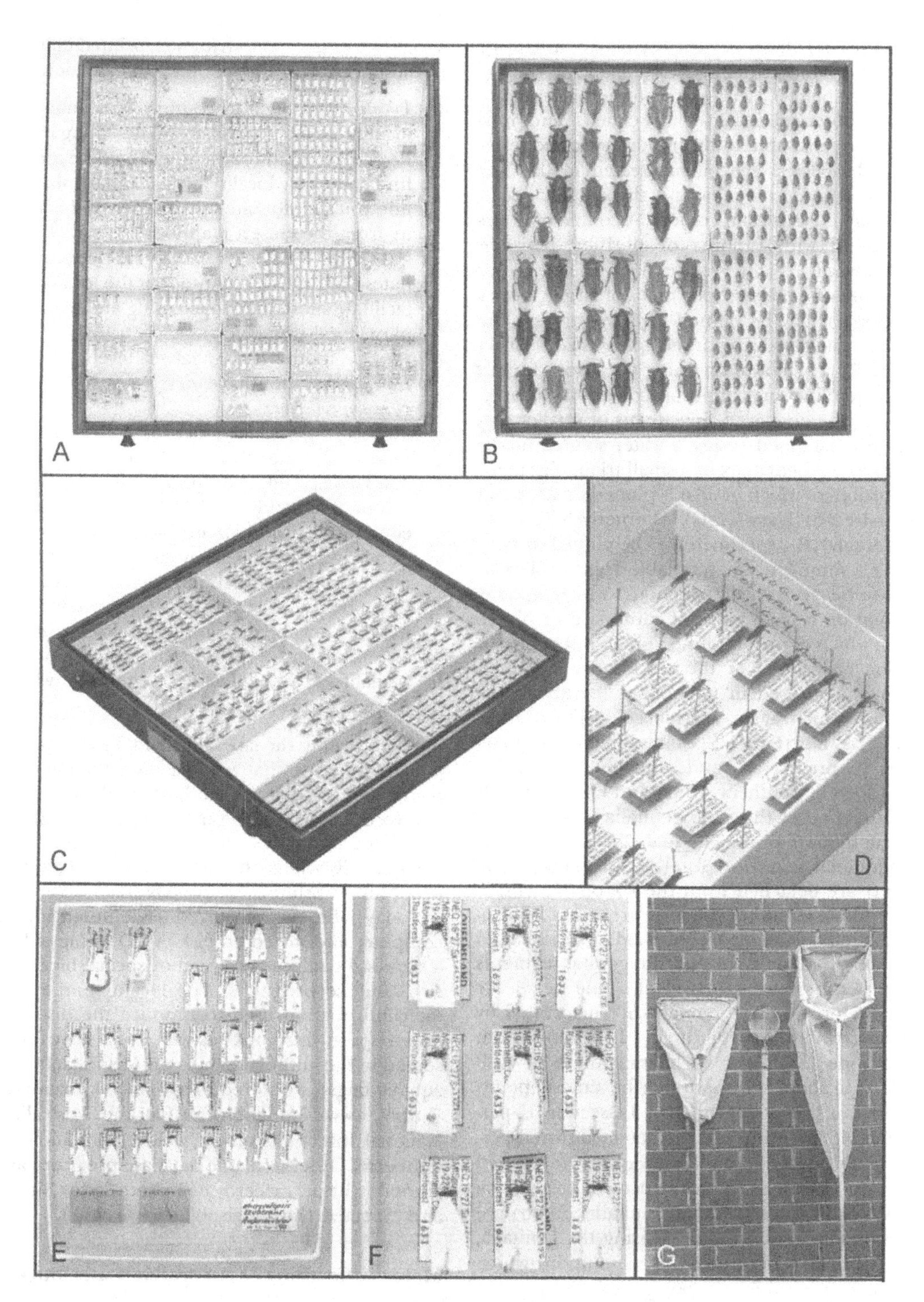

Figure 5.1. Storage of dry mounted collections of water bugs in Australian National Insect Collection, CSIRO Entomology, Canberra. A-C, drawers with unit trays containing: A, Australian species of Veliidae. B, Australian species of Belostomatidae. C, Australian *Limnogonus* spp. (Gerridae). D, unit tray with *Limnogonus fossarum gilguy*. E-F, *Microvelopsis exuberans:* E, unit box with holotype and paratypes. F, point-mounted specimens. G, different types of aquatic nets. A-F, photographs by David McClenaghan (CSIRO Entomology). G, reproduced from Gooderham & Tsyrlin (2002).

mens should always be clearly labelled, usually by placing a hand-written or printed label inside the glass (see below). Certain structures, however, are better observed in dried specimens and it is therefore advisable to dry mount synoptic series of each species. Specimens treated in this way may become greasy, loosing their dull appearance as well as certain colours. A couple of hours treatment with vapours of purified benzene (toxic!) will usually restore the velvety surface structure of the bugs. Specimens of larger species (>10 mm) may be pinned directly through the thorax using ordinary insect pins (size 0 or 1) (Plate 5). Try to avoid pinning the insect directly in the midline since important diagnostic characters may be destroyed. Specimens of smaller species may be glued (using a water soluble hobby glue) to the bent apex of a small triangular piece of cardboard which in turn is pinned on an insect pin (size 3-5) (Plate 5, E-G). Alternatively (but not preferably), the specimen may be pinned directly using a minutien pin (available from dealers in accessories for insect collections) which again is pinned on a small piece of cardboard or plastic foam which in turn is pinned on an insect pin. In major collections of the world, dry mounted insects are stored in unit boxes (cardboard or plastic) which again are placed inside drawers with glass lids (Fig. 5.1). General methods of collecting, preserving, and studying insects are described by Smithers (1981) and Upton (1991).

Dissection of the male genitalia is performed in the following way: The male genital segments (abdominal segments distal of the seventh segment) are detached from the abdomen. This is easily done if the specimen is preserved in alcohol. Dried specimens are softened in 70% alcohol for about half an hour and the genital segments then detached with the aid of a fine needle or pair of forceps. The genitalia are macerated by placing them overnight in 10% potassium hydroxide (KOH) or by heating them in 10% KOH for about ten minutes (the time depends upon the size of specimens). Dissection is performed in glycerine and the genitalia stored permanently in a microvial with glycerine pinned with the specimen. Some structures (e.g., parameres or sclerites of the male phallic organ) are most easily observed after clearing the genitalia, e.g., in lactic acid (50% aqueous solution) which also may be used for temporary storage.

Labelling

Specimens of water bugs are practically useless without attached information on where and when the specimens were collected. Labels should be either hand-written or printed and at a minimum contain the following information:

(1) *Locality data*, such as country, state, name of locality (e.g., Australia, NSW, Mongarlowe River E of Braidwood). You should be able to find the exact locality on a standard map. If not, you should state the distance of the locality from the closest map locality (e.g., 15 km N by E of Smallville). Giving the geocoordinates (latitude & longitude) of the locality will facilitate subsequent databasing and mapping of material (see below).
(2) *Altitude*, if other than near sea level.
(3) *Date*, as day, month, year.
(4) *Name of collector.*
(5) *Accessory information*, such as collecting method (e.g., water sweep, light trap), details of habitat (e.g., edge of flowing river, large shallow pool in sandy & rocky river bed, grass at edge, algal growth), and biological information (e.g., pairs taken in copula, feeding on mosquito larva). Such information may eventually be put on extra labels. Alternatively, a number may be put on the label referring to an excursion log or journal (which must be available at the collection or museum where the specimens are deposited).

When writing the labels by hand, be sure to use pencils or pens that will guarantee long time readability. If labels are put in alcohol, always use waterproof pens. If the labels are printed using a computer, remember that most ink-jet and laser print usually will be dissolved in alcohol (waterproof ink-jet printers are available).

Today, label data are routinely entered into a computer and stored on disk or CD, using either a standard database program or special biodiversity database programs. The locality and specimen information used to produce the distribution maps in this book (as well as the maps presented in the authors' papers on Australian semiaquatic bugs) are stored in a "BioLink" database (http://www.csiro.biolink.au) at the Australian National Insect Collection, CSIRO Entomology, Canberra. Other available database programs are "Biota" (http://viceroy.eeb.uconn.edu/Biota) and "Specify" (http://usobi.org/specify/).

Major collections of Australian water bugs

This section contains an alphabetical list of the most important collections (with acronyms) containing Australian water bugs, including types of Australian species described by major workers on Heteroptera (see also Schuh & Slater 1995). Collections within Australia are listed first.

Australian Museum, Sydney (AMS). Reasonable collection of water bugs, mainly from NSW and Queensland, but good holdings of Gelastocoridae from all over Australia. Also wet collections of EPA river surveys of NSW. Types of species of water bugs described by Frederick A. A. Skuse (1864-1896), N. M. Andersen & T. A. Weir and G. Cassis & R. Silveira.

Australian National Insect Collection, CSIRO Entomology, Canberra (ANIC). The major collection of Australian water bugs, including most of the important collections by Ivor Lansbury, Oxford, and material gathered by various expeditions by ANIC staff in Australia, including the extensive material collected by the second author during fieldwork in Queensland, Northern Territory, and Western Australia. Types of species of water bugs described within the past 20 years are deposited in ANIC, in particular those described by the present authors as well as some described by I. Lansbury, J. T. Polhemus & I. Lansbury, J. T. Polhemus, J. Polhemus & L. Cheng, J. T. Polhemus & D. A. Polhemus, J. T. Polhemus & P. B. Karunaratne, G. Cassis & R. Silveira, and C. V. Reichardt.

Museum of Victoria, Melbourne (MVM) Large collection of waterbugs, mainly from surveys in Victoria. Mostly wet collections with strength in Corixidae, Notonectidae, Mesoveliidae and Veliidae. Types of species of water bugs described by M. Malipatil, I. Lansbury, J. N. Knowles, I. M. King and E. L. Todd.

Northern Territory Museum, Darwin (NTMD). Moderate collection of water bugs, mainly from NT and a lot collected by M. B. Malipatil. Good representation of marine species. Type species of water bugs described by I. Lansbury and N. M. Andersen & T. A. Weir.

Queensland Museum, Brisbane (QMB). Moderate collection of water bugs, mainly from Queensland and chiefly collected by Geoff B. Monteith. Types of species of water bugs described by I. Lansbury, N. M. Andersen & T. A. Weir, A. Menke, M. Baehr, and G. Cassis & R. Silveira.

South Australian Museum, Adelaide (SAMA). Moderate collection of water bugs with strength in NT, Queensland and SA. Types of species of water bugs described by Herbert M. Hale (1895-1963), I. Lansbury, and E. L. Todd.

University of Queensland Insect Collection, Brisbane (UQIC). Large collection of water bugs with strength in Queensland material and mainly collected by Thomas E. Woodward (1918-1985), G. Monteith, and T. A. Weir. A few types of species of water bugs described by N. M. Andersen & T. A. Weir.

Western Australian Museum, Perth (WAM). Small collection, almost solely from Western Australia. Types of species of water bugs described by I. Lansbury and M. Baehr.

Ivor Lansbury collection, Hope Museum of Zoology, Oxford, U.K. (ILC). Collections of Australian water bugs. Types in ANIC.

John T. Polhemus collection, Englewood, Colorado, U.S.A. (JTPC). Extensive collection of aquatic and semiaquatic bugs from all over the World, including Australia and New Guinea. Types deposited in USNM (see below).

Natural History Museum (formerly *British Museum, Natural History)*, London, U.K. (BMNH). Large collections of chiefly historical interest, including numerous types of species described by William L. Distant (1845-1922), William E. China (1895-1979), and other workers.

National Museum of Natural History, Smithsonian Institution, Washington, D.C., U.S.A. (USNM). Extensive collections of water bugs, in particular the collections of Carl J. Drake (1885-1965) and John T. Polhemus (although still maintained by him). Many types of species described by C. J. Drake, J. T. Polhemus, and Dan A. Polhemus.

Bernice P. Bishop Museum, Honolulu, U.S.A. (BPBM). Large collections from the Indo-Australian region, in particular New Guinea.

Natural History Museum, Budapest, Hungary (HNHM). Important collections including the types of species described by Geza Horváth (1847-1937).

Naturhistoriska Riksmuseet (*Swedish Museum of Natural History*), Stockholm, Sweden (NRS). Repository for the collections studied by Carl Stål (1833-1878).

Snow Entomological Museum, University of Kansas, Lawrence, Kansas, U.S.A. (SEMC). Perhaps the largest collection of aquatic and semiaquatic bugs gathered by Herbert B. Hungerford (1885-1963). Many types of Australian taxa described by Hungerford and his co-workers (G. T. Brooks, L. C. Chen, E. L. Todd and others).

Zoologisk Museum (Zoological Museum), University of Copenhagen, Copenhagen, Denmark (ZMUC). Large collections of semiaquatic bugs gathered by Nils M. Andersen, including the L. Cheng collection of *Halobates*. Type collection of Iohan Christian Fabricius (1745-1808) as well as numerous types of species described by N. M. Andersen.

6. Key to Adults of Australian Families of Water Bugs

1. Antennae longer than head, inserted in front of eyes and plainly visible from above (Figs 6.1B, D, H, J) 2
- Antennae shorter than head and folded beneath eyes, not visible from above (Figs 6.2F, H), at most apex of antenna exposed beyond margin of large eyes (Figs 6.2E, G). True aquatic bugs, except Gelastocoridae and Ochteridae (NEPOMORPHA) 11
2. Head with at least 3 pairs of conspicuous trichobothria (long, slender setae) located near inner margins of eyes (Figs 4.2A; ct, 7.3C), inserted in deep pits (Fig. 7.3D). Fore wings, when present, lacking claval commissure and not divided into a well demarcated corium-clavus and membrane (Fig. 4.5A). Part or most of body always covered with a distinct layer of microtrichia. Semiaquatic bugs (GERROMORPHA) 3
- Head without trichobothria, or if one or more pairs of trichobothria present on head, these never inserted in deep pits. Fore wings, when present, with or without claval commissure, usually divided into a well demarcated corium-clavus and membrane. Venter infrequently with a layer of microtrichia (some Leptopodomorpha) .... Terrestrial Hemiptera-Heteroptera
3. Head prolonged, much longer than greatest width; eyes (Fig. 6.1A; ey) far removed from anterior margin of prothorax (pn) .... HYDROMETRIDAE
- Head at most twice as long as greatest width; eyes situated very close to anterior margin of prothorax (Fig. 6.1B; ey) 4
4. Macropterous (long-winged) adult form (Figs 6.1C, I) 5
- Apterous (wingless) adult form (Figs 6.1B, H, J) 7
5. Meso-scutellum exposed, forming rounded or transverse plate behind pronotal lobe (Figs 6.1C, D; sc) 6
- Meso-scutellum not exposed, covered by posteriorly prolonged pronotal lobe (Fig. 6.1I) 10
6. Ventral surface of head with a pair of prominent, vertical plates (bucculae) covering base of rostrum (Fig. 6.1E; bu). Tarsi 2-segmented, first segment very short .... HEBRIDAE
- Ventral surface of head not modified as above. Tarsi 3-segmented .... MESOVELIIDAE
7. Abdominal scent orifice present on tergum 4 (Fig. 4.1A; so). Claws of middle and hind logs inserted apically on last tarsal segment(Fig. 4.4; un) 8
- Abdominal scent orifice absent. Claws of middle and hind legs inserted distinctly before apex of last tarsal segment (Figs 4.4; un, 6.1G; cl) 10
8. Pronotum long (Fig. 6.1D; pn), at least covering mesonotum. Ventral surface of head with prominent bucculae covering base of rostrum (Fig. 6.1E; bu)........ HEBRIDAE
- Pronotum very short (Fig. 6.1B; pn), exposing both mesonotum (ms) and metanotum (mt). Ventral surface of head not modified as above 9
9. Head longer than wide, porrect (Fig. 6.1B). Fore femora only moderately thickened. Claws of all legs inserted apically on last tarsal segment........ MESOVELIIDAE
- Head shorter than wide, subvertical (Fig. 6.1J). Fore femora distinctly thickened. Claws of fore legs inserted before apex of

---

Figure 6.1. A, dorsal view of head and thorax of *Hydrometra* (Hydrometridae) (ey, ey; fv, reduced fore wing; mt, metathorax). B, apterous ♂ of *Mesovelia* (Mesoveliidae) (ey, eye; ms, mesonotum; mt, metanotum; pn, pronotum). C, macropterous ♂ of *Mesovelia* (Mesoveliidae) (fw, fore wing; mt, metanotal elevation; sc, meso-scutellum). D, dorsal view of head and thorax of *Hebrus* (Hebridae) (mt, metanotal elevation; pn, pronotum; sc, meso-scutellum). E, lateral view of head and thorax of *Hebrus* (Hebridae) (bu, bucculum). F, fore leg of *Halovelia* ♂ (Veliidae) (gr, tibial grasping comb). G, distal segment of middle tarsus of *Phoreticovelia* (Veliidae) (cl, claw; va, ventral arolium). H, apterous ♂ of *Drepanovelia* (Veliidae) (ms, mesontum; mt, metanotum; pn, pronotum). I, macropterous ♂ of *Drepanovelia* (Veliidae) (fw, fore wing). J, apterous ♂ of *Hermatobates* (Hermatobatidae) (he, head). K, fore tarsus of *Hermatobates* (Hermatobatidae). Reproduced with modification from: A-E, Andersen & Weir (in press); F, Andersen & Weir (1999); G-I, Andersen & Weir (2001); J-K, F, Andersen & Weir (2000).

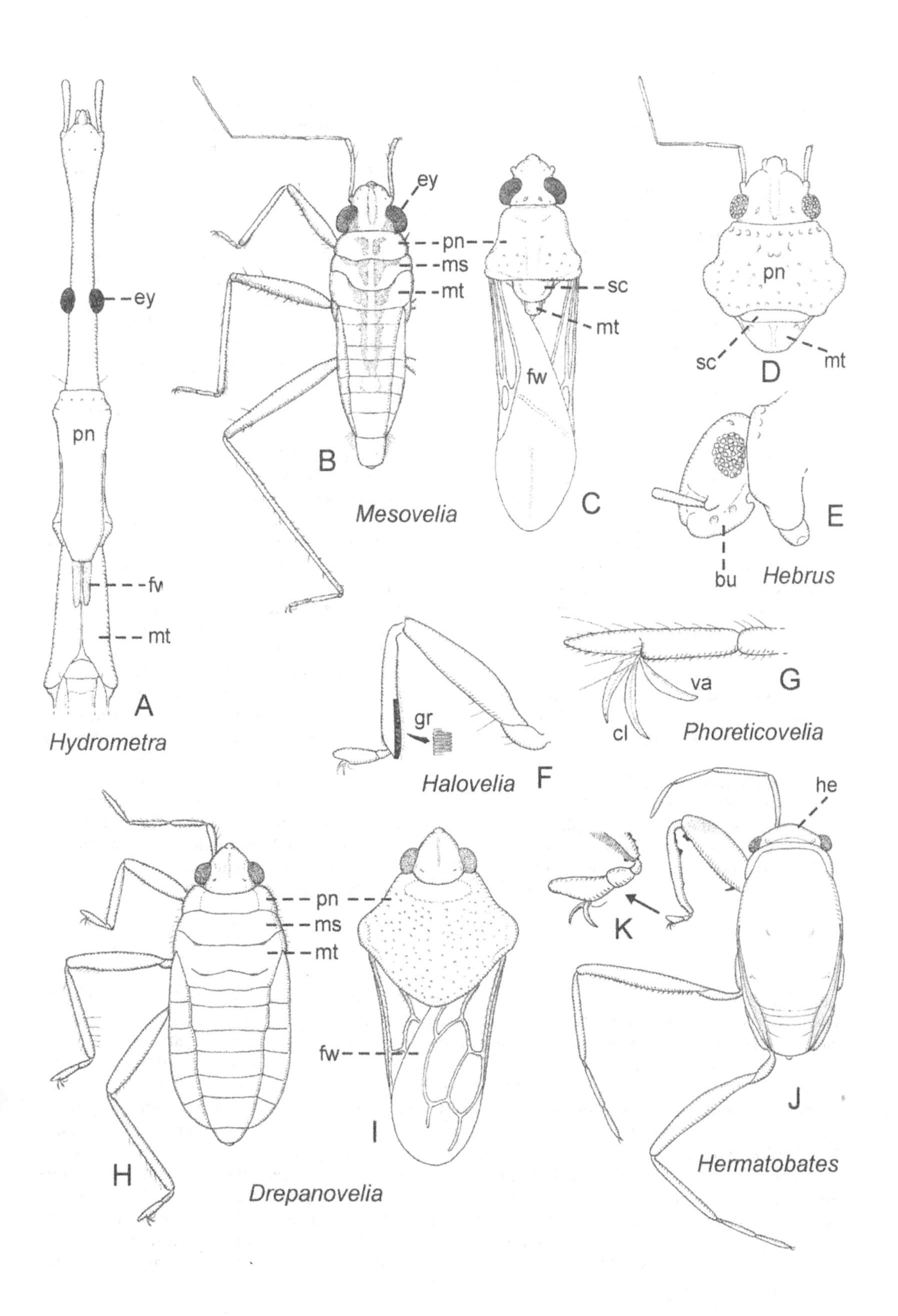
ey
pn
fv
mt
A
Hydrometra
ey
pn
ms
mt
B
Mesovelia
sc
mt
fw
C
pn
sc
mt
D
bu
E
Hebrus
gr
Halovelia
F
va
cl
G
Phoreticovelia
pn
ms
mt
fw
H
I
Drepanovelia
he
K
J
Hermatobates

last tarsal segment (Fig. 6.1K), those of middle and hind legs inserted apically on last tarsal segment........... HERMATOBATIDAE

10. Head with a distinct longitudinal, median impressed line on dorsal surface (Fig. 11.3B). Fore tibiae of male usually with a distal grasping comb of short spines along inner margin (Fig. 6.1F; gr) (except in *Xenobates*). Hind femora usually stouter than middle femora. Coxal cavity (acetabulum) of metathorax with scent evaporatorium (Figs 11.1C, G, ev) .... VELIIDAE
 - Head without a longitudinal, median impressed line on dorsal surface. Fore tibiae of male never with a distal grasping comb. Hind femora usually more slender than middle femora. Coxal cavity of metathorax never with scent evaporatorium [Australian species only] ... GERRIDAE

11. Rostrum short, broadly triangular (Fig. 6.2A; ro), non-segmented, with transverse sulcation (except Cymatiainae). Fore tarsus 1-segmented (sometimes fused with tibia), spoon-, scoop-, or sickle-shaped (Fig. 6.2B; ta), rarely cylindrical (*Cnethocymatia*), with ventral fringe of long hairs. Head overlapping pronotum antero-dorsally................................ CORIXIDAE
 - Rostrum cylindrical to conical, distinctly segmented, without transverse sulcation (Fig. 6.2F-H; ro). Fore tarsus cylindrical, with one or more segments, without a fringe of long hairs. Head never overlap ping pronotum antero-dorsally................... 12

12. Apex of abdomen with paired processes, either in the shape of a long siphon (Fig. 6.2C; si) or a pair of short air-straps (Fig. 6.2D; as). Body cylindrical or ovoid and flat. Medium-sized to gigantic bugs. Fore legs raptorial.. 13
 - Apex of abdomen without respiratory processes. Body never cylindrical, dorsum flat to extremely convex. Very small to medium-sized bugs. Fore legs raptorial or cursorial, often modified in males 14

13. Respiratory siphon non-retractile, usually long and filiform (Fig.6.2C; si), sometimes relatively short (Figs 13.6A, 13.8A). All tarsi 1-segmented. Hind tibiae simple. Hind coxae short, free, not united with metapleuron............................... NEPIDAE
 - Respiratory siphon retractile, short and strap-like, often only apices visible (Fig. 6.2D, 14.4A; as). Tarsi 2- or 3-segmented. Hind tibiae usually flattened, with swimming hairs. Hind coxae conical, firmly united with metapleuron ..... BELOSTOMATIDAE

14. Ocelli present (Fig. 6.2E; oc), if obsolete or absent, then head transverse and eyes pedunculate or subpedunculate (Fig. 6,2F; ey) (submacropterous forms of Gelastocoridae) 15
 - Ocelli absent; eyes sessile. Hind and/or middle legs usually flattened, fringed with hairs (except minute PLEIDAE) 16

15. Antennae relatively long, filiform, partially visible in dorsal view (Fig. 6.2E; an). Head moderately transverse; eyes sessile. Scutellum flat (Fig. 6.2E; sc). Fore femora (fl) not incrassate, anterior surface not modified................................ OCHTERIDAE
 - Antennae short and stout, concealed in cavity between head and prothorax (Fig. 6.1F; an). Head strongly transverse; eyes pedunculate to subpedunculate (Fig. 6.2F; ey). Scutellum tumid. Fore femora (Fig. 6.2F; fe) incrassate, anterior surface sulcate to carinate .... GELASTOCORIDAE

Figure 6.2. A, frontal view of head of *Sigara* (Corixidae) showing rostrum (ro). B, right fore leg of ♂ of *Sigara* (Corixidae) (fe, femur; ta, tarsus; ti, tibia). C, ventral view of abdominal end of *Ranatra* (Nepidae) (si, respiratory siphon; s6, s7. abdominal sterna 6-7). D, dorsal view of abdominal end of *Diplonychus* (Belostomatidae) (as, air-straps; t7, abdominal tergum 7). E, dorsal view of head and prothorax of *Ochterus* (Ochteridae) (an, antenna; fl, fore leg; pn, pronotum; sc, scutellum). F, ventral view of head and prothorax of *Nerthra* (Gelastocoridae) (an, antenna; ey, eye; fe, fore femur; ps, prosternum). G, oblique ventral view of head of *Aphelocheirus* (Aphelocheiridae) (an, antenna; ro, rostrum). H, ventral view of head and prothorax of *Naucoris* (Naucoridae) (an, antenna; fe, fore femur; ps, prosternum; ro, rostrum). I, dorsal view of head and prothorax of *Anisops* (Notonectidae) (pn, pronotum; sc, meso-scutellum; ve, vertex). J, dorsal view of *Paraplea* (Plaeidae) (fw, fore wing (hemelytron); pn, pronotum; ve, vertex). K-L, *Trephotomas* (Helotrephidae): K, dorsal view (cp, cephalonotum; fw, fore wing (hemelytron); sc, mesoscutellum). L, lateral view. Reproduced with modification from: A-B, Lansbury (1970); C, Menke & Stange (1964); F, Todd (1955); G, Parsons (1969b); J, Lansbury & Lake (2002); K-L, Papacek et al. (1988). D-E, H-I, original.

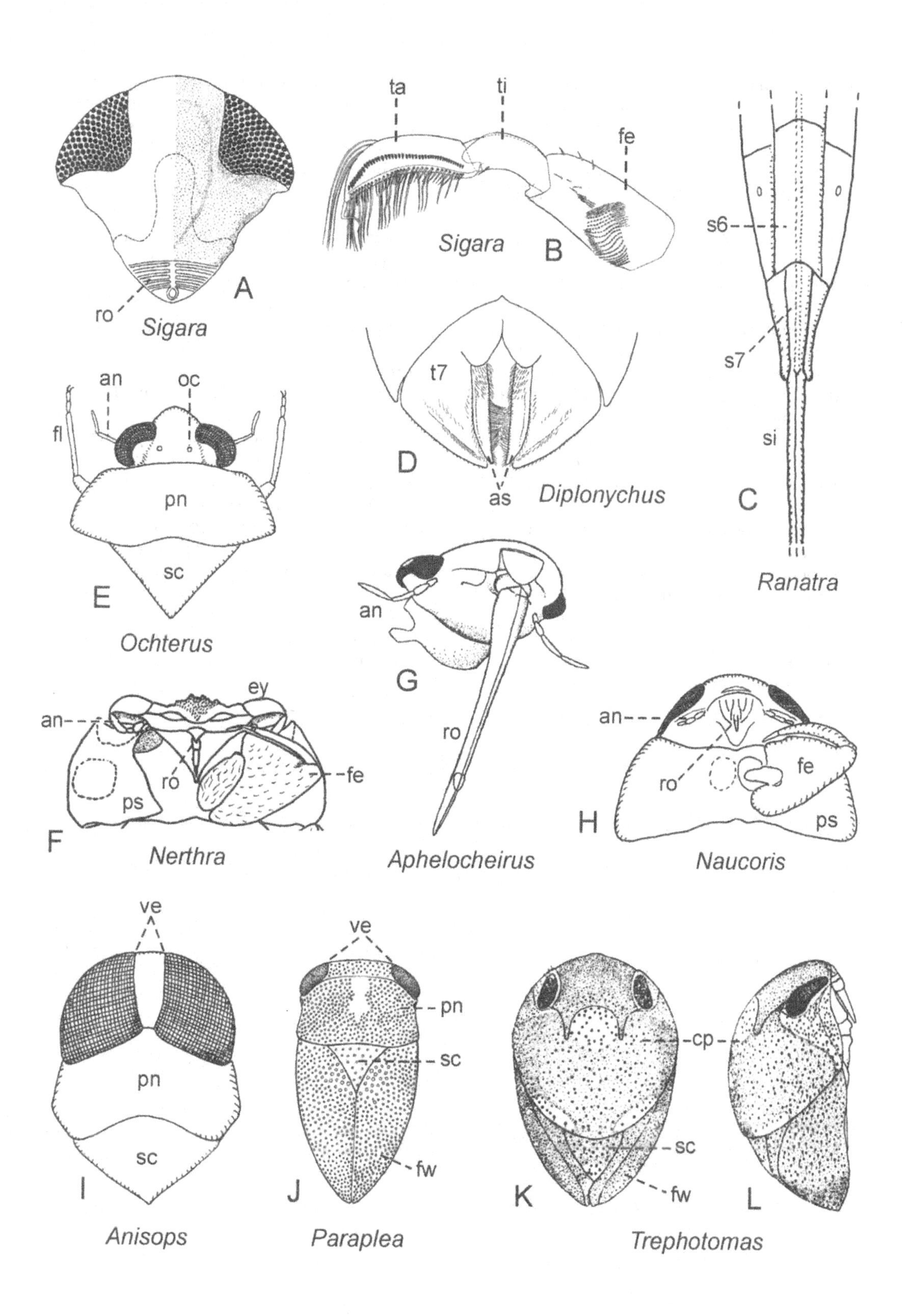

ro
Sigara
A
ta
ti
fe
Sigara
B
s6
s7
si
C
Ranatra
t7
D
as
Diplonychus
an
oc
fl
pn
sc
E
Ochterus
an
G
ro
Aphelocheirus
ey
an
ro
fe
ps
F
Nerthra
an
ro
fe
ps
H
Naucoris
ve
pn
sc
I
Anisops
ve
pn
sc
fw
J
Paraplea
cp
sc
fw
K
L
Trephotomas

16. Dorsum flat to moderately convex. Fore legs strikingly raptorial or antennae relatively long, protruding beyond outline of head 17
- Dorsum usually strongly convex, inversely boat-shaped to tectiform. Fore legs never raptorial, although sometimes modified in ♂ 18

17. Apex of antennae extending beyond lateral margins of head (Fig. 6.2G; an). Rostrum (ro) slender, reaching to mesosternum. Head narrow, elongate, strongly produced in front of eyes. Fore tarsus 3-segmented, mobile, with two well-developed claws .... APHELOCHEIRIDAE
- Apex of antennae not extending beyond lateral margins of head, usually not visible from above (Fig. 6.2H; an). Rostrum (ro) short and thick, not surpassing prosternum (ps). Head usually transverse, anteocular part only slightly produced in front of eyes. Fore tarsus 2- or 1-segmented, immobile, with one or two small claws or without claws .... NAUCORIDAE

18. Elongate, wedge-shaped species, usually over 4 mm long. Eyes large, vertex narrow (Fig. 6.2I; ve). Leg pairs often markedly different; hind legs long, oar-shaped, with reduced or inconspicuous claws. Head not firmly and immovably associated with prothorax. Midline of meso- and metasternum without carina; midline of abdominal venter with a flattened to sharp keel, never with an irregular carina .... NOTONECTIDAE
- Broadly oval, robustly built species, under 4 mm long. Eyes small to medium-sized, vertex broad (Figs 6.2J-K; ve). Leg pairs quite similar to each other; hind legs not oar-shaped, usually with two claws. Head firmly and immovably associated with prothorax. Thoracic sterna and basal abdominal segments with an irregularly shaped carina 19

19. Dorsum of head not fused with prothorax; head-pronotum suture distinct, straight (Fig. 6.2J). Antennae 3-segmented. Dorsum strongly convex .... PLEIDAE
- Dorsum of head completely fused with pronotum, forming a cephalonotum; head-pronotum suture not straight (Figs 6.2K, L). Antennae 2-segmented or non-segmented, Dorsum rather flat to extremely convex. [Not yet recorded from Australia] .... HELOTREPHIDAE

Infraorder GERROMORPHA

Semiaquatic Bugs

Identification

Usually small to medium-sized, rarely large insects inhabiting the surface of both freshwater and saline waters. A few species live in humid terrestrial habitats. Gerromorphans are characterised by: (1) dorsal head surface with three pairs of trichobothria (long slender setae) inserted in deep cuticular pits (Figs 7.3C, D); (2) mouth-parts with quadrangular (instead of triangular) mandibular lever (Andersen 1982: fig.26); (3) pretarsus (distal part of last tarsal segment) with two aroliae, one dorsal and one ventral (Fig. 4.4G; da, va); and (4) female genital tract with gynatrial complex, including a long tubular and entirely glandular spermatheca and a secondary fecundation canal (Fig. 4.5G).

Overview

The semiaquatic bugs are formally classified as an infraorder, Gerromorpha, of heteropterous bugs. This infraorder was formally proposed by Popov (1971b), but previously known by the name Amphibiocorisae proposed by Dufour (1833) as part of his classification of true bugs. Dufour's two other groups were the aquatic bugs, Hydrocorisae, and the terrestrial bugs, Geocorisae. At its current taxonomical state, the Gerromorpha comprises about 1,800 described species classified in eight families (Andersen 1982; Andersen & Weir in press).

7. Family MESOVELIIDAE

Water Treaders, Pondweed Bugs

Identification

Mesoveliids are small or very small bugs, length 1.2-4.4 mm. Usually pale yellowish or brownish, with a greenish tinge when alive. Body surface with layer of microtrichia restricted to head and thorax. Ocelli usually present (macropterous adults of the subfamily Mesoveliinae) or absent (apterous adults). Antennae long and slender (Fig. 7.1B), sometimes flagelliform, with very long and slender distal segments (Fig. 7.1A). Rostrum long and slender, basal segments of labium not obscured by bucculae (Fig. 7.3B). Pronotum of macropterous form truncate behind (Fig. 7.1C; pn); meso-scutellum (Fig. 7.1D; ms) well developed and exposed behind pronotum, either subtriangular (in the Afrotropical and Neotropical subfamily Madeoveliinae; Andersen 1982) or forming a transverse plate (Fig. 7.1C; ms); a small oval lobe behind scutellum belongs to metanotum (mt); pronotum of apterous form very short, exposing mesonotum (Fig. 7.1E; pn). Legs inserted on thoracic sterna, close to the ventral midline of each segment (Fig. 4.3C); tarsi 3-segmented, claws inserted apically (subapically in the subfamily Madeoveliinae). Forewing venation reduced distally, longitudinal veins enclosing 2-4 cells (Fig. 7.1C; fw). Male phallus with specialised ejaculatory pump; parameres small but distinct, symmetrical (Figs 7.4C, D; pa). Female ovipositor well-developed, laciniate, with serrate gonapophyses. Anterior end of egg truncate, with an egg cap developed in many species (Fig. 2.7G); eclosion by means of an embryonic bladder rather than an egg burster (Cobben 1968).

Overview

This small family includes 11 genera and about 40 species distributed worldwide. All species are predators on other arthropods. These yellowish green bugs live in moist surroundings, either marginal aquatic habitats, on water surfaces extensively covered with floating leaves of water plants, or humid terrestrial habitats (moss, litter). They overwinter as eggs which are inserted into plant tissue. There are five (in some species four) larval instars. Most species are wing dimorphic, but winged adults are usually rare. Jell & Duncan (1986) described a fossil mesoveliid, *Duncanovelia extensa*, from the Lower Cretaceous Koonwarra Formation, Victoria (Figs 3.4A, B). Two extant genera and 7 species are known to occur in Australia (Andersen & Weir in press).

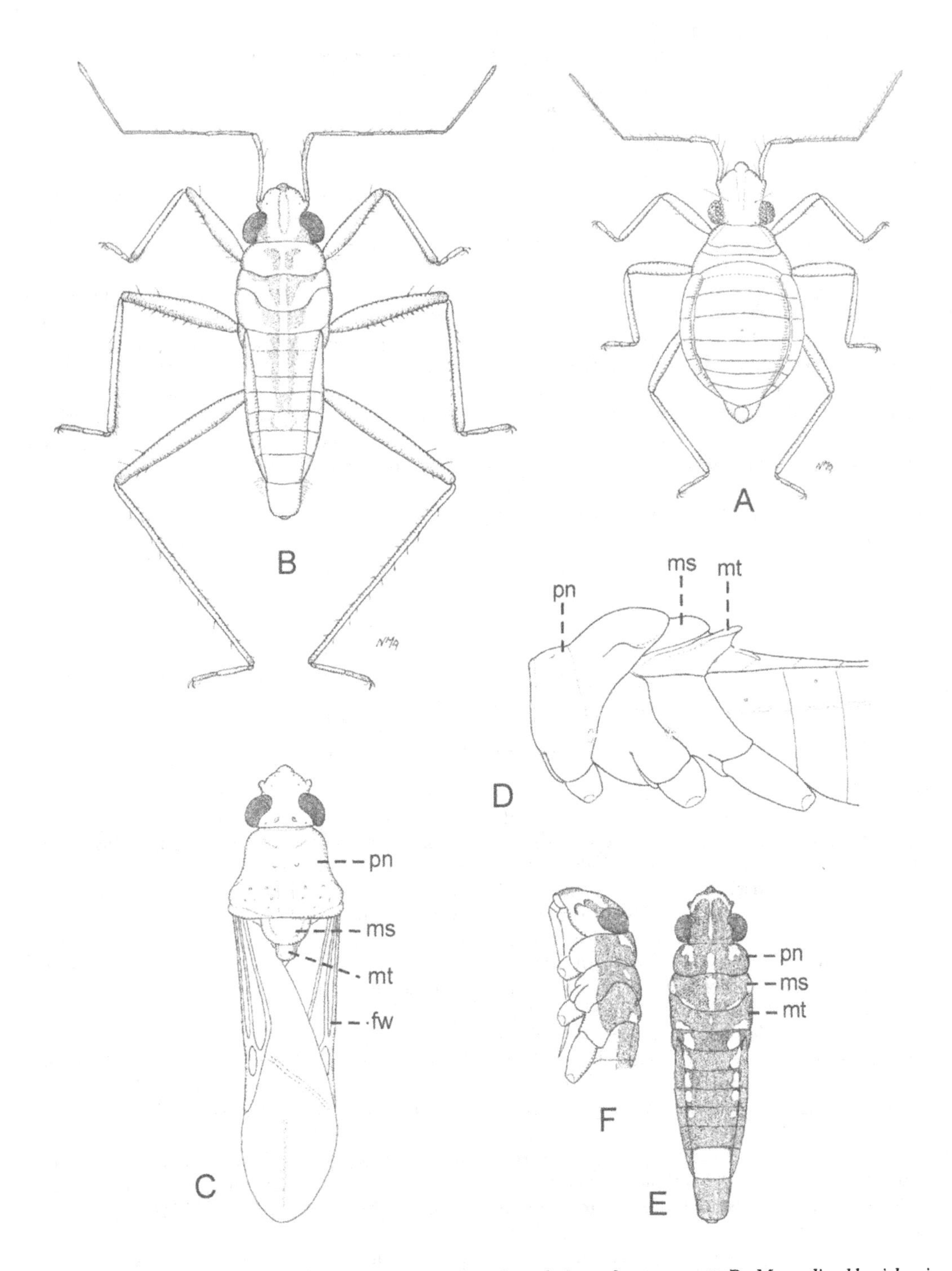

Figure 7.1. MESOVELIIDAE. A, *Austrovelia queenslandiae*: dorsal view of apterous ♂. B, *Mesovelia ebbenielseni*, apterous ♂: dorsal view. C, *M. stysi*, macropterous ♂: dorsal view, antennae and legs omitted (fw, fore wing; ms, meso-scutellum; mt, metanotal elevation; pn, pronotum). D, *Mesovelia* sp.: lateral view of thorax of macropterous form (pn, pronotum; ms, mesoscutellum; mt, metanotal elevation). E-F, *M. hackeri*, apterous ♂: E, dorsal view, antennae and legs omitted (ms, mesoscutellum; mt, metanotal lobe; pn, pronotum). F, lateral view of thorax, antennae and distal segments of legs omitted. Reproduced with modification from: A-C, E-F, Andersen & Weir (in press); D, Andersen (1982).

Key to the Australian genera of the family Mesoveliidae

1. Mesonotum of apterous form prolonged in middle, distinctly longer than pronotum (Figs 7.1B, E; ms). Both apterous and macropterous forms known for most species (Figs 7.1B, C) .... *Mesovelia*
- Mesonotum of apterous form not as above, subequal in length or shorter than pronotum (Fig. 7.1A). Only apterous form known.................................... *Austrovelia*

AUSTROVELIA

Identification

Very small, broadly oval bugs (Fig. 7.1A), length 1.2-1.5 mm. Head slightly longer than wide across eyes; three pairs of trichobothria almost equidistantly located on dorsal head surface. Ocelli absent (only apterous adults known). Eyes spherical, ommatidia large, varying in number from about 10 to more than 20. Antennae 0.7-0.9x total length of insect, flagelliform, segments 3-4 much longer and thinner than segments 1-2. Ventral surface of head with a pair of plate-like elevated longitudinal ridges behind rostral base, gradually widened and fading towards base of head (Malipatil & Monteith 1983: fig. 23). Rostrum extending to hind coxae or beyond. Pronotum of apterous form subequal in length to meso-and metanotum combined; posterior margin of metanotum curved. Femora moderately incrassate, without dark bristles on anterior surface and without black spines beneath. First and second segments of all tarsi subequal in length, third segment distinctly longer; claws and aroliae simple (Malipatil & Monteith 1983: fig. 24). Abdominal mediotergite 1 not subdivided into one median and two lateral parts. Abdominal scent gland opening distinct, located in middle of tergite 4.

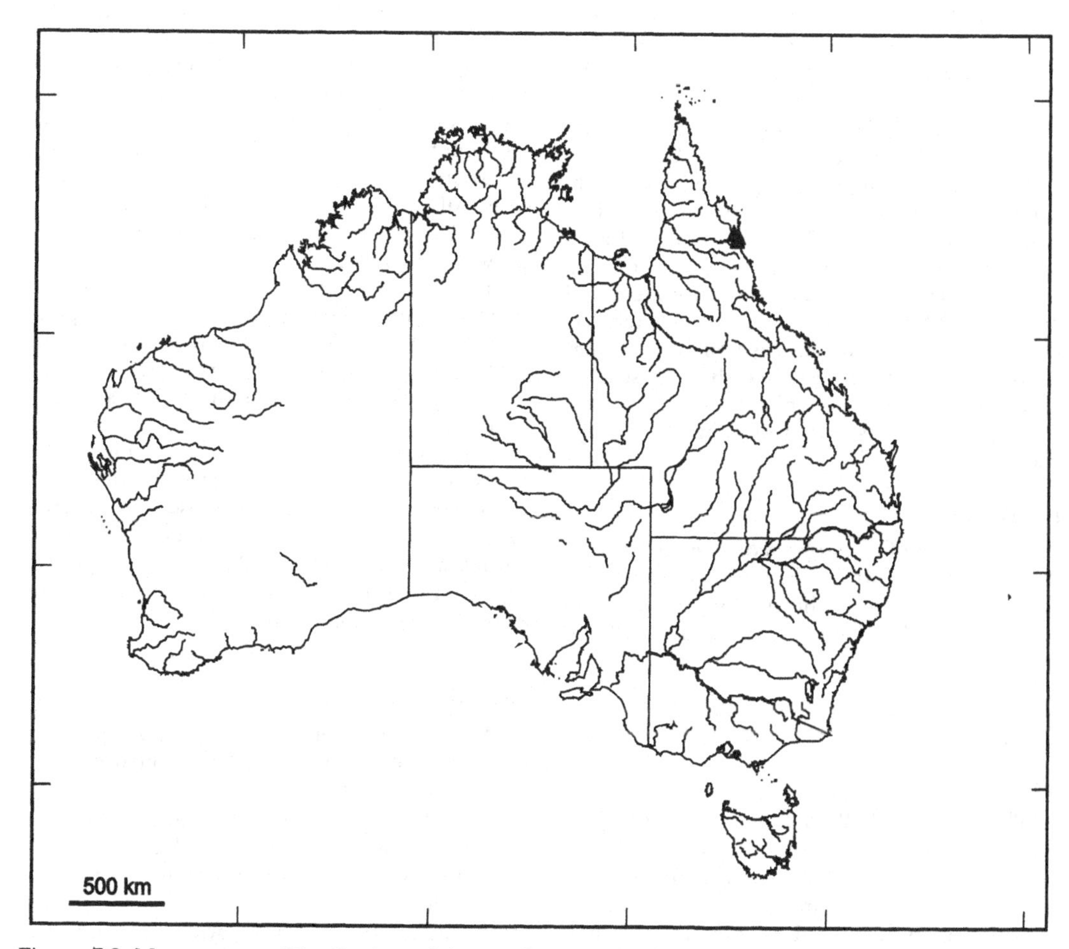

Figure 7.2. MESOVELIIDAE. Distribution of *Austrovelia queenslandiae* in Australia.

Male genital segments relatively large; dorsal hind margin of segment 8 slightly concave; proctiger simple, without lateral extensions; parameres small, shallow falciform. Female ovipositor large, gonapophyses laciniate and strongly serrated (Malipatil & Monteith 1983: figs 11-15).

The genus *Austrovelia* was erected by Malipatil & Monteith (1983) for two species, *A. queenslandica* (Queensland, Australia) and *A. caledonica* Malipatil & Monteith (New Caledonia). It is related to the genus *Phrynovelia* Horvath (Papua New Guinea, New Caledonia), but differs in having a pair of distinct ridges on the ventral head surface and in having a simple instead of tripartite first abdominal tergite.

Biology

These tiny bugs were extracted from sieved ground litter in rainforests (Malipatil & Monteith 1983). In systematic collections of litter at 13 sites along a transect from Cape Tribulation at sea level, across the Mt Sorrow tableland, to the slopes of Mt Pieter Botte at 780 m, *A. queenslandica* was taken in numbers at all sites above 500 m except one (a rocky ridge top). This distribution correlates closely with the extent of "Simple Microphyll Vine - Fern Thicket" along the transect, a type of vegetation typical of "cloudy wet and moist windswept of topslopes and highlands". Cape Tribulation is the focus of a high-rainfall area with over 3750 mm a year and rainfall on the mountain tops can be expected to be much greater where cloud cover is frequent and persistent. Adults are recorded from September-October and April, but no nymphs were found during the first, predominantly dry season. Breeding probably takes places during the wet season, and nymphs have been found in January and April.

Distribution

The type series of *A. queenslandiae* was collected in the Mt. Sorrow - Mt Pieter Botte tableland near Cape Tribulation, northern Queensland. Also known from Mt Hemmant, 6 km SW of Cape Tribulation (Fig. 7.2).

MESOVELIA

Identification

Small bugs, length 2.0-4.4 mm; body elongate oval to almost parallel-sided (Figs 7.1B, E). Colour chiefly yellowish (in live specimens with a greenish tinge) with more or less extensive brownish patterns. Head slightly longer than wide across eyes; basal pair of cephalic trichobothria inserted towards the base of head, two other pairs inserted on anterior part of head, before the level of eyes. Ocelli present in macropterous adults (Fig. 7.1C), absent in apterous adults (Fig. 7.1B). Eyes large, spherical, with many ommatidia. Antennae 0.6-1.0x total length of insect, subflagelliform; segment 1 with 2-3 dark bristles before apex. Ventral surface of head simple. Rostrum extending to middle coxae or beyond. Pronotum of macropterous form subquadrangular (Fig. 7.1C, D; pn), exposing mesoscutellum (ms) and metanotal lobe (mt). Pronotum much shorter than head in apterous adults (Fig. 7.1E; pn), exposing both mesonotum (ms) and metanotum (mt); mesonotum distinctly longer than pronotum, with posterior margin produced in middle. Legs long and slender, with hind leg longest; femora moderately incrassate, usually with 1-2 dark bristles on anterior surface, just before apex, and with a varying number of black spines beneath (Fig. 7.1B). Hind tibia with numerous dark bristles. Segment 1 of all tarsi relatively short, segment 2 longer, and segment 3 longest except in hind tarsus. Fore wings of macropterous form largely membranous, usually with thickened veins forming 3 closed cells in the antero-proximal part of the wing (Fig. 7.1C; fw). Abdomen long, usually more or less shiny. Abdominal scent gland opening distinct, located in middle of tergite 4. Male genital segments relatively large; segment 8 usually modified ventrally, with tubercles, patches of dark bristles, or spines; proctiger distally widened, with a pair of lateral, spinous extensions (Fig. 7.4C; pr); parameres (pa) relatively small and robust, usually with recurved and pointed apices (Fig. 7.4D).

The genus *Mesovelia* Mulsant & Rey is distributed world-wide with 27 species. The fauna of New Guinea was revised by J. Polhemus & D. Polhemus (2000b) who described or redescribed 5 species, three of which are also known from Australia. Andersen & Weir (in press) revised the Australian fauna and described one new species. Out of a total of six Australian species, two are endemic to the continent.

Key to Australian species

1. Middle femur ventrally with at most one or two black spines. Antenna relatively long, 0.8-1.1x total length 2
– Middle femur ventrally with many (usually more than 12) black spines (Fig. 7.1B). Antenna shorter, less than 0.9x total length 3
2. Middle femur ventrally without black spines. Apterous form chiefly brownish

Figure 7.3. MESOVELIIDAE. A-F, *Mesovelia vittigera*, apterous ♂: A, dorsal view. B, lateral view of head and thorax. C, dorsal view of head and pronotum; arrows mark the insertions of cephalic trichobothria. D, posterior cephalic trichobothrium. E, cuticular peg-plate. F, cuticular pit. Scales: A-C, 0.5 mm; D, 0.01 mm, E-F, 0.001 mm. Scanning Electron Micrographs (CSIRO Entomology, Canberra).

on thoracic and abdominal dorsum (Figs 7.1E, F). Abdominal sternum 8 of ♂ simple, without groups of black hairs and spines (Fig. 7.4E). Length 2.3-3.2 mm. New South Wales, Norfolk Island, Queensland .... *hackeri* Harris & Drake

– Middle femur ventrally with at least one or two medium-sized black spines distally. Apterous form chiefly brownish yellow above. Abdominal sternum 8 of ♂ basally with two tufts of black hairs, normally (genital segments not distended) originating under the hind margin of sternum 7 (Fig. 7.4G). Length 2.0-2.6 mm. Northern Territory, Queensland, Western Australia .... *horvathi* Lundblad

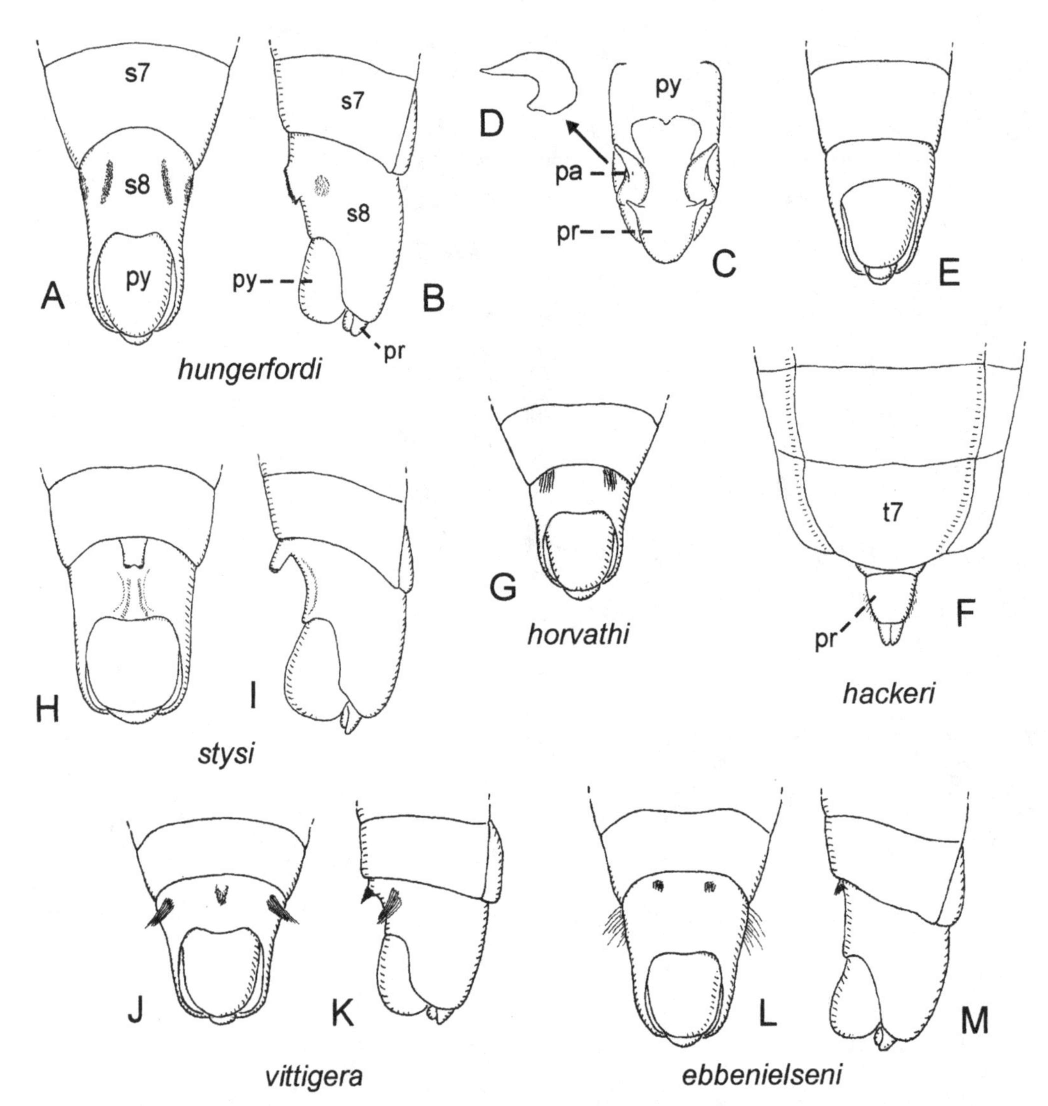

Figure 7.4. MESOVELIIDAE. A-D, *Mesovelia hungerfordi*, apterous ♂: A, ventral view of abdominal end. B, lateral view of abdominal end (pr, proctiger; py, pygophore; s7, sternum 7; s8, segment 8). C, dorsal view of pygophore and proctiger (pa, paramere; pr, proctiger; py, pygophore). D, left paramere at higher magnification. E-F, *M. hackeri*: ventral view of abdominal end of apterous ♂. F, dorsal view of abdominal end of apterous ♀ (pr, proctiger; t7, tergum 7). G, *M. horvathi*, apterous ♂: ventral view of abdominal end. H-I, *M. stysi*, apterous ♂: ventral view of abdominal end. I, lateral view of abdominal end. J-K, *M. vittigera*, apterous ♂: J, ventral view of abdominal end. K, lateral view of abdominal end. L-M, *M. ebbenielseni*, apterous ♂: L, ventral view of abdominal end. M, lateral view of abdominal end. Reproduced with modification from Andersen & Weir (in press).

3. Middle femur ventrally with 23-26 black spines (15-16 medium-length spines, interspersed with a 8-10 smaller spines). Abdominal sternum 8 of ♂ basally with a median quadrate process bearing two tufts of very short black hairs (Figs 7.4H, I). Length 2.8-4.0 mm. Northern Territory, Queensland ....*stysi* J. Polhemus & D. Polhemus
– Middle femur ventrally with less 20 black

spines. Abdominal sternum 8 of ♂ otherwise modified 4

4. Middle femur ventrally with 17-18 black spines (12-13 medium-length spines, interspersed with about 5 smaller spines). Abdominal sternum 8 of ♂ basally with a small median process bearing short stout black hairs and a tuft of pale hairs laterally (Figs 7.4J, K). Length 2.1-3.3 mm. Northern Territory, Queensland, Western Australia..........*vittigera* Horvath
- Middle femur ventrally with less than 17 black spines. Abdominal sternum 8 of ♂ otherwise modified... 5

5. Larger species, length 3.1-4.4 mm. Middle femur ventrally with 12-16 black spines (5-7 medium-length spines, interspersed with 7-9 smaller spines). Abdominal sternum 8 of ♂ basally with a pair of oblique processes bearing numerous very short black hairs (Figs 7.4A, B). New South Wales, Northern Territory, Queensland, South Australia, Tasmania, Victoria, Western Australia .... *hungerfordi* Hale
- Smaller species. length 2.1-3.3 mm. Middle femur ventrally with 12-13 black spines (6 medium-length spines, interspersed with 6-7 smaller spines). Abdominal sternum 8 of ♂ basally with a two groups of densely set, short black hairs (Figs 7.4L, M). Northern Territory, Queensland........ *ebbenielseni* Andersen & Weir

Biology

Water treaders or pondweed bugs are mainly found in stagnant water with plenty of vegetation along the edges, in permanent ponds with water lilies, billabongs, and in pools in river beds, as well as in temporary pools with algal growth. A few records are from brackish water and mangrove swamps. Adults are recorded throughout most of the year, nymphs from March-April and October-November (southeastern and southern Australia), March-June, August, October-November (northeastern and northern Australia). Unlike other Australian *Mesovelia* species, *M. horvathi* is usually cryptic, living amongst vegetation on the banks of flowing streams, stagnant pools in stream beds, and billabongs. Winged individuals are frequently attracted to light.

Mesovelia walks or runs in the way normal for insects: the three leg pairs are moved as alternating tripods of support composed of the fore and hind legs of one side of the body and the middle leg of the other side (Fig. 2.3A). Andersen (1976, 1982) studied locomotion in the European species *M. furcata*. This species moves rapidly both across floating vegetation and areas of free water surface. Where solid matter and water surface meet, a meniscus is usually formed which because it is steep and smooth may present an obstacle for surface bugs. *Mesovelia* is able to overcome this obstacle by a mechanism known as ascension of a meniscus. Prior to the ascension, the bug takes a characteristic body and leg posture (Figs 2.5A, B): the fore and hind legs are stretched out symmetrically, forwards and backwards, respectively, and pull up the surface film. The middle legs are extended laterally and carry the weight of the insect. The forces of attraction between menisci of the same shape causes the insect to be drawn up the slope of the larger meniscus (Andersen 1976, 1982).

Mesovelia species are predators or scavengers, feeding on micro-crustaceans (ostracods, cladocerans) or dead or half-dead culicid or chironomid midges, but never attempting to capture larger moving prey (Ekblom 1930; J. Polhemus & Chapman 1979a). Despite this *Mesovelia furcata* seems to be strictly carnivorous, Cobben (1965, 1968) found alleged endosymbionts in the gut this species.

Mating in *Mesovelia* species often takes place when the bugs are resting on the water surface. The mating behaviour is quite simple. The male clasps his fore legs around the mesothorax of the female, rests the middle legs on the water surface or on some part of the female's body, and holds the hind legs in the air. Copulation usually lasts a few minutes, after which the sexes separate (Hungerford 1920; Ekblom 1930). During oviposition, the female drills a hole in plant stems with her long, concave and serrate ovipositor. A number of eggs are embedded in the same plant stem, with each egg in a separate hole. The anterior end of the egg is truncate and provided with a cap through which the nymph hatches. The life history of *Mesovelia* species may be different from that of other Australian gerromorphans. In northern America and Europe, eggs laid by *Mesovelia* species in the autumn overwinter in a state of embryonic diapause which last until the following spring where the small nymphs emerge (Ekblom 1930; Galbreath 1973, 1976). The lifecycle of the European *M. furcata* only involves four nymphal instars (both wingless and winged specimens). This species is trivoltine in Switzerland which also may be the case in Australian *Mesovelia* species (Zimmermann 1984).

Distribution

Mesovelia species are distributed throughout Australia except for the dry interior and larger

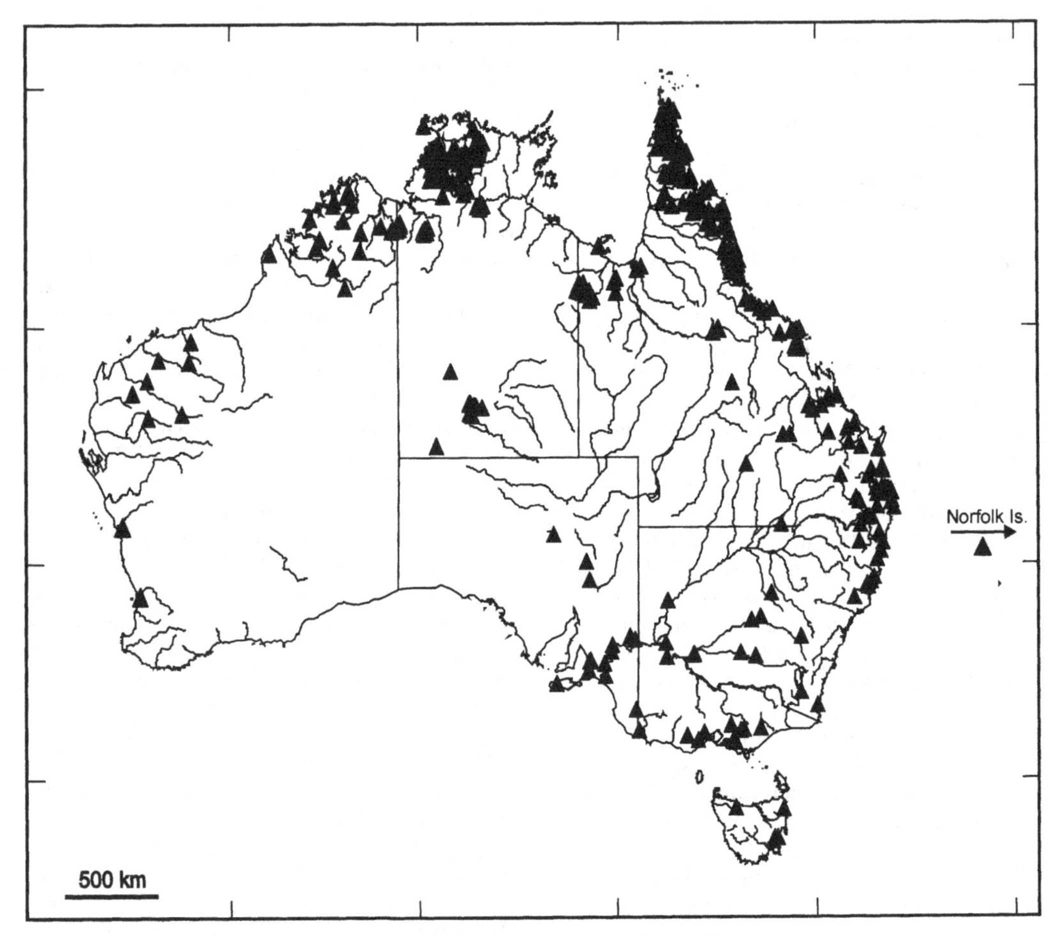

Figure 7.5. MESOVELIIDAE. Distribution of *Mesovelia* species in Australia.

parts of South Australia and Western Australia (Fig. 7.5). The most common and widespread species, *M. hungerfordi*, is recorded from all states except A.C.T., but apparently absent from most parts of Northern Territory and Western Australia, where *M. vittigera* is the most common species. The latter species is globally one of the most widespread species of *Mesovelia*, recorded from larger parts of Asia, Africa, and southern Europe (Andersen 1967, 1995; J. Polhemus and D. Polhemus 2000). In Australia, *M. hackeri* is limited to northern New South Wales and southern Queensland, but this species is also known from Norfolk Island and New Zealand where it is one of very few gerromorphan bugs (Andersen & Weir in press).

8. Family HEBRIDAE

Velvet Water Bugs

Identification

Hebrids are stout-bodied, small or very small bugs, length 1.4-2.0 mm. Body usually dull-coloured, densely covered with a velvety hydrofuge hair-pile, interspersed with small peg-plates and mushroom-shaped micro-hairs (Fig. 8.4D). Adults usually macropterous or brachypterous; micropterous and apterous adults rare. Two ocelli present on head (absent or vestigial in *Austrohebrus*). Antennae variable in length, usually slender (Figs 8.1A, 8.3A). Rostrum long and slender, resting in a groove on the underside of head and thoracic sternum; basal segments of labium obscured by a pair of plate-shaped bucculae (Figs 8.1C, 8.3G, H); labial segments 1-2 very small, segment 3 longest, segment 4 relatively long, tapering in width. Pronotum of macropterous form emarginated behind (Figs 8.3C, G; pn); meso-scutellum (ms) exposed behind pronotum, forming a transverse plate; a subtriangular lobe behind that (commonly termed "scutellum") is actually a median elevation of metanotum (mt). Legs stout, equally spaced on thorax, somewhat removed from the ventral midline of each segment; all tarsi 2-segmented, basal segment very short; claws well-developed, inserted apically on tarsus. Forewing venation reduced distally, longitudinal veins enclosing 1-2 cells (Figs 8.3A, 8.4A, E). Abdomen relatively long, mediotergite 4 with a median scent orifice. Genital segments of both sexes relatively small, seemingly inserted on the abdominal venter, slightly before the apex of abdomen (Fig. 8.3B). Male parameres usually small but distinct, symmetrical (Figs 8.3D, E; pa), rarely slender and long (Fig. 8.3B; pa). Female ovipositor short, gonapophyses plate-shaped and non-serrated. Eggs relatively large, usually ovoid, with a single micropyle at the anterior end (Fig. 2.7B); eclosion through a longitudinal split caused by a median, clypeal egg burster (Andersen 1982: fig. 11).

Overview

The family includes two subfamilies, 9 genera, and about 160 species distributed worldwide. All species are predators on small insects and other arthropods. These small bugs live in moist surroundings, either humid terrestrial (litter, moss), marginal aquatic habitats (sphagnum bogs), rarely on water surfaces covered with floating plants. Overwinter as adults. The relatively large eggs are deposited superficially on plant leaves. There are five nymphal instars. Most species are macropterous or brachypterous, rarely micropterous or apterous. A fossil hebrid was described by J. Polhemus (1995) from the Oligocene-Miocene Mexican amber. Three extant genera and 6 species are recorded from Australia, all belonging to the subfamily Hebrinae (the subfamily Hyrcaninae is confined to the Oriental region).

Key to the Australian genera of the family Hebridae

1. Antennae distinctly shorter than the greatest width of pronotum (Figs 8.3I, 8.4E-F). Antennal segments 3 and 4 short and club-like; segment 4 not subdivided .... *Merragata*
– Antennae subequal in length to or longer than the greatest width of pronotum (Figs 8.3A, C). Antennal segments 3 and 4 long and slender; segment 4 subdivided in middle by a membranous zone (Fig. 8.3C; is, 8.4C), appearing 5-segmented 2
2. Macropterous (Figs 8.3A, 8.4A), brachypterous, or micropterous with distinct wing-pads. Ocelli present and distinct (Figs 8.3C, G; oc). Meso-scutellum (Fig. 8.3C; ms) and metanotal elevation (mt) well-developed .... *Hebrus*
– Apterous, no traces of wings (Fig. 8.1A). Ocelli absent or vestigial (Figs 8.1B, C). Meso-scutellum and metanotal elevation not developed .... *Austrohebrus*

AUSTROHEBRUS

Identification

Small, suboval bugs, length 1.6-1.7 mm (Fig. 8.1A), dull coloured. Head elongate, distinctly extended in front of eyes. Ocelli reduced to a pair of semicircular pits close to the inner margins of eyes (Figs 8.1B, C). Eyes relatively small and spherical, with about 20 large ommatidia. Antennae long and subflagelliform; segments 1 and 2 relatively short and stout, segments 3 and 4 long and slender; segment 4 subdivided in middle by a constricted, membranous zone. Ventral surface of head with a pair of prominent bucculae (Fig. 8.1C; bu). Rostrum long and slender, reaching base of abdomen. Pronotum short, length less than half the length of head (Figs 8.1B, C; pn). Mesonotum not visible behind pronotal lobe. Metanotum (Figs 8:1B, C; mt) short, with no trace of a median elevation. Legs relatively long,

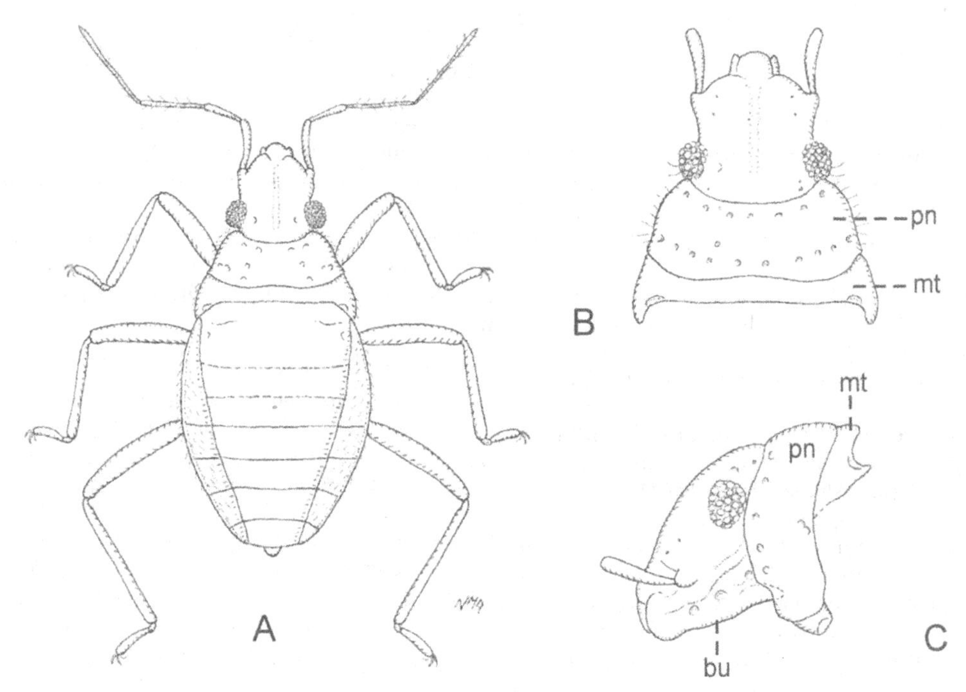

Figure 8.1. HEBRIDAE. A-C, *Austrohebrus apterus*, apterous ♀: A, dorsal view. B, dorsal view of head and thorax; distal antennal segments and legs omitted (pn, pronotum; mt, metanotum). C, lateral view of head and thorax; distal antennal segments and legs omitted (bu, buccula; pn, pronotum; mt, metanotum. Reproduced with modification from Andersen & Weir (in press).

femora moderately incrassate. Wings completely absent. Abdomen relatively long, mediotergite 4 with a median scent orifice. Genital segments of female relatively large, inserted slightly before the apex of pregenital abdomen; proctiger small, suboval, protruding from abdominal end. Male characters unknown.

The genus *Austrohebrus* was described by Andersen & Weir (in prep.) with one species, *A. apterus* Andersen & Weir, length 1.6-1.7 mm. Western Australia.

Biology

Nothing is known about the biology of these small obscure, wingless bugs.

Distribution

Only known from the holotype and single paratype from Broome, Western Australia (Fig. 8.2).

HEBRUS

Identification

Small or very small bugs, length 1.4-2.0 mm. Usually dull coloured, forewings with pale spots. Head elongate, distinctly extended in front of eyes (Figs 8.3A, C, F, 8.4F). Eyes relatively small, spherical, with coarse ommatidia. Antennae long and subflagelliform; segments 1 and 2 relatively short and stout, segments 3 and 4 long and slender; segment 4 subdivided in middle by a constricted, membranous zone (Fig. 8.3C; is, 8.4C) with ring-like cuticular thickenings (Andersen 1982: fig. 114). Ventral surface of head with a pair of prominent bucculae (Figs 8.G, H; bu, 8.4B), which form a deep groove concealing the basal rostral segments. Rostrum long and slender, usually reaching hind coxae and base of abdomen. Pronotum relatively short and broad (Fig. 8.3C, G; pn). Meso-scutellum (ms) narrow, transverse; metanotal elevation (mt) usually subtriangular, somewhat reduced in brachypterous and micropterous adults, but never completely absent; posterior margin more or less rounded (Figs 8.3C, I), sometimes bifid (Fig. 8.3F). Legs moder-

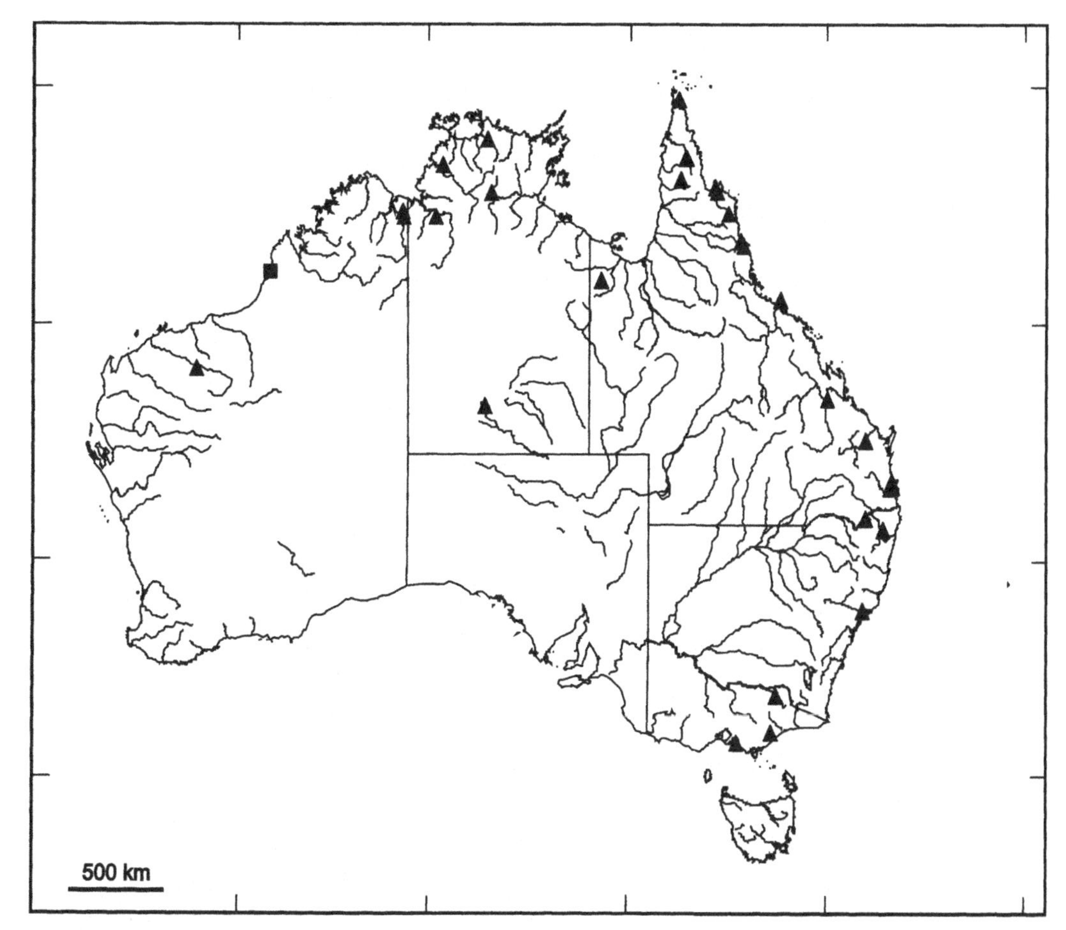

Figure 8.2. HEBRIDAE. Distribution of *Austrohebrus apterus* (square) and *Merragata hackeri* (triangles) in Australia.

ately prolonged, hind leg longest; all tarsi with two segments of which the basal segment is very small. Fore wings with two thickened veins in basal part forming one closed cell (Fig. 8.3A); distal part without veins. Parameres of male small, symmetrical, and robust (Figs 8.3D, E; pa) or long and slender (Fig. 8.3B; pa).

The large genus *Hebrus* Curtis (with about 120 species) is found in all continents and on many islands. Four species are recorded from Australia (Lansbury 1990; Andersen & Weir in prep.).

Key to Australian species

1. Dorsal surface of head and thorax, bases of fore wings, and ventral thorax and abdomen strongly pilose (Fig. 8.3H). Length 1.8-1.9 mm. Northern Territory .... *pilosus* Andersen & Weir
- Head and body not strongly pilose 2
2. Apex of metanotal elevation ("scutellum") bifurcate (Fig. 8.3F). Posterior corners of buccula with blunt or hook-shaped process (Fig. 8.3G; bu). Length 1.4-1.8 mm. New South Wales, Northern Territory, Queensland .... *nourlangiei* Lansbury
- Apex of metanotal elevation rounded or acuminate (Figs 8.3A, C; mt), rarely with a small median, round incision. Buccula otherwise shaped 3
3. Parameres long and slender, projecting beyond the abdominal end (Fig. 8.3B; pa). Membrane of fore wings not reaching end of abdomen, especially in ♂ (Fig. 8.3A). Length 1.9-2.0 mm. Queensland .... *monteithi* Lansbury
- Parameres short and robust, not project-

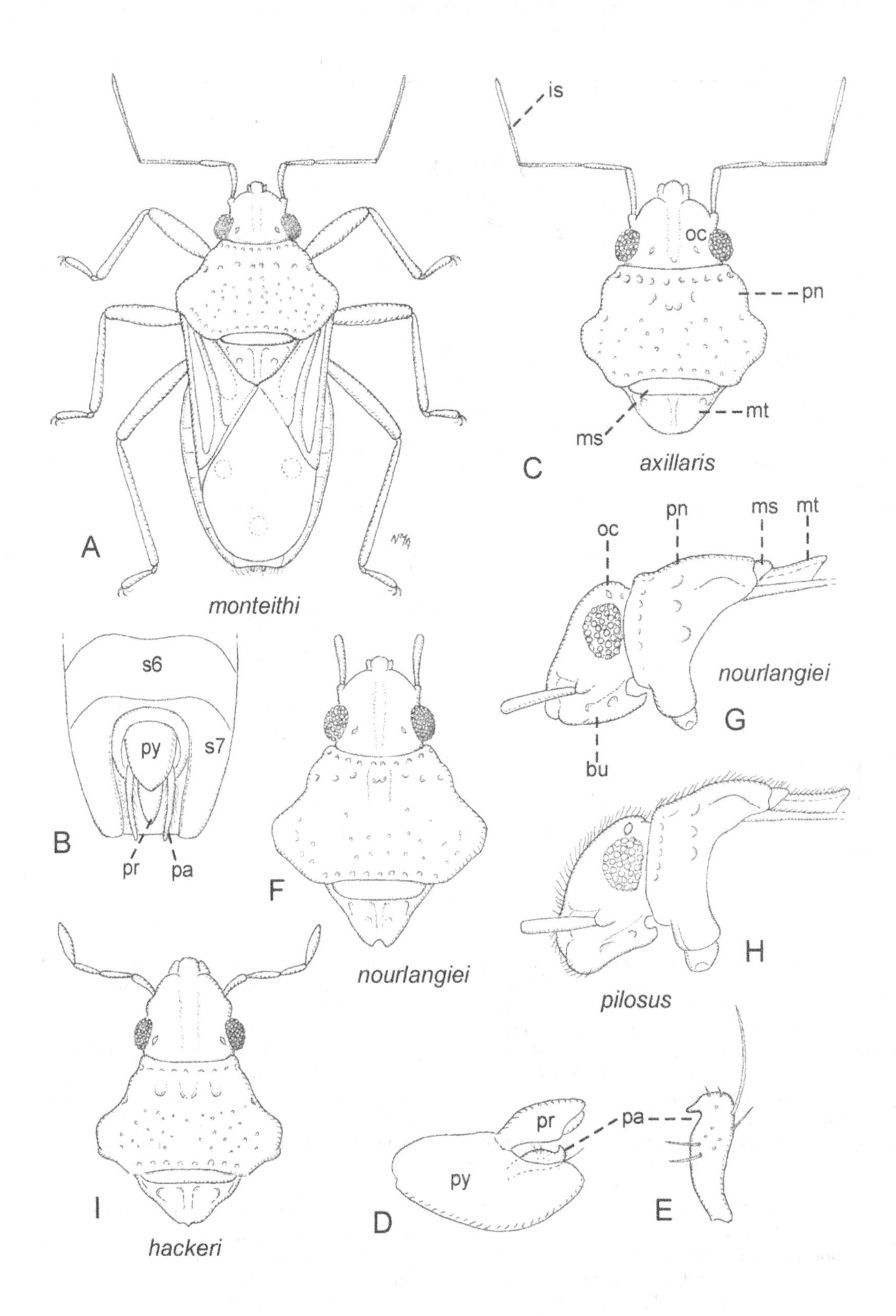

A
monteithi
is
oc
pn
ms
mt
C
axillaris
s6
py
s7
B
pr
pa
F
nourlangiei
oc
pn
ms
mt
bu
nourlangiei
G
H
pilosus
I
hackeri
pr
pa
py
D
E

ing beyond the abdominal end (Figs 8.3D, E; pa). Membrane of fore wings usually reaching or surpassing end of abdomen in macropterous adults. Length 1.65-2.0 mm. New South Wales, Northern Territory, Queensland, South Australia, Tasmania, Victoria, Western Australia .... *axillaris* Horvath

Biology

The habitats of *Hebrus* species are not the water surface, but weeds, litter, and debris, normally close to either stagnant or flowing freshwater bodies; occasionally also muddy or sandy soil, more rarely in mangrove swamps. *H. monteithi* was collected by sieving litter and brushing sticks in rainforests. Species living in *Sphagnum* bogs (such as the European *H. ruficeps*) may be collected by stepping on the floating mats of *Sphagnum* so that clear water appears above the moss; the bugs will then float to the surface and can be picked up with a small dipper. In Europe and North America *Hebrus* species overwinter as adults although they may be active during winter in milder climates. In Australia, adults may be collected throughout the year, but most frequently in March-June and September-December; nymphs are recorded from November and January. Adults usually macropterous and occasionally taken at light or by flight intercept traps.

The powerful claws of *Hebrus* species are well suited to walk and climb on stems and leaves of plants, but these velvety water bugs move very slowly on water surfaces. All observations unequivocally point to a carnivorous habit for *Hebrus* which may feed on Collembola and other small insects (Polhemus & Chapman 1979b). Reports by Jordan (1952) that he had observed the European *Hebrus ruficeps* sucking out leaves of *Sphagnum* is probably erroneous, since the probing and insertion of the rostrum into plant material may have been part of the search for prey. The life history, unknown for Australian species, has been studied in European and North American species (Hungerford 1920; Southwood & Leston 1959; Polhemus & Chapman 1979b). The eggs are very large relative to the size of the female, elongate oval, with a single micropyle (Fig. 2.7B). They are laid on moss, usually in groups in leaf axils or between closely spaced leaves (Fig. 2.7A). Incubation takes from 8-12 days and the five nymphal instars are completed in about 20 days (Lebrun 1960; Cobben 1968; Polhemus & Chapman 1979b). The functional morphology of the male and female genital organs have been extensively documented for the European species *H. pusillus* and *ruficeps* (Heming-van Battum & Heming 1986, 1989).

Distribution

Hebrus species are distributed throughout Australia, except for most of the dry interior and larger parts of South Australia and Western Australia (Fig. 8.5). The most common and widespread species, *H. axillaris*, is recorded from all states except A. C. T.

Merragata

Identification

Very small, elongate oval bugs (Fig. 8.4E), length 1.4-1.7 mm. Head elongate (Figs 8.3I, 8.4F), distinctly extended in front of eyes. Two ocelli situated at the base of head. Eyes relatively small, globular, with coarse ommatidia. Antennae short and robust, length much less than width of pronotum; segments 1-3 relatively short and slender, segment 4 suboval, not subdivided in middle (Figs 8.3I, 8.4F). Ventral surface of head with a pair of prominent bucculae. Rostrum usually reaching base of abdomen. Pronotum relatively short and broad, hind margin medially concave or straight. Meso-scutellum narrow, transverse; metanotal elevation subtriangular (Fig. 8.3I). Fore wings with two thickened veins in basal part forming one closed cell (Fig. 8.4E). Genital segments of both sexes relatively small; parameres of

---

Figure 8.3. HEBRIDAE. A-B, *Hebrus monteithi*, macropterous ♂: A, dorsal view. B, ventral view of abdominal end (pa, paramere; pr, proctiger; s6, s7, sterna 6 and 7). C-E, *H. axillaris*, macropterus ♂: C, dorsal view of head and thorax (oc, ocellus; pn, pronotum; ms, meso-scutellum; mt, metanotal elevation). D, lateral view of pygophore (py) and proctiger (pr) with paramere (pa). E, left paramere in higher magnification. F-G, *H. nourlangei*, macropterus ♂: F, dorsal view of head and thorax; distal antennal segments omitted. G, lateral view of head and thorax; distal antennal segments omitted (bu, buccula; oc, ocellus; pn, pronotum; ms, meso-scutellum; mt, metanotal elevation). H, *H. pilosus*, macropterus ♀: lateral view of head and thorax; distal antennal segments omitted. I, *Merragata hackeri*, macropterus ♂: dorsal view of head and thorax. Reproduced with modification from Andersen & Weir (in press).

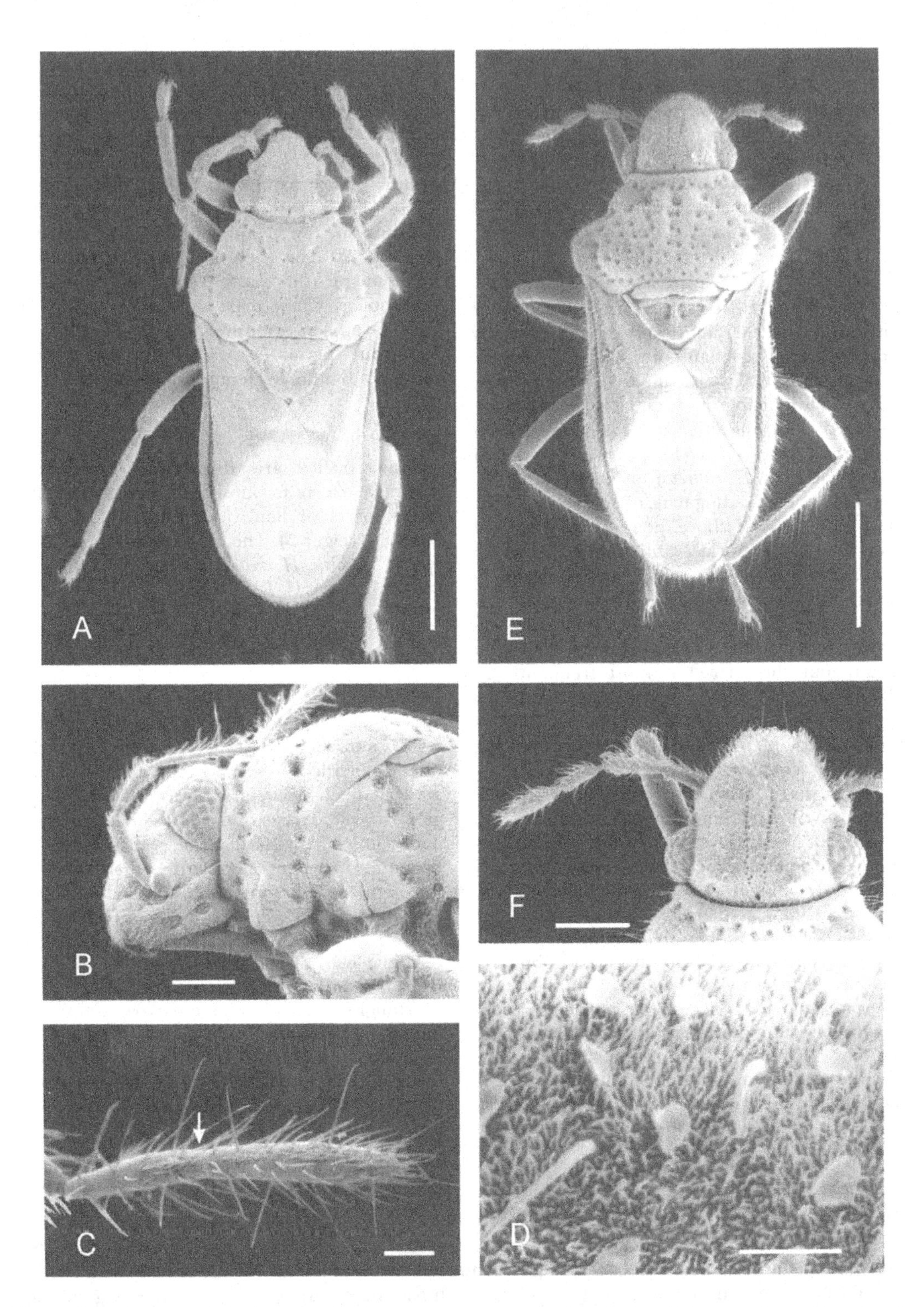

Figure 8.4. HEBRIDAE. A-D, *Hebrus axillaris*, macropterous ♀: A, dorsal view. B, lateral view of head and thorax. C, fourth antennal segment with median, membranous zone (arrow). D, surface structure of pronotum. E-F, *Merragata hackeri*, macropterous ♀: E, dorsal view. F, dorsal view of head. Scales: A, E, 0.5 mm; B, F, 0.1 mm; C, 0.02 mm; D, 0.01 mm. Scanning Electron Micrographs (CSIRO Entomology, Canberra).

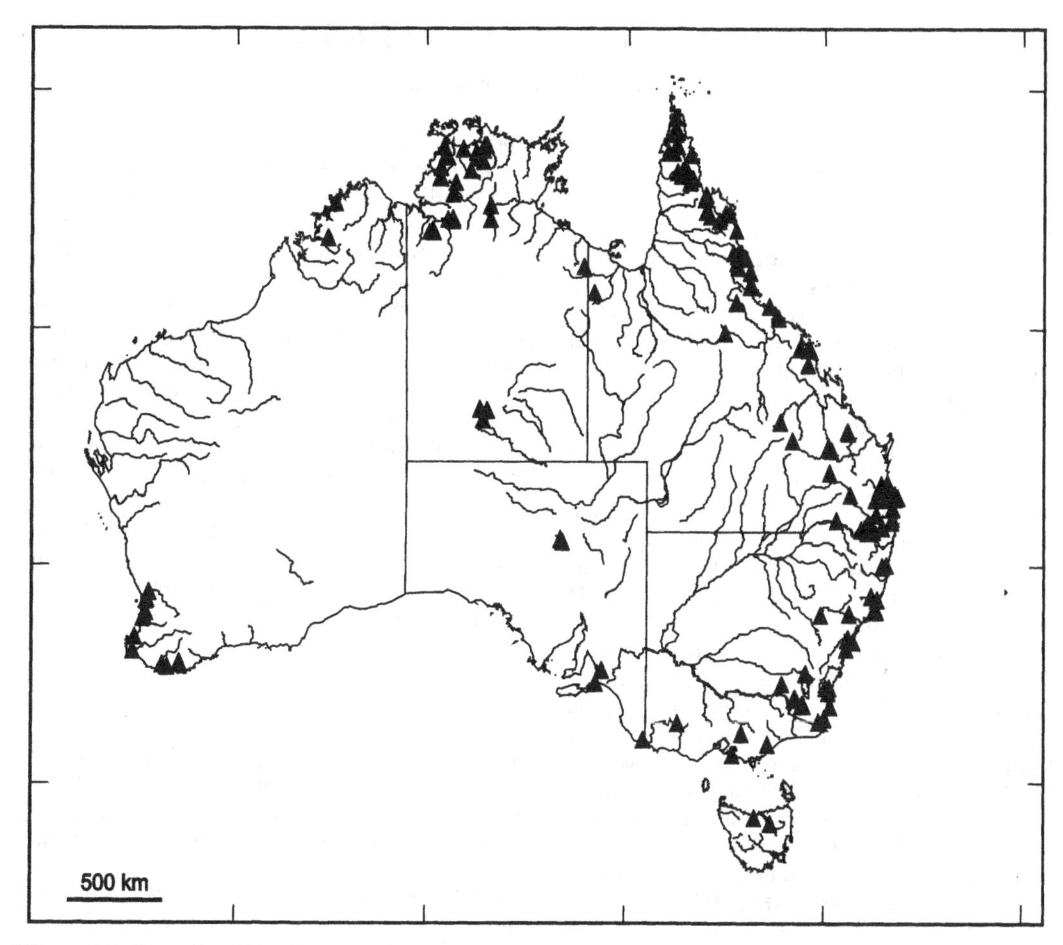

Figure 8.5. HEBRIDAE. Distribution of *Hebrus* species in Australia.

male small, stout, symmetrical. Other characters as in *Hebrus*.

Merragata White is a small genus with only 7 species, most of which occur in the Nearctic and Neotropical regions. There is one species in the Oriental region and one in Australia (Lundblad 1933; Lansbury 1990): *M. hackeri* Hungerford, length 1.4-1.7 mm, New South Wales, Northern Territory, Queensland, Victoria, Western Australia.

Biology

Merragata hackeri lives on the plant-covered water surfaces along margins of ponds, pools, and farm dams, usually with plenty of floating vegetation. Adults are always macropterous and found throughout the year; nymphs are recorded from June-August, October, and December. Drake (1917) characterised *Merragata* species as true surface dwellers, capable of running on the surface film. They were also seen walking around on submerged plants, staying below the water surface for about half an hour.

Distribution

Merragata hackeri is sparsely distributed in Northern Territory and Western Australia and along the East coast of Queensland, New South Wales, and Victoria (Fig. 8.2).

9. Family HYDROMETRIDAE

Water Measurers, Marsh Treaders

Identification

Hydrometrids are slender-bodied, medium-sized to large bugs, length 7.8-18 mm (Fig. 9.1A; Plate 6, A). Body usually dull-coloured, densely covered with a velvety hydrofuge hair-pile. Adults usually macropterous or brachypterous; micropterous and apterous adults rare. Head more or less elongate, usually very slender, with the compound eyes distinctly removed from anterior margin of thorax (Fig. 9.1I; ey, 9.2B); ocelli usually absent (only present in the subfamily Heterocleptinae, not in Australia). Antennae long, antennal segments slender, in particular the two distal ones (Fig. 9.1A). Rostrum long and slender, basal segments of labium obscured by a pair of semicircular bucculae (Fig. 9.1C; bu). Thorax more or less prolonged (Fig. 9.1I); pronotum (pn) posteriorly extended to cover mesonotum; metanotum (mt) usually exposed. Legs slender and long, equally spaced on thorax, inserted ventro-laterally or laterally on body; all tarsi 3-segmented, basal segment very short; claws usually small, inserted apically on tarsus (Fig. 9.2D). Wings relatively narrow; forewing venation relatively simple. Abdomen usually long and slender, with or without a scent orifice. Genital segments of both sexes protruding from the apex of the pregenital abdomen (Figs 9.2E, 9.3E, F). Male parameres small, symmetrical. Female ovipositor short, gonapophyses plate-shaped and non-serrated. Eggs relatively large, usually spindle-shaped, with a basal pedicel and a single micropyle at the anterior end (Fig. 9.4); eclosion through a longitudinal split caused by a median, frontal egg burster (Andersen 1982: fig. 167).

Overview

This family includes 3 subfamilies, 7 genera, and about 130 species distributed world-wide. All species are predators on insects and other arthropods. These extremely slender bugs live in moist surroundings, either humid terrestrial (litter), marginal aquatic habitats, or on water surfaces covered with floating plants. Overwinter as adults. The relatively large eggs are deposited in an upright position, above water on plant leaves or stems, etc. There are five nymphal instars. Many species are wing dimorphic. The unique design of head and body sets the Hydrometridae apart from other water bugs. Yet, this design is very old. Fossil hydrometrids are known from Baltic amber (50-40 Myr), Burmese amber (100-90 Myr), and from the lower Cretaceous of Brazil (Andersen 1998; Andersen & Grimaldi 2001). Only one extant genus with 9 species is found in Australia (belonging to the subfamily Hydrometrinae).

HYDROMETRA

Identification

Medium-sized to large bugs, length 7.8-18.1 mm; head, thorax, and abdomen elongate and very slender. Head extremely prolonged, with semiglobular eyes inserted on the sides of the head; ratio between the anteocular and postocular portions of the head varies between 1.2:1 and 2.8:1. Anterior part of head swollen, with anteclypeus protruding in middle (Figs 9.1B, C; ac) together with the large maxillary plates (mx). Ocelli absent. Antennae very long and slender; segment 1 relatively stout, much shorter than any of the following segments of which segment 3 is the longest. Rostrum usually reaching backwards to base of head (Fig. 9.2A). Thorax, and in particular metathorax prolonged, with coxae inserted in acetabula projecting from the sides of thorax. Pronotum long, almost parallel-sided. Metathor-

Figure 9.1. HYDROMETRIDAE. A-C, *Hydrometra feta*, macropterous ♂: A, dorsal view. B, dorsal view of apex of head; antenna omitted (ac, anteclypeus; mx, maxillary plate). C, lateral view of apex of head; antenna and distal part of rostrum omitted (ac, anteclypeus; bu, buccula; ct, cephalic trichobothrium; mx, maxillary plate). D, *H. illingworthi*, macropterous ♂: lateral view of apex of head; antenna and rostrum omitted. E-F, *H. papuana*, macropterous ♂: lateral view of abdominal end. F, ventral view of abdominal end (py, pygophore; s7, sternum 7; s8, segment 8. G-H, *H. papuana*, macropterous ♀: G, dorsal view of abdominal end. H, lateral view of abdominal end (gx, gonocoxa; pr, proctiger; s7, sternum 7; t8, tergum 8. I, *H. strigosa*, micropterous ♀: dorsal view of head and thorax; distal antennal segments and legs omitted (ey, eye; fw, forewings; mt, metathorax; pn, pronotum). J, *H. strigosa*: dorsal view of apex of head; antenna omitted (ac, anteclypeus). K, *H. novaehollandiae*: dorsal view of apex of head. Reproduced with modification from Andersen & Weir (in press).

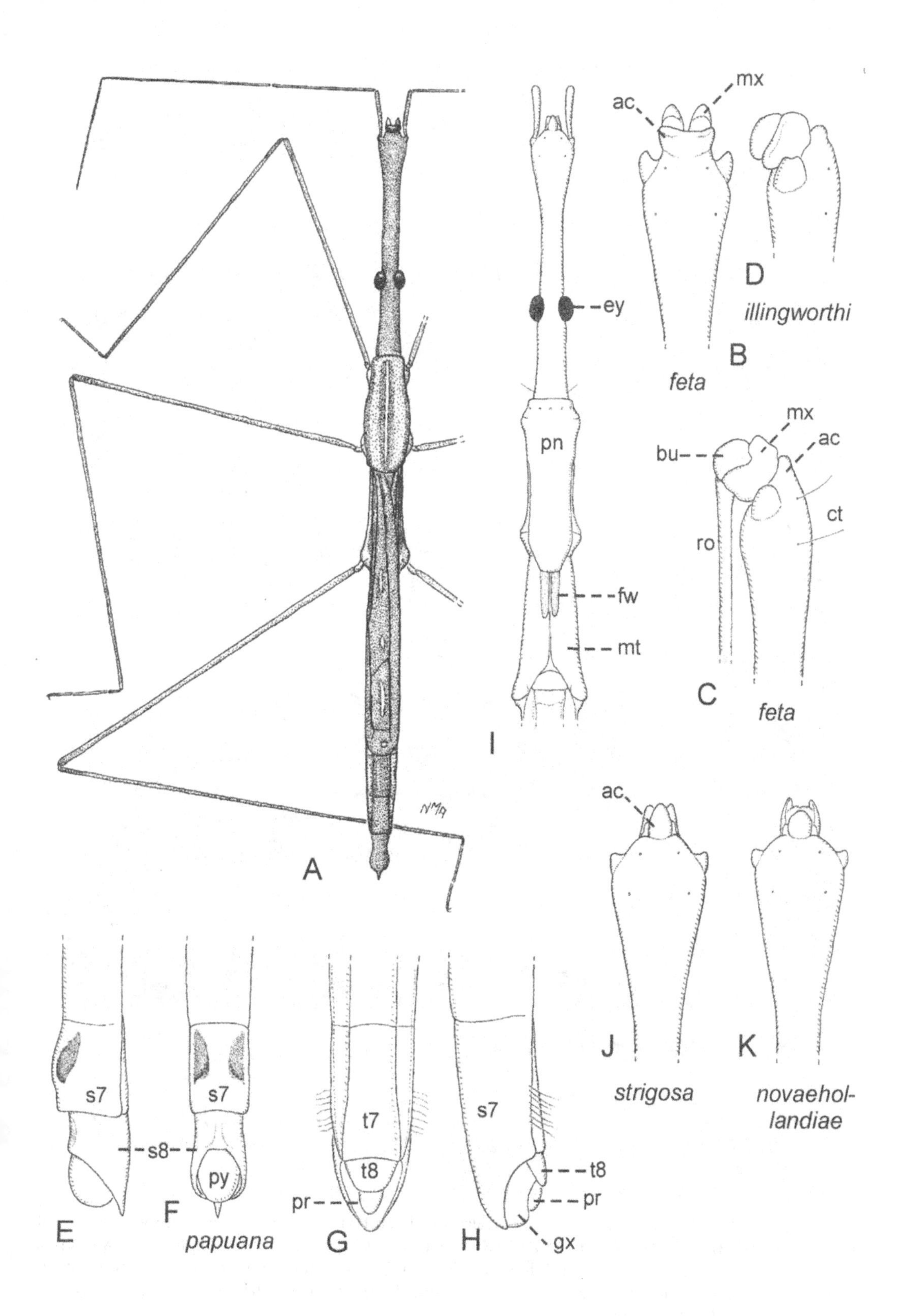

mx
ac
D
illingworthi
ey
B
feta
mx
ac
bu
ct
ro
C
feta
pn
fw
mt
I
ac
A
J
K
strigosa
novaehol-
landiae
s7
s7
s8
py
t7
t8
pr
s7
t8
pr
gx
E
F
papuana
G
H

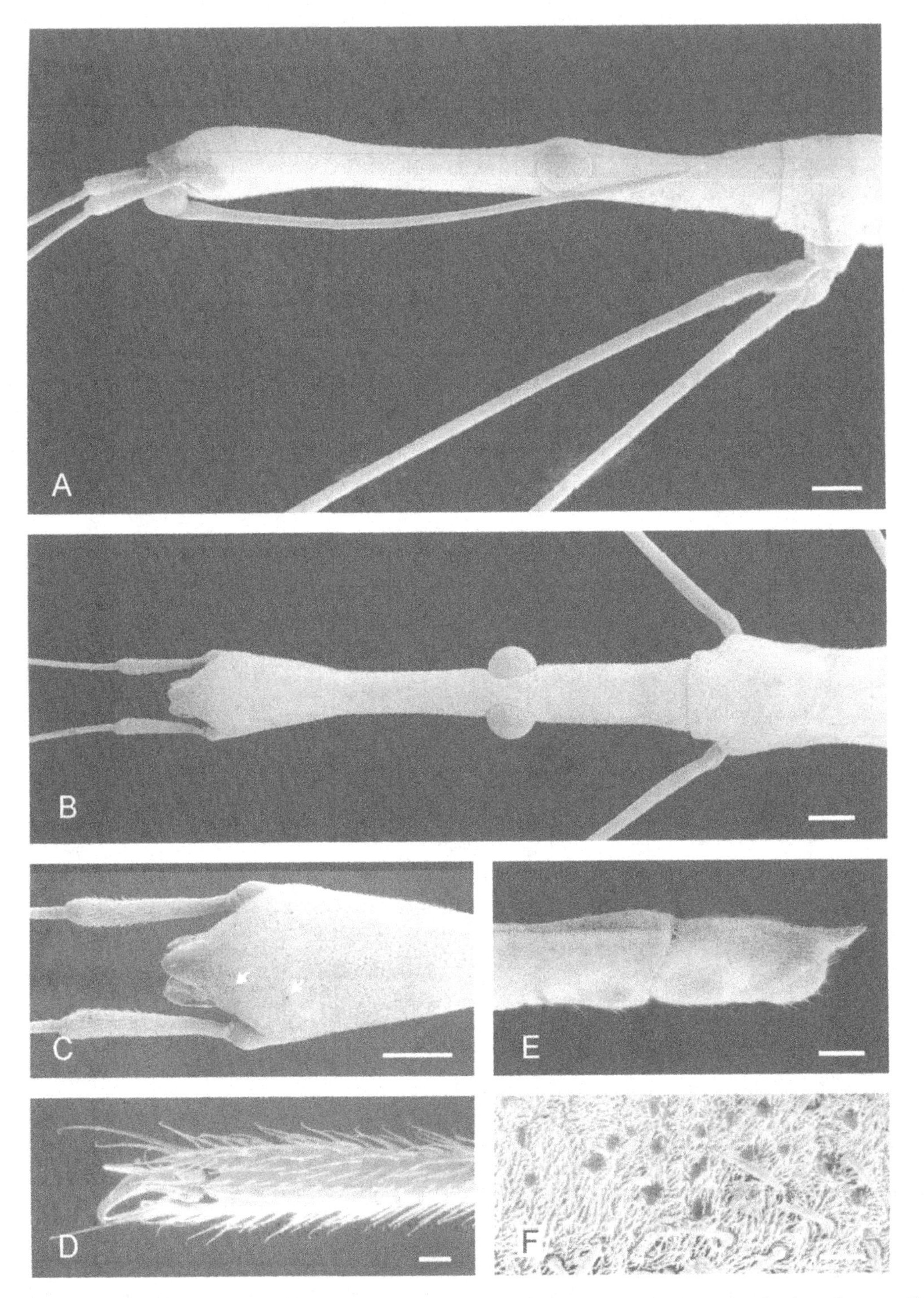

Figure 9.2. HYDROMETRIDAE. A-F, *Hydrometra strigosa*: lateral view of head and anterior thorax of micropterous ♂. B, dorsal view of head and anterior thorax of micropterous ♀. C, dorsal view of apex of head. D, ventral view of apex of middle tarsus. E, lateral view of abdominal end of micropterous ♂. F, surface structure of head. Scales: A-C, E, 0.5 mm; D, 0.1 mm; F, 0.01 mm. Scanning Electron Micrographs (CSIRO Entomology, Canberra).

acic as well as abdominal scent glands absent. Legs very long and slender, increasing in length from fore to hind legs; femora not thickened Macropterous adults with relatively narrow fore wings at least reaching abdominal tergite 6 (Fig. 9.1A), with two longitudinal veins running approximately parallel, connected by two cross-veins; wings reduced in brachypterous (wings reaching between tergites 1 and 3) and micropterous adults (wings rudimentary, not reaching tergite 1; Fig. 9.1I, fw). Abdomen relatively long and slender, sometimes modified ventrally in ♂. Genital segments relatively large; posterior margin of tergum 8 with a median extension in both male (Figs 9.3A, B; s8) and female (Fig. 9.3E, F; t8); ventral surface of segment 8 usually with species-specific modifications in ♂.

Hydrometra Latreille is a large genus distributed worldwide with about 120 species. Hungerford & Evans (1934) gave the first extensive revision of the genus and described or recorded 5 species from Australia. Recent revisions by J. Polhemus & D. Polhemus (1995a), J. Polhemus & Lansbury (1997), and Andersen & Weir (in press) have increased this number to 9 Australian species.

Key to Australian species

1. Anteclypeus of head broad, truncate, sometimes medially concave (Fig. 9.1B; ac).. 2
- Anteclypeus of head conical, narrowly to broadly rounded anteriorly (Fig. 9.1J; ac) 6
2. Sternum 7 of ♂ with a pair of prominent pad-like lateral projections covered with stiff black hairs (Figs 9.1E, F). Tergum 8 of ♀ without a prominent distal process (Figs 9.1G, H). Length 11.9-15.7 mm. Northern Territory, Queensland, Western Australia .... *papuana* Kirkaldy
- Sternum 7 of ♂ without a pair of prominent pad-like lateral projections covered with stiff black hairs. Tergum 8 of ♀ with a prominent distal process (Figs 9.3E, F)...... 3
3. Sternum 6 of ♂ with a pair of prominent blunt processes on either side of the midline adjacent to the posterior margin (Figs 9.3A, B, G, H; arrows) 4
- Sternum 6 of ♂ without a pair of prominent blunt processes on either side of the midline adjacent to the posterior margin..... 5
4. Sternum 8 of ♂ with two tightly packed clusters of short, stout black hairs (Figs 9.3A, B). Length 12.0-13.8 mm. Queensland .... *illingworthi* Hungerford & Evans
- Sternum 8 of ♂ with two elongate clusters of short black hairs ventrally (Figs 9.3G, H). Length 13.0-14.0 mm. Queensland... .... *claudie* J. Polhemus & Lansbury
5. Sternum 8 of ♂ essentially smooth, with two small barely visible clusters of very short, golden hairs on either side of the midline (Fig. 9.3I). Length 13.0-13.4 mm. Queensland .... *jourama* J. Polhemus & Lansbury
- Sternum 8 of ♂ with two tightly packed clusters of short, stout brown hairs on either side of the midline (Figs 9.3C, D). Length 12.0-14.0 mm. New South Wales, Northern Territory, Queensland, Western Australia...................................... *feta* Hale
6. Sternum 7 of ♂ with a broad transverse sulcus, deepest medially; anterior and posterior margins densely pilose, posterior margin declivant (Figs 9.3O, P). Anterior half of sternum 6 of ♂ with scattered long hairs, distal half moderately to densely pilose, less so medially. Female connexiva with short tufts of inwardly pointing black hairs at ends; tergite 8 with tip deflexed. Length 9.4-11.9 mm. Northern Territory, Western Australia..................... .... *darwiniana* J. Polhemus & Lansbury
- Sternum 7 of ♂ without a broad transverse sulcus; anterior and posterior margins not densely pilose, posterior margin not declivant. Sterna 6 and 7 of ♂ variably pilose. Female connexiva without short tufts of black hairs; tergite 8 with tip not deflexed... 7
7. Sternum 7 of ♂ smooth, without prominent fields of stout, dark hairs, distinctly flattened medially (Fig. 9.3Q). Macropterous ♀ with distinct whitish, longitudinal stripe on fore wings. Tergite 7 of ♀ parallel-sided, same width anteriorly as posteriorly. Length 9.8-13.4 mm. Northern Territory, Queensland....... *orientalis* Lundblad
- Sternum 7 of ♂ not distinctly flattened medially, either with two prominent fields of stout, dark hairs (Figs 9.3J, K), or two groups of very small peg fields (Fig. 9.3M). Macropterous ♀ at most with indistinct whitish longitudinal stripe on fore wings. Tergite 7 of ♀ distinctly widening posteriorly 8
8. Sternum 7 of ♂ proximally with two groups of stout hairs superficially resembling spines, each set on a weak tumescence; distally with two small fields of pegs or very short hairs on either side of midline (Figs 9.3J. K). Sternum 8 of ♂ anteriorly tumid, ventral margin weakly convex in lateral view; not pilose basally. Anteclypeus projecting anteriorly, reaching anterior margin of maxillary plates

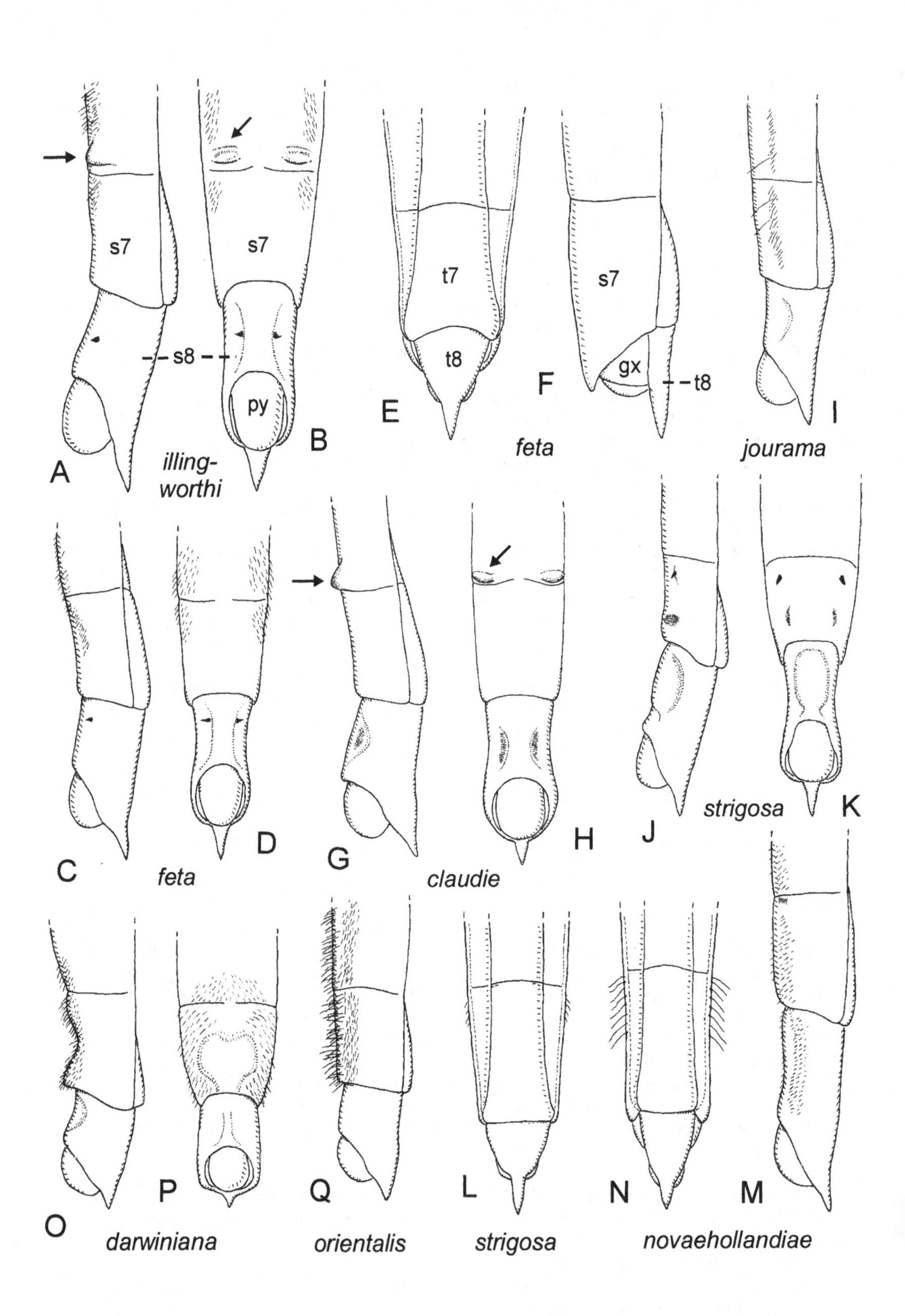
s7
s7
s8
py
A
illing-
worthi
B
t7
t8
E
s7
gx
t8
F
feta
I
jourama
C
feta
D
G
claudie
H
J
strigosa
K
O
darwiniana
P
Q
orientalis
L
strigosa
N
M
novaehollandiae

(Fig. 9.1J). Macropterous or micropterous, wing pads of the latter form reaching almost halfway to base of abdomen. Length 7.8-13.8 mm. New South Wales, Norfolk Island, Northern Territory, Queensland, South Australia, Tasmania, Victoria, Western Australia...... *strigosa* (Skuse)

– Sternum 7 of ♂ proximally with two groups of very small peg fields, sometimes weakly developed, thus appearing as raised brown patches, not set on tumescences; distally without fields of pegs or stiff dark hairs on either side of midline; pilose ventrally (Fig. 9.3M). Sternum 8 of ♂ proximally pilose. Anteclypeus not projecting anteriorly, not reaching anterior margin of maxillary plates (Fig. 9.1K). Micropterous, wing pads short to barely visible, not reaching one fourth of distance to base of abdomen. Length 9.7-11.2 mm. New South Wales, Queensland ....
.... *novaehollandiae* J. Polhemus & Lansbury

Biology

These stick-like bugs are most frequent in quiet waters with abundant vegetation such as pools and ponds, permanent or temporary, edge of canals and streams, pools and billabongs in river and stream beds. Hydrometrids spend most of their lives in the vegetation quite close to the water margin. Only occasionally do they cross larger stretches of free water surface where their movements are much slower compared to that of other surface bugs ("Hydrometra" means water measurer and refers to the slow measured gait of these bugs). In Europe and North America, *Hydrometra* species migrate to land in the autumn and are found in moist, protected places. They emerge in the spring where they can be seen in numbers in the vegetation near the water's edge (Sprague 1956; Andersen 1982). In tropical species, numerous individuals can be found under stones in dry stream beds, indicating a kind of aestivation (Andersen 1982).

Hydrometra species are predators or scavengers on dead and half-dead insects found on the water surface, but are also able to spear living mosquito larvae, cladocerans, and ostracods through the surface film (Hungerford 1920; Ekblom 1926; J. Polhemus & Chapman 1979c). Water measurers are "timid predators" (Andersen 1982) which approach their prey with great care, waving their long antennae from side to side, presumably to perceive the smell of potential prey items (the tip of the antenna has specialized sense organs; Andersen 1982: fig. 182). After careful examination, the prey is speared with the long and slender rostrum and usually carried to land where it is sucked out. When describing the feeding act of *Hydrometra*, Cobben (1978) observed that due to their extreme flexibility, the maxillary stylets were able to reach every part of the prey (a fruit-fly) from the point of insertion of the rostrum (Fig. 2.6C).

The life history is not known for any of the Australian *Hydrometra* species. In the most common species, *H. feta* and *strigosa*, adults are active throughout the year whereas nymphs are recorded from January, March-July, August, and October- December (Andersen & Weir, in press), thus suggesting multiple generations per year, at least in the subtropical and tropical areas. Most Australian species are wing dimorphic, with the macropterous (long-winged) form being the most frequent adult form in two thirds of the species. The flight activity may be restricted to the day, however, since hydrometrids are not common in light-trap catches. The flightless adult form is either brachypterous or micropterous.

The structure of the unique egg of *Hydrometra* is well known through the work by Cobben (1968). The egg shell has a complex, multi-layered structure (Fig. 9.4A). The single micropyle (Fig. 9.4B, mp) traverses a tube-like anterior elongation of the egg whereas the base of the egg

Figure 9.3. HYDROMETRIDAE. A-B, *Hydrometra illingworthi*, ♂: A, lateral view of abdominal end. B, ventral view of abdominal end (py, pygophore; s7, sternum 7; s8, segment 8). C-D, *H. feta*, ♂: C, lateral view of abdominal end. D, ventral view of abdominal end. E-F, *H. feta*, ♀: E, dorsal view of abdominal end. F, lateral view of abdominal end (gx, gonocoxa; s7, sternum 7; t7, t8, terga 7 and 8). G-H, *H. claudie*, ♂: G, lateral view of abdominal end. H, ventral view of abdominal end. I, *H. jourama*, ♂: lateral view of abdominal end. J-K, *H. strigosa*, ♂: J, lateral view of abdominal end. K, ventral view of abdominal end. L, *H. strigosa*, ♀: dorsal view of abdominal end. M, *H. novaehollandiae*, ♂: lateral view of abdominal end. N, *H. novaehollandiae*, ♀: dorsal view of abdominal end. O-P, *H. darwiniana*, ♂: O, lateral view of abdominal end. P, ventral view of abdominal end. Q, *H. orientalis*, ♂: lateral view of abdominal end. Reproduced with modification from Andersen & Weir (in press).

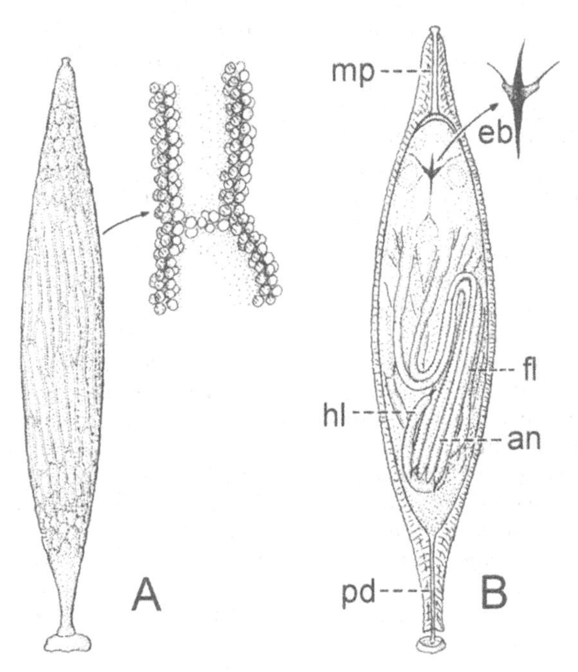

Figure 9.4. HYDROMETRIDAE. A-B, *Hydrometra gracilenta* Horvath: A, structure of deposited egg with detail of surface sculpture shown at higher magnification. B, egg showing micropyle, pedicel, and embryo (an, antenna; eb, egg burster; fl, fore leg; hl, hindleg; mp, micropyle; pd, pedicel). Reproduced from Andersen (1982).

forms a distinct pedicel (pd). The egg is normally deposited on vertical objects (like plant stems) and attached in an upright position by its pedicel (Fig. 2.7J). Unlike other surface bugs, the egg of *Hydrometra* has a greater tolerance for a dry environment. The oviposition is always terrestrial, the eggs being laid up to several centimeters above water level. The egg shell is obviously designed to trap air, but the egg develops normally under water without any air store in the shell. The embryo has a frontal egg-burster or ruptor ovi (Fig. 9.4B; eb) which is used to produce a longitudinal split prior to eclosion

Both adults and nymphs of *Hydrometra* lack scent glands and being relatively slow they seem to be an easy catch for predators. The slender body shape combined with stilt legs possibly affords some camouflage protection. Besides, *Hydrometra* are often seen to raise and lower its body rhythmically, which tends to obscure the outline of the insect (Rensing 1962). As a last resort, hydrometrids may "play dead". When handled roughly they become immobile, the antennae are stretched forward and the legs pressed against the body (Andersen 1982).

Distribution

Hydrometra species are distributed throughout Australia, except for the dry interior and larger parts of South Australia and Western Australia (Fig. 9.5). The most common and widespread species, *H. strigosa*, is recorded from all states, but apparently absent from most northern parts of Northern Territory and Western Australia, where *H. feta, papuana*, and *orientalis* are the most common species. The two latter species are widespread in the Oriental region and New Guinca (J. Polhemus & Lansbury 1997). *H. strigosa* is also known from Norfolk Island and New Zealand where it is one of very few gerromorphan bugs (Wise 1990; Andersen & Weir in press).

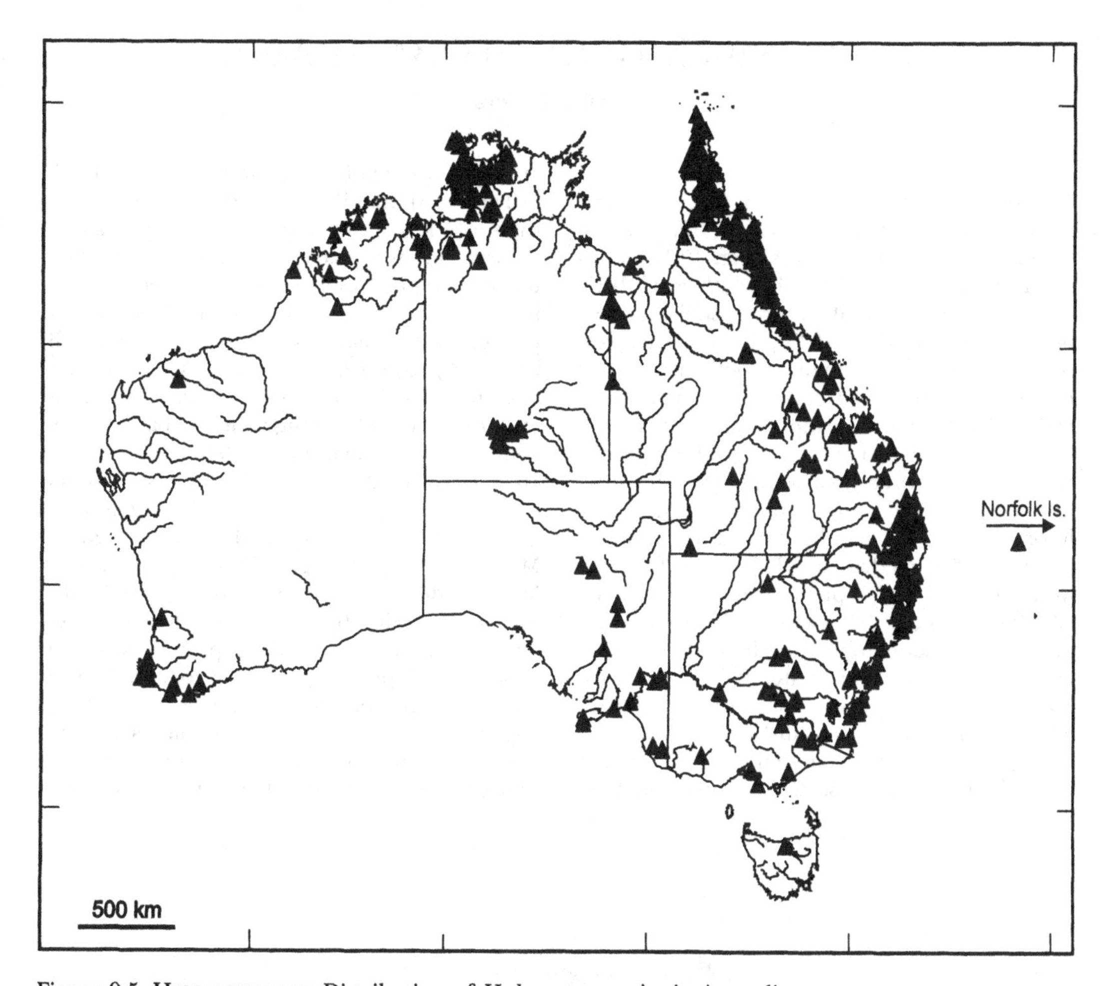

Figure 9.5. HYDROMETRIDAE. Distribution of *Hydrometra* species in Australia.

10. Family HERMATOBATIDAE

Coral Treaders

Identification

This small family is one of very few exclusively marine families of insects. Coral treaders are small, apterous water striders (Fig. 10.1) with a short, broad and declivent head (Figs 10.2E, F, 10.3B), a strongly modified thorax subject to sexual dimorphism (Figs 10.2A, B), an extremely short pregenital abdomen in both sexes, and enlarged and modified genital segments in the male (Figs 10.3C). The thorax structure resembles that of the true water striders, family Gerridae, but all tarsi are 3-segmented, with claws inserted preapically only on the fore tarsi, apically on the middle and hind tarsi. Head width across eyes 3-5x median length. Basal part of head with three pairs of cephalic trichobothria (Fig. 10.2E; ct) and a transverse line which turns forward along inner margin of each eye. Middle and hind pairs of acetabula laterally displaced on body (Figs 10.2C, F); mesosternum (Fig. 10.2C; ms) and metasternum (mt) thereby greatly enlarged, indistinctly separated from each other. Middle and hind legs inserted on body so that the enlarged coxae point straight caudad (Fig. 10.2F; cx2 and cx3); middle and hind legs much longer than fore leg. A metathoracic scent apparatus is present, but its orifice is hidden beneath the posterior margin of metasternum (Andersen 1982: fig. 499). Abdomen much reduced, subject to pronounced sexual dimorphism (Figs 10.2A, B; ab, 10.3C). Intersegmental sutures between tergites more or less obsolete, especially in ♀; laterotergites fused, forming a narrow band-shaped connexivum on each side of abdomen in ♂ (Fig. 10.2; cn), greatly expanded and fused to the metanotum in ♀ (Fig. 10.2A; cn). A vestigial scent gland opening is found on abdominal tergite 4. Male genital segments greatly enlarged and peculiarly modified (Andersen 1982: figs 505-508). Segment 8 (Fig. 10.4A; s8) cylindrical, turned downward, provided with a pair of styliform processes (st) pointing obliquely forward, almost reaching posterior margin of metasternum (mt); pygophore (py) ovate, rotated so that its opening has retained its dorsal orientation; proctiger (pr) small, facing caudad, situated in a slit on the dor-

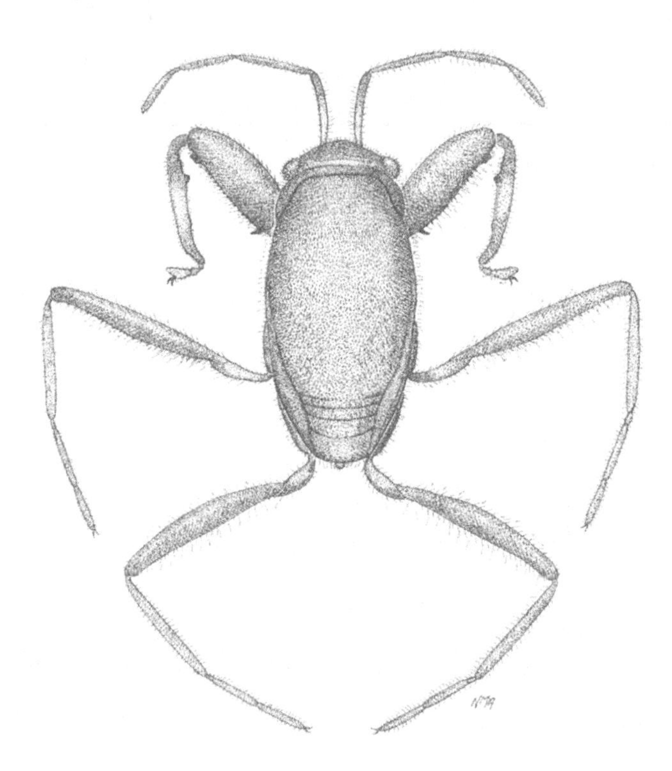

Figure 10.1. HERMATOBATIDAE. *Hermatobates marchei*, dorsal view of apterous ♂.

sal side of segment 8. Female genital segments including ovipositor reduced. Eggs are broadly oval (Fig. 10.3G), with outer sculpture of numerous fine peg-like processes; anterior end pointed, with small blunt process carrying the single micropyle.

Overview

Only one genus, *Hermatobates* Carpenter, originally classified in the family Gerridae. Coutière & Martin (1901) proposed a separate subfamily, Hermatobatinae, for the genus. Matsuda (1960) excluded *Hermatobates* from the Gerridae, based on its unique characteristics, e.g., presence of an abdominal scent gland (absent in gerrids), 3-segmented tarsi (2-segmented in gerrids), and extremely reduced and modified pregenital abdomen. This probably lead Poisson (1965b) to place *Hermatobates* in a family of its own. This was confirmed by Andersen (1982) who classified the Hermatobatidae with the families Veliidae and Gerridae in the superfamily Gerroidea.

HERMATOBATES

Identification

Small water striders, length 2.5-4.0 mm; adults always apterous. Body fusiform to ovate (Figs 10.1, 10.2A, B, D), chiefly dark coloured, but covered by a thick greyish pilosity. Eyes relatively small, globular. Antennae long, with dense, erect pilosity; segment 2 subequal to or slightly longer than segment 1, segments 3 and 4 shortest. Rostrum short and stout, apex reaching anterior part of mesosternum (Fig. 10.2F; ro). Pronotum very short in middle (Figs 10.2A, B, E; pn), laterally enlarged and covering antero-lateral parts of mesonotum (ms) which is prolonged and indistinguishably fused with metanotum. Thoracic dorsum of ♂ without visible sutures (Fig. 10.2A) while that of ♀ has a deep median, longitudinal furrow (Fig. 10.2B). Hind margin of metasternum of ♂ usually modified in middle (with a process or groups of pegs). Fore femur more or less thickened, especially in ♂ (Figs 10.2A, D), usually armed ventrally with a basal and apical tooth and a row of smaller teeth in between (the latter also present in ♀). Middle and hind femora moderately thickened, the former armed beneath with a row of spines which is denser and longer in ♂ (Figs 10.2A, D). Fore tarsus short with claws inserted near middle of last segment; middle and hind tarsi subequal in length to tibiae with a very short segment 1 and much longer segments 2 and 3; claws inserted apically on last segment. Abdominal venter extremely short in ♂ and covered by the enormously enlarged genital segments (Figs 10.3A; ab, 10.4C). Abdominal sterna 2-7 fused to each other in ♀ (Fig. 10.2C; s2-s7).

Carpenter (1892) described *Hermatobates haddoni* from Australia (Mabuiag Island, Torres Strait), the first species of this genus. Andersen & Weir (2000), the most recent revision of the genus, recognised 8 species. *H. weddi* described by China (1957) from Monte Bello Islands, Western Australia, was synonymised with the widespread *H. marchei* (Coutière & Martin). Previously, J. Polhemus (1982) synonymised *H. walkeri* China with *H. haddoni* Carpenter. Andersen & Weir (2000) recorded two species from Australia and described *H. armatus* from Chesterfield Islands (belonging to New Caledonia) in the Coral Sea. The latter species is included in the key below since it may be found inside the territorial waters of Australia.

Key to Australian species

1. Head width across eyes distinctly less than 4x median length (Figs 10.2A, B). Posterior margin of metasternum of ♂ straight or slightly emarginated (Fig. 10.4A; mt), without median lobe or process, only a number of minute, dark teeth. Middle femur of ♀ with a ventral row of short but distinct, dark spines (Fig. 10.2B). Length 2.6-3.2 mm (♂), 2.6-2.7 mm (♀). Northern Territory, Queensland, Western Australia...... *haddoni* Carpenter
- Head width across eyes 4-6x median length (Fig. 10.2D, E). Posterior margin of metasternum of ♂ with a median process (Figs 10.4D, G). Middle femur of ♀ at most with scattered, very small black teeth on ventral surface 2
2. Metasternum of ♂ not depressed posteriorly (Fig. 10.4C), posterior margin with a small, rounded process in middle (Fig. 10.4D). Basal third of first antennal segment pale. Length 3.4-4.0 (♂), 3.3-3.5 mm (♀). Northern Territory, Queensland, Western Australia .... *marchei* (Coutière & Martin)
- Metasternum of ♂ depressed in posterior part (Fig. 10.4F), with a prominent, angular process in middle (Fig. 10.4G). First antennal segment dark except for extreme base. Length 3.4-3.7 (♂), 3.5-3.7 mm (♀). Chesterfield Islands (Coral Sea)......................... *armatus* Andersen & Weir

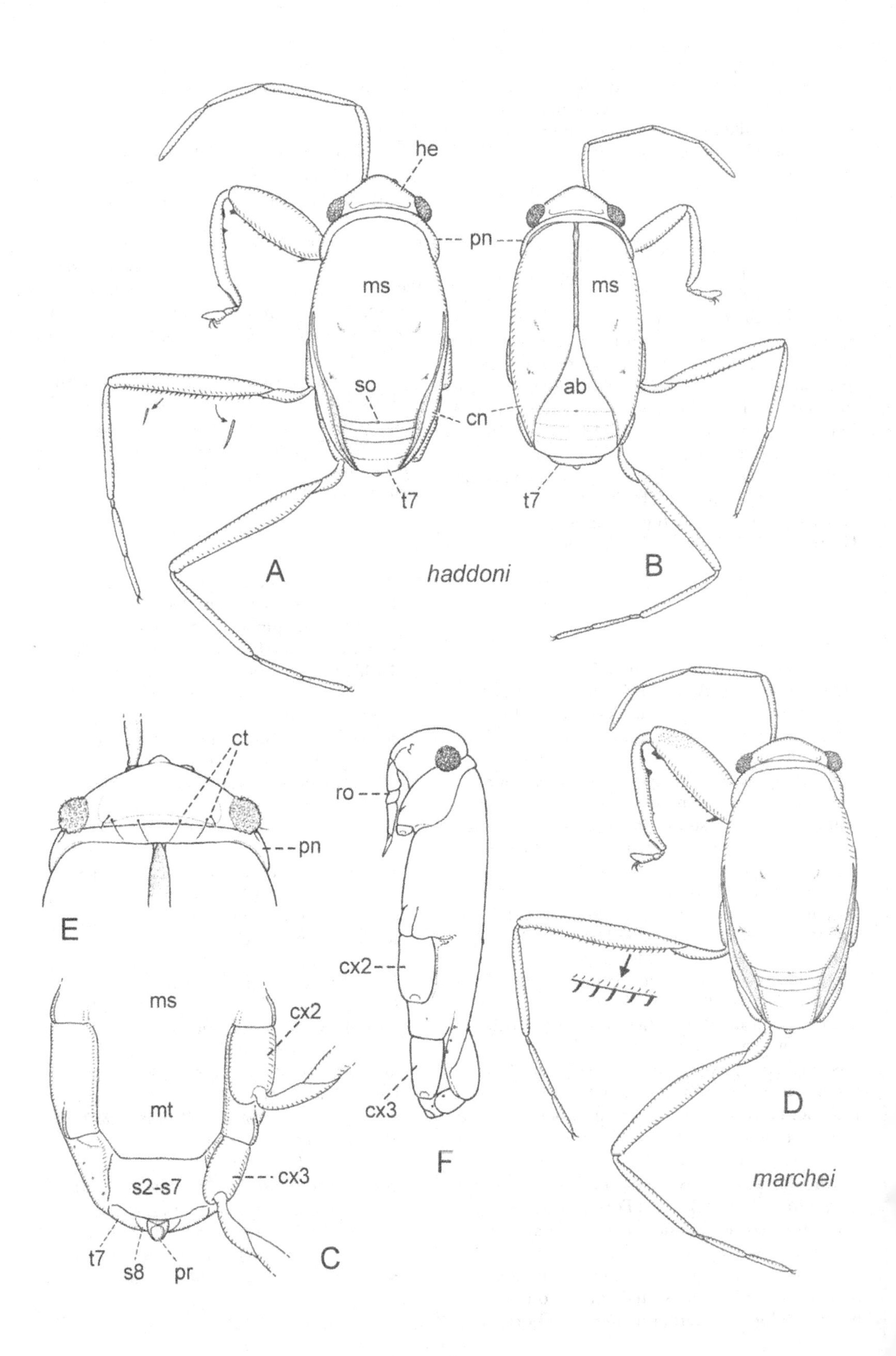
he
pn
ms
so
cn
t7
A
haddoni
B
ms
ab
t7
ct
pn
E
ms
cx2
mt
s2-s7
cx3
t7
s8
pr
C
ro
cx2
cx3
F
D
marchei

Biology

Knowledge of the biology and ecology of coral treaders was for many years impeded by the lack of sufficient material and observations. Generally, neither entomologists nor marine biologists pay much attention to marine insects and *Hermatobates* species turned out to be particularly elusive. By understanding the habits and habitats of coral treaders, Lanna Cheng and her husband, Ralph A. Lewin, from Scripps Institute of Oceanography, California, were able to collect far more specimens than all previous collections combined. They also were able to confirm that *Hermatobates* species live in the intertidal zone of coral reefs and are active at low tide whereas they recede to holes in rocks and blocks of coral during high tide (Cheng 1977). Specimens taken on the sea surface at high tide and on the open ocean were characterised as strays (Cheng & Leis 1980; J. Polhemus 1982). Foster (1989) found *H. marchei* (as *H. weddi*) abundant on shores that provide rocks with cavities that remain air-filled during high tide. The zone of suitable habitats is restricted to a narrow, about 20 cm wide band, roughly centered around Mean Low Water Neaps. Foster conducted transplant experiments showing that the insects are unable to survive above or below this zone. Adults and nymphs emerge from their crevices and holes about an hour before low tide and skate and forage on the surface of tidal pools of the reef crest or on protected areas of sea immediately seaward. They return to their hiding places about an hour after the tide has turned.

In Australia, *H. haddoni* has only been collected on reef flats and under coral rubble at low tide (J. Polhemus 1982). Of 19 samples of *H. marchei*, 52% were taken in the same type of intertidal habitats while 48% were taken on the sea surface outside coral reefs, usually at light (Cheng 1977; Cheng & Schmitt 1982). Andersen & Weir (2000) hypothesised that the behaviour pattern for *H. marchei* (and possibly other coral treaders) has a build-in flexibility that allows populations to exploit resources both on reef flats, where the activity is controlled by the tidal cycle, and on the open sea outside reefs, especially when the sea surface is relatively calm. In habitats of the first category, *H. marchei* is frequently found together with species of coral bugs, *Halovelia* (Andersen & Weir 1999). On the sea surface coral treaders keep company with several species of sea skaters, *Halobates.*

Coral treaders are uniquely adapted to life in the intertidal zone of rocky and coral coasts, biologically as well as structurally. Most of the body surface is clothed in a layer of microtrichia, each composed of a tapering, 0.4-2.5 mm long shaft tipped with a sphere. Such microtrichia occur at very high densities on the surface of the body (up to 3.5×10^6 mm^{-1}), and the hair layer makes connection with the thoracic spiracles. Cheng (1977) and Foster (1989) conducted experiments showing that *Hermatobates* is able to survive submergence in recirculated seawater for many hours and suggested that the dense layer of microtrichia may act as a plastron. Coral treaders are undoubtedly predators and scavengers, probably feeding on other arthropods like marine chironomid midges, collembola, etc. (Foster 1989).

Mating pairs have been observed in tidal pools (Foster 1989; D. Polhemus 1990). The incrassate fore legs probably help the male to control the female during copulation, but are not associated with prolonged mate-guarding since the copulation only lasts for a few minutes. Males of both Australian species show a relatively wide range of size variation and the thickness and armature of the fore legs vary even more, probably as a result of allometric growth (Andersen & Weir 2000).

Figure 10.2. HERMATOBATIDAE. A, *Hermatobates haddoni*, apterous ♂: dorsal view; antennae and legs of right side omitted (cn, connexivum; he, head; ms, mesonotum; pn, pronotum; so, scent orifice; t7, tergum 7). B-C, *H. haddoni*, apterous ♀: B, dorsal view; antennae and legs of left side omitted (ab, abdomen; cn, connexivum; ms, mesonotum; pn, pronotum; t7, tergum 7). C, ventral view of posterior thorax and abdomen; coxae of right side omitted (cx2, cx3, middle and hind coxae; ms, mesosternum; mt, metasternum; pr, proctiger; s2-s7, abdominal sterna 2-7; s8, segment 8; t7, tergum 7). D, *H. marchei*, apterous ♂: dorsal view; antennae and legs of right side omitted; ventral spines of middle femur shown at higher magnification. E-F, *H. marchei*, apterous ♀: E, dorsal view of head and anterior thorax (ct, cephalic trichobothria; pn, pronotum). F, lateral view; antennae and legs distal of coxae omitted (cx2, cx3, middle and hind coxae; ro, rostrum). Reproduced with modification from: A-D, Andersen & Weir (2000); E-F, Andersen (1982).

Figure 10.3. HERMATOBATIDAE. A, *Hermatobates marchei*, apterous ♂: ventral view. B, *H. marchei*, apterous ♀: dorsal view of head and anterior thorax. C, *H. marchei*, apterous ♂: ventral view of posterior thorax and abdomen. D, *H. marchei*, apterous ♀: apex of middle tarsus. E, *H. marchei*, apterous ♀: surface structure of head. F, *Hermatobates* sp.: piece of dead coral with batch of eggs (Phuket, Thailand). G, *Hermatobates* sp.: eggs with mature embryos (Phuket, Thailand). Scales: A, mm; B-C, 0.2 mm; D, 0.1 mm; E, 0.01 mm; F, 5 mm; G, 1 mm. A-E, Scanning Electron Micrographs (CSIRO Entomology, Canberra); F-G, photographs (G. Brovad, Zoological Museum, Copenhagen).

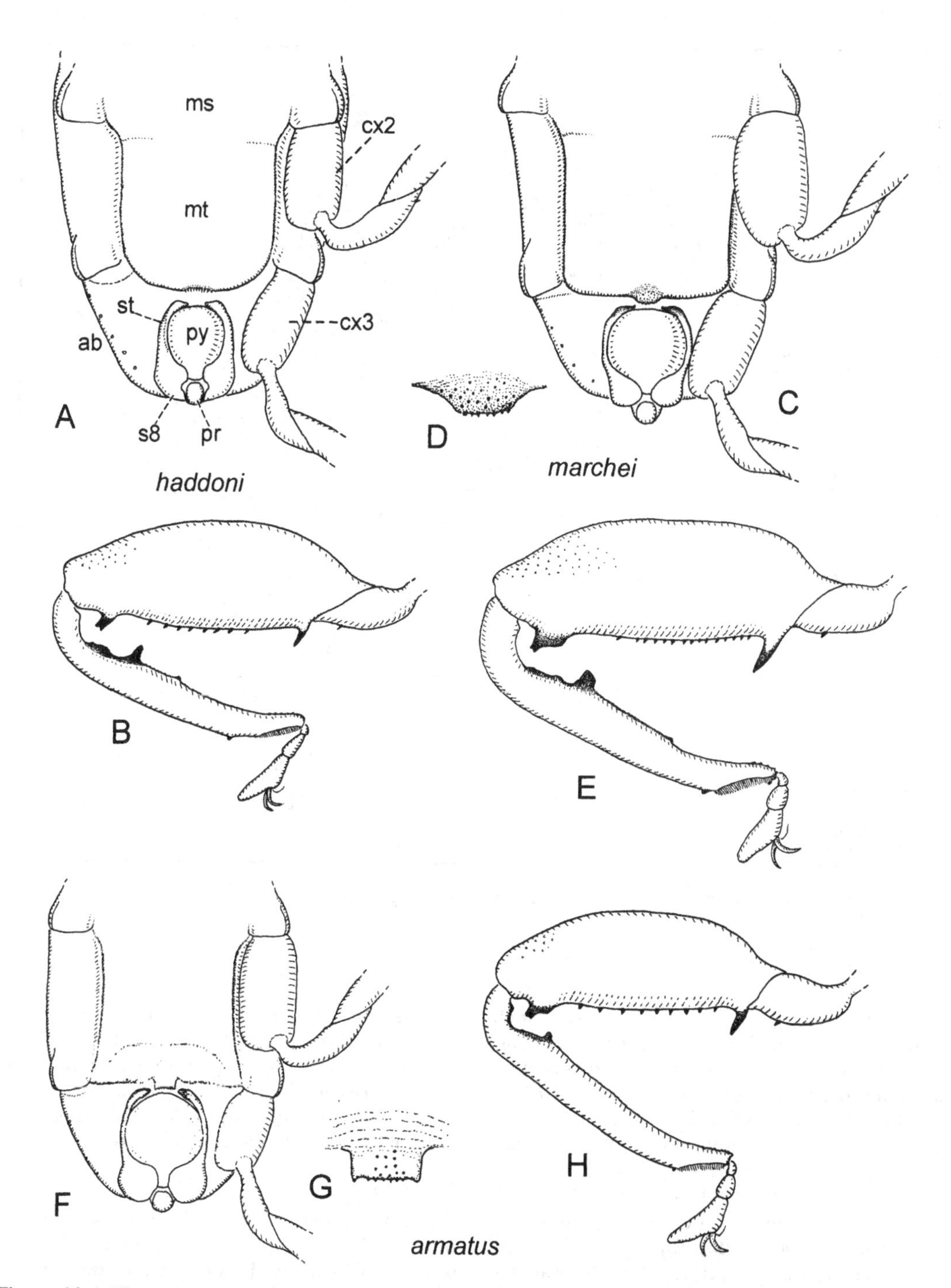

Figure 10.4. HERMATOBATIDAE. A-B, *Hermatobates haddoni*, apterous ♂: A, ventral view of posterior thorax and abdomen; coxae of right side removed (ab, abdomen; cx2, cx3, middle and hind coxae; ms, mesosternum; mt, metasternum; pr, proctiger; py, pygophore; st, styliform process of segment 8; s8, segment 8). B, right fore leg. C-E, *H. marchei*, apterous ♂: C, ventral view of posterior thorax and abdomen; coxae of right side removed. D, median process of metasternum shown at higher magnification. E, apterous ♂: right fore leg. F-H, *H. armatus*, apterous ♂: F, ventral view of posterior thorax and abdomen; coxae of right side removed. G, median process of metasternum shown at higher magnification. H, right fore leg. Reproduced with modification from Andersen & Weir (2000).

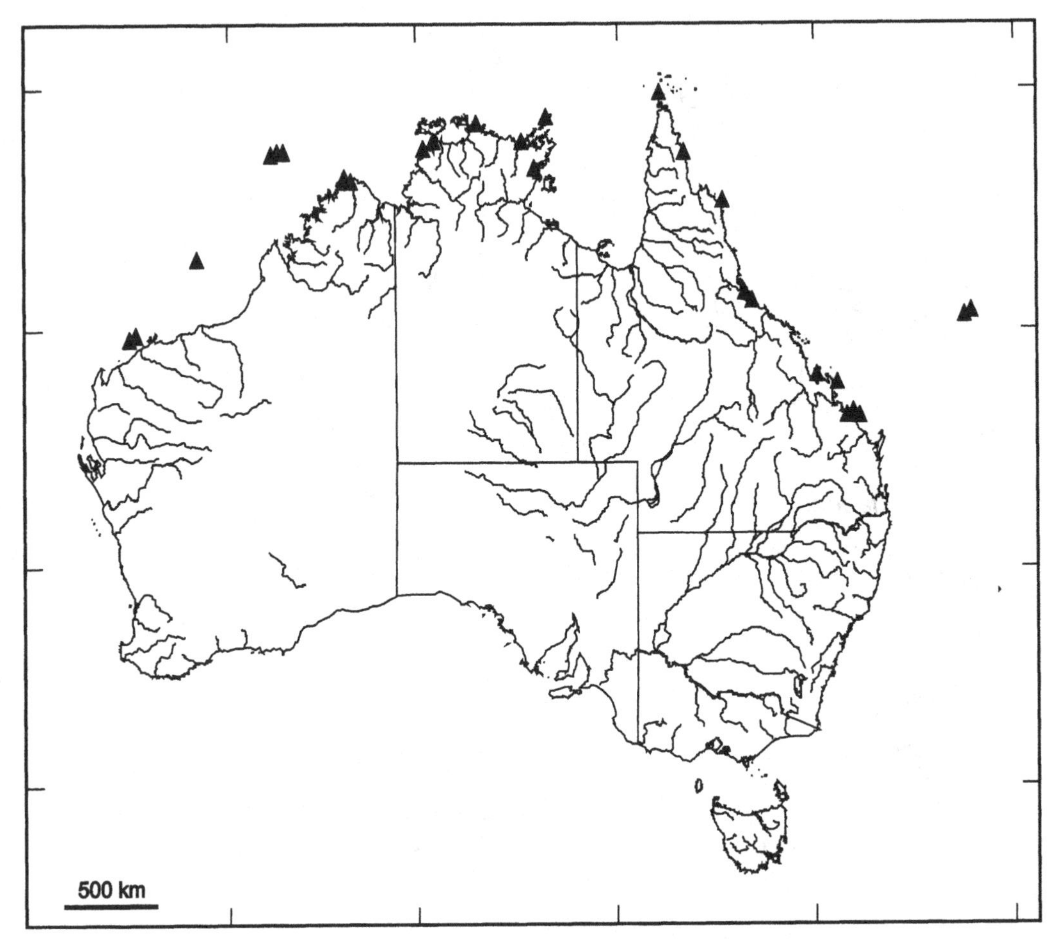

Figure 10.5. HERMATOBATIDAE. Distribution of *Hermatobates* species in Australia.

Oviposition in *Hermatobates* has never been observed in the field, but the eggs are probably deposited in the same holes and crevices as used as hiding places by the insects. A piece of *Porites* coral collected in Phuket, Thailand (Fig. 10.3F), contains a batch of about 20 eggs of similar size and shape as ripe ovarian eggs dissected from female *Hermatobates* (Andersen 1982: fig. 483). Each egg is attached in an upright position (Fig. 10.3G) to the coral surface by a gelatinous substance. The egg (Fig. 2.7H, I) is broadly oval, 0.8-0.9 mm long, isodiametrical, with outer sculpture of numerous fine peg-like processes; anterior end pointed, with small blunt process carrying the single micropyle. Based upon morphometrical analyses, Foster (1989) found only four nymphal instars in *Hermatobates*. Moulted cuticles, of all nymphal instars, are frequently found in holes and crevices in intertidal rocks indicating that coral treaders moult in these dry, air-filled spaces (Foster 1989; Andersen & Weir 2000).

Distribution

Hermatobates species are widely distributed in the Indo-Pacific area with a single species on islands of the Caribbean Sea. In Australia, the two species are sparsely distributed along the coasts of Western Australia, Northern Territory, and Queensland (Fig. 10.5).

Plate 1. Habitats of Australian water bugs. A, small, stagnant, temporary pool with long fringing grass; Jarrnarm Walks, Keep River National Park, Northern Territory (July 1994). B, small deep sided narrow spring fed billabong fringed by long grass, *Pandanus* and trees, water lillies on surface; Giyamungkurr (Black Jungle Spring), Kakadu National Park, Northern Territory (July 1994). C, pool on rocky base with grass and fringing palms; Palm Valley, Finke Gorge National Park, Northern Territory (March 1995). D, spring fed waterhole, partly shaded, with water lillies on surface; Murrays Spring, Musselbrook area, Lawn Hill National Park, Queensland (May 1995). E, temporary, murky pool in rocky creek bed; "Amphitheatre Camp", Musselbrook area, Lawn Hill National Park, Queensland (May 1995). F, permanent lagoon with water lillies and fringing grass; Cockatoo Lagoon, Keep River National Park, Northern Territory (July 1994). Photographs by Tom A. Weir

Plate 2. Habitats of Australian water bugs. A, pool in gorge, with sandy bottom, rocky edges and no vegetation; Simpsons Gap, West Macdonnells National Park, Northern Territory (March 1995). B, still shaded pools in sandy creek bed, with lots of floating leaves; Canoe Creek, Cape York Peninsula, Queensland (October 1992). C, shaded end of billabong; Policeman Waterhole (Dilirba), Keep River National Park, Northern Territory (July 1994). D, deep billabong in Keep River, with steep muddy banks and floating leaves; Keep River Gorge, Keep River National Park, Northern Territory (July 1994). E, large billabong in Magela Creek, with water lillies and reeds; Georgetown Billabong, Magela Creek, Kakadu National Park, Northern Territory (July 1994). F, large deep billabong with steep sides and *Pandanus* only; Muirella Park, Kakadu National Park, Northern Territory (July 1994). Photographs by: A, C-F, Tom A. Weir; B, Paul Zborowski.

Plate 3. Habitats of Australian water bugs. A, large permanent lake with sandy bottom in open forest; Lake Wicheura, Cape York Peninsula, Queensland (October 1992). B, large permanent waterhole, with water lillies and other aquatic vegetation; Python Waterhole, Lakefield National Park, Cape York Peninsula, Queensland (October 1992). C, shaded pools in running creek in monsoon forest remnant; Gubara (Baroalba Springs), Kakadu National Park, Northern Territory (July 1994). D, permanent spring fed waterhole with water lillies and reeds; Bitter Springs, Roper River National Park, Northern Territory (July 1994). E, creek polluted by green/blue seepages of mining waste, with algal growth; Murrays Creek, Musselbrook area, Lawn Hill National Park, Queensland (May 1995). F, small deep sided narrow spring fed billabong fringed by long grass, with *Pandanus* and trees, water lillies on surface; Giyamungkurr (Black Jungle Spring), Kakadu National Park, Northern Territory (July 1994). Photographs by: A-B, Paul Zborowski; C-F, Tom A. Weir.

Plate 4. Habitats of Australian water bugs. A, rock pool near river; Davies Creek Falls area, SW of Kuranda, Queensland. B, river with rocks, Murrumbidgee River at Cotter, Australian Capital Territory (November 1997). C, pool on rockshelf with mangroves ahead of incoming tide; East Point, Darwin, Northern Territory (July 1994). D, intertidal reef flat at low tide with coral rocks; Cape Tribulation, Queensland (August 1990). E, mangroves and nearshore sea, Roonga Point, Cape York Peninsula, Queensland (October 1992). F, neuston net in operation on ocean surface; one way of catching oceanic *Halobates* species. Photographs by: A, E, Paul Zborowski; B, D, Annemarie M. Andersen; C, Tom A. Weir; F, Lanna Cheng.

11. Family VELIIDAE

Small Water Striders

Identification

The veliids are generally small or very small water striders with relatively short thorax and robust legs. The head is generally rather short and broad, more or less deflected in front of the large, globular eyes (Fig. 11.2B); ocelli absent (only present in the South African genus *Ocellovelia*); the dorsal head surface has typically a slender median impressed line. Most species are wing dimorphic, with macropterous and apterous adult forms (Figs 11.1E, F). Pronotum of apterous form variable in length; pronotum of macropterous form large, pentagonal (Fig. 11.1F; pn). Fore wings (fw) largely membranous, usually with three longitudinal, branching veins forming four closed cells. Metasternal scent gland apparatus always present. Metasternum with a midventral scent orifice (Fig. 11.C; so) and scent grooves leading laterally onto the metacetabula where they end in evaporatoria each provided with a tuft of hairs (Figs 11.1C, G; ev). Leg segments variable but typically rathcr short and robust. Male fore usually tibia provided with a distal *grasping comb* composed of a row of minute, stout hairs (Figs 11.1D; gr, 11.2C, D). Number of tarsal segments variable, from one to three; claws inserted preapically in a cleft on last tarsal segment. Abdomen variable in length; abdominal scent gland absent in both nymphs and adults. Male genital segments usually large; parameres of variable size; phallic vesica with well-defined sclerites. Female ovipositor plate-shaped, gonapophyses short and non-serrate.

Overview

Small water striders of the family Veliidae move along the water surface in two different ways: (1) walking or running, where the three pairs of legs are moved as alternating tripods of support (see Chapter 2) and (2) rowing by simultaneous strokes of the middle and (to a lesser degree) the hind legs. The two ways of locomotion is reflected in the leg structures. The relative leg length increases from the fore to the hind legs in veliids moving by walking or running (most Microveliinae), whereas the middle legs are prolonged in veliids using the rowing rhythm of leg movements (Haloveliinae, Rhagoveliinae) (Fig. 2.3B). Various modifications of the pretarsal structures (claws and arolia) to form "swimming fans" is most likely associated with the latter (Andersen 1982). The most elaborate structures are found in the genus *Rhagovelia* (which includes one Australian species) where the last segment of the middle tarsus is deeply cleft (Fig. 11.1A). From the bottom of this cleft arises two very long and flattened claws and a "swimming fan" composed of about 20 plumose structures which branch off from a common stem representing a dorsal extension of the ventral arolium (Andersen 1976, 1982).

A peculiar way of locomotion called "expansion skating" or "skimming" has been described for nymphs and adults of some European *Microvelia* and *Velia* species (Andersen 1976, 1982), but most likely apply to other veliids (see also Chapter 2). The movements are initiated by the spread of a detergent fluid (probably saliva ejected by the salivary pump through the rostrum) which lowers the surface tension and creates a smaller or larger "hole" in the surface film. This effect is easily demonstrated on a water surface covered by dust or a thin layer of *Lycopodium* spores (Fig. 2.5E). "Skimming" is probably an anti-predator adaptation. The release of a fluid for lowering the surface tension may propel the bug with a sudden speed in almost any direction. If the predator is a surface feeder (for example another water strider), it will furthermore be expelled in the opposite direction.

Veliid water striders are common inhabitants of freshwater bodies, both stagnant and flowing, throughout the world; a few groups (including Australian species) have extended their habitats into the marine environment. Veliids are predators or scavengers feeding on emerging aquatic insects, terrestrial insects accidentally caught in the surface film, etc. (J. Polhemus & Chapman 1979d; Andersen 1982). The European *Velia caprai* assemble in two different kinds of groups: mixed groups composed of adults and larger nymphs (instars 4-5) and juvenile groups (instars 1-3). A greater proportion of the available prey are captured by the mixed groups whereas individuals in large groups have a lower capture rate than individuals in small groups (Erlandsson & Giller 1992). Although *Velia* seems to be distasteful to predators such as trout, living in groups or "schools" probably lowers the risk of predation because frequent predator-prey encounters may assist in retention of the avoidance learned by the predator. Individual behavioural responses to attacks are variable and the frequency of expansion skating (see above) and thanatosis (death feigning) seems temperature dependent (Brönmark et al. 1984).

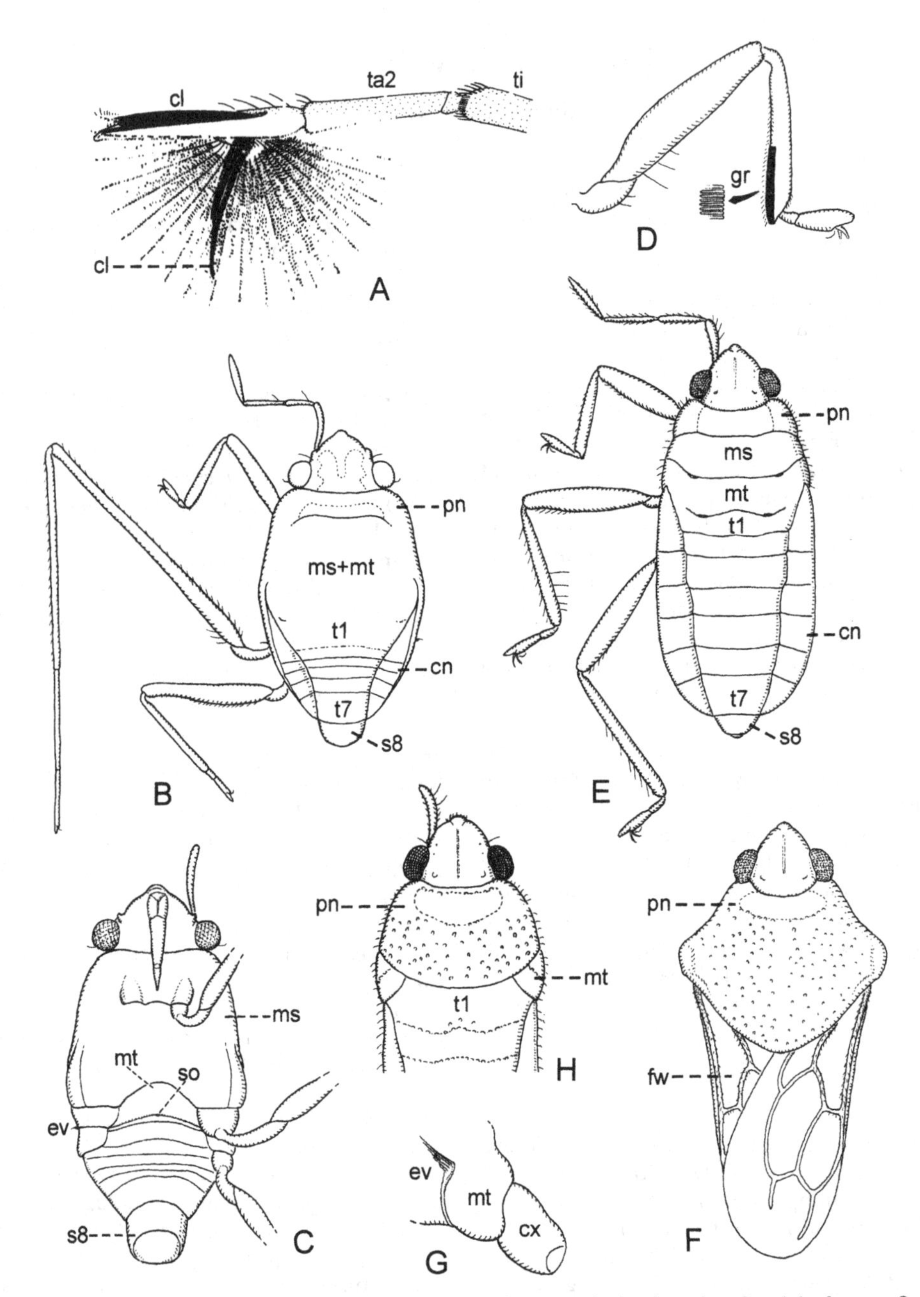

Figure 11.1. VELIIDAE. A, *Rhagovelia* sp.: middle tarsus with expanded swimming fan (cl, claws; ta2, tarsal segment 2; ti, tibia). B-C, *Xenobates mangrove*, apterous ♂: B, dorsal view; antennae and legs of right side omitted (cn, connexivum; ms, mesonotum; mt, metanotum; pn, pronotum; t1, abdominal tergite 1; s8, segment 8). C, ventral view; most of antennae and legs omitted (ev, evaporatorium; ms, mesosternum; mt, metansternum; so, scent orifice; s8, segment 8). D, *Halovelia hilli*: fore leg of ♂ with tibial grasping comb (gr) shown at higher magnification. E-G, *Drepanovelia dubia*: E, dorsal view of apterous ♂; antennae and legs of right side omitted (cn, connexivum; ms, mesonotum; mt, metanotum; pn, pronotum; t1, t7, abdominal tergites 1 and 7; s8, segment 8). F, dorsal view of macropterous ♂; antennae and legs omitted (fw, forewing; pn, pronotum). G, metacetabula with evaporatorium (cx, coxa; ev, evaporatorium; mt, metacetabulum). H, *Microvelia* (*Austromicrovelia*) *queenslandiae*: dorsal view of head and thorax (mt, metanotum; pn, pronotum; t1, abdominal tergite 1). Reproduced with modification from: A, Andersen (1976); B-D, Andersen & Weir (1999); E-H, Andersen & Weir (2001).

Veliid males are usually slightly smaller than the females, but more pronounced sexual size dimorphism is for example found in *Halovelia hilli* (Fig. 11.4B) and species of *Phoreticovelia* (Fig. 11.2E, 11.35A) where it is associated with extreme phoresy (male rides on the females back for a long time). Veliid eggs are relatively small (0.4-1.0 mm), with a porous shell penetrated by 2-4 micropyles, rarely only by one micropyle (*Halovelia*) (Cobben 1968; Andersen 1982). As far as we know, the eggs are always deposited below water or at the water surface, glued lengthwise to emergent or floating water plants. The embryo has a median, unpaired clypeal egg-burster (Fig. 2.7C; eb) which is used to produce a longitudinal split in the egg shell (es) through which the first instar nymph hatches.

The Veliidae is the largest family of semiaquatic bugs with about 840 described species in 57 genera worldwide. Andersen (1982) classified the Veliidae in 5 subfamilies: Haloveliinae, Microveliinae, Ocelloveliinae, Perittopinae, Rhagoveliinae, and Veliinae. The last subfamily is represented by fossils from the Lower Eocene Baltic amber (Andersen 1998, 2001a). Only the subfamilies Haloveliinae, Microveliinae, and Rhagoveliinae are represented in the Australian fauna (Cassis & Gross 1995; Andersen & Weir 1999, 2001, 2003b). A fossil veliid of unknown affinity is recorded from the Lower Cretaceous Koonwarra Formation, Victoria (Jell & Duncan 1986).

Key to the Australian genera of the family Veliidae

1. All tarsi with three segments. Middle tarsi deeply cleft, with leaf-like claws and plumose swimming fans arising from base of the cleft (Fig. 11.1A). (RHAGOVELIINAE) .... *Rhagovelia*
- Tarsi at most with two segments. Middle tarsi not deeply cleft, without plumose swimming fan 2
2. All tarsi two-segmented (basal segment of fore tarsi short). Middle femora distinctly longer than hind femora (Fig. 11.1B). Middle tarsi more than twice as long as hind tarsi. Marine. (HALOVELIINAE) 3
- Fore tarsi with only one segment, middle and hind tarsi two-segmented (Fig. 11.1E). Middle femora subequal or shorter than hind femora. Middle tarsi rarely more than twice as long as hind tarsi. Freshwater. (MICROVELIINAE) 4
3. Head with extensive pale markings. Pronotum with pale transverse mark or paired spots. Male fore tibia without grasping comb (Fig. 11.7B). Male genital segments distinctly protruding from abdominal end (Fig. 11.1B, C, 11.7C).. *Xenobates*
- Head and pronotum with very limited pale markings. Male fore tibia with grasping comb (Fig. 11.3D; gr). Male genital segments withdrawn into pregenital abdomen, largely concealed from dorsal view (Fig. 11.1D) .... *Halovelia*
4. Middle tarsus short, less than 0.8x length of middle tibia 5
- Middle tarsus long, at least 0.8x length of middle tibia 12
5. Apterous form 6
- Macropterous form 9
6. Pronotum of apterous form very short, median length much less than head length (Fig. 11.1E; pn) 7
- Pronotum of apterous form long, median length at least subequal to head length (Fig. 11.1H; pn) 8
7. Male fore femora thickened but not strongly incrassate. Pilosity of body and appendages chiefly short. Segment 8 of male with either one midventral, vertical process (Figs 11.10E-G), or a pair of posteriorly directed processes (Figs 11.10B-D, H-K) .... *Drepanovelia*
- Male fore femora strongly incrassate (Fig. 11.12A). Pilosity of body and appendages long. Segment 8 of male simple, without ventral processes.. *Lacertovelia*
8. Large species, length 2.9-3.0 mm. Male segment 8 very long, with a pair of triangular processes at ventral, posterior margin (Fig. 11.32B)............................. *Nesidovelia*
- Small species, length less than 2.9 mm (except for a few species in which male segment 8 is otherwise modified).... *Microvelia*
9. Fore wings uniformly dark, without pale stripes or spots .... *Drepanovelia*
- Fore wings at least with pale longitudinal stripes in basal part..................................... 10
10. Inner margin of male fore tibiae sinuated (Fig. 11.37A), grasping comb long, about 0.55x tibial length .... *Tarsoveloides*
- Male fore tibiae not modified as above 11
11. Anterior part of pronotum distinctly narrowed (Fig. 11.30A); anterior margin deeply concave. Female abdomen constricted in basal part .... *Microvelopsis* (part)
- Anterior part of pronotum not distinctly narrowed; anterior margin only slightly concave. Female abdomen usually not constricted in basal part.................. *Microvelia*
12. Large species, length 3.1-3.8 mm .... *Microvelopsis* (part)

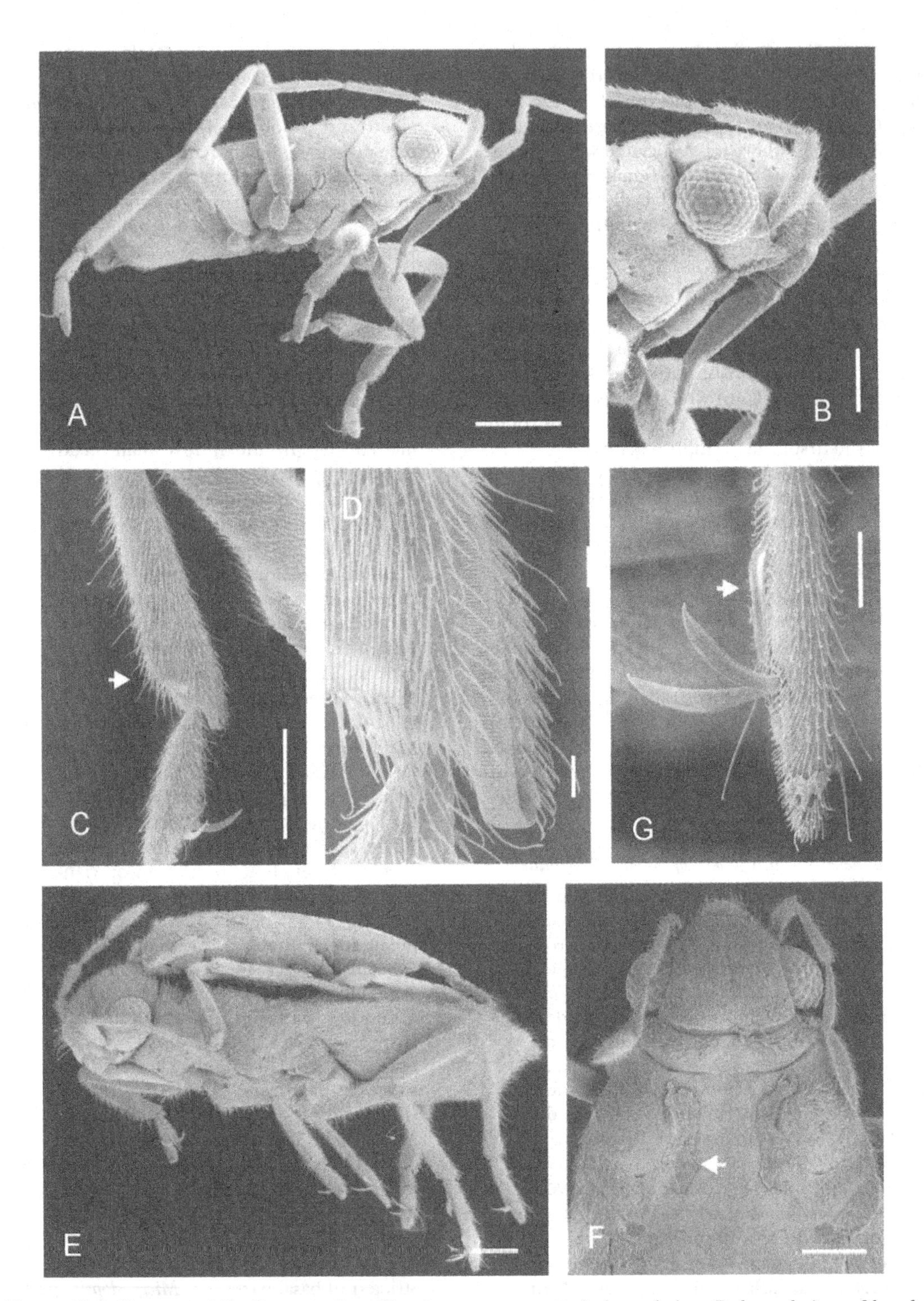

Figure 11.2. VELIIDAE. A-D, *Drepanovelia millennium*, apterous ♂: A, lateral view. B, lateral view of head and thorax. C, fore leg with grasping comb (arrow). D, grasping comb. E-G, *Phoreticovelia rotunda*: E, lateral view of ♀ with small ♂ on her back. F, dorsal view of head and thorax of ♀ with patches of waxy substance (arrow). G, middle tarsus with blade-like ventral arolium (arrow). Scales: A, E, 0.5 mm; B-F, 0.1 mm; G, 0.05 mm. Scanning Electron Micrographs. Reproduced with modification from Andersen & Weir (2001).

- Small species, length less than 3.0 mm....... 13

13. Middle tarsi with long slender, falcate claws (Fig. 11.33A); ventral arolium bristle-like. Grasping comb of male fore tibiae about 0.4x tibial length. Length 2.0-2.8 mm (♂) or 2.0-2.9 mm (♀); male and female thus of about same size .... *Petrovelia*

- Middle tarsi with claws blade-like expanded, ventral arolium long, blade-like (Fig. 11.35D; va). Grasping comb of male fore tibiae less than 0.25x tibial length. Length 1.0-1.2 (♂) or 1.7-2.0 mm (♀); male thus distinctly smaller than female (Fig. 11.35A). . . . . . *Phoreticovelia*

Subfamily Haloveliinae

Identification

Small or very small surface bugs with slender middle legs which are distinctly longer than hind legs (Fig. 11.1B) and all tarsi 2-segmented. All Australian species marine. Adults always apterous (wingless). Body suboval, chiefly dark coloured and covered by a thick pilosity. Head much shorter than wide, moderately deflected in front of eyes; dorsal surface at most with an indistinct median groove and without pseudocellar pits. Eyes globular but relatively small. Pronotum much shorter than head length (Fig. 11.B; pn); indistinctly separated from mesonotum laterally. Dorsal boundaries between meso- and metathorax and between metathorax and abdominal terga indistinct. Ventral sutures of thorax and abdomen distinct; metathoracic scent channels extending laterally and obliquely backward, distinctly separated from hind margin of metasternum. Fore legs short and relatively robust. Middle very long and slender, with very long mesotrochanters; middle tarsus long and slender. Hind legs distinctly shorter than middle legs. Abdomen shortened, especially in males. Genital segments of male relatively large, with a pair of large, symmetrically developed parameres. Genital segments of female usually covered by abdominal sternum 7 from beneath.

Overview

When describing the genus *Halovelia*, Bergroth (1893) placed this genus in the Veliidae. Hale (1926) supported this classification by pointing out the presence of a specialised grasping comb on the male fore tibia of *Halovelia*, a feature shared with the Australian species of *Microvelia* Westwood (Veliidae). Despite this, Esaki (1924, 1926) transferred *Halovelia* to the family Gerridae because of superficial similarities among species belonging to these two groups, e.g. points of insertion of the middle and hind legs distinctly removed from that of the fore legs, long and slender middle femora. China (1957) and Andersen (1982, 1989b) extensively discussed the relationships of the haloveliine water striders, confirming their classification in the family Veliidae.

The subfamily Haloveliinae was erected by Esaki (1930) for the genera *Halovelia* Bergroth, *Xenobates* Esaki, *Entomovelia* Esaki and *Strongylovelia* Esaki. The two last-mentioned genera contain freshwater species distributed in the Oriental region, including New Guinea, but are not represented in the Australian fauna. All Australian species belonging to this subfamily are marine, but the two genera usually occupy different types of habitats. *Xenobates* species are confined to estuaries and mangrove swamps with tidal influence, whereas *Halovelia* species usually inhabit intertidal reef flats with coral rubble and blocks of coral, but may also be found in mangroves adjacent to coral reefs.

Halovelia

Coral Bugs

Identification

Small or very small marine water striders. Body usually suboval, length 1.7-2.6x greatest width, dark coloured and covered by a thick pilosity which is more or less greyish, especially on abdomen. Head and pronotum without extensive pale markings. Thoracic and abdominal dorsum without definite spots of silvery hairs. Eyes globular but relatively small, width of each eye less than 0.3x interocular width. Antennae long and slender, usually 0.6-0.8x total length of insect; segment 1 usually as long as head, subequal to or slightly shorter than segment 4; segment 3 longer than segment 2. Pronotum much shorter than head (Fig. 11.4). Fore tibia of male with a distal grasping comb (Fig. 11.1D; gr) composed of a compact row of short spines along the inner margin. Middle femur very long, usually more than 0.5x total length of insect; middle tibia and tarsus very slender and long. Hind femur relatively short, usually thickened proximally; second tarsal segment much longer than first segment. Abdomen short with broadly rounded sides in male, longer and usually with more straight sides in female. Abdominal venter of male usually simple. Male genital segments relatively large but withdrawn into pregenital abdomen and only slightly protruding from abdominal end (Figs 11.3D); segment 8 and pygophore usually simple; parameres large and symmetrically developed. Hind margin of female sternum 7 usually produced in

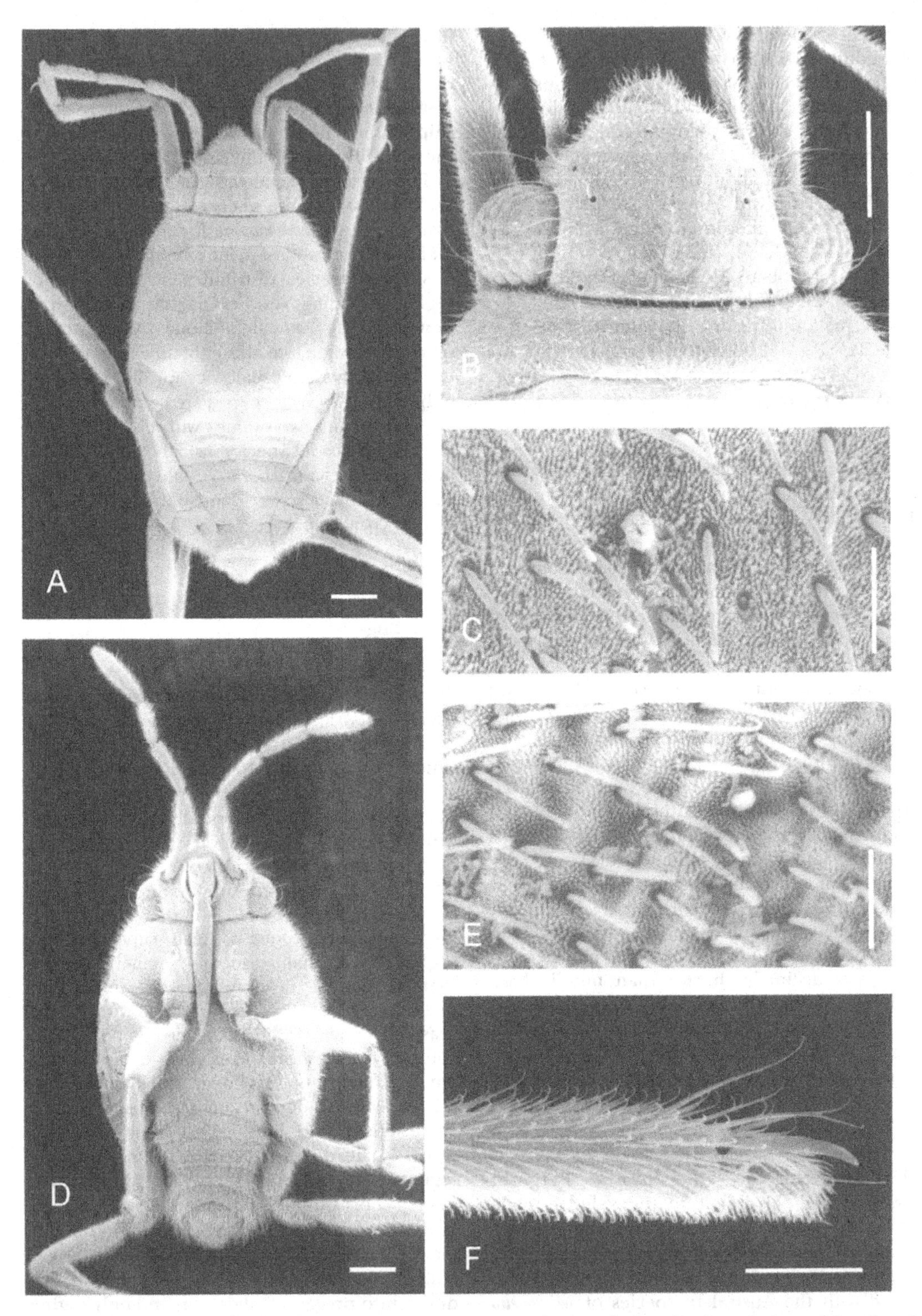

Figure 11.3. VELIIDAE. A-C, *Xenobates mangrove*, apterous ♀: A, dorsal view. B, dorsal view of head and pronotum. C, body surface. D-F, *Halovelia* hilli, apterous ♂: D, ventral view. E, body surface. F, apex of middle tarsus. Scales: A, D, 0.2 mm; B, F, 0.1 mm; C, E, 0.01 mm. Scanning Electron Micrographs (CSIRO Entomology, Canberra).

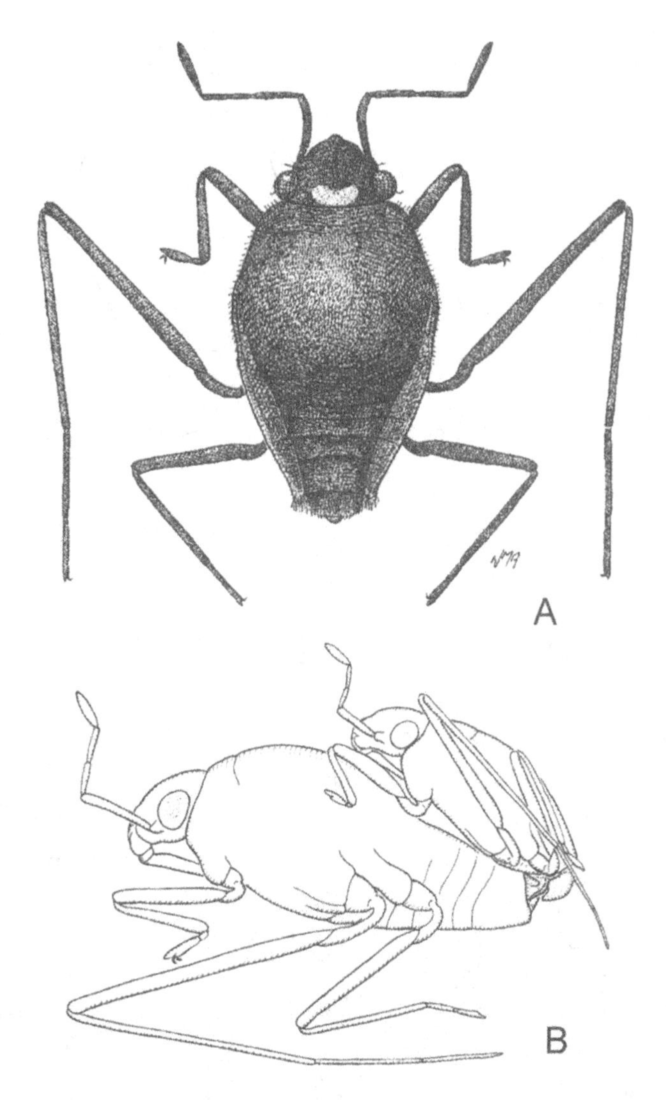

Figure 11.4. VELIIDAE. A-B, *Halovelia hilli*: A, apterous ♀. B, copulating pair. Reproduced with modification from Andersen & Weir (1999).

middle. Female genital segments clearly visible behind tergum 7; proctiger cone- or button-shaped, usually concealed beneath tergum 8.

The type species of this genus, *Halovelia maritima* Bergroth, 1893, was described from specimens collected by the British naturalist J. J. Walker on the small Cartier Island in the Timor Sea, about 250 km NW of Australia. Additional species were described by China (1957), J. Polhemus (1982), Andersen (1989b, 1989c), and Lansbury (1989). In his monographic revision of the genus *Halovelia*, Andersen (1989b, 1989c) presented the results of a phylogenetic analysis of relationships between species. A number of monophyletic species-groups can be defined upon apomorphic characters. The Australian species of *Halovelia* belong to several of these groups. *H. hilli* and *heron* belong to the *bergrothi* Esaki group; *H. polhemi* is closely related to *H. esakii* Andersen. Finally, *H. maritima* and *H. corallia* both assume a basal position on the reconstructed phylogeny.

Key to Australian species

1. Antennal segment 3 subequal to or only slightly longer than segment 2 (ratio less than 1.2). Grasping comb of ♂ fore tibia more than half length of tibia (Fig.

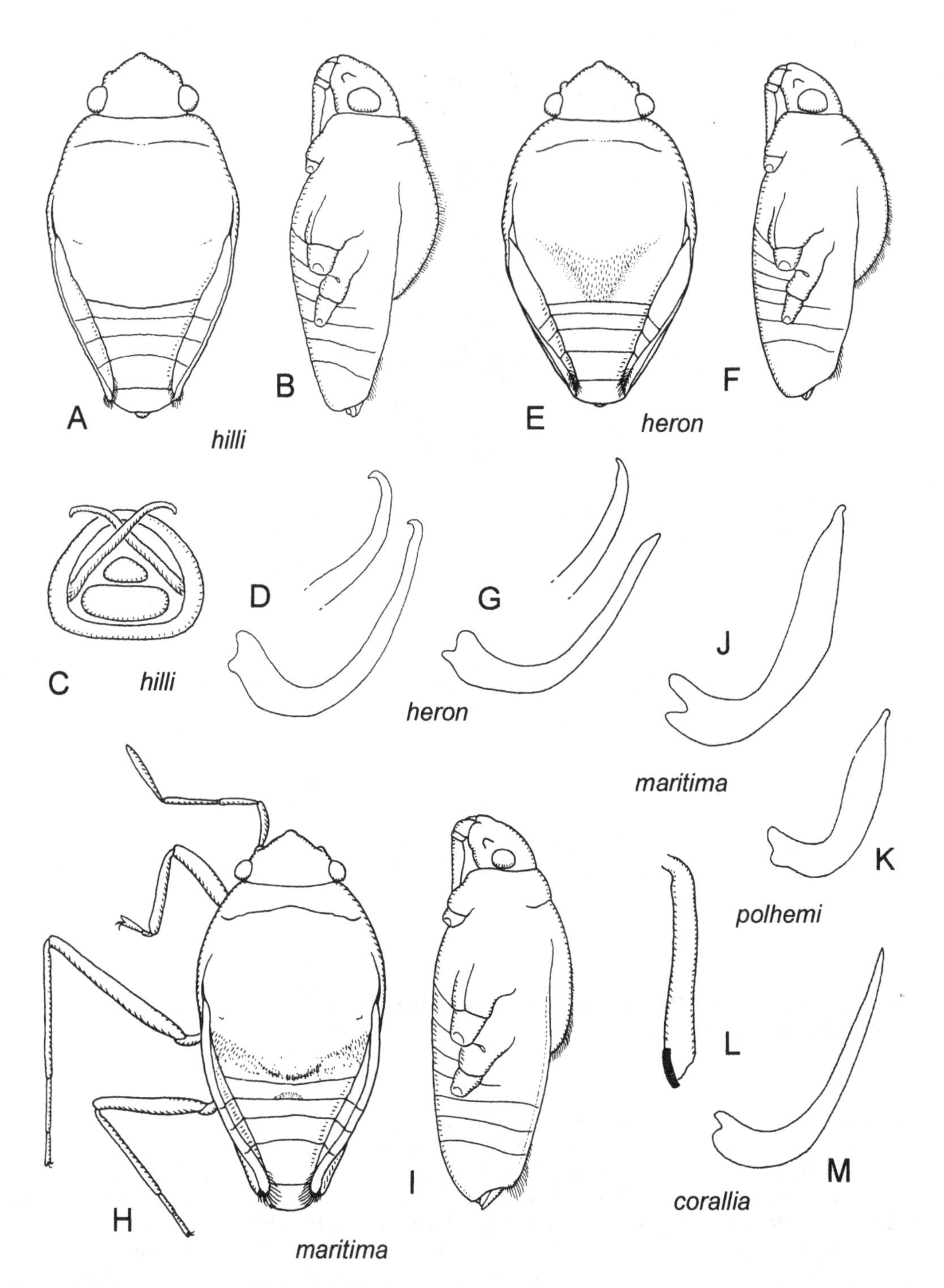

Figure 11.5. VELIIDAE. A-D, *Halovelia hilli*: A, dorsal view of apterous ♀; antennae and legs omitted. B, lateral view of apterous ♀. C, caudal view of abdominal end of apterous ♂. D, paramere, with distal part in different view. E-G, *H. heron*: E, dorsal view of apterous ♀; antennae and legs omitted. F, lateral view of apterous ♀. G, paramere, with distal part in different view. H-J, *H. maritima*: H, dorsal view of apterous ♀; antennae and legs of right side omitted. I, lateral view of apterous ♀. J, paramere. K, *H. polhemi*: paramere. L-M, *H. corallia*, fore tibia of apterous ♂. M, paramere. Reproduced with modification from Andersen & Weir (1999).

11.1D; gr). Parameres slender and very long (length subequal to or more than half width of head across eyes), in resting position distinctly crossing each other above genital segments (Fig. 11.5C). Thoracic dorsum of ♀ distinctly swollen (Figs 11.5B, F) 2
- Antennal segment 3 distinctly longer than segment 2 (ratio 1.2: 1 or more). Grasping comb of ♂ fore tibia less than half length of tibia (Fig. 11.5L). Parameres relatively short (length less than half width of head across eyes), in resting position reaching or barely crossing each other above genital segments. Thoracic dorsum of ♀ only moderately swollen (Figs 11.5I) 3
2. Extreme apex of each paramere hook-shaped (at least in some angles of view, Fig. 11.5D). Length of ♀ about 2.0x greatest width across thorax. Length 1.4-1.6 mm (♂) or 1.8-2.2 mm (♀). Northern Territory, Queensland, Western Australia .... *hilli* China
- Extreme apex of each paramere almost straight (Fig. 11.5G). Length of ♀ about 1.8x greatest width across thorax. Length 1.4-1.5 mm (♂) or 1.9-2.0 mm (♀). Queensland .... *heron* Andersen
3. Eyes small, width of an eye distinctly less than one fourth interocular width. Middle femur of ♀ less than half total length (Fig. 11.5H). [Paramere (Fig. 11.5J)]. Length 1.7 mm (♂) or 2.1-2.2 mm (♀). Northern Territory.................... .... *maritima* Bergroth
- Eyes larger, width of an eye subequal to or more than one fourth interocular width. Middle femur of ♀ subequal to or longer than half total length......................... 4
4. Abdominal venter of ♂ with basal tumescence furnished with long hairs. Parameres relatively short and broad (Fig. 11.5K). Grasping comb of male fore tibia a little more than 0.25x length of tibia. Thoracic and basal abdominal dorsum of ♀ with markings of greyish pubescence. Smaller species, length 1.3-1.4 mm (♂) or 1.5-1.6 mm (♀). Northern Territory .... *polhemi* Andersen
- Abdominal venter of ♂ not modified as above; paramere longer and more slender (Fig. 11.5M). Grasping comb of male fore tibia very short, less than 0.20x length of tibia (Fig. 11.5L). Thoracic and basal abdominal dorsum of ♀ without markings of greyish pubescence. Larger species, length 1.6 mm (♂) or 1.9 mm (♀). Queensland .... *corallia* Andersen

Biology

Kellen (1959) and Andersen (1989c) gave comprehensive accounts of the biology of *Halovelia* species which were given the trivial name *coral bugs* because of their association with coral reefs. Most species are found on intertidal reef flats, on the surface of tidal pools among stands of *Porites* and *Acropora* corals on the mid-reef flat, but only rarely on the inner reef flat or towards the outer reef margin. Species of *Halovelia* have also been recorded from rocky coasts without corals. The coral bugs usually appear when the tide recedes, in particular upon the surface of shallow pools around or beneath blocks of coral or porous rock. When the tide rises, both adults and nymphs retreat to cavities and holes in such blocks and stay submerged during high tide, surrounded by an air bubble. Coral bugs probably deposit their eggs on blocks of dead coral and the ecdysis of nymphs take place in the cavities and holes in such blocks. Occasionally, *Halovelia* species are found in mangrove habitats, especially when these border coasts fringed by coral reefs.

In *Halovelia hilli*, the most common and widespread Australian species, adults were collected in February, April, June-August, October, and December. Copulating pairs were found in August and October, nymphs in February and August (Andersen & Weir 1999). The male of *H. hilli* is much smaller than the female (female/male size ratio 1.4-1.5) and its very long parameres interlock with the genital segments of the female during mating (Fig. 11.4B). Males are so persistent in guarding their mate that they remain seated on the female's back even after being killed by submersion in 70% alcohol.

The eggs of *Halovelia* species are elongate oval, 0.6-0.7 mm long, 0.3 mm wide, with a small tubercle at the anterior pole, traversed by a single micropyle (Andersen 1989c: fig. 72-73). The are five nymphal instars (illustrated for *Halovelia bergrothi* Esaki by Kellen 1959: fig. 5). Fifth instar with extensive brownish sclerotisations dorsally; head with distinct U-shaped ecdysial line; thoracic nota uniformly brownish; abdomen with large, paired sclerites. Antennal segments 2 and 3 with short pilosity. Middle femora and tibia without distinct row of hairs along anterior margin.

Distribution

In Australia, *Halovelia* species are widely distributed along the northeastern, northern, and northwestern coast of the mainland (Fig. 11.6) as

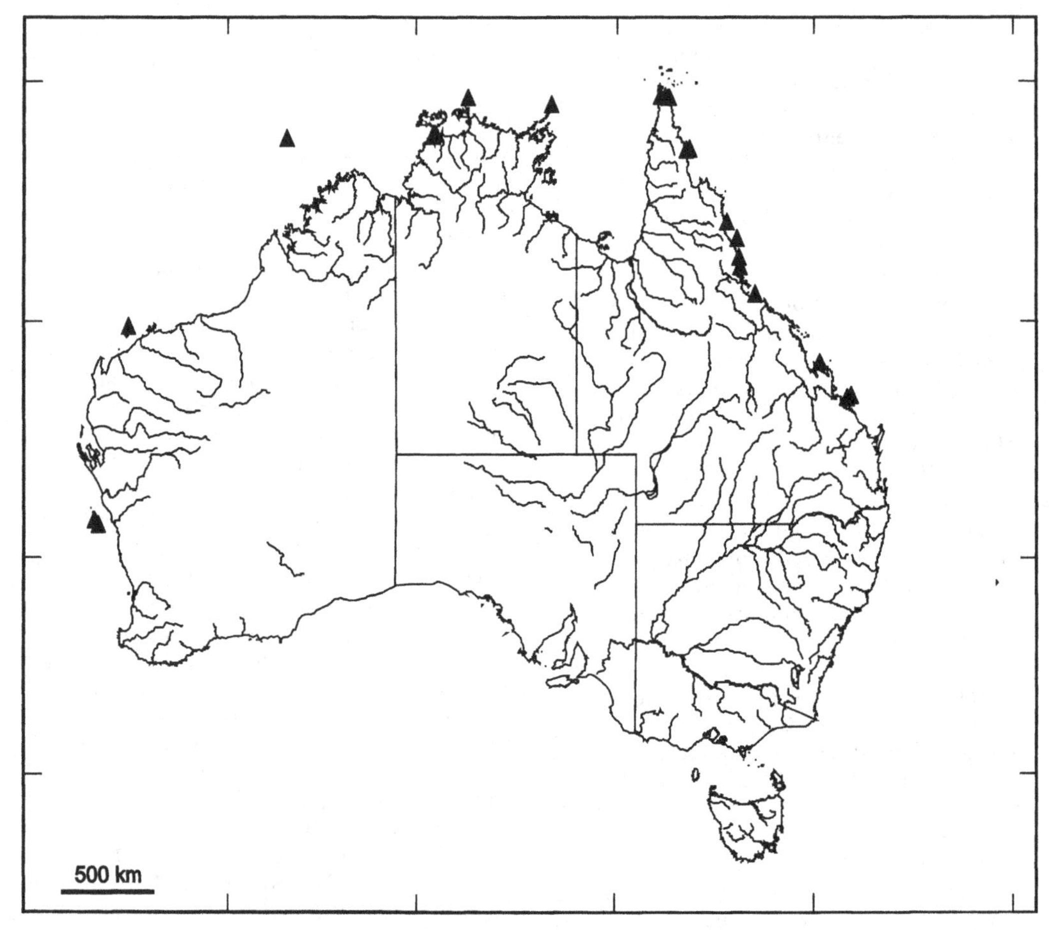

Figure 11.6. VELIIDAE. Distribution of *Halovelia* species in Australia.

well as along islands of the Timor Sea (e.g. Cartier Island), Torres Strait, and the Coral Sea (e.g. Heron Island), and some islands along the Indian Ocean coast of Australia (e.g. Hermite Island and Pelsart group). The genus is widely distributed in the Indo-West Pacific area as far east as Samoa (Andersen 1989b, 1989c). The present distribution, however, may only be part of a much wider, ancient distribution since a fossil *Halovelia* species is known from Dominican amber found on the island of Hispaniola in the Caribbean Sea (Andersen & Poinar 1998; Andersen 1998).

XENOBATES

Mangrove Bugs

Identification

Small or very small marine water striders; adults always apterous (wingless). Body fusiform to oval (Figs 11.3A, 11.7D), 1.8-2.2x greatest width across mesothorax, dark coloured, covered by a thick pilosity, usually forming definite spots of greyish or silvery hairs on thoracic and abdominal dorsum. Head largely pale; pronotum with transverse pale marking(s) in middle. Eyes relatively large, globular, width of each eye 0.3-0.5x interocular width. Antennae slender, 0.5-0.7x total length of insect; segment 1 always shorter than head, subequal in length to segment 4; segment 3 slightly longer than segment 2. Pronotum much shorter than head (Fig. 11.1B; pn). Fore tibia of male without grasping comb (Fig. 11.7B), but usually with a row of scattered, spine-like hairs along inner surface before apex. Middle femur very long, 0.7-0.9x total length of insect, distinctly thickened in proximal part; femur usually with a row of bristle-like hairs along anterior margin (Fig. 11.7E), continuing on tibia and tarsus. Hind femur relatively short, more or less thickened

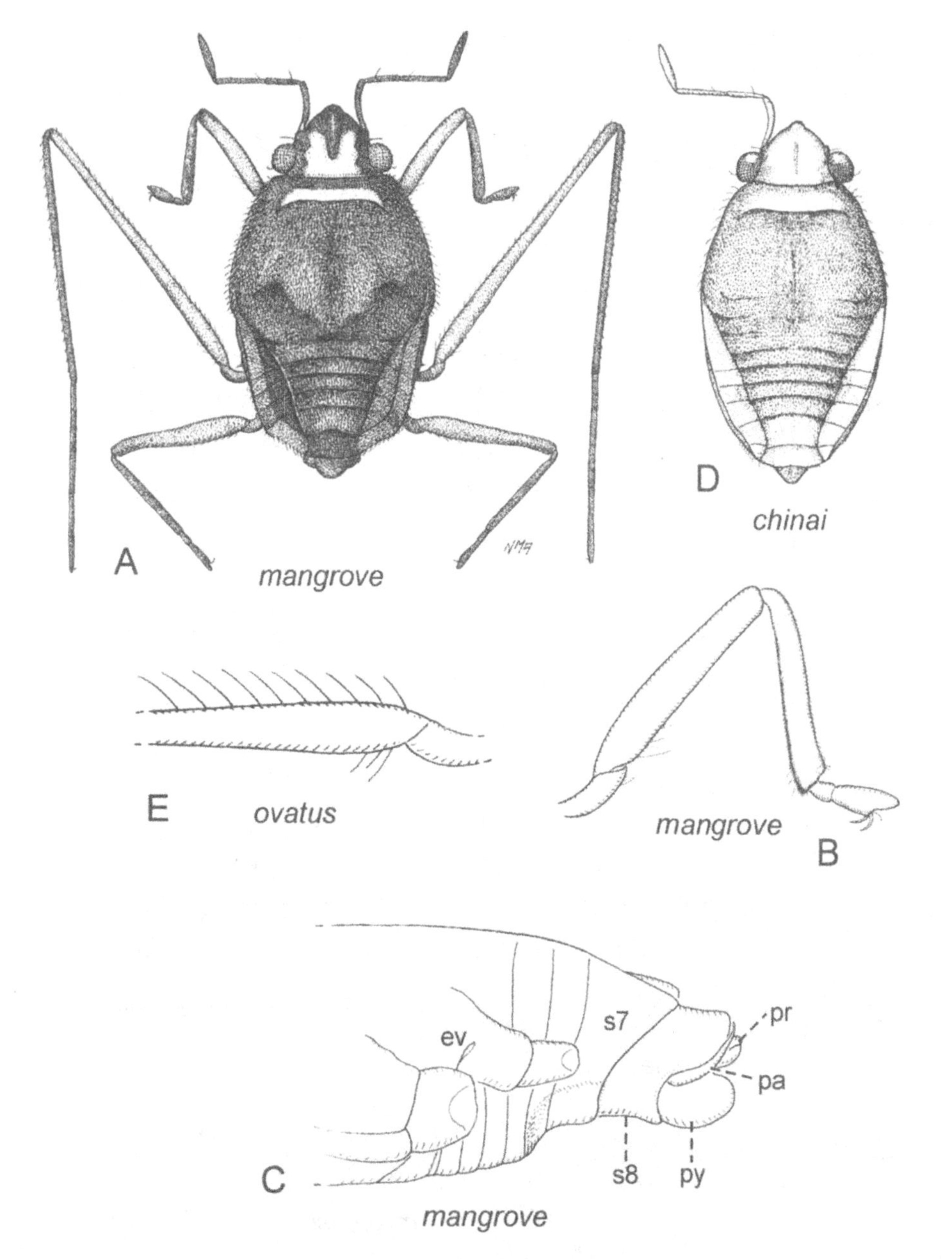

Figure 11.7. VELIIDAE. A-C, *Xenobates mangrove*: apterous ♀. B, fore leg of apterous ♂. C, lateral view of abdomen of apterous ♂ (ev, evaporatorium; pa, paramere; pr, proctiger; py, pygophore; s7, abdominal sternum 7; s8, segment 8). D, *X. chinai*, dorsal view of apterous ♀; most of antennae and legs omitted. E, *X. ovatus*, base of middle femur. Reproduced with modification from Andersen & Weir (1999).

proximally, especially in male. Abdomen relatively short and tapering in width posteriorly in both sexes, sides almost straight (♂) or more or less rounded (♀). Abdominal venter of male simple or modified (with basal tumescence and median ridge; Fig. 11.7E); male genital segments large, conspicuous and distinctly protruding from pregenital abdomen (Figs 11.1C, 11.7E); segment 8 (s8) simple or depressed ventrally; pygophore (py) simple; parameres (pa) large, symmetrically

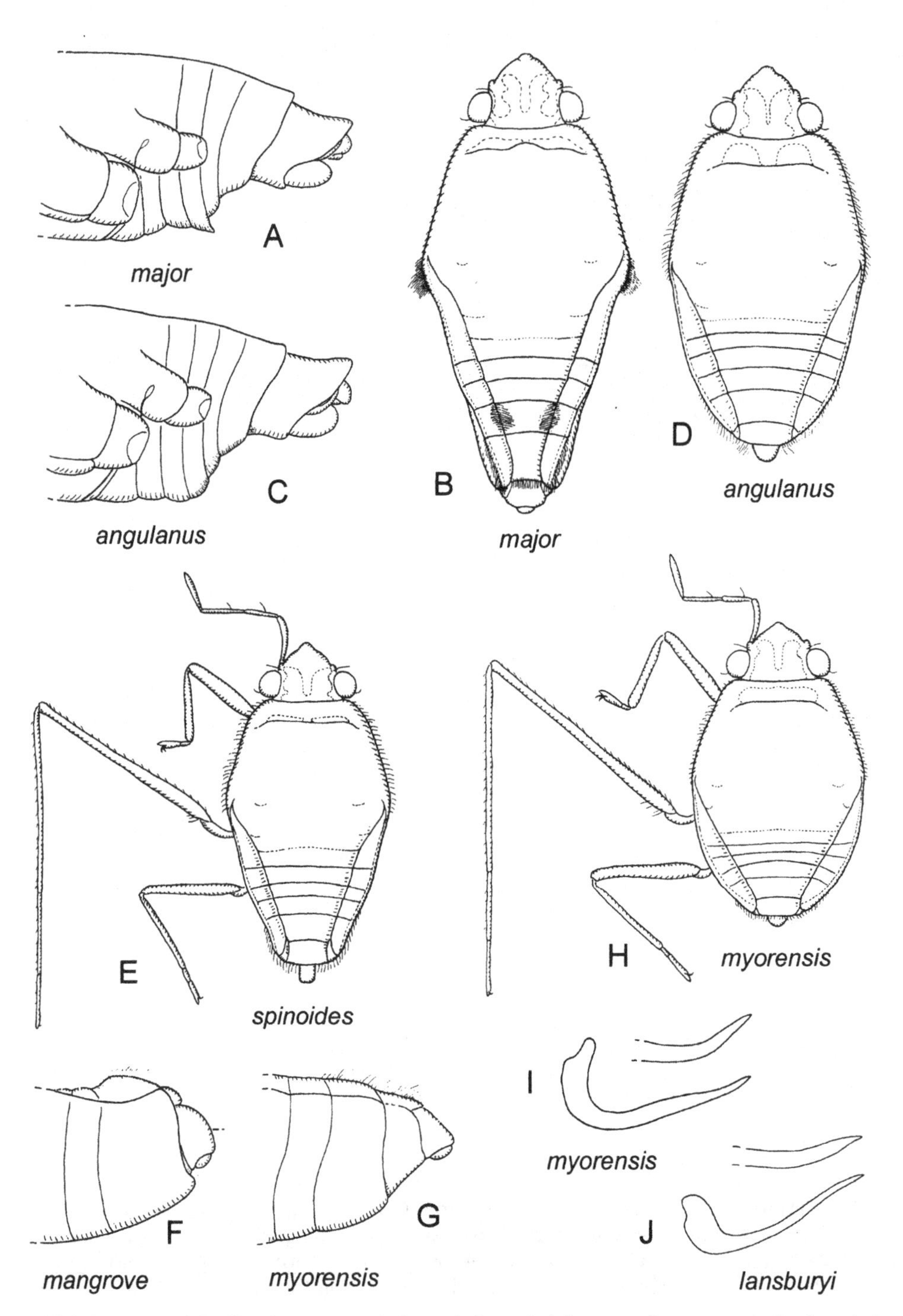

Figure 11.8. VELIIDAE. A-B, *Xenobates major*: A, lateral view of abdomen of apterous ♂. B, dorsal view of apterous ♀; antennae and legs omitted. C-D, *Xenobates angulanus*: C, lateral view of abdomen of apterous ♂. D, dorsal view of apterous ♀; antennae and legs omitted. E, *X. spinoides*: dorsal view of apterous ♀; antennae and legs of right side omitted. F, *X. mangrove*, apterous ♀: lateral view of abdominal end. G, *X. myorensis*, apterous ♀: lateral view of abdominal end. H-I, *X. myorensis*: H, dorsal view of apterous ♀; antennae and legs of right side omitted. I, paramere, with distal part shown at a different angle. J, *X. lansburyi*: paramere, with distal part shown at a different angle. Reproduced with modification from Andersen & Weir (1999).

developed, and usually falciform. Hind margin of sternum 7 of female usually straight; female proctiger variable in shape, protruding from segment 8 or more or less deflected (Figs 11.8F, G).

Esaki (1926) described this genus under the name "*Microbates*", which was preoccupied (in Aves) and subsequently changed to *Xenobates* (Esaki 1927). J. Polhemus (1982) described the subgenus *Colpovelia* stating that it was "similar to *Halovelia s.s*, but without definite sclerotised comb on fore tibia". Lansbury (1989) described several new species of haloveliines, placing some of these in the genus *Halovelia* (including the Australian *Xenobates myorensis*), others in the genus *Xenobates*. At present this genus contains 18 described species, including 8 from Australia, but several undescribed species are known from the Indo-Australian area (N. M. Andersen, unpublished).

Key to Australian species

1. Large species, length more than 1.9 mm (♂) or 2.4 mm (♀).. 2
- Small species, length less than 1.8 mm (♂) or 1.9 mm (♀).. 3
2. Basal tumescence of male abdominal venter forming a spine before the depressed sterna 6-7 (Fig. 11.8A). Abdomen of ♀ relatively long, with distinct lateral hair tufts anteriorly (Fig. 11.8B). Length 2.1-2.3 mm (♂), 2.6-2.9 mm (♀). Queensland................ *major* Andersen & Weir
- Basal tumescence of male abdominal venter forming a steep angle, but no spine, before depressed sterna 6-7 (Fig. 11.8C). Abdomen of ♀ relatively short, without distinct hair tufts (Fig. 11.8D). Length 2.2-2.4 mm (♂), 2.4-2.6 mm (♀). Northern Territory...... *angulanus* (Polhemus)
3. Abdominal end of ♀ with distinctly protruding, button-like proctiger (Figs 11.8E). Abdominal venter of ♂ with narrow basal tumescence. Length 1.6-1.8 mm (♂), 1.9-2.0 mm (♀). Queensland .... *spinoides* Andersen & Weir
- Abdominal end of ♀ not modified as above, proctiger cone-shaped and more or less deflected. Abdominal venter of ♂ without basal tumescence.............................. 4
4. Abdomen of ♀ broad, trough-shaped (Figs 11.3A, 11.7A), proctiger widened and strongly deflected (Figs 11.8F). Abdominal venter of ♂ distinctly depressed on sterna 6-7. Length 1.6-1.7 mm (♂), 1.7-1.8 mm (♀). Queensland.................... ....*mangrove* Andersen & Weir
- Abdomen of ♀ not trough-shaped, proctiger cone-shaped and only slightly deflected (Fig. 11.8G). Abdominal venter of ♂ simple 5
5. Larger species, length more than 1.6 mm (♂) or 1.7 mm (♀).. 6
- Smaller species, length 1.3-1.6 mm (♂), 1.5-1.7 mm (♀).. 7
6. Hind femora of ♀ slender, not as thick as middle femora (Fig. 11.8H). Paramere as in Fig. 11.8I. Length 1.6-1.8 mm (♂), 1.7-1.9 mm (♀). Queensland.................... .... *myorensis* (Lansbury)
- Hind femora of ♀ distinctly thickened, as thick as middle femora. Paramere (Fig. 11.8J). Length 1.6 mm (♂), 1.7-1.8 mm (♀).Northern Territory .... *lansburyi* Andersen & Weir
7. Chiefly dark brownish or black above, antenna, legs, and connexiva dark. Length 1.4-1.6 mm (♂), 1.5-1.7 mm (♀). Queensland .... *ovatus* Andersen & Weir
- Chiefly brownish above, antenna, legs, and connexiva yellowish (Fig. 11.7B). Length 1.3-1.4 mm (♂), 1.5-1.6 mm (♀). Northern Territory.... *chinai* Andersen & Weir

Biology

Species belonging to the genus *Xenobates* are typical inhabitants of mangrove swamps where they occur singly or in groups amongst mangrove shoots and in tidal streams (hence their trivial name *mangrove bugs*). Very little is known about the biology of these insects. The present records for Australian *Xenobates* indicate that species may breed throughout most of the year (Andersen & Weir 1999). Adults and nymphs belonging to various instars are usually collected together and samples often contain more than one species. Lansbury (1996) used mercury vapour lights to attract *X. solomonensis* Lansbury and *X. pilosellus* Lansbury in Papua New Guinea, and often found these species in habitats where saline water and freshwater mingle.

Lansbury (1989: 99) characterises the habitat of *X. myorensis* as follows: "Collected from mangrove swamp, water shallow about 1-2 cms deep. The bugs occurring singly amongst mangrove shoots, their cryptic coloration making them difficult to see on the greyish-black ooze of the swamp". Adults of this species were collected in February, May, June - August, and November, nymphs in May and November. The eggs of *X. mangrove* are elongate oval, 0.6-0.7 mm long, 0.2-0.25 mm wide, with a small tubercle at the anterior pole, traversed by one micropyle (Andersen & Weir 1999: figs 16-17). Probably five nymphal

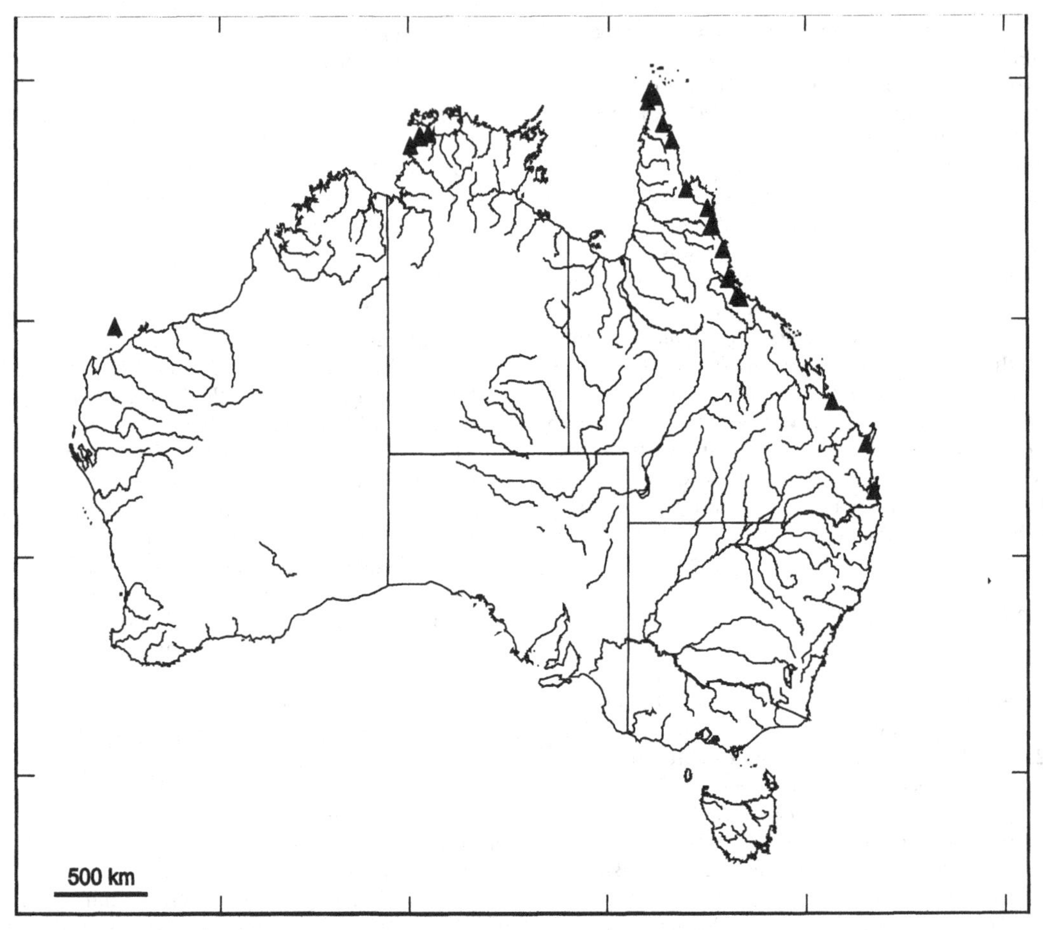

Figure 11.9. VELIIDAE. Distribution of *Xenobates* species in Australia.

instars. Fifth instar largely pale (Andersen & Weir 1999: fig. 15), with some brownish sclerotisations dorsally; head without distinct ecdysial line; thoracic nota brownish with pale markings; abdomen with paired sclerites. Antennal segments 2 and 3 with short pilosity and a few longer hairs. Middle femora and tibia with short but distinct row of hairs along anterior margin.

Distribution

In Australia, *Xenobates* species are sparsely but widely distributed along the northeastern and northern coast of the mainland (Fig. 11.9). An undescribed species is also known from Hermite Island, Western Australia (Andersen & Weir 1999). The genus is widely distributed in the Indo-Australian region, including New Guinea, Solomon Islands, and New Caledonia.

SUBFAMILY MICROVELIINAE

Identification

Small or very small water striders which above all are characterised by their unique state of tarsal segmentation: fore tarsi with only one segment, middle and hind tarsi with two segments each, the two-segmented state derived from a fusion of the first and second segment of the normally three-segmented veliid tarsus. Adults are either apterous (the most common adult form in most species) or macropterous. Pronotum of apterous form either shorter than head or prolonged, covering mesonotum. Ventral sutures of thorax and abdomen distinct; metathoracic scent channels extending laterally from median scent orifice, running close to hind margin of metasternum; small evaporative areas and hair tufts present on metacetabula. Forewings usually dark with pale spots and stripes. Femora of all legs usually stout.

Fore tibiae and sometimes also middle tibiae of male with grasping comb of varying length. Middle legs are usually intermediate in length between the fore and hind legs. Middle tibia usually with row of long bristles on inner side. Middle and hind tarsus usually shorter than tibiae; claws distinct, slender and falcate, usually with bristle-like aroliae. Male abdomen long, often modified ventrally. Genital segments of male may be variously modified. Female abdomen long, connexiva usually raised and sometimes modified distally.

Overview

The subfamily Microveliinae is distributed worldwide, comprising 29 genera with a total of about 300 described species. Whereas the Microveliinae is undoubtedly a monophyletic group, the phylogenetic relationships among the microveliine genera remain largely unsettled. China & Usinger (1949) gave a key to genera, but several new genera have been added since. The most recent key was provided by Andersen (1982: 419-420) and includes 21 genera. Additional genera were described by Zimmermann (1984), Andersen (1989a), J. Polhemus & D. Polhemus (1988, 1994a), J. Polhemus & Copeland (1996), D. Polhemus & J. Polhemus (2000), and Andersen & Weir (2001). The majority of microveliines, however, are still classified in the large genus *Microvelia* (with about 170 species). Structurally, these species are quite uniform except for various modifications of the male genital segments, characters which are insufficiently known in many species. As presently defined, the genus *Microvelia* is most probably not a monophyletic group.

The Microveliinae include some of the smallest and most common Australian water striders. Most species inhabit the nearshore, plant-covered zone of stagnant fresh water. Some species of *Microvelia* are frequently found in temporary habitats, such as puddles and pools filled with rainwater. This also includes small natural as well as artificial containers such as treeholes, coconut-shells, and tin cans (Laird 1956). A few Neotropical *Microvelia* species have become specialised to live in the water-filled leaf axils of epiphytic Bromeliaceae (Andersen 1982). Other microveliines live in water-filled internodes of bamboo (Kovac & Yang 2000). Microveliines are scavengers and predators feeding on other small arthropods, such as ostracods, cladocerans, collemboles, eggs and newly hatched larvae of mosquitoes (Laird 1956). They normally walk or run on the water surface (see above). The eggs of microveliine bugs are 0.5-0.8 mm, ovoid, with 2-3 micropyles (Cobben 1968; Andersen 1982). They are usually laid on floating material at the edge of the water, glued lengthwise to the substrate.

Drepanovelia

Millennium Bugs

Identification

Small water striders which are either apterous (Fig. 11.1E, 11.10A) or macropterous (Fig. 11.1F). Body elongate oval to oval, 1.9-2.8x greatest width across mesothorax, chiefly dark coloured with scarce pilosity of long, semierect hairs (Plate 6, B). Antennae slender and long, between one half and two thirds length of body; segment 1 always shorter than head, subequal in length to segment 3; segment 2 slightly shorter than segments 1 and 3, segment 4 longest. Pronotum of apterous form distinctly shorter than head (Fig. 11.1E; pn), with large, transverse pale marking in middle. Mesonotum exposed, distinctly longer than pronotum. Pronotum of macropterous form large (Fig. 11.1F; pn), pentagonal, with a large, transverse pale marking along anterior margin and numerous punctures on pronotal lobe, between raised humeral angles. Fore wings uniformly dark, without pale spots or stripes. Fore femur thickened (especially in male), but otherwise not modified; grasping comb of male fore tibia short (Fig. 11.2C, D), about one fourth of tibial length. Middle tarsus relatively short, about half length of tibia, with first tarsal segment about half length of second segment. Hind femur slightly thicker than middle femur. Abdomen relatively long, sides almost straight (♂) or more or less rounded (♀); connexiva broad, obliquely raised or sometimes vertical in posterior parts; connexival margins of female not modified. Abdominal venter of male not modified; male genital segments conspicuous, segment 8 modified on ventral surface, either carrying one median, chisel-shaped process (Figs 11.10E-G) or a pair of more or less conspicuous, pointed processes (Figs 11.10B-D, H-K). Posterior margin of sternum 7 of female straight, completely covering gonocoxae from beneath; proctiger broad and deflected.

The genus *Drepanovelia* Andersen & Weir is endemic to Australia with 4 described species. The type species, *D. dubia*, was originally described by Hale (1926) in the genus *Microvelia*. Andersen & Weir (2001) named one of the Australian species for the new millennium to make the point that this bug is not harmful, that we have a very large proportion of our biodiversity to inventory in the new millennium and that water striders are potentially useful indicators of

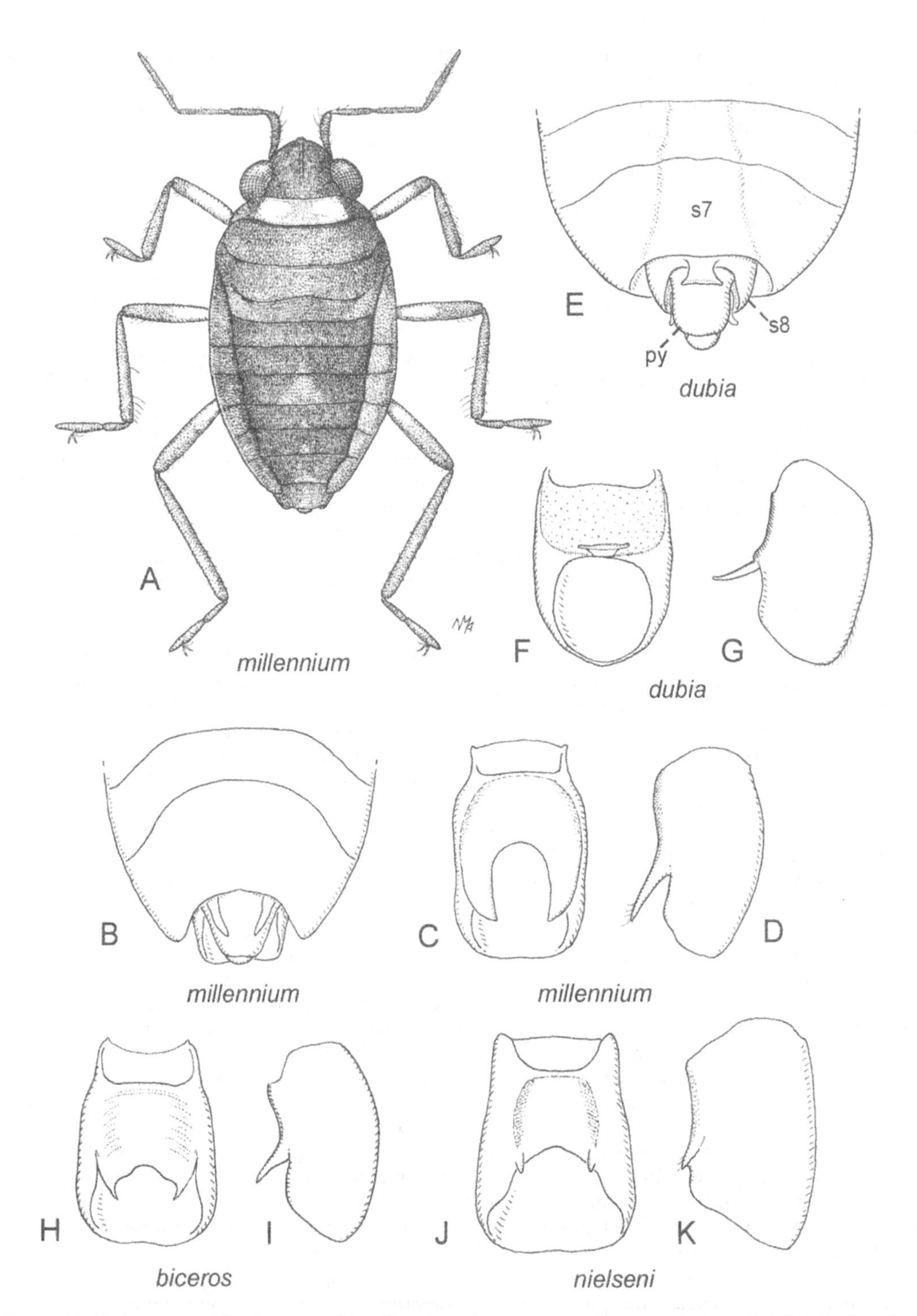

Figure 11.10. VELIIDAE. A-D, *Drepanovelia millennium*: A, apterous ♀. B, ventral view of abdominal end of apterous ♂. C, ventral view of segment 8 of apterous ♂. D, lateral view of segment 8 of apterous ♂. E-G, *D. dubia*, apterous ♂: E, ventral view of abdominal end (py, pygophore; s7, sternum 7; s8, segment 8). F, ventral view of segment 8. G, lateral view of segment 8. H-I, *D. biceros*, apterous ♂: H, ventral view of segment 8. I, lateral view of segment 8. J-K, *D. nielseni*, apterous ♂: J, ventral view of segment 8. K, lateral view of segment 8. Reproduced with modification from Andersen & Weir (2001).

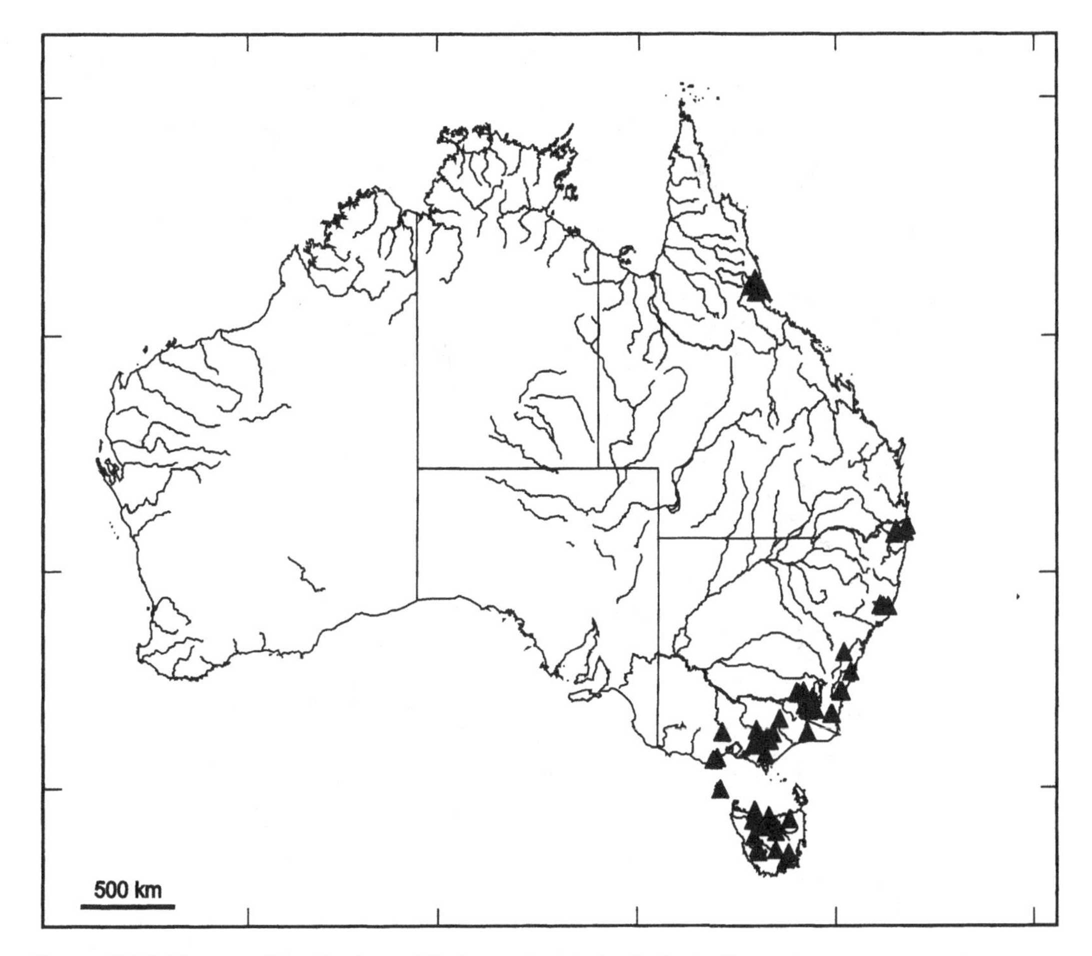

Figure 11.11. VELIIDAE. Distribution of *Drepanovelia* species in Australia.

the quality of inland waters. With reference to the imminent formal description of *D. millennium* (Figs 11.2A, 11.10A), CSIRO issued a press release at midnight of 1 January 2000 (http://www.csiro.au/) that the "real" Millennium Bug had been found. The term 'The Millenium Bug' was widely used for the "Y2K" concern about what impact the Year 2000 date change could have on businesses, utilities and other organisations that rely on computerised and automated systems around the world. The discovery of the real Millennium Bug was widely reported in the media during early 2000.

Key to Australian species

1. Segment 8 of ♂ with one prominent, chisel-shaped process at ventral, posterior margin, directed obliquely downwards (Figs 11.10E-F). Length 2.2-2.8 mm. New South Wales, Tasmania, Victoria... *dubia* (Hale)
- Segment 8 of ♂ with a pair of more or less conspicuous, pointed processes at ventral, posterior margin.............................. 2
2. Segment 8 of ♂ with a pair of prominent, sickle-shaped processes at ventral, posterior margin, directed obliquely downwards and backwards (Figs 11.10B-D). Abdomen of ♀ broad, with rounded sides (Fig. 11.10A); connexiva obliquely raised. Length 1.9-2.2 mm. New South Wales, Queensland........ *millennium* Andersen & Weir
- Segment 8 of ♂ otherwise modified ventrally 3
3. Segment 8 of ♂ with a pair of conspicuous, pointed processes arising from at ventral, posterior margin (Figs 11.10H, I). Length 1.9-2.0 mm (apterous ♂; female unknown). New South Wales.................. .... *biceros* Andersen & Weir
- Segment 8 of ♂ with a pair of small pro-

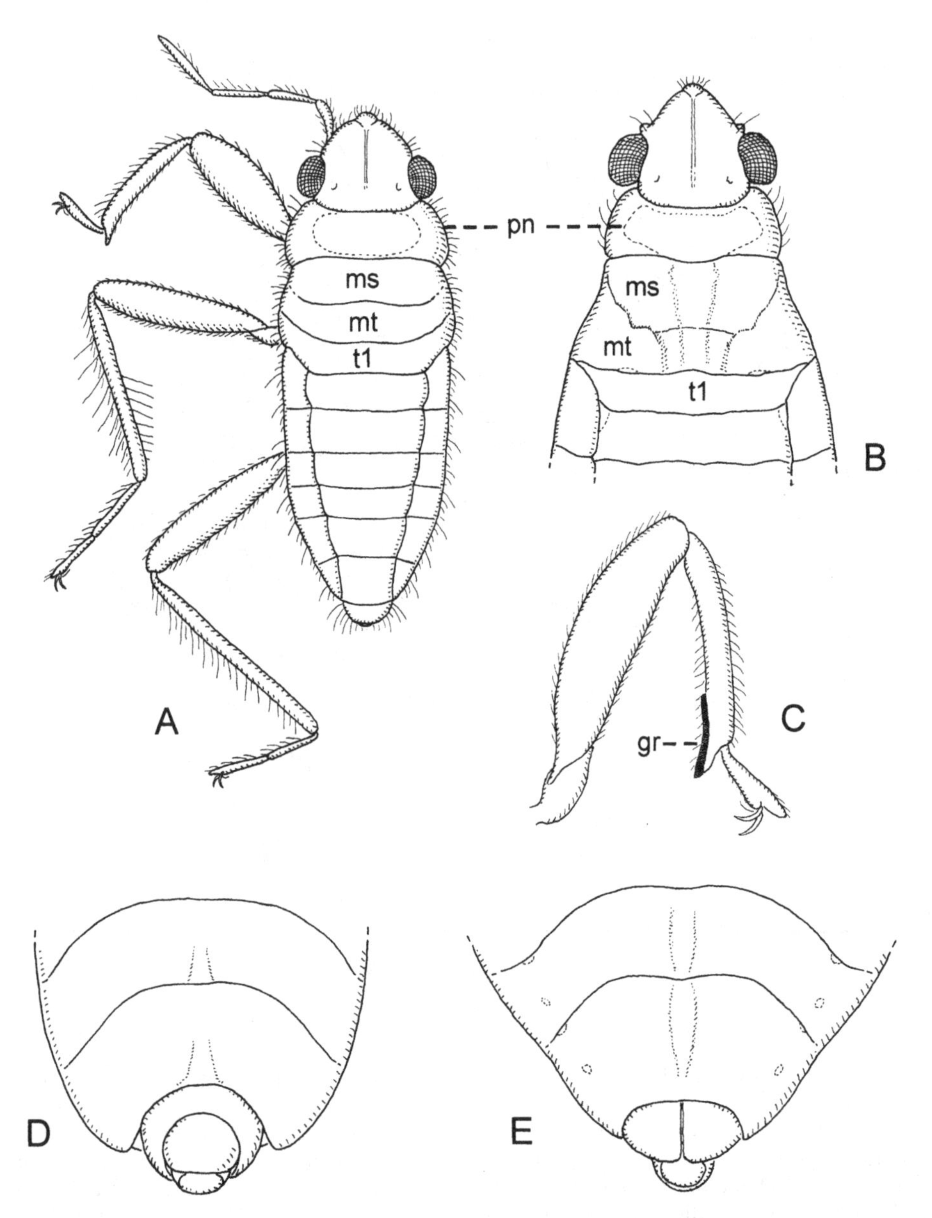

Figure 11.12. VELIIDAE. A-E, *Lacertovelia hirsuta*: A, dorsal view of apterous ♂; antennae and legs of right side omitted (ms, mesonotum; mt, metanotum; pn, pronotum; t1, abdominal tergite 1). B, dorsal view of head, thorax, and anterior abdomen of apterous ♀. C, fore leg of apterous ♂ with grasping comb (gr). D, ventral view of abdominal end of apterous ♂. E, ventral view of abdominal end of apterous ♀. Reproduced with modification from Andersen & Weir (2001).

cesses arising from posterior margin (Figs 11.10J, K). Abdomen of ♀ slightly constricted at base; connexiva almost vertically raised. Length 2.2-2.6 mm. Queensland.......................... *nielseni* Andersen & Weir

Biology

The most common and widespread species, *Drepanovelia dubia*, lives at the edge of streams and rivers, usually lined with forests, with low to moderate flow, from sea level up to 1000 m (Andersen & Weir 2001). Occasionally found in road-

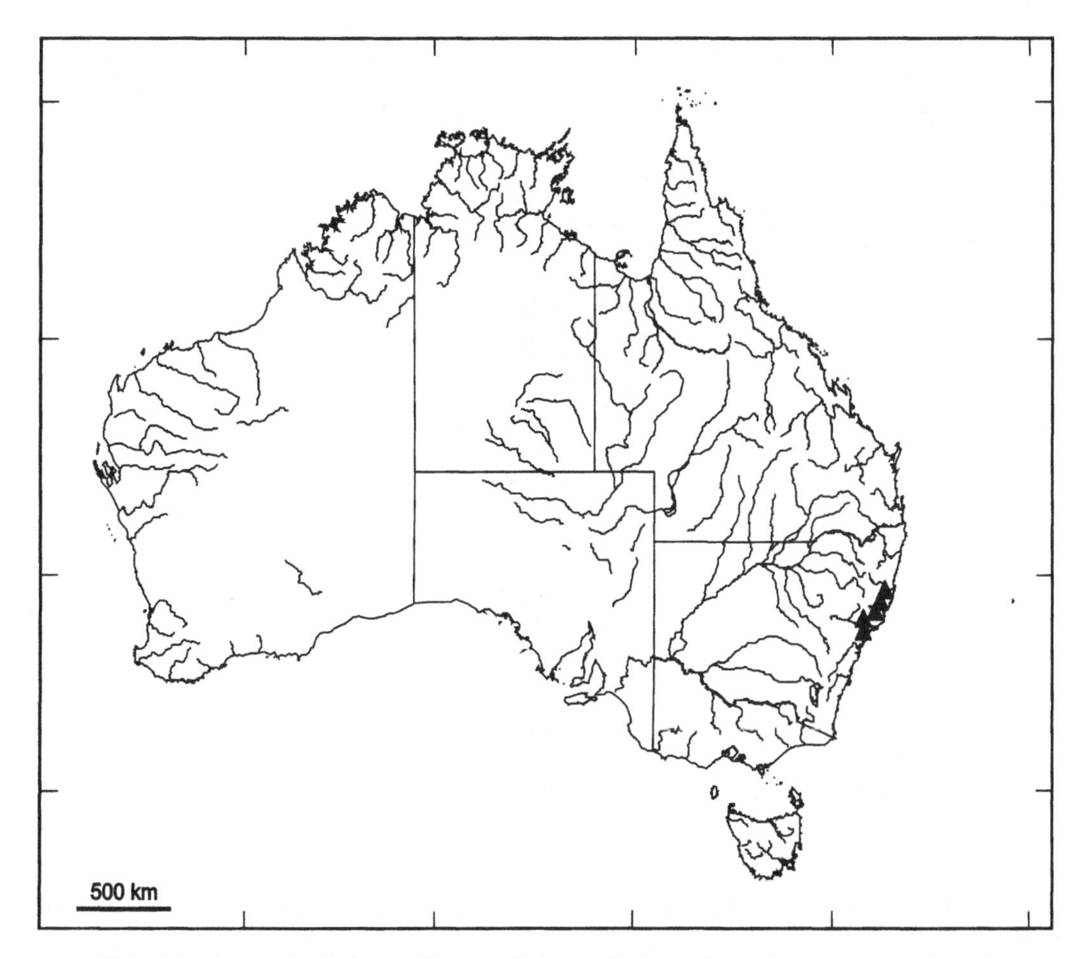

Figure 11.13. VELIIDAE. Distribution of *Lacertovelia hirsuta* in Australia.

side ponds and pools. Adults, which usually are apterous (wingless), are recorded from all months except June and July, nymphs from December-April, with most records from February and March.

Distribution

Endemic to eastern Australia with 4 described species of which three species are confined to the southeastern states incl. Tasmania, *D. nielseni* to northern Queensland (Fig. 11.11).

LACERTOVELIA

Identification

Small water striders which are either apterous (Figs 11.12A) or macropterous. Body elongate (♂), 2.8x greatest width across mesothorax, or suboval (♀), 2.2x greatest width across abdomen, chiefly dark coloured, with dense pilosity of long black, erect hairs (♂, pilosity more sparse in ♀). Antennae slender and long, about 0.5x total length of insect; segment 1 much shorter than head, surpassing tip of head with less than half of its own length, subequal in length to segment 2; segment 3 longer than segments 1 and 2, segment 4 longest. Pronotum of apterous form (Figs 11.12A, B; pn) distinctly shorter than head, with large, transverse pale marking in middle. Mesonotum exposed, as long as pronotum. Pronotum of macropterous form large, pentagonal, with a large, transverse pale marking along anterior margin and numerous punctures on pronotal lobe, between raised humeral angles. Fore wings uniformly dark, without pale spots or stripes; most posterior of two distal, free veins reduced. Fore femora of male incrassate (Fig. 11.12A), but otherwise not modified; fore tibiae of male with a

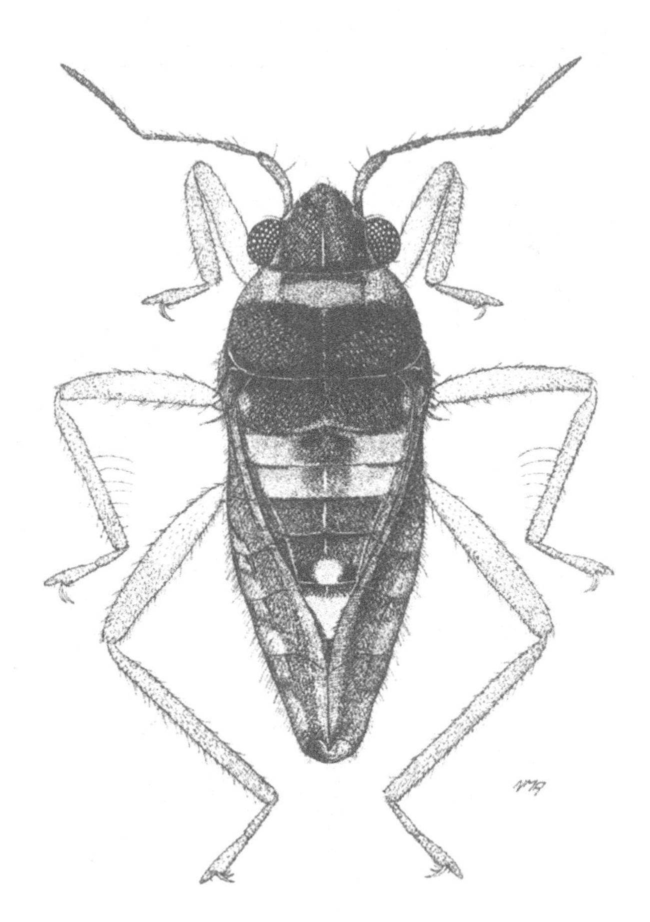

Figure 11.14. VELIIDAE. *Microvelia* (*Austromicrovelia*) *childi*, apterous ♀. Reproduced from Andersen (1969).

relatively short grasping comb (Fig. 11.12C; gr) on inner surface before apex; each comb about 0.3x length of tibia. Middle femur thickened, middle tarsus relatively short, about 0.6x length of tibia; first tarsal segment slightly shorter than second segment. Hind femur slightly thicker than middle femora. Abdomen relatively long, sides almost straight (♂) or distinctly rounded (♀). Connexiva broad, obliquely raised; connexival margins of female not modified. Abdominal venter of male not modified; male genital segments relatively small, protruding from pregenital abdomen (Figs 11.12D); segment 8 rectangular, simple. Posterior margin of sternum 7 of female emarginated (Fig. 11.12E), exposing gonocoxae from beneath; proctiger broadly conical, protruding.

The genus *Lacertovelia* was described by Andersen & Weir (2001) with a single species, *L. hirsuta* Andersen & Weir, length 2.3-2.8 mm (♂), 2.9-3.1 mm (♀). New South Wales.

Biology

Lacertovelia hirsuta has been collected at the edge of forest lined streams at times of low to moderate flow at altitudes of between 410 m and 1160 m (Andersen & Weir 2001). Adults, which usually are apterous (wingless), are recorded from March-May and October-November, the last month together with nymphs. The densely pilose males are distinctly smaller than females which, in combination with the incrassate male fore legs, suggest that the males exhibit postcopulatory guarding behaviour, remaining with the female for some time after copulation.

Distribution

Confined to central and northern New South Wales (Fig. 11.13). So far recorded from the catchment areas of the Clarence, Bellinger, Hastings and Manning rivers.

Microvelia

Identification

Small or very small water striders which are either apterous (Figs 11.14, 11.15A, B) or macropterous (Fig. 11.15C). Body shape variable, usually elongate oval or suboval, 2.4-3.0x greatest width across mesothorax, chiefly black or dark brownish above with yellowish brown markings (Plate 6, C); pilosity variable but usually relatively short often arranged in pruinose and silvery patches. Antennae relatively long, 0.4-0.6x total length of insect; segment 1 always shorter than head, surpassing tip of head with less than half of its own length; segment 4 longer than segment 2 and 3, fusiform. Pronotum of apterous form usually much longer than head (Figs 11.15A, B; pn), usually with large, transverse pale marking or paired spots anteriorly. Mesonotum usually completely covered by the pronotal lobe; metanotum only exposed laterally (Figs 11.15A, B; mt). Pronotum of macropterous form large, pentagonal, with distinctly raised humeral angles (Fig. 11.15C; pn); anterior collar never formed. Fore femora moderately thickened, but otherwise not modified; fore tibia of male usually with a grasping comb on inner surface before apex (Fig. 11.15D; gr). Middle femur slender; middle tibia of male with or without a short grasping comb; middle tarsus relatively short, 0.55-0.7x length of tibia; first tarsal segment distinctly shorter than second segment. Forewings of macropterous form usually dark with whitish stripes and spots (Plate 6, D). Abdomen relatively long, sides usually evenly rounded; connexiva broad, obliquely raised throughout (♂) or sometimes almost vertical and converging to meet each other posteriorly (♀). Abdominal venter of male variable, simple or modified to some extent, depressed or with hair tufts, tubercles, etc.; genital segments of variable shape and relative length, but always protruding from pregenital abdomen (Figs 11.15E, 11.16A); segment 8 sometimes modified ventrally; parameres variable in size and shape, usually relatively small, symmetrical (Fig. 11.15I; pa); but sometimes large and more or less asymmetrical (Figs 11.26C, 11.28G; pa). Tergum 8 of female often prolonged above proctiger; sternum 7 more or less tubular (Figs 11.18A, B, s7), usually covering bases of gonocoxae from beneath; proctiger either button-shaped and protruding (Fig. 11.23G) or suboval and deflected (Fig. 11.18K).

The large genus *Microvelia* Westwood (with about 170 species) is distributed world-wide. It is in need of generic and subgeneric revision. Andersen & Weir (2003b), revising the Australian species, erected three new subgenera and reinstated another subgenus, which are keyed out below.

Key to Australian subgenera

1. Large species, total length more than 2.0 mm (with a few exceptions). Grasping comb of male fore tibia more than 0.4x tibial length (Fig. 11.15D; gr). Connexiva of ♀ (apterous form) vertically raised and converging towards each other posteriorly (Fig. 11.15B, cn) 2
- Small species, total length less than 2.0 mm. Grasping comb of male fore tibia usually shorter. Connexiva of ♀ (apterous form) variable in shape and orientation....... 3
2. Anterior margin of male abdominal sternum 7 with a pair of barb-like patches of hairs (Figs 11.15K, 11.22D). Segment 8 of ♂ with two long and slender processes on ventral surface (Figs 11.22E, 11.24D, E) .... *Microvelia* (*Barbivelia*)
- Male abdominal sternum 7 and segment 8 not modified as above .... *Microvelia* (*Austromicrovelia*) [part]
3. Male with grasping comb only on fore tibia (Fig. 11.26B; gr). Male genitalia almost symmetrical, parameres subequal in size, small or large 4
- Male with short grasping comb on both fore and middle tibiae (Figs 11.28A, B; gr). Male genitalia asymmetrical, right paramere much larger than left paramere (Figs 11.28C) .... *Microvelia* (*Picaultia*)
4. Abdominal venter of ♂ modified, either with a pair of tooth-like processes on sternum 7 (Fig. 11.15J) or segment 8 with a low, transverse symmetrical process at its ventral, posterior margin (Fig. 11.22C). Grasping comb of male fore tibia more than 0.25x tibial length. Hind margin of pronotum dark .... *Microvelia* (*Austromicrovelia*) [part]
- Male abdominal venter not modified. Grasping comb of male fore tibia less than 0.25x tibial length. Hind margin of pronotum usually pale .... *Microvelia* (*Pacificovelia*)

Microvelia (*Austromicrovelia*)

Identification

Medium-sized to large *Microvelia* species. Antennae long and slender, about half as long as body (Fig. 11.15A); antennal segment 4 not distinctly thicker than segment 3. Fore tibia of male with a well-developed grasping comb usually occupying the distal half of tibia or more (Fig. 11.15D; gr). Fore wings of macropterous form dark, usually with two broad stripes in basal parts and 3-4 spots in distal part, whitish. Abdominal venter and genital segments of male usually mod-

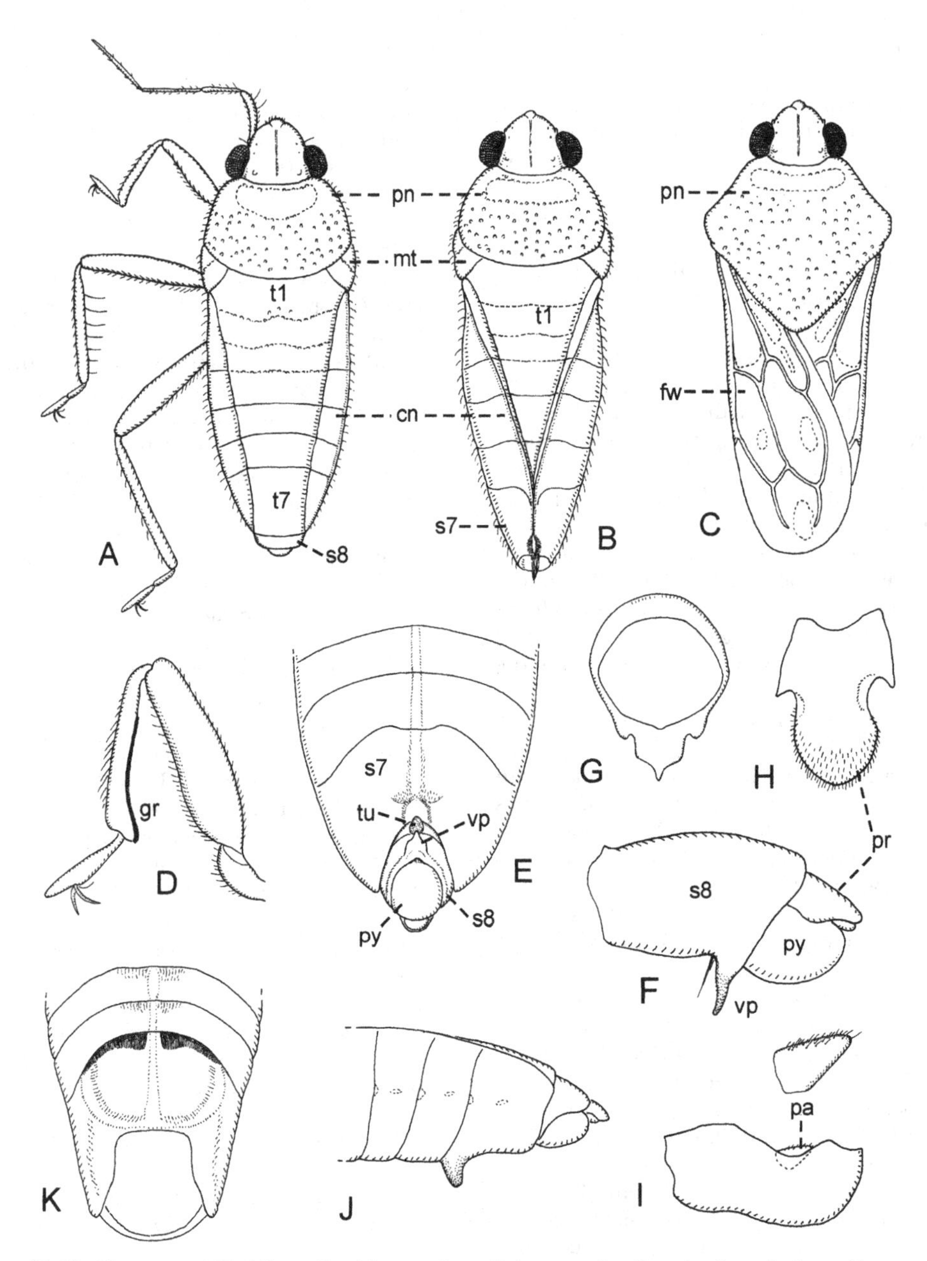

Figure 11.15. VELIIDAE. A-F, *Microvelia (Austromicrovelia) queenslandiae*. A, dorsal view of apterous ♂; antennae and legs of right side omitted (cn, connexivum; mt, metanotum; pn, pronotum; s8, abdominal segment 8; t1, t7, abdominal tergites 1 and 7). B, dorsal view of apterous ♀; antennae and legs omitted (s7, abdominal segment 7). C, dorsal view of macropterous ♂; antennae and legs omitted (fw, forewing; pn, pronotum). D, fore leg of apterous ♂ with grasping comb (gr). E, ventral view of abdominal end of apterous ♂ (py, pygophore; s7, sternum 7; s8, segment 8; tu, ventral tubercle of sternum 7; vp, ventral plate-like process on segment 8). F, lateral view of genital segments of apterous ♂ (pr, proctiger; py, pygophore; s8, segment 8; vp, ventral plate-like process of segment 8). G, *M. (A.) mjobergi*, apterous ♂: caudal view of segment 8. H-I, *M. (A.) childi*: apterous ♂: H, dorsal view of proctiger (pr). I, lateral view of pygophore of apterous ♂ with paramere (pa) shown at higher magnification. J, *M. (A.) odontogaster*, apterous ♂: lateral view of abdominal end. K, *M. (Barbivelia) falcifer*, apterous ♂: ventral view of abdominal end exclusive of genital segments. Reproduced with modification from Andersen & Weir (2003b).

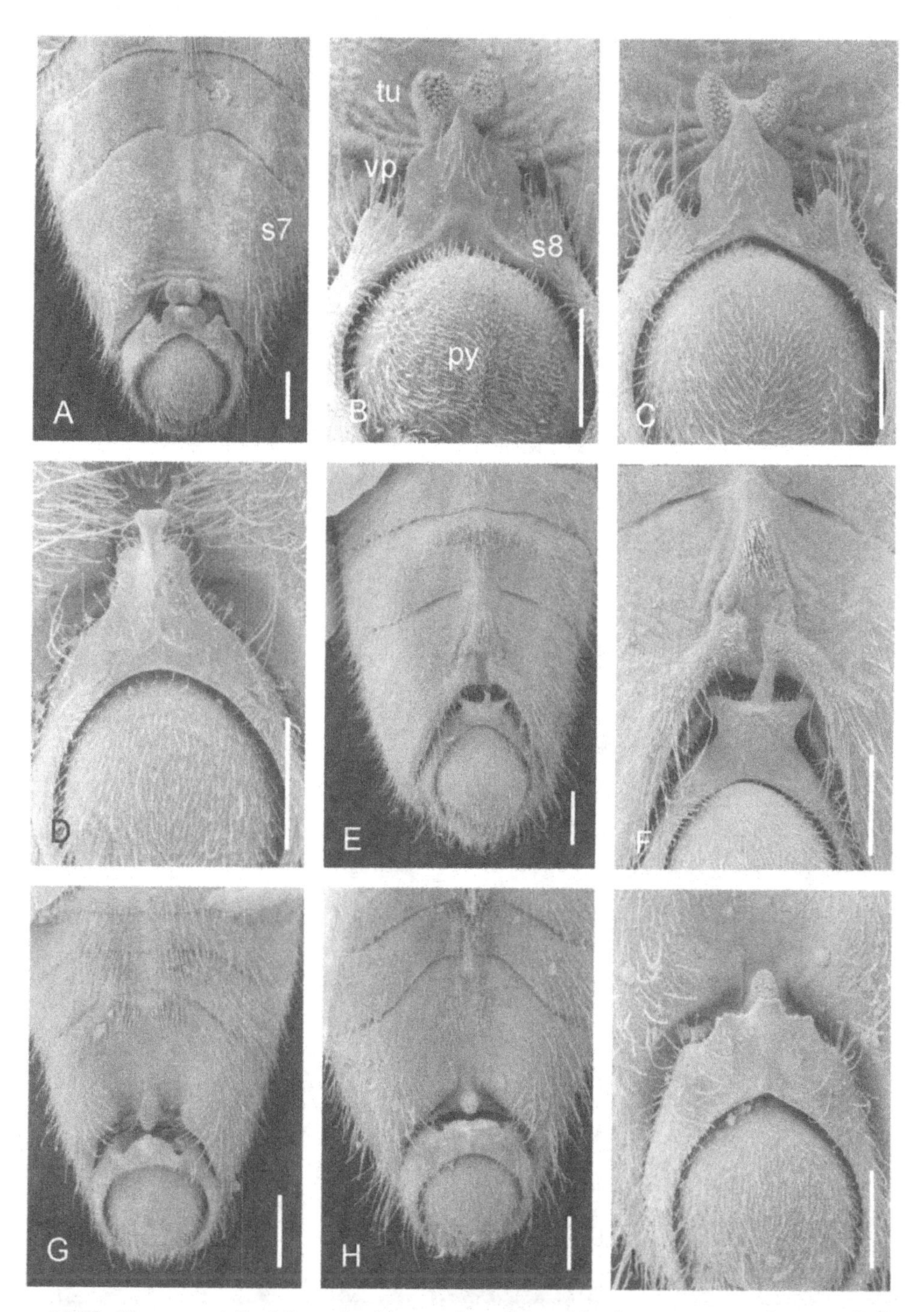

Figure 11.16. VELIIDAE. A-B, *Microvelia* (*Austromicrovelia*) *mjobergi*, apterous ♂: A, ventral view of abdomen (s7, sternum 7). B, ventral view of genital segments (py, pygophore; s8, segment 8; tu, ventral tubercle of sternum 7; vp, ventral plate-like process of segment 8). C, *M.* (*A.*) *spurgeon*: ventral view of genital segments. D, *M.* (*A.*) *carnavon*: ventral view of genital segments. E-F, *M.* (*A.*) *distincta*, apterous ♂: E, ventral view of abdominal end. F, ventral view of genital segments. G, *M.* (*A.*) *hypipamee*, apterous ♂: ventral view of abdominal end.. H-I, *M.* (*A.*) *margaretae*, apterous ♂: H, ventral view of abdominal end. I, ventral view of genital segments. All scales 0.1 mm. Scanning Electron Micrographs. Reproduced with modification from Andersen & Weir (2003b).

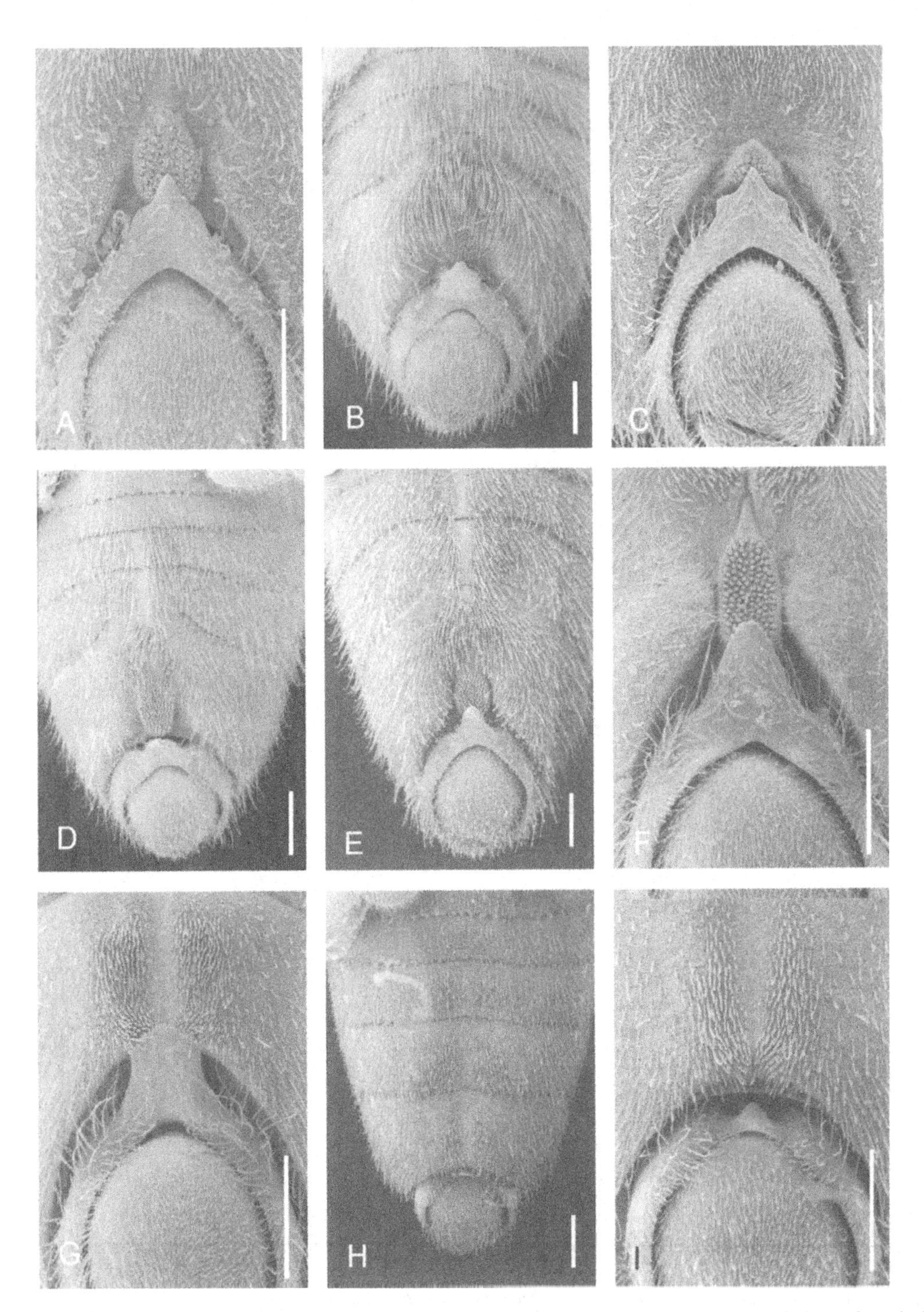

Figure 11.17. VELIIDAE. A, *Microvelia* (*Austromicrovelia*) *queenslandiae*, apterous ♂: ventral view of genital segments. B, *M.* (*A.*) *woodwardi*, apterous ♂: ventral view of abdominal end. C, *M.* (*A.*) *myorensis*: ventral view of genital segments. D, *M.* (*A.*) *tuberculata*, ventral view of abdominal end. E, *M.* (*A.*) *monteithi*: ventral view of abdominal end. F, *M.* (*A.*) *childi*: ventral view of genital segments. G, *M.* (*A.*) *annemarieae*: ventral view of genital segments. H-I, *M.* (*A.*) *mossmann*, apterous ♂: H, ventral view of abdominal end. I, ventral view of genital segments. All scales 0.1 mm. Scanning Electron Micrographs. Reproduced with modification from Andersen & Weir (2003b).

ified; segment 8 with one or two plate-like processes arising from its ventral, posterior margin (Figs 11.15E-G; s8); parameres small, symmetrical and cone-shaped; proctiger widened basally, with a notch on each side forming lateral, usually pointed processes (Figs 11.15H, 11.19G, 11.23D). Female connexiva usually vertically raised and converging posteriorly (Fig. 11.15B; cn).

Key to Australian species

1. Sternum 7 of ♂ with a pair of stout, tooth-like processes (Figs 11.15J). Chestnut brown or dark brown above with conspicuous patches of silvery pubescence on anterior pronotum and abdominal terga (Fig. 11.23E). Small species, total length 1.7-2.0 mm. Northern Territory...... *M. odontogaster* Andersen & Weir
- Sternum 7 of ♂ not modified as above. Colouration not as above. Usually larger species, total length more than 2.0 mm 2
2. Segment 8 of ♂ with one large process at ventral, posterior margin (Figs 11.15E, G) 3
- Segment 8 of ♂ otherwise modified 18
3. Entire anterior part of pronotum pale (Fig. 11.14). Segment 8 of ♂ with a large, plate-like process at ventral, posterior margin (Figs 11.15E; vp, 11.16B). Connexiva of ♀ (apterous form) converging posteriorly and usually meeting each other above abdominal tergum (Fig. 11.15B; cn). Fore wings of macropterous form dark with broad, whitish stripes in basal parts and 3-4 pale spots in distal parts 4
- Anterior part of pronotum with a pair of pale spots (Figs 11.19A, F). Segment 8 of ♂ with one large, triangular or spatulate, process at ventral, posterior margin (Figs 11.21A, D). Connexiva of ♀ (apterous form) converging posteriorly, at most meeting each other at the end of abdomen (Figs 11.19B, C). Fore wings of macropterous form dark with broad, whitish stripes in basal parts; distal parts without pale spots 17
4. Large or medium-sized species, total length 2.2-3.4 mm. Median process in posterior part of male sternum 7 prominent, distally widened into a pair of suboval pads (Figs 11.16B, C); hind margin of segment 8 with a transverse, plate-like process which is broad basally and abruptly narrowing to a pointed apex 5
- Smaller species, total length less than 2.8 mm. Median process in posterior part of male sternum 7 not as above 6
5. Ventral process of male segment 8 as wide as long basally (Figs 11.15G, 11.16B); distal part distinctly bent in lateral view. Length 3.1-3.4 mm. Queensland....................................... *M. mjobergi* Hale
- Ventral process of male segment 8 not as wide as long basally (Fig. 11.16C); distal part almost straight in lateral view. Length 2.9-3.2 mm. Queensland .... *M. spurgeon* Andersen & Weir
6. Ventral process of male segment 8 broad at base, abruptly narrowing to slender and long distal part (Figs 11.16D-I). 7
- Ventral process of male segment 8 not as above 10
7. Ventral process of male segment 8 broad at base, abruptly narrowing to a slender and long apical part which is widened distally (Fig. 11.16D). [Connexiva of ♀ (apterous form) converging posteriorly, meeting each other above terga 6-7; connexival margins rarely with tufts of black bristles; posterior corners of laterotergites 7 distinctly produced, with tufts of long, black bristles]. Length 2.3-2.9 mm. Queensland..... *M. carnavon* Andersen & Weir
- Ventral process of male segment 8 not as above 8
8. Posterior part of male sternum 7 with a median ridge carrying a pair of hairy pads (Figs 11.16E, F). Ventral process of male segment 8 transverse plate-like at base, with a slender and pointed distal part (Fig. 11.16F). Length 2.3-3.0 mm. Victoria .... *M. distincta* Malipatil
- Posterior part of male sternum 7 with a median process, distally widened into a suboval pad furnished with minute denticles (Figs 11.16G, H). Ventral process of male segment 8 broad at base, with a median, blunt distal part (Fig. 11.16I).......... 9
9. Sterna 4 and 5 of ♂ each with a pair of dark hair patches in middle (Fig. 11.16G). Connexiva of ♀ converging posteriorly, but barely meeting each other above tergum 7. Small species, length 2.1-2.4 mm. Queensland.......................... .... *M. hypipamee* Andersen & Weir
- Sterna 4 and 5 of ♂ each with a with a pile of long, brownish hairs in middle (Fig. 11.16H). Connexiva of ♀ converging posteriorly, meeting each other above terga 6-7 (Figs 11.18A, B). Large species, length 2.7 mm. New South Wales, Queensland, Victoria .... *M. margaretae* Andersen & Weir

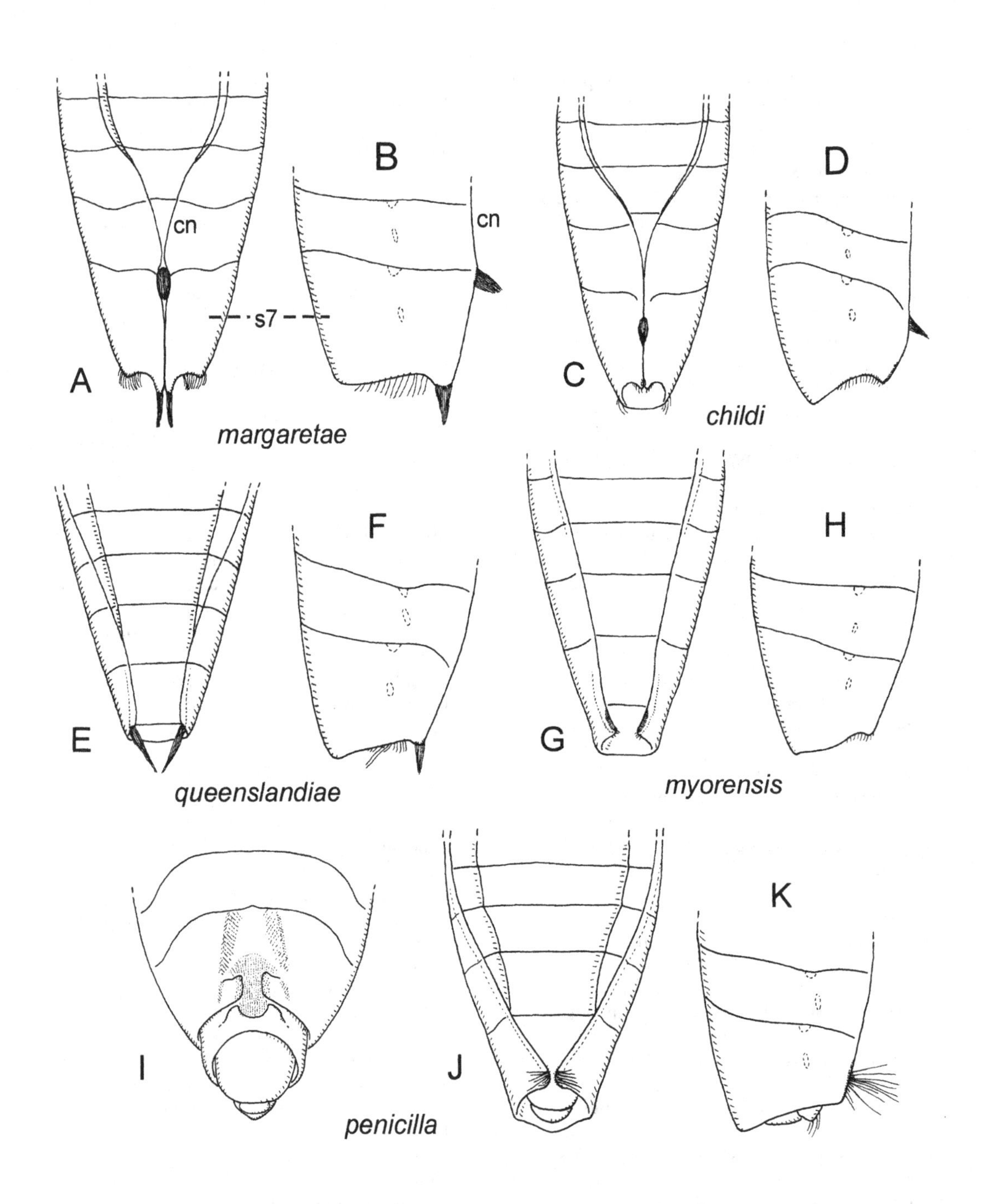

Figure 11.18. VELIIDAE. A-B, *Microvelia* (*Austromicrovelia*) *margaretae*, apterous ♀: A, dorsal view of abdomen. B, lateral view of abdominal end (cn, connexivum). C-D, *M.* (*A.*) *childi*, apterous ♀: C, dorsal view of abdomen. D, lateral view of abdominal end. E-F, *M.* (*A.*) *queenslandiae*, apterous ♀: E, dorsal view of abdomen. F, lateral view of abdominal end. G-H, *M.* (*A.*) *myorensis*, apterous ♀: H, dorsal view of abdomen. H, lateral view of abdominal end. I-K, *M.* (*A.*) *penicilla*: I, ventral view of abdomen of apterous ♂. J, dorsal view of abdomen of apterous ♀. K, lateral view of abdominal end of apterous ♀. Reproduced with modification from Andersen & Weir (2003b).

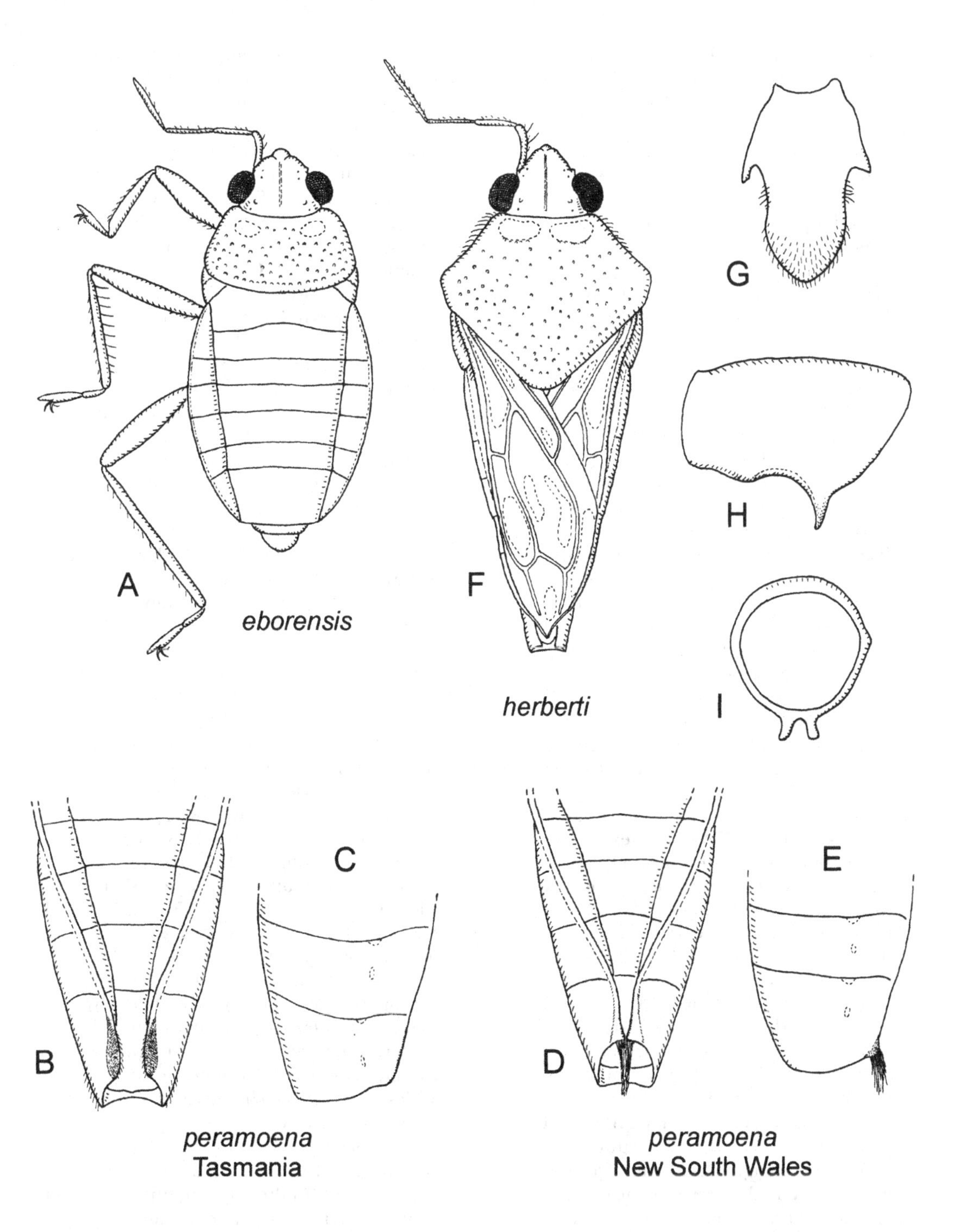

Figure 11.19. VELIIDAE. A, *Microvelia* (*Austromicrovelia*) *eborensis*, apterous ♂: dorsal view; antenna and legs of right side omitted. B-C, *M.* (*A.*) *peramoena*, apterous ♀ from Tasmania: B, dorsal view of abdomen. C, lateral view of abdominal end. D-E, *M.* (*A.*) *peramoena*, apterous ♀ from New South Wales: D, dorsal view of abdomen. E, lateral view of abdominal end. F-I, *M.* (*A.*) *herberti*: F, dorsal view of macropterous ♀; right antenna and all legs omitted. G, dorsal view of proctiger of apterous ♂. H, lateral view of segment 8 of apterous ♂. I, caudal view of segment 8 of apterous ♂. Reproduced with modification from Andersen & Weir (2003b).

10. Posterior part of male sternum 7 with a low ovate, median tubercle carrying minute denticles (Figs 11.17A-F). Connexiva of ♀ converging posteriorly, meeting above tergum 7; connexival margins and posterior corners with or without hair tufts 11
- Posterior part of male sternum 7 without a median tubercle (Figs 11.17G-I). Connexiva of ♀ converging posteriorly, but never meeting above tergum 7 (as in Fig. 11.18G); connexival margins and posterior corners without hair tufts 16
11. Ventral process of male segment 8 with square base and triangular apex (Figs 11.17A-C) 12
- Ventral process of male segment 8 triangular (Figs 11.17D-F) 14
12. Sterna 5-7 of ♂ each with a transverse pile of long, blackish hairs (Figs 11.17B); median tubercle of sternum 7 ovate. [Connexiva of ♀ converging posteriorly, meeting above tergum 7; connexival margins usually with a suberect tuft of black bristles at the intersegmental limit between laterotergites 6 and 7 (as in Figs 11.18A, B); posterior corners of connexiva without long, black bristles]. Length 2.6-3.0 mm. Queensland.......................... .... *M. woodwardi* Andersen & Weir
- Sterna 5-7 of ♂ without piles of long, blackish hairs 13
13. Median tubercle of male sternum 7 drop-like (Fig. 11.17A). Connexiva of ♀ converging posteriorly, meeting above tergum 7; connexival margins usually with a suberect tuft of black bristles in middle of laterotergite 7; posterior corners of connexiva usually with tufts of long, black bristles (as in Figs 11.18E, F). Length 2.2-2.8 mm. Queensland .... *M. queenslandiae* Andersen & Weir
- Median tubercle of male sternum 7 suboval (Fig. 11.17C). Connexiva of ♀ converging posteriorly, but never meeting above tergum 7; connexival margins and posterior corners without hair tufts (Figs 11.18G, H). Length 2.3-2.7 mm. Queensland................. *M. myorensis* Andersen & Weir
14. Median tubercle of male sternum 7 very large, occupying more than half of the length of the segment (Fig. 11.17D). Length 2.3-2.7 mm. Queensland .... *M. tuberculata* Andersen & Weir
- Median tubercle of male sternum 7 smaller, less than half length of segment ... 15
15. Sternum 7 of ♂ with a patch of brownish hairs at anterior margin (Fig. 11.17E). Connexiva of ♀ converging posteriorly, meeting above tergum 7; connexival margins usually with a suberect tuft of black bristles at the intersegmental limit between laterotergites 6 and 7 (as in Figs 11.18A, B). Length 2.5-2.8 mm. Queensland .... *M. monteithi* Andersen & Weir
- Sternum 7 of ♂ without a patch of hairs at anterior margin. Connexiva of ♀ converging posteriorly, meeting above tergum 7; connexival margins usually with a suberect tuft of black bristles in middle of laterotergite 7 (Figs 11.18C, D). Length 2.4-2.8 mm. New South Wales..... .... *M. childi* Andersen
16. Ventral process of male segment 8 large, rectangular (Figs 11.17G); sternum 7 with a pair of hair patches in middle, but without tubercle at posterior margin. Length 2.1-2.4 mm. Queensland .... *M. annemarieae* Andersen & Weir
- Ventral process of male segment 8 small, triangular (Figs 11.17H); sternum 7 with a small group of minute, dark denticles at posterior margin. Length 2.2-2.4 mm. (Queensland)... *M. mossman* Andersen & Weir
17. Segment 8 of ♂ with a triangular, dark process ventrally (Fig. 11.21A). Connexiva of ♀ almost vertically raised throughout, converging but not meeting each other posteriorly; margins of laterotergites 7 with elongate shiny pads; posterior margin of female tergum 8 more or less produced (Fig. 11.21B). Length 2.1-2.6 mm. New South Wales, Queensland, Victoria .... *M. fluvialis* Malipatil
- Segment 8 of ♂ with a spatulate, slightly asymmetrical, pale process ventrally (Fig. 11.21D). Connexiva of ♀ obliquely raised throughout, not converging posteriorly; margins of laterotergites 7 not modified; posterior margin of female tergum 8 straight. Length 2.1-2.8 mm. New South Wales .... *M. eborensis* Andersen & Weir
18. Segment 8 of ♂ with a pair of processes or one bifid process at its ventral, posterior margin.. 19
- Segment 8 of ♂ otherwise modified 24
19. Entire anterior part of pronotum pale. Sternum 7 of ♂ with two widely separated tubercles at its posterior margin, each carrying numerous denticles (Figs 11.18I, 11.21E); segment 8 with two widely separated, slender processes at ventral, posterior margin. Small species, length 1.8-2.3 mm....................................... 20

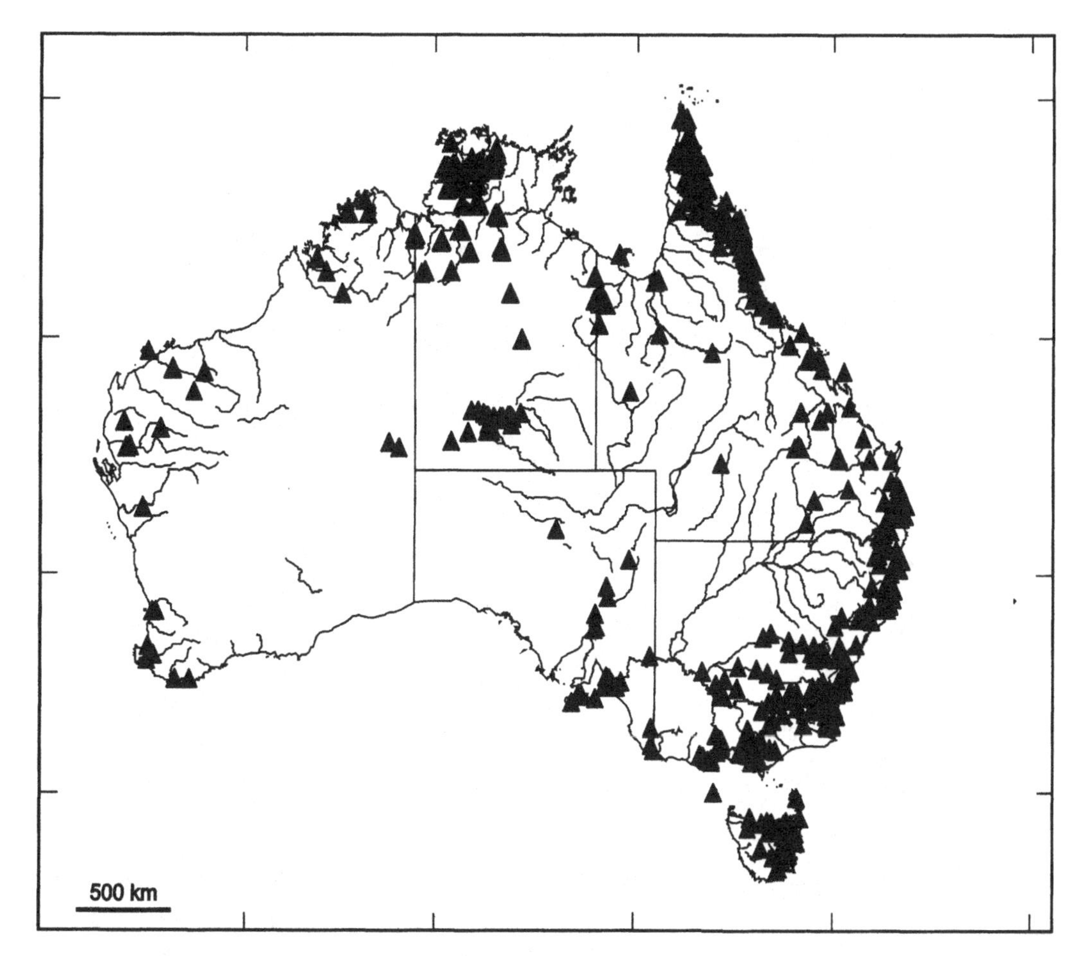

Figure 11.20. VELIIDAE. Distribution of *Microvelia* (*Austromicrovelia*) species in Australia.

- Anterior part of pronotum with a pair of pale spots or a transverse pale band. Sternum 7 and segment of ♂ 8 not modified as above. Large species, length 2.1-2.8 mm 21

20. Sternum 7 of ♂ deeply depressed posteriorly (Fig. 11.18I). Posterior corners of female laterotergites 7 with tufts of long, black hairs (Figs 11.18J, K). Length 1.8-2.1 mm. Northern Territory, Western Australia............ *M. penicilla* Andersen & Weir

- Sternum 7 of ♂ not deeply depressed posteriorly (Fig. 11.21E). Posterior corners of female laterotergites 7 without hair tufts. Length 2.0-2.3 mm. Northern Territory .... *M. angelesi* Andersen & Weir

21. Segment 8 of ♂ with a pair of distinctly separated processes at its ventral, posterior margin (Figs 11.21F, G) 22

- Segment 8 of ♂ with a bifid process at its ventral, posterior margin (Figs 11.21H, I).. 23

22. Pronotum with a pair of distinctly separated, pale spots in anterior part. Ventral processes of the male segment 8 triangular (Fig. 11.21F). Margins of female laterotergites 7 with elongate, shiny pads (Fig. 11.19B) or carrying tufts of hairs (Fig. 11.19D, E). Length 2.1-2.7 mm. Australian Capital Territory, New South Wales, Northern Territory, Queensland, South Australia, Tasmania, Victoria, Western Australia............... *M. peramoena* Hale

- Pronotum with a transverse pale band in anterior part (rarely dissolved into two separate spots). Ventral processes of the male segment 8 slender, slightly curved (Figs 11.19H, I, 11.21G). Margins of female laterotergites 7 not modified ex-

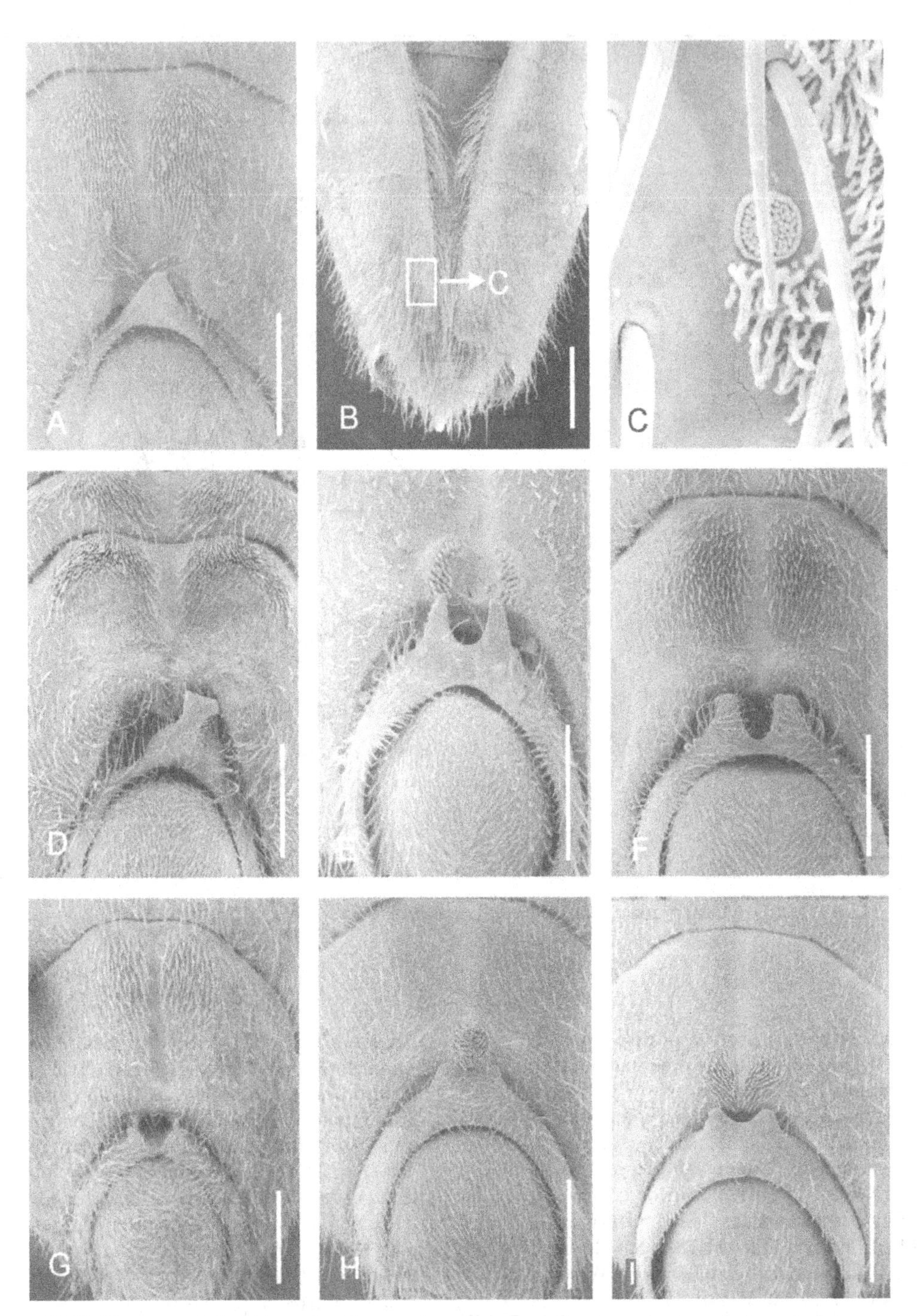

Figure 11.21. VELIIDAE. A-C, *Microvelia* (*Austromicrovelia*) *fluvialis*: A, ventral view of genital segments of apterous ♂. B, dorsal view of abdominal end of apterous ♀. C, surface structure of connexival margin of apterous ♀ (marked area in B). D, *M.* (*A.*) *eborensis*, apterous ♂: ventral view of genital segments. E, *M.* (*A.*) *angelesi*, apterous ♂: ventral view of genital segments. F, *M.* (*A.*) *peramoena*, apterous ♂: ventral view of genital segments. G, *M.* (*A.*) *herberti*, apterous ♂: ventral view of genital segments. H, *M.* (*A.*) *malipatili*, apterous ♂: ventral view of genital segments. I, *M.* (*A.*) *alisonae*, apterous ♂: ventral view of genital segments. Scales: A, B, D-I, 0.1 mm; C, 0.05 mm. Scanning Electron Micrographs. Reproduced with modification from Andersen & Weir (2003b).

cept for occasional tufts of hairs. Length 2.3-2.8. Northern Territory, Queensland, Western Australia ....
.... *M. herberti* Andersen & Weir

23. Sternum 7 of ♂ with a small, ovate group of minute, dark hairs (Fig. 11.21H); segment 8 with a narrowly bifid process at its ventral, posterior margin. Length 2.2-2.5 mm. Northern Territory ....
.... *M. malipatili* Andersen & Weir
- Sternum 7 of ♂ with two small groups of minute, dark hairs (Fig. 11.21I); segment 8 with a broadly bifid process at its ventral, posterior margin. Length 2.1-2.7 mm. Northern Territory ....
.... *M. alisonae* Andersen & Weir

24. Sternum 6 of ♂ with a median process at its posterior margin (Figs 11.23F, H, I). [Connexiva of ♀ obliquely raised, not converging posteriorly] 25
- Sternum 6 of ♂ without a median process 26

25. Median process of male sternum 6 spine-like (Fig. 11.23F). Proctiger of ♀ slightly deflected (Fig. 11.23G). Fore wings dark with two broad stripes in basal part and one small spot in distal part, whitish. Length 2.0-2.4 mm. New South Wales, Queensland ....
.... *M. ventrospinosa* Andersen & Weir
- Median process of male sternum 6 tuberculate (Figs 11.23H, I). Proctiger of ♀ almost horizontally produced. Length 2.25-2.7 mm. New South Wales...............
..............................*M. milleri* Andersen & Weir

26. Posterior margin of pronotal lobe of macropterous form yellowish brown; fore wings uniformly dark without pale spots or stripes. Process at ventral, posterior margin of male segment 8 broad, abruptly narrowing to a pointed distal part (Fig. 11.22B). Length 2.0-2.5 mm. Northern Territory, Queensland ....
........................*M. apunctata* Andersen & Weir
- Posterior margin of pronotal lobe of macropterous form dark; fore wings dark with pale stripes and spots. Process at ventral, posterior margin of male segment 8 not as above 27

27. Segment 8 of ♂ with a small asymmetrical cone-shaped process at its ventral, posterior margin (Figs 11.22A, 11.23B, C). Length 2.3-2.8 mm. Northern Territory, Queensland, Western Australia ....
.... *M. torresiana* Andersen & Weir
- Segment 8 of ♂ with a low, transverse symmetrical process at its ventral, posterior margin (Fig. 11.22C). Length 1.9-2.3 mm. Northern Territory, Queensland, Western Australia ....
.... *M. australiensis* Andersen & Weir

Biology

Hale (1926) and Malipatil (1980) gave notes on the habitats and feeding biology of the common Australian species, *Microvelia peramoena.* It lives in a variety of habitats, such as roadside pools, farm dams, backwater pools of streams, edges of lakes and stagnant pools in stream beds. This species seems to be able to colonise the most temporary of habitats. The closely related *M. herberti* lives along the edges of flowing rivers and streams and in stream pools with or without rocks, but is only occasionally found in temporary pools. Other species, in particular those belonging to the *M. mjobergi* group, are most frequently found in habitats along the edges of slow flowing rivers, streams, and in stream pools, from sea level up to 1000 m (Andersen & Weir 2003b). Adults of *Microvelia* are found on the water throughout the year, although the relative abundance varies. In the southeast, nymphs are chiefly found in February-March and October-December. In the tropical north, breeding seems to be more continuous and nymphs are observed throughout most of the year. Most species are predominantly apterous (wingless), but macropterous (winged) specimens are not rare in some species (for example *M. peramoena*) and may even be the dominant form in others (for example *M. australiensis*). Such species are often collected at light or in flight intercept traps. Whereas most *Microvelia* species are semiaquatic, *M. ventrospinosa* may be hygropetric, living in moist leaf litter and debris rather than on the water surface.

Distribution

Most Australian species of *Microvelia* belong to the subgenus *Austromicrovelia* (Andersen & Weir 2003b) which also includes one species from Bali and probably many undescribed *Microvelia* species from New Guinea (Andersen & Weir, unpublished; D. Polhemus, unpublished). Most Australian species can be placed in one of four species groups: (1) *M. (Austromicrovelia) mjobergi* Hale group with 14 species distributed along the entire East coast of Australia; (2) *M. (Austromicrovelia) angelesi* Andersen & Weir group, with two species sparsely distributed along the coast of Northern Territory and northwestern Australia; (3) *M. (Austromicrovelia) peramoena* Hale group with 4 species, including the most widely distributed species known from all Australian states and from the interior (Fig.

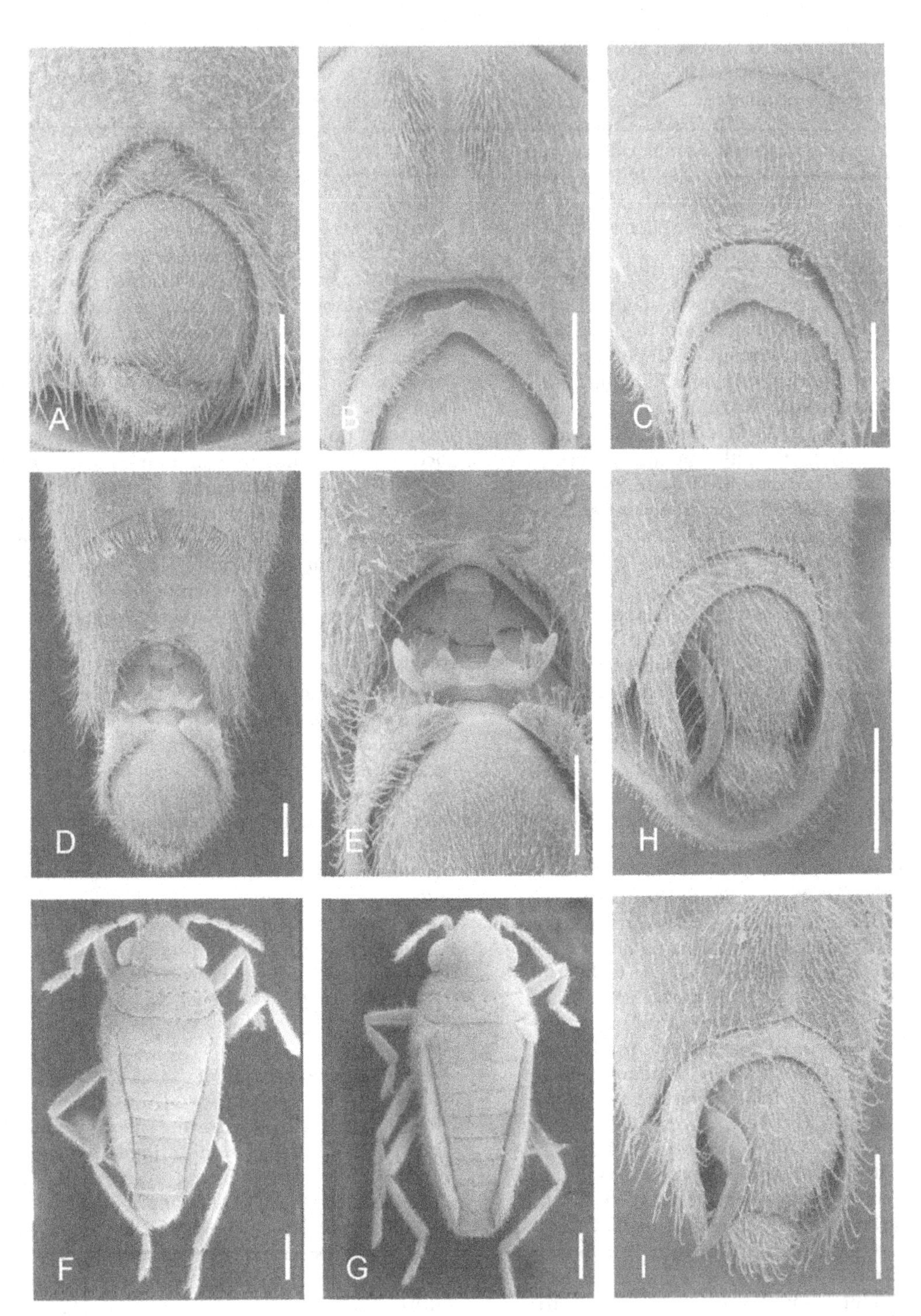

Figure 11.22. VELIIDAE. A, *Microvelia* (*Austromicrovelia*) *torresiana*, apterous ♂: ventral view of genital segments. B, *M.* (*A.*) *apunctata*, apterous ♂: ventral view of genital segments. C, *M.* (*A.*) *australiensis*, macropterous ♂: ventral view of genital segments. D-E, *M.* (*Barbivelia*) *barbifer*, apterous ♂: D, ventral view of abdominal end. E, ventral view of genital segments. F-G, *M.* (*Pacificovelia*) *lilliput*: F, apterous ♂. G, apterous ♀. H, *M.* (*Picaultia*) *justi*, macropterous ♂: ventral view of genital segments. I, *M.* (*P.*) *paramega*, apterous ♂: ventral view of genital segments. Scales: A-E, H, I, 0.1 mm; F, G, 0.2 mm. Scanning Electron Micrographs. Reproduced with modification from Andersen & Weir (2003b).

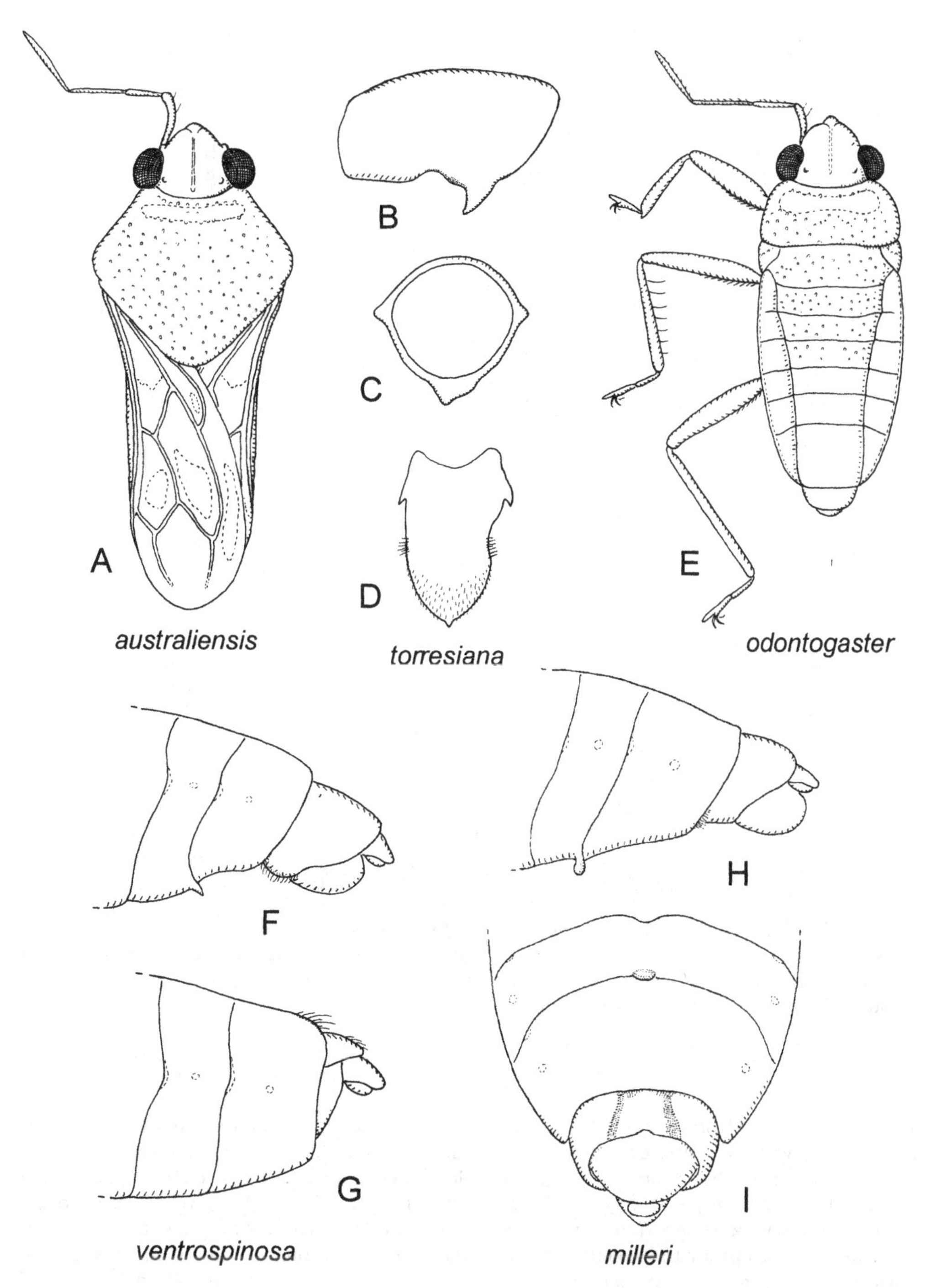

Figure 11.23. VELIIDAE. A, *Microvelia* (*Austromicrovelia*) *australiensis*, macropterous ♂: dorsal view; right antenna and all legs omitted. B-D, *M.* (*A.*) *torresiana*, apterous ♂: B, lateral view of abdominal segment 8. C, caudal view of segment 8. D, dorsal view of proctiger. E, *M.* (*A.*) *odontogaster*, apterous ♂: dorsal view; antenna and legs of right side omitted. F-G, *M.* (*A.*) *ventrospinosa*: F, lateral view of abdominal end of apterous ♂. G, lateral view of abdominal end of apterous ♀. H-I, *M.* (*A.*) *milleri*, apterous ♂: H, lateral view of abdominal end. I, ventral view of abdominal end. Reproduced with modification from Andersen & Weir (2003b).

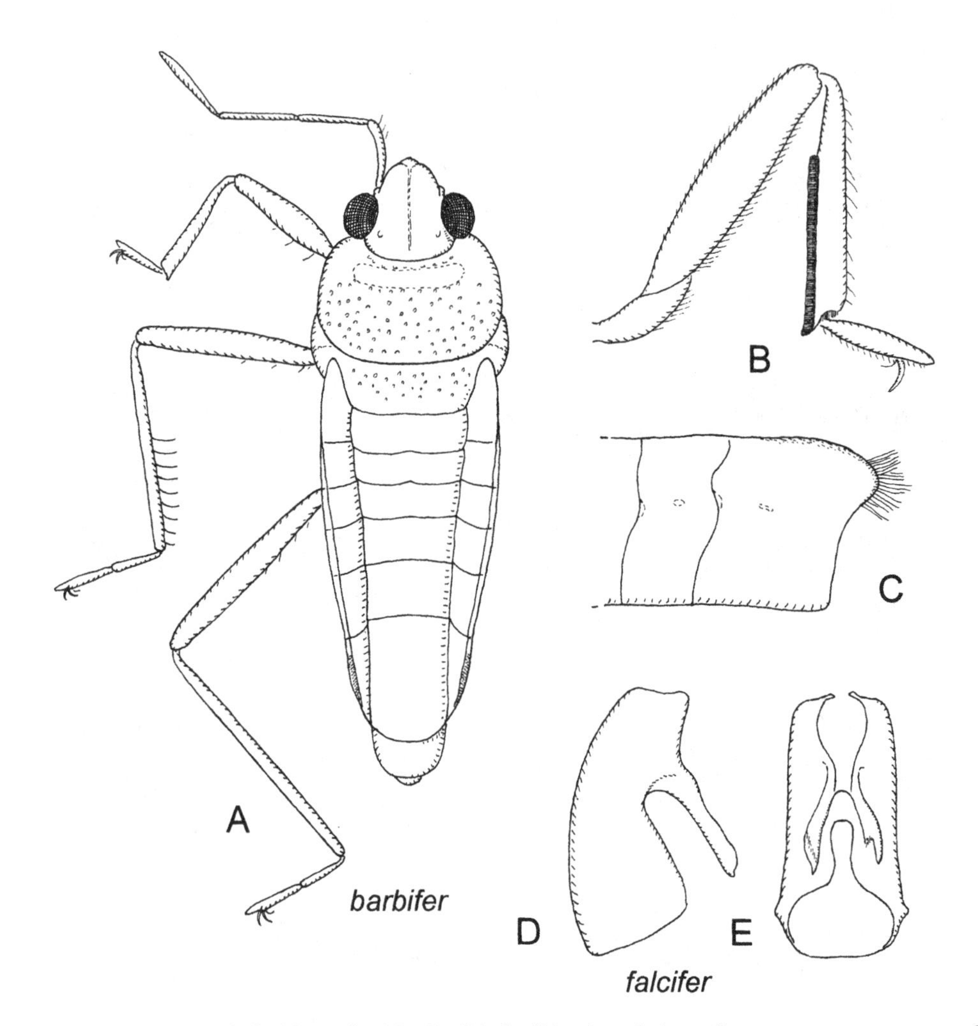

Figure 11.24. VELIIDAE. A-C, *Microvelia* (*Barbivelia*) *barbifer*: dorsal view of apterous ♂; antenna and legs of right side omitted. B, fore leg of apterous ♂ with grasping comb. C, lateral view of abdominal end of apterous ♀. D-E, *M.* (*B.*) *falcifer*, apterous ♂: D, ventral view of segment 8. E, lateral view of segment 8. Reproduced with modification from Andersen & Weir (2003b).

11.20); and (4) *M.* (*Austromicrovelia*) *fluvialis* Malipatil group, with two species confined to southeastern Australia. The remaining 6 species are found in the northern parts of Queensland, Northern Territory, and northern part of Western Australia, except for the species-pair *M.* (*Austromicrovelia*) *ventrospinosa* and *milleri*, which are confined to New South Wales.

MICROVELIA (BARBIVELIA)

Identification

Medium-sized *Microvelia* species (Fig. 11.24A). Antennae long and slender, almost half as long as body; antennal segment 4 not distinctly thicker than segment 3. Fore tibia of male with a well-developed grasping comb, length almost 3/4 of tibia (Fig. 11.24B; gr). Fore wings of macropterous form dark brown with paler stripes in intervenal areas. Abdominal venter and genital segments of male strongly modified; sternum 7 broadly depressed in middle with a pair of transverse, barb-like patches of dark hairs along anterior margin (Figs 11.15K, 11.22D); segment 8 with deep, median incision from behind and a pair of very long processes on ventral surface (Figs 11.22E, 11.24D, E). Connexiva of apterous female almost vertically raised anteriorly, con-

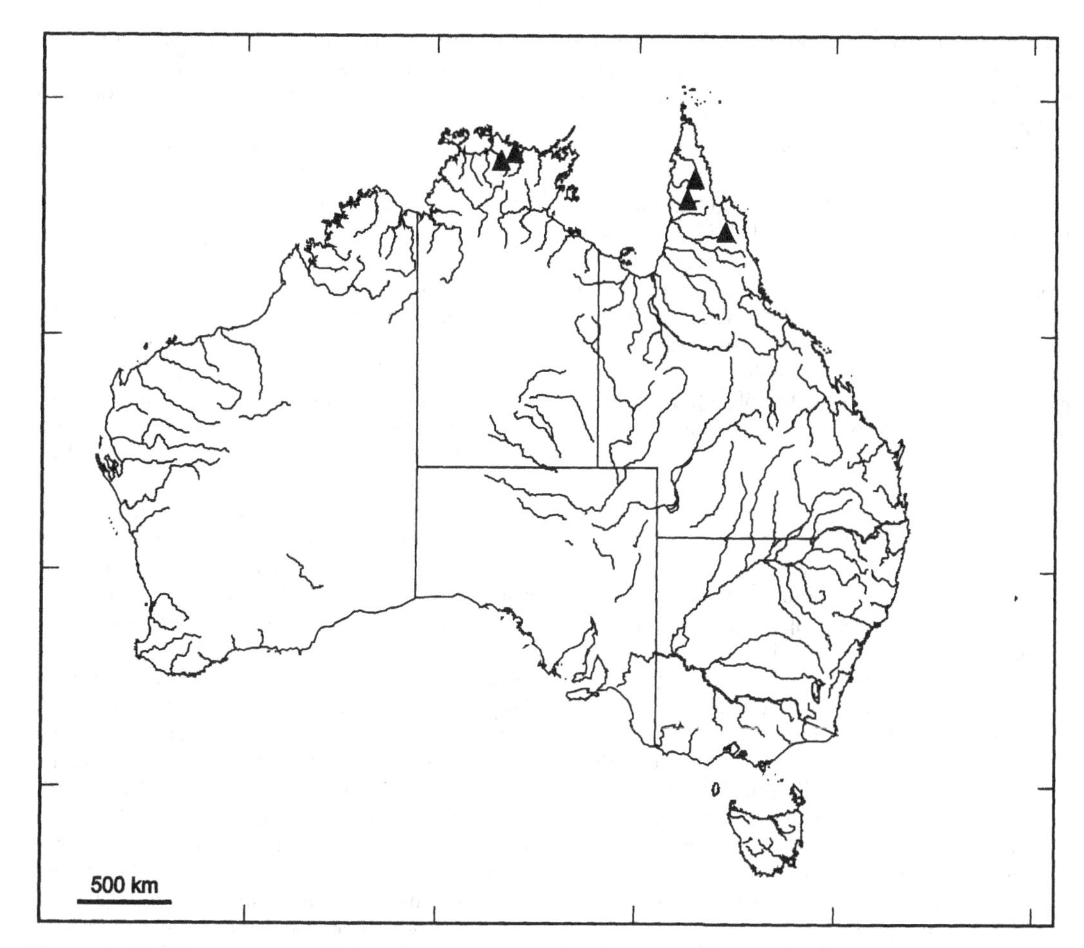

Figure 11.25. VELIIDAE. Distribution of *Microvelia* (*Barbivelia*) species in Australia.

verging posteriorly and meeting each other above terga 6 and 7; posterior corners of laterotergites 7 produced (Fig. 11.24C).

Key to Australian species

1. Male segment 8 a pair of large, symmetrical, antler-like processes on ventral surface (Figs 11.22D, E). Posterior corners of female laterotergites 7 broadly produced, furnished with a tuft of long, dark bristles (Fig. 11.24C). Length 2.4-2.8 mm Queensland......... *M. barbifer* Andersen & Weir
– Male segment 8 with a pair of asymmetrical, sword-like processes on ventral surface (Figs 11.24D, E). Posterior corners of female laterotergites 7 slightly produced, without hair tufts. Length 2.7-2.8 mm Northern Territory............................ .... *M. falcifer* Andersen & Weir

Biology

Almost nothing is known about the biology of these rare *Microvelia* species. Both *M. barbifer* and *falcifer* were found in deep, permanent billabongs with vegetation. Adults were recorded from May-July and in September (Andersen & Weir 2003b).

Distribution

The two species of *Microvelia* subgenus *Barbivelia* are only know from a few localities in northern Queensland and Northern Territory (Fig. 11.25).

MICROVELIA (PACIFICOVELIA)

Identification

Very small to medium-sized *Microvelia* species (Figs 11.22F, G, 11.26A, G). Antennae slender, 0.4-0.55x total length of insect; antennal segment

4 distinctly thicker than segment 3. Male fore tibial comb less than 1/4 tibial length (Fig. 11.26B); grasping comb not present on middle tibia. Fore wings of macropterous form usually dark with broad stripes in basal parts and 3-4 spots or stripes in distal parts. Male abdominal venter simple; genital segments relatively large, almost symmetrical; parameres large (Fig. 11.26C; pa), slightly asymmetrically developed (right paramere largest); proctiger with small, lateral processes (Figs 11.26E, J, L). Female connexiva vertically raised, usually converging and sometimes meeting above abdominal end, with or without marginal tufts of hairs (Fig. 11.26F, H, I); female proctiger deflected to cover gonocoxae.

Key to Australian species

1. Smaller species, total length 1.0-1.5 mm...... 2
- Larger species, total length more than 1.5 mm 3
2. Antenna about 0.4x total length. Connexiva of ♀ with pale spots, converging posteriorly, but not meeting each other at abdominal end (Fig. 11.22G); tergum 8 without hair tuft. Length 1.0-1.4 mm. Northern Territory, Queensland, Western Australia .... *lilliput* Andersen & Weir
- Antenna 0.5-0.6x total length. Connexiva of ♀ usually dark, converging and sometimes meeting above abdominal end; tergum 8 with a tuft of long hairs. Length 1.2-1.5 mm. Northern Territory, Queensland, Western Australia .... *kakadu* Andersen & Weir
3. Grasping comb of fore tibia of ♂ about 0.2x tibial length (Fig. 11.26B, gr). Length 1.5-2.2 mm (apterous), 2.1-2.5 mm (macropterous). Australian Capital Territory, Norfolk Island, New South Wales incl. Lord Howe Island, Northern Territory, Queensland, South Australia, Tasmania, Victoria, Western Australia .... *oceanica* Distant
- Grasping comb of fore tibia of ♂ less than 0.2x tibial length... 4
4. Grasping comb of fore tibia of ♂ about 0.1x tibial length. Female connexiva strongly converging posteriorly, meeting each other above terga 6 and 7 (Fig. 11.26H); connexival margins with hair tufts at the intersegmental limits between segment 6 and 7 (Fig. 11.26I); posterior corners of laterotergites 7 rounded, with tufts of long bristles. Length 1.9-2.2 mm. Norfolk Island .... *macgregori* (Kirkaldy)
- Grasping comb of fore tibia of ♂ extremely short, only about 0.03x tibial length. Female connexiva slightly converging posteriorly, not meeting each other above terga 6 and 7; connexival margins and posterior corners of connexiva without prominent hair tufts. bristles. Length 1.7-2.35 mm. Tasmania......... .... *tasmaniensis* Andersen & Weir

Biology

Hale (1926) gave notes on the habitat, food, and copulatory behaviour of *Microvelia oceanica* (as *M. australica*) in South Australia and Malipatil (1980) recorded its habitats in Gippsland, Victoria. Apterous and macropterous adults were found in all seasons, normally in still or slow-flowing backwaters and along banks partly shaded by overhanging vegetation, mainly grass and sedge, in billabongs and swamps, and in permanent or semi-permanent farm ponds with a variety of emergent vegetation. Habitats recorded by Andersen & Weir (2003b) vary from the most temporary of roadside pools and pools in dry river beds, to edges of canals, rivers, and lakes. Adults were recorded from all months, but most frequently from the periods March-May and October-December (only few records from June-August in southeastern and southern Australian), nymphs from March and November (New South Wales, South Australia, Tasmania, and Victoria) and March and May-July (Northern Territory, Queensland, and Western Australia). Macropterous specimens are often taken at light and in intercept traps. The life history of *Microvelia macgregori* was studied by Don (1967) in New Zealand. He reared 225 apterous adults from nymphs that were exposed to various temperatures, and in each case there were four instars. All other species of *Microvelia* which have been reared in the laboratory go through five nymphal instars as is the usual number for semiaquatic bugs (Andersen 1982). Jackson & Walls (1998) investigated the behaviour of prey location in the same species. Experiments indicate that vibratory cues from both the struggling prey and from conspecifics are important. Optical cues are also used, but there is no evidence that chemical cues have a role.

Distribution

Microvelia subgenus *Pacificovelia* includes *M. oceanica*, the most common and widespread Australian *Microvelia* species found in all states of mainland Australia, in Tasmania, and in Norfolk Island (Fig. 11.27). It is furthermore recorded from New Caledonia, Loyalty Islands, Vanuatu (New Hebrides), and Fiji. The four other Australian species have more restricted distribu-

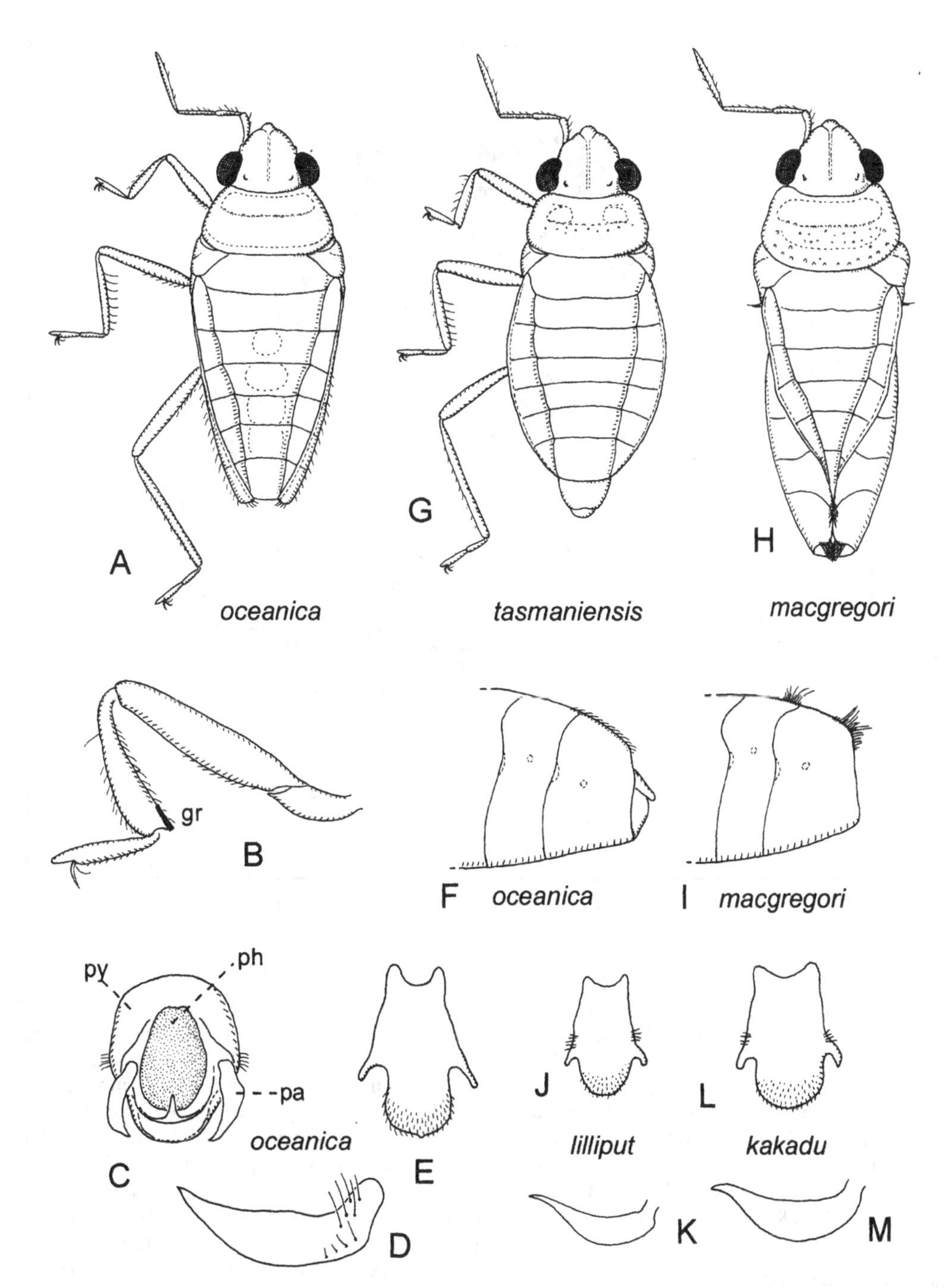

Figure 11.26. VELIIDAE. A-F, *Microvelia* (*Pacificovelia*) *oceanica*: A, dorsal view of apterous ♀; antenna and legs of right side omitted. B, fore leg of apterous ♂ with grasping comb (gr) C, dorsal view of pygophore of apterous ♂ (pa, paramere; ph, phallus; py, pygophore. D, dorsal view of proctiger of apterous ♂. E, right paramere of apterous ♂. F, lateral view of abdominal end of apterous ♀. G, *M.* (*P.*) *tasmaniensis*, apterous ♂: dorsal view, antenna and legs of right side omitted. H-I, *M.* (*P.*) *macgregori*, apterous ♀: H, dorsal view; antenna and legs omitted. I, lateral view of abdominal end. J-K, *M.* (*P.*) *lilliput*, apterous ♂: J, dorsal view of proctiger. K, right paramere. L-M, *M.* (*P.*) *kakadu*, apterous ♂: L, dorsal view of proctiger. M, right paramere. Reproduced with modification from Andersen & Weir (2003b).

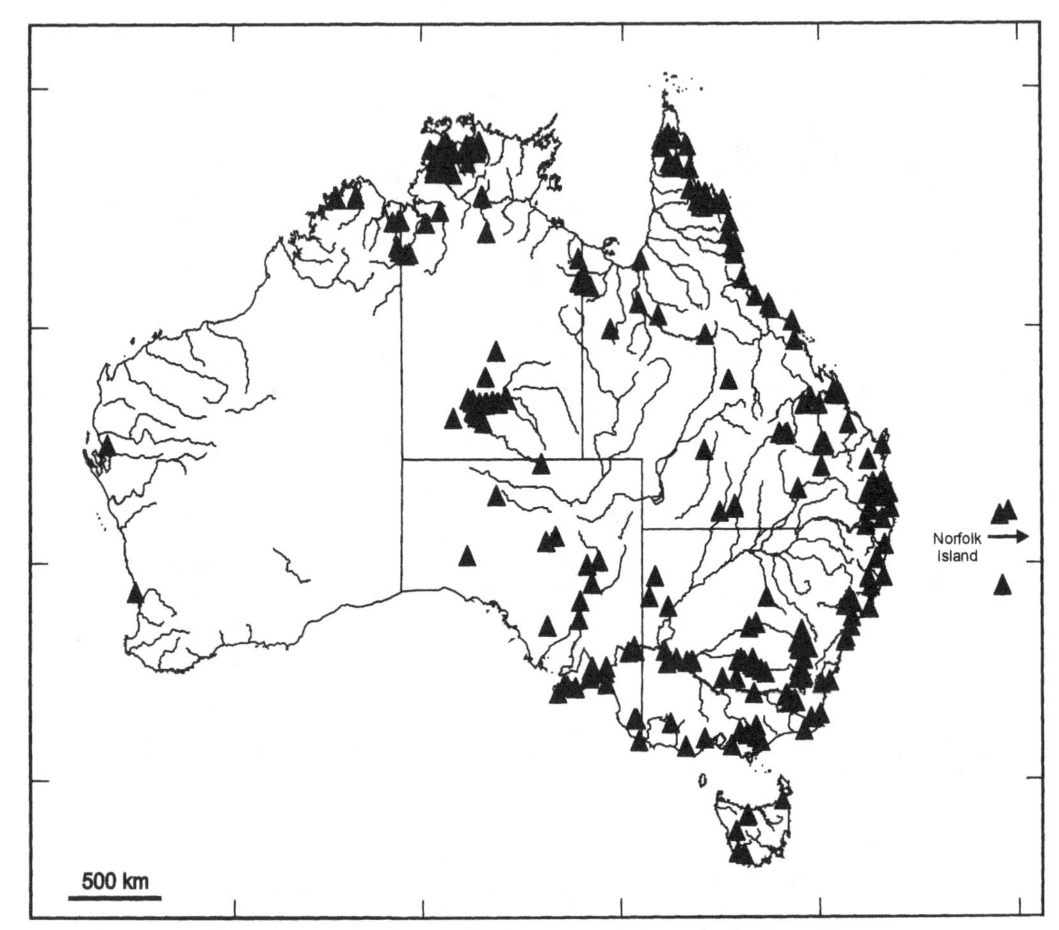

Figure 11.27. VELIIDAE. Distribution of *Microvelia* (*Pacificovelia*) species in Australia.

tion. In Australian territory, *M.* (*Pacificovelia*) *macgregori* is only known from Norfolk Island, but otherwise widely distributed and common in New Zealand where it is the only species of the Veliidae (Wise 1965, 1990; Towns 1978). Other non-Australian species of the subgenus are widespread in the Pacific area including Hawaii and also found in adjacent areas of East and Southeast Asia (Andersen & Weir 2003b).

MICROVELIA (PICAULTIA)

Identification

Small to very small *Microvelia* species (Fig. 11.28J) Antennae slender, about 0.5x total length of insect; antennal segment 4 distinctly thicker than segment 3. Male fore tibial comb less than 1/4 of tibial length (Fig. 11.28A; gr); small grasping comb also present on middle tibia (Fig. 11.28B, gr). Fore wings of macropterous forms dark, usually with two broad stripes in basal parts and 3-5 spots or stripes in distal parts, whitish (Fig. 11.28F). Male abdominal venter simple. Male genital segments large; parameres asymmetrically developed, right paramere very large (Figs 11.28C, D; pa), left paramere distinctly smaller. Female connexiva obliquely or vertically raised, usually converging posteriorly and sometimes meeting each other above abdominal end (Figs 11.28E, L); proctiger broad, deflected.

Andersen & Weir (2003b) reinstated *Picaultia* (type species *Picaultia pronotalis* Distant, 1913, from the Seychelles) as a subgenus of *Microvelia*. Here belong about 20 species and subspecies of Palearctic, Afrotropical, and Oriental *Microvelia*. In Australia there are 4 species.

Key to Australian species

1. Genital segments of ♂ very large (Fig. 11.22I), median length about 2.5x length

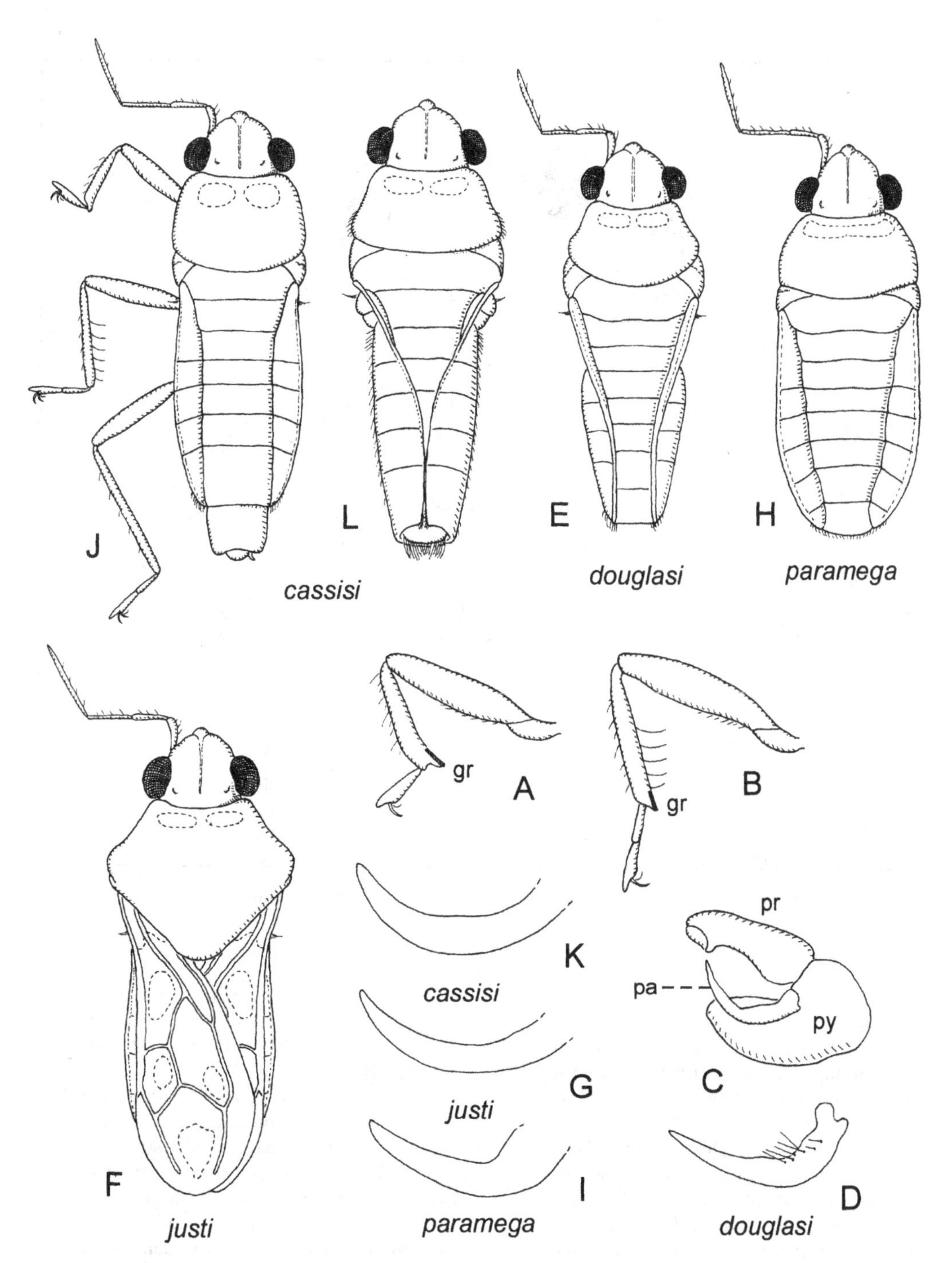

Figure 11.28. VELIIDAE. A-E, *Microvelia* (*Picaultia*) *douglasi*: fore leg of apterous ♂ with grasping comb (gr). B, middle leg of apterous ♂ with grasping comb (gr). C, lateral view of pygophore (py) and proctiger (pr) of apterous ♂ showing right paramere (pa). D, right paramere. E, dorsal view of apterous ♀; right antenna and all legs omitted. F-G, *M.* (*P.*) *justi*: dorsal view of macropterous ♀; right antenna and all legs omitted. G, right paramere of macropterous ♂. H-I, *M.* (*P.*) *paramega*: dorsal view of apterous ♀; right antenna and all legs omitted. I, right paramere of apterous ♂. J, *M.* (*P.*) *cassisi*, apterous ♂: dorsal view, antenna and legs of right side omitted. K-L, *M.* (*P.*) *cassisi*, right paramere of apterous ♂. L, dorsal view of apterous ♀; antenna and legs omitted. Reproduced with modification from Andersen & Weir (2003b).

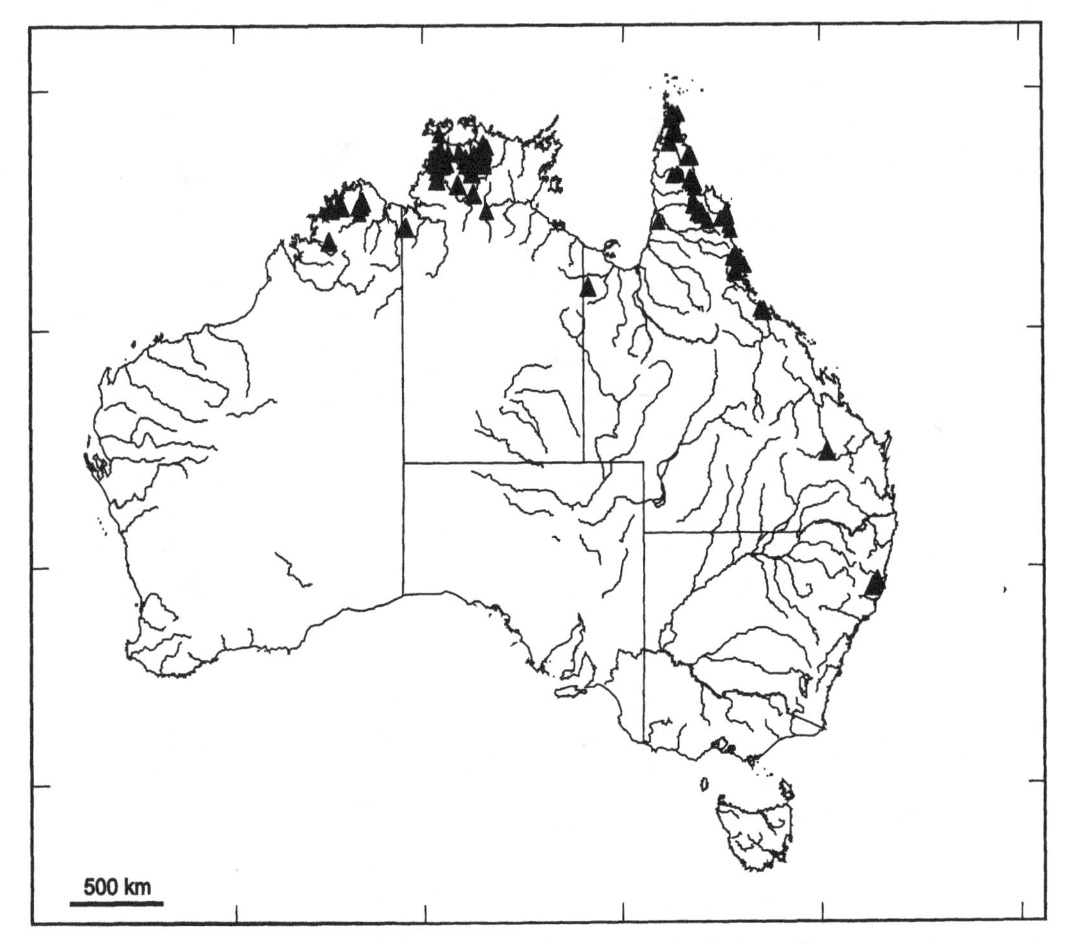

Figure 11.29. VELIIDAE. Distribution of *Microvelia* (*Picaultia*) species in Australia.

of sternum 7; right paramere very large, protruding behind genital segments (Fig. 11.22I, 11.28I). Connexiva of apterous ♀ obliquely raised throughout, not converging posteriorly (Fig. 11.28H). Length 1.3-1.7 (♂), 1.4-1.8 mm (♀). Northern Territory, Queensland, Western Australia.... *M. paramega* Andersen & Weir

- Genital segments of ♂ moderately large, median length less than 2x length of sternum 7; right paramere large, but not distinctly protruding behind genital segments. Connexiva of apterous female almost vertically raised throughout, converging posteriorly (Figs 11.28E, L) 2

2. Apterous form 3
- Macropterous form 4

3. Distal part of right paramere of male almost as wide as basal part (Fig. 11.28K). Female connexiva vertically raised throughout, converging posteriorly, meeting each other above terga 6 and terga 7 (Fig. 11.28L). Length 1.6-1.7 mm (♂, ♀). New South Wales..........................
.............................*M. cassisi* Andersen & Weir

- Distal part of right paramere of male more slender than basal part (Fig. 11.28D). Female connexiva vertically raised throughout, converging posteriorly, sometimes meeting each other at abdominal end (Fig. 11.28E). Length 1.2-1.4 mm (♂, ♀). Northern Territory, Queensland, Western Australia ....
.... *M. douglasi* Scott

4. Larger species, length 1.7-1.8 mm. Pale markings of fore wings well-defined, distinctly separated from each other (Fig. 11.28F). Northern Territory, Queensland, Western Australia ....
.... *M. justi* Andersen & Weir

- Smaller species, length 1.3-1.6 mm. Pale markings of fore wings more or less confluent or obscured. Northern Territory, Queensland, Western Australia .... *M. douglasi* Scott

Biology

Habitats in pools and billabongs in river and stream beds; also in slow flowing streams and temporary pools. Adults collected throughout the period March to December, most frequently from May to July. In Australia, macropterous (winged) individuals are seemingly more frequent than apterous (wingless) individuals, at least in some species. The former may disperse between habitats by flight (Fernando 1961, 1964) and are often collected at light or by car traps. Many aspects of the biology of *Microvelia douglasi* have been studied by Muraji & Nakasuji (1988, 1990) and Muraji et al. (1989a, 1989b). In the Philippines, this minute water strider is considered one of the most important natural enemies of the brown planthopper, *Nilaparvata lugens* Stål, an important pest in paddy fields (Nakasuji & Dyck 1984).

Distribution

Species of *Microvelia* subgenus *Picaultia* are found in New South Wales, Northern Territory, Queensland, and Western Australia (Fig. 11.29). *M.* (*Picaultia*) *douglasi* is widely distributed in the Oriental Region and also recorded from New Guinea, Marianna Islands (Guam), and Samoa. The other three Australian species are endemic to the continent (Andersen & Weir 2003b).

MICROVELOPSIS

Identification

Small or very small water striders, length 1.9-3.3 mm (♂) or 2.0-3.8 mm (♀); adults always macropterous (Figs. 11.30A, C). Body elongate, 2.5-3.0x greatest width across pronotum, chiefly dark coloured, with scarce pilosity of long, semierect hairs. Head produced posteriorly, extending well behind hind margin of eyes. Antennae slender and long, 0.45-0.55x total length of insect; segment 1 much shorter than head, barely surpassing tip of head, shorter than all other segments; segment 2 slightly shorter than segment 3, segment 4 slightly longer than segment 3. Pronotum of macropterous form large, pentagonal, without anterior collar (Figs 11.30A, C); a large, transverse pale marking along anterior margin and numerous punctures on pronotal lobe, between distinctly raised humeral angles. Middle legs slightly longer than fore legs, but shorter than hind legs. Fore femur thickened (especially in male), but otherwise not modified; fore tibiae of male with a relatively short grasping comb on inner surface before apex (Fig. 11.30B; gr); each comb 0.2-0.3x length of tibia. Middle femur slender, middle tarsus prolonged, more than 0.6x length of tibia. Hind femur as thick as middle femur; hind tarsus as long as middle tarsus. First segment of both middle and hind tarsi longer than second segment (*melancholica*, Fig. 11.30A) or first segment subequal to or slightly shorter than second segment (*exuberans* and *minor*, Fig. 11.20C). Wings long, reaching abdominal end when folded; fore wings dark, with pale stripes at base and four spots on the remaining wing. Abdomen relatively long, sides almost straight (♂) or abdomen distinctly constricted in basal parts with sides more or less rounded (♀). Connexiva broad, vertical in posterior parts. Abdominal venter of male not modified; male genital segments relatively small (Fig. 11.30D); segment 8 elongate and simple, pygophore elongate, slightly modified distally; parameres small, symmetrically developed, with slender distal part. Posterior margin of sternum 7 of female slightly produced, partly covering gonocoxae from beneath; proctiger narrow, slightly deflected (Fig. 11.30E).

The genus *Microvelopsis* was erected by Andersen & Weir (2001) for *Microvelia melancholica* Hale (1926) and two related, but smaller species.

Key to Australian species

1. Middle tarsus about as long as middle tibia (Fig. 11.30A); first tarsal segment slightly longer than second segment. Larger species, length 3.1-3.3 (♂) or 3.5-3.8 mm (♀). Queensland... *melancholica* (Hale)
- Middle tarsus distinctly shorter than middle tibia (Fig. 11.30C); first tarsal segment shorter than second segment. Smaller species, length less than 2.7 (♂) or 3.3 mm (♀) 2
2. Male fore tibial grasping comb 0.3-0.4x tibial length. Female abdomen constricted beyond the level of abdominal segment 3 (Fig. 11.30C). Larger species, length 2.5-2.6 (♂) or 2.7-3.3 mm (♀). Queensland .... *exuberans* Andersen & Weir
- Male fore tibial grasping comb 0.2x tibial length. Female abdomen constricted at base. Smaller species, length 1.9-2.0 (♂) or 2.0-2.2 mm (♀). Queensland............... ....*minor* Andersen & Weir

Biology

Habitats in half shaded creeks and stream pools in rainforests, water usually clear and with sandy

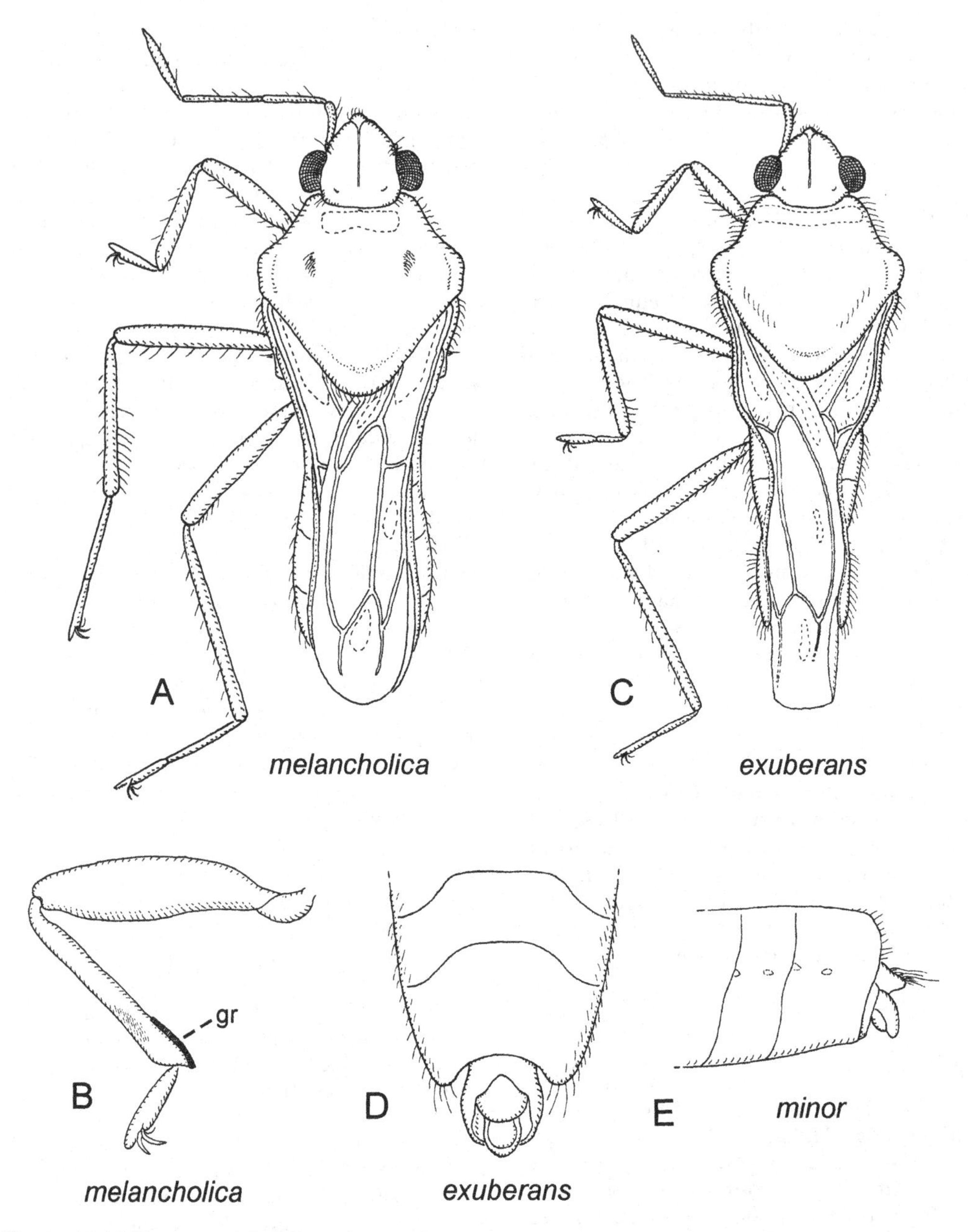

Figure 11.30. VELIIDAE. A-B, *Microvelopsis melancholica*, macropterous ♂: A, dorsal view; antenna and legs of right side omitted. B, fore leg with grasping comb (gr). C-D, *M. exuberans*: C, dorsal view of macropterous ♀; antenna and legs of right side omitted. D, ventral view of abdominal end of macropterous ♂. E, *M. minor*, macropterous ♀: lateral view of abdominal end. Reproduced with modification from Andersen & Weir (2001).

or stony bottom (Andersen & Weir 2001). Always macropterous (winged) and many specimens have been collected by flight intercept traps in mountain rainforests at altitudes 700-1100 m, suggesting a vigorous flight activity. Adults recorded from September-March.

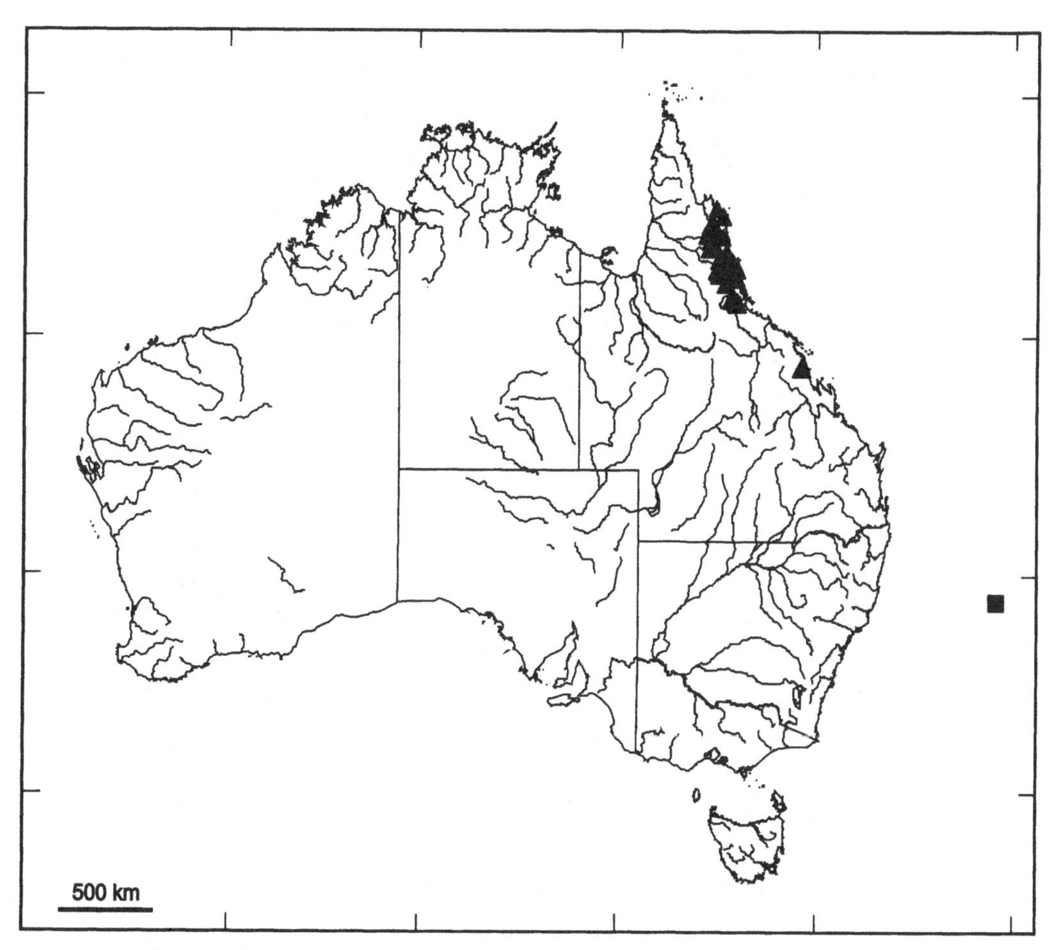

Figure 11.31. VELIIDAE. Distribution of *Microvelopsis* species (triangles) and *Nesidovelia howensis* (square) in Australia.

Distribution

This genus is endemic to Australia and the three decscribed species are so far only known from northern Queensland (Fig. 11.31).

NESIDOVELIA

Identification

Small water striders, length 2.9-3.0 mm (♂) or 2.9 mm (♀); adults apterous (Fig. 11.32A). Body elongate, 2.9-3.0x greatest width across mesothorax, chiefly black or dark brownish above with yellowish brown markings and dense pilosity of long, semierect hairs. Head produced posteriorly, extending well behind hind margin of eyes (Fig. 11.32A). Antennae slender and long, about 0.5x total length of insect; segment 1 shorter than head, surpassing tip of head with about half of its own length. Pronotum of apterous form (Fig. 11.32A; pn) long, median length subequal to head length, with large, transverse pale marking anteriorly; anterior margin distinctly concave. Mesonotum completely covered by pronotal lobe; metanotum only exposed laterally. Middle legs distinctly longer than fore legs, but shorter than hind legs. Fore femora moderately thickened, but otherwise not modified; fore tibiae of male with a relatively long grasping comb on inner surface before apex; each comb about 0.4x length of tibia. Middle femur slender, middle tarsus relatively short, about 0.5x length of tibia; first tarsal segment distinctly shorter than second segment. Abdomen relatively long, sides relatively straight. Abdominal venter of male relatively simple, sternum 7 depressed; male genital segments relatively large, protruding from pregenital abdomen (Fig. 11.32B); segment 8 (s8) very long, ventrally depressed and with a pair of triangular processes

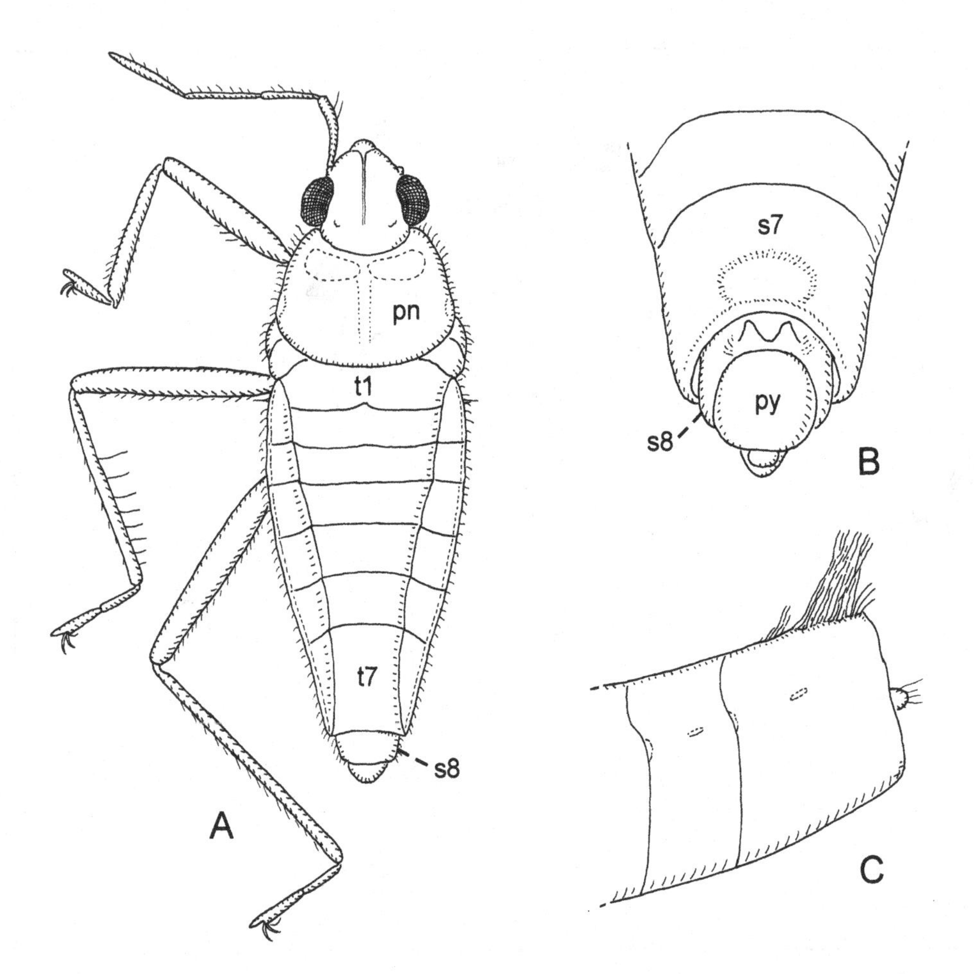

Figure 11.32. VELIIDAE. A-C, *Nesidovelia howensis*: A, dorsal view of apterous ♂; antenna and legs of right side omitted (pn, pronotum; s8, segment 8; t1, t7, abdominal tergites 1 and 7). B, ventral view of abdominal end of apterous ♂ (py, pygophore; s7, sternum 7; s8, segment 8). C, lateral view of abdominal end of apterous ♀. Reproduced with modification from Andersen & Weir (2001).

arising from posterior margin; parameres small, symmetrically developed. Connexiva of female almost vertical and converging to meet each other above tergum 7, with a brush of long dark hairs towards abdominal end (Fig. 11.32C); sternum 7 tubular; proctiger suboval and deflected. Macropterous adult form unknown.

The genus *Nesidovelia* was erected by Andersen & Weir (2001) for *Microvelia howensis* Hale (1926). Only one species, *Nesidovelia howensis* (Hale), length 2.9-3.1 mm (apterous form). Lord Howe Island (New South Wales).

Biology

Habitats in creeks and rock-pools (Andersen & Weir 2001).

Distribution

Endemic to Lord Howe Island (Fig. 11.31).

PETROVELIA

Identification

Small water striders, length 2.0-2.8 mm (♂) or 2.0-2.9 mm (♀); adults apterous (Fig. 11.33A) or macropterous (Fig. 11.33B). Body elongate oval to suboval, 2.1-2.7x greatest width across mesothorax, chiefly blackish and shiny above (♂) or blackish with yellowish brown markings (♀), with pilosity of short, semierect hairs. Antennae slender and long, about 0.6x total length of insect; segment 1 always shorter than head, surpassing tip of head with about half of its own

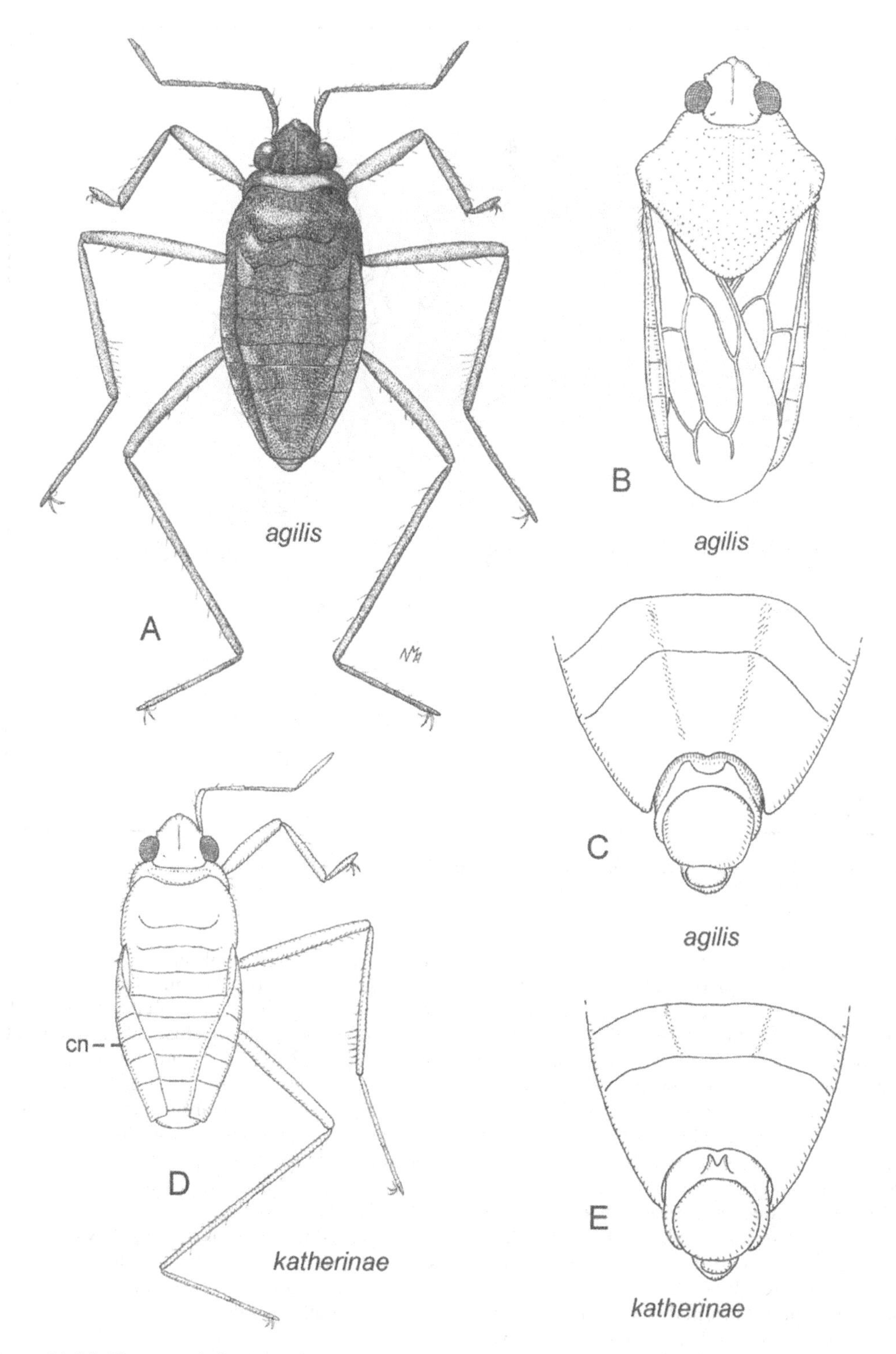

Figure 11.33. VELIIDAE. A-C, *Petrovelia agilis*: A, apterous ♂. B, dorsal view of macropterous ♀; antennae and legs omitted. C, ventral view of abdominal end of apterous ♂. D-E, *P. katherinae*: dorsal view of apterous ♀; antenna and legs of left side omitted (cn, connexivum). E, ventral view of abdominal end of apterous ♂. Reproduced with modification from Andersen & Weir (2001).

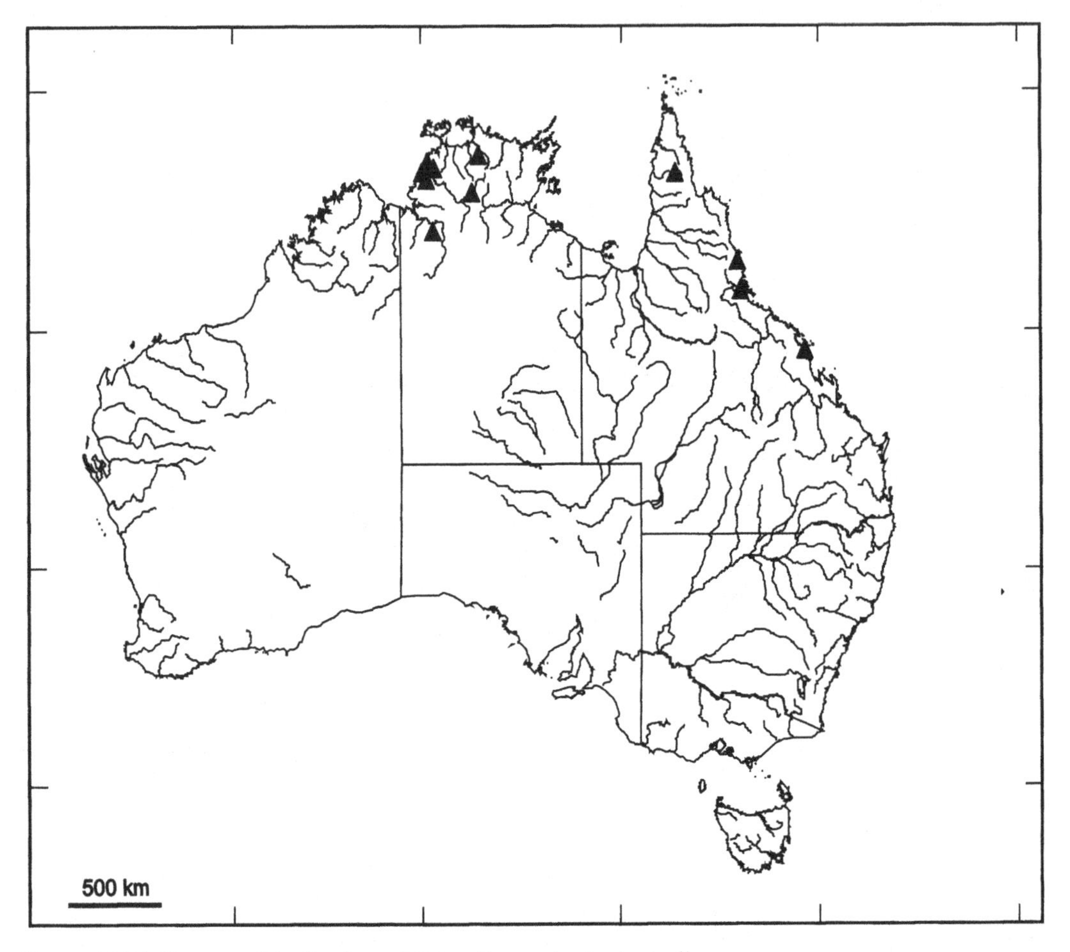

Figure 11.34. VELIIDAE. Distribution of *Petrovelia* species in Australia.

length. Pronotum of apterous form very short (Figs 11.33A, D), median length only about 0.25x head length, with large, transverse pale marking in middle. Mesonotum exposed, distinctly longer than pronotum; metanotum also exposed and much shorter than mesonotum. Pronotum of macropterous form large, pentagonal, without anterior collar (Fig. 11.33B) Middle legs distinctly longer than fore legs, but shorter than hind legs. Fore femora moderately thickened, but otherwise not modified; fore tibiae of male with a relatively long grasping comb, each comb about 0.4x length of tibia. Middle femur slender, middle tarsus relatively long, 0.8-0-9x length of tibia; first tarsal segment distinctly longer than second segment. Claws of middle and hind legs relatively long, falcate. Fore wings uniformly dark, without pale spots or stripes. Abdomen relatively long, sides more or less rounded. Connexiva of female broad (Fig. 11.33D, cn), obliquely raised or inflexed upon abdominal tergum (except when abdomen is inflated). Abdominal venter of male depressed in middle; male genital segments relatively small, protruding from pregenital abdomen (Figs 11.33C, E); segment 8 depressed ventrally and with a pair of triangular processes arising from posterior margin; parameres small, symmetrically developed. Posterior margin of sternum 7 of female straight, covering bases of gonocoxae from beneath; proctiger (pr) suboval and deflected.

The genus *Petrovelia* was described by Andersen & Weir (2001) to hold two Australian microvelines with extraordinary long middle and hind legs and peculiar habits (see below).

Key to Australian species

1. Triangular processes at ventral, posterior margin of male segment 8 widely separat-

ed (Fig. 11.33C). Female connexiva obliquely raised, broad but simple. Larger species, 2.2-2.9 mm. Queensland .... *agilis* Andersen & Weir
- Triangular processes at ventral, posterior margin of male segment 8 more closely set (Fig. 11.33E). Female connexiva almost erect anteriorly (Fig. 11.33D), inflexed upon abdominal tergum posteriorly (except when abdomen is inflated). Smaller species, 2.0-2.7 mm. Northern Territory........ *katherinae* Andersen & Weir

Biology

Habitats in billabongs or slow flowing rivers (Andersen & Weir 2001). *Petrovelia* species tends to spend time resting on moist rocks and logs just above the water surface and in the slash zone of water falls. When disturbed, they rush on to the water surface before returning to their rocks or logs. Adults and nymphs are recorded from January, June-August, and October-December.

Distribution

This genus is possibly endemic to tropical, northern Australia (Fig. 11.34) although its presence in southern New Guinea can be predicted.

Phoreticovelia

Zeus Bugs

Identification

Very small water striders, length 1.0-1.2 mm (♂) or 1.7-2.0 mm (♀); male thus distinctly smaller than female (Figs 11.2E, 11.35A). Dimorphic with respect to wing development, apterous (Figs 11.35A, E) or macropterous. Body suboval to oval, chiefly brownish, moderately pilose; mesonotum and abdominal dorsum (apterous form) with patches of silvery hairs. Head strongly deflected in front of eyes (Fig. 11.35B). Antennae relatively short, less than half of body length; first antennal segment barely reaching apex of head. Pronotum of macropterous form large, pentagonal in outline, without anterior collar. Pronotum of apterous form very short, median length less than width of an eye (Fig. 11.35E; pn). Meso- and metanotum indistinctly separated, modified in female, depressed (for reception of small male during copulation; Figs 11.2E, 11.35A); anterior part of mesonotum with lateral, punctate swellings and an elongate patch of whitish, waxy secretion on each side of midline (Figs 11.2F, 11.35E; wa). Middle leg subequal in length to hind leg, fore leg distinctly shorter. Fore leg of male with thickened and curved femora (Fig. 11.35C); fore tibiae curved with a short and broad grasping comb (gr), less than one fifth of tibial length. Middle tarsi long, subequal in length to middle tibiae; first segment slightly longer than second segment; claws long, falcate and blade-like expanded (Figs 11.2G, 11.35D; cl), inserted towards middle of last tarsal segment; ventral arolium (va) long, blade-like. Hind leg claws long, falcate. Wings of macropterous form brownish with basal and distal whitish markings. Male abdomen almost parallel-sided; abdominal venter not modified. Genital segments small, simple. Parameres small, symmetrically developed, elongate and straight, with blunt apices. Female abdomen very broad (Fig. 11.35E), with rounded sides; connexiva broad, not converging, margins simple. Ventral margin of sternum 7 almost straight. Genital segments exposed; gonocoxae large, proctiger conical, protruding.

The genus *Phoreticovelia* was erected by D. Polhemus & J. Polhemus (2000) and redescribed by Andersen & Weir (2001). Apart from the two species recorded from Australia, this genus also includes *P. nigra* D. Polhemus & J. Polhemus (2000) from Biak and Salawati, Irian Jaya, and *P. notophora* (Esaki) from the Palau Islands.

Key to Australian species

1. Ground colour chiefly pale brownish; mesonotum of male without dark punctures (Fig 11.35A). Patches of silvery hairs present on male mesonotum (except medially), all of male abdominal terga 3-4, and laterally on female terga 3 and 5-6. Length 1.0-1.1 mm (♂), 1.7-2.05 mm (♀). New South Wales, Northern Territory, Queensland, Victoria .... *rotunda* D. Polhemus & J. Polhemus
- Ground colour chiefly dark brownish or blackish; mesonotum of male with numerous dark punctures (Fig. 11.35F). Patches of silvery hairs present on most of male mesonotum (including medially), all of male abdominal terga 3-4, and female laterotergites 3 and mediotergite 6. Length 1.2-1.25 mm (♂), 2.0-2.05 mm (♀). Queensland ....*disparata* D. Polhemus & J. Polhemus

Biology

Species of *Phoreticovelia* are found (often in large numbers) at the edges of streams and billabongs, especially at night (Andersen & Weir 2001; D. Polhemus & J. Polhemus 2000). Adults are recorded from March-April (New South Wales, Victoria) and May-December and January (Northern Territory and Queensland). Nymphs

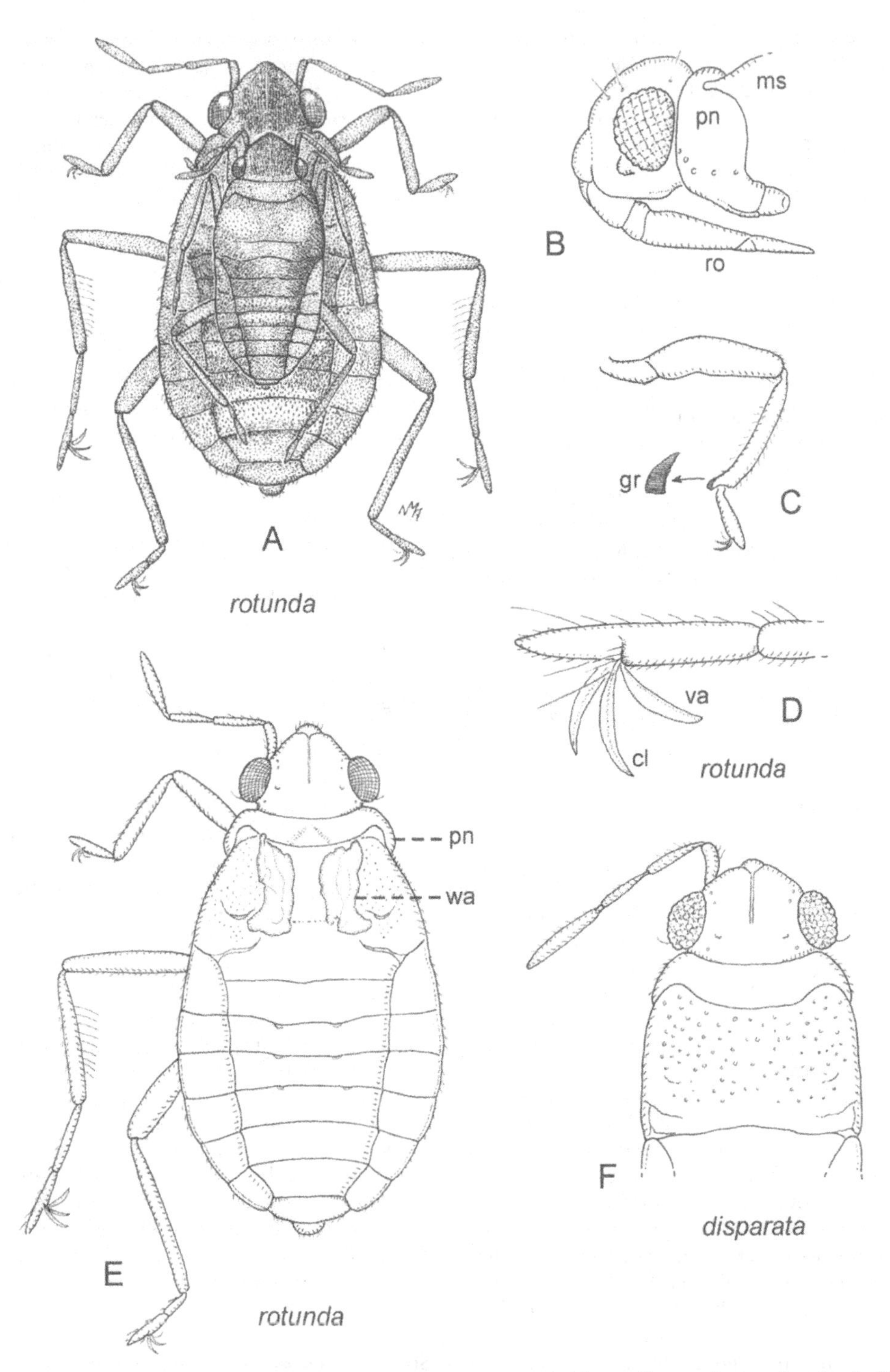

Figure 11.35. VELIIDAE. A-E, *Phoreticovelia rotunda*: A, apterous ♀ with small ♂ resting on her back. B, lateral view of head and prothorax of apterous ♀ (ms, mesonotum; pn, pronotum; ro, rostrum). C, fore leg of apterous ♂ with grasping comb (gr) shown at higher magnification. D, distal part of middle tarsus (cl, claw; va, ventral arolium). E, dorsal view of apterous ♀; antenna and legs of right side omitted (pn, pronotum; wa, patch of waxy substance). F, *P. disparata*, apterous ♂: dorsal view of head and thorax. Reproduced with modification from Andersen & Weir (2001).

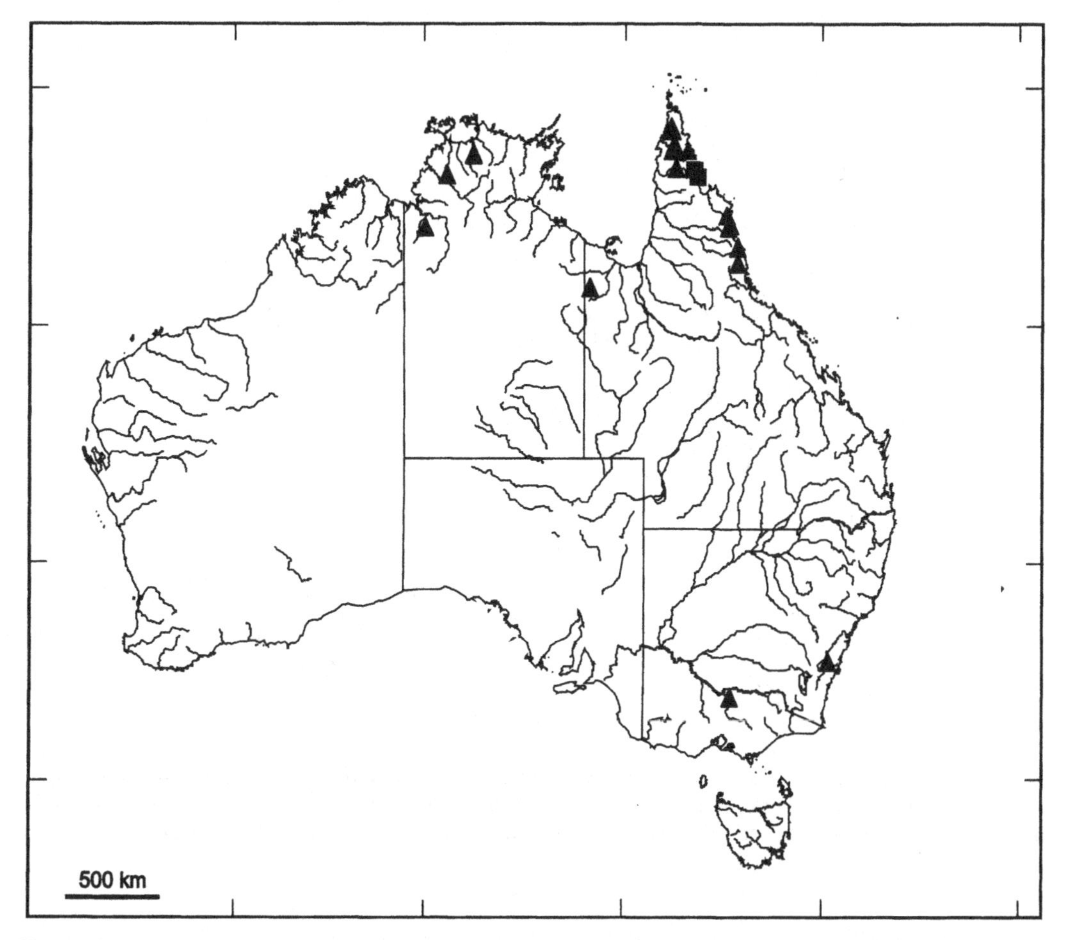

Figure 11.36. VELIIDAE. Distribution of *Phoreticovelia* species (triangles) and *Tarsoveloides brevitarsus* (squares) in Australia.

from March-April (New South Wales, Victoria) and May-August and January (Northern Territory and Queensland).

This genus is unique in several respects: (1) The presence of large patches of waxy secretion on each side of the midline of the female mesonotum; these secretions are already present in the last (fifth?) instar female nymphs where they seemingly are formed by secretion from a longitudinal row of pores on each side of the mesonotal midline; (2) the pronounced sexual size dimorphism where the female is between 1.7 and 2 times the size of the female. The male is phoretic, sitting in a depression on the females back with his head situated between the patches of waxy secretions (Figs 11.2E, F, 11.35A); and (3) the ventral arolium of the middle tarsus is flattened, leaf-like instead of bristle-like, forming a swimming fan together with the two claws (Figs 11.2G, 11.35D). Similar swimming fans are found in a few other Indo-Australian genera of Microveliinae (Andersen 1982; D. Polhemus & J. Polhemus 2000).

It has recently been demonstrated that males may feed on the female glandular secretions during mating (Arnqvist et al. 2003). Experimentally, females were kept isolated from males and fed on radio-labelled food (*Drosophila* flies). Males were then brought together with the females and fed only non-labelled food. After four days males contained labels indicating that material had been passed by the female to the male. Females only produce the secretions when ridden by a male, but the presence of males apparently do not affect the life-span and fecundity of females. Females do not need to mate more than every two-three weeks in order to maintain full fecundity. Arnqvist et al. (2003) suggest the trivial name

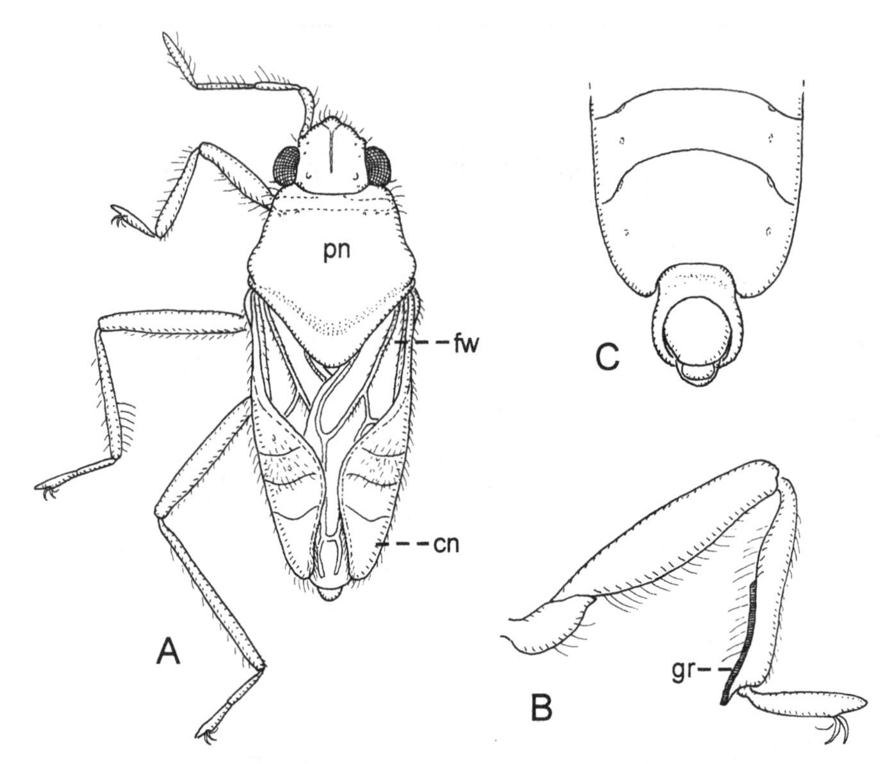

Figure 11.37. VELIIDAE. A-C, *Tarsoveloides brevitarsus*: dorsal view of macropterous ♀; antenna and legs of right side omitted. B, fore leg of macropterous ♂ with grasping comb (gr). C, ventral view of abdominal end of macropterous ♂. Reproduced with modification from Andersen & Weir (2001).

"Zeus bugs" for *Phoreticovelia*. In Greek mythology, Zeus consumed his first wife Metis.

Distribution

In Australia, the two species of the genus *Phoreticovelia* are sparsely distributed in northern Queensland and the top end of Northern Territory (Fig. 11.36), with geographically widely separated distributions in Victoria and New South Wales (*P. rotunda*).

TARSOVELOIDES

Identification

Very small water striders, length 1.8-1.9 mm (♂) or 1.95-2.1 mm (♀); adults macropterous (Fig. 11.37A). Body elongate, 2.8-3.0x greatest width across pronotum, chiefly dark coloured, with pilosity of long black, semierect hairs. Antennae slender and long, 0.5x total length of insect; segment 1 much shorter than head, surpassing tip of head with about half its own length. Pronotum of macropterous form subpentagonal, without anterior collar (Fig. 11.37A; pn); an ill-defined, transverse pale marking along anterior margin and numerous punctures on pronotal lobe. Middle legs slightly longer than fore legs, but shorter than hind legs. Fore femur slightly thickened, but otherwise not modified; inner margin of fore tibiae of male sinuated, with a relatively long grasping comb before apex (Fig. 11.37B, gr); each comb about 0.55x length of tibia. Middle femur slender, middle tarsus only 0.7x length of tibia. First segment of both middle and hind tarsi subequal to or slightly shorter than second segment. Fore wings (Fig. 11.37A; fw) dark with pale stripes at base and some longitudinal pale stripes or spots in distal part. Abdomen relatively long, sides almost straight (♂) or relatively broad anteriorly, narrowing posteriorly (♀). Connexiva of female vertically raised anteriorly, inflexed and depressed medially (Fig. 11.37A; cn), with marginal hair fringes posteriorly. Abdominal venter of male not modified; male genital segments relatively large; segment 8 simple, without ventral modifications; pygophore elongate, posterior margin slightly modified; parameres very small, symmetrically developed. Posterior margin of sternum 7 of female slightly concave, exposing gonocoxae; proctiger cone-shaped, protruding.

The genus *Tarsoveloides* was described by Andersen & Weir (2001) and contains only one species, *T. brevitarsus* Andersen & Weir, length 1.8-1.9 mm (macr. ♂), length 1.9-2.1 mm (macr. ♀). Queensland.

Biology

A specimen from near Bald Hill, McIlwraith Range, was taken at the edge of a slow flowing rainforest stream in half shade. D. Polhemus (personal communication) has collected a related species in foam at the edge of streams in New Guinea. Adults recorded from June-July.

Distribution

In Australia only known from northern Queensland (Fig. 11.36), but species of this genus probably also occur in New Guinea (D. Polhemus, unpublished).

SUBFAMILY RHAGOVELIINAE

Identification

Small water striders which above all are characterised by the unique structure of their middle tarsi, which are deeply cleft and provided with an elaborate swimming fan (Fig. 11.1A). The head is relatively short and steeply declivent in front. Antennae relatively long, with long and slightly curved segment 1 and segments 2 and 3 longer than segment 4. Rostrum short and robust. All legs with 3-segmented tarsi, basal segments of fore tarsus very short. Adults are either apterous (the most common adult form in most species) or macropterous. Pronotum of apterous form either shorter than head or prolonged, covering mesonotum. Ventral sutures of thorax and abdomen distinct; metathoracic scent channels almost straight, extending laterally from median scent orifice, close to hind margin of metasternum; small evaporative areas and hair tufts present on metacetabula. Forewings usually dark, with well-developed veins (at least in basal half). Hind femora more or less incrassate and usually armed ventrally with numerous teeth of varying size (most prominent in ♂). Fore tibiae of male with well developed grasping comb. Middle legs distinctly longer than fore and hind legs; middle tarsus as long as or longer than tibia; last tarsal segment split for about 3/4 of its length, with blade-like flattened claws, bristle-like dorsal arolium, and a hairy or plumose swimming fan arising from ventral arolium (Fig. 11.1A). A similar but smaller fan is present on the hind tarsus in the Oriental genus *Tetraripis* Lundblad (Andersen 2000a). Male abdomen moderately prolonged; genital segments usually simple, parameres large and symmetrically developed. Female abdomen relatively long, connexiva usually raised and sometimes modified.

Overview

The subfamily Rhagoveliinae is distributed worldwide, comprising four genera with a total of about 270 described species. In his revision of the genus *Rhagovelia* Mayr from the western hemisphere, D. Polhemus (1997) suggested that the genus *Trochopus* Carpenter (with five marine species) is nothing but a specialised branch of some Central and South American species groups of *Rhagovelia*. Furthermore, D. Polhemus (1997) proposed that *Tetraripis* should be reassigned to the subfamily Veliinae, regarding the presence of a pretarsal swimming fan in this genus and *Rhagovelia* as a case of parallelism (convergent evolution), based on differences in structural details of the fan (hairy vs. plumose) and presence of such structures on both middle and hind legs in *Tetraripis*. The reclassification by D. Polhemus (1997) was opposed by Andersen (2000a) who pointed out that *Tetraripis* shares several unique synapomorphies (structural details of the middle tarsus) with *Rhagovelia*, which inevitably leads to the conclusion that these genera are members of the same monophyletic taxon, the subfamily Rhagoveliinae. Only the genus *Rhagovelia* is represented in Australia.

RHAGOVELIA

Riffle Bugs

Identification

Small water striders which are either apterous (Figs 11.38A, B) or (very rarely) macropterous. Body elongate oval, 2.0-2.9x greatest width across mesothorax, dark coloured or with yellowish and reddish-brown pattern; pilosity dense but mostly short. Antennae slender and long, between one half and two thirds length of body; segment 1 distinctly longer than head and other segments, segments 2 and 3 subequal in length, segment 4 shortest. Pronotum of apterous form long, covering all but a narrow band of mesonotum (in Australian species; Figs 11.38B), yellowish with two large dark markings in middle. Pronotum of macropterous form large, pentagonal, with raised humeral angles. Fore wings uniformly dark, without pale spots or stripes. Fore femur simple; grasping comb of male fore tibia short, about one fifth length of tibia. Middle leg long, femur about one third body length; middle tarsus long, sube-

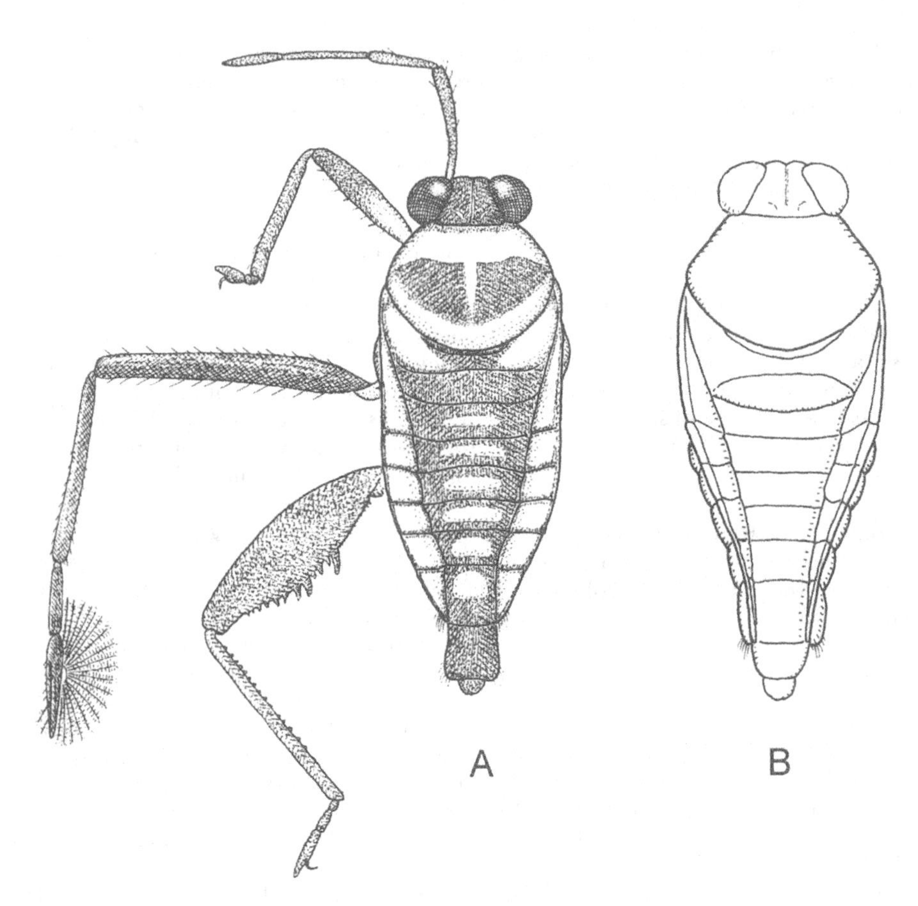

Figure 11.38. VELIIDAE. A-B, *Rhagovelia australica*: A, apterous ♂; antenna and legs of right side omitted. B, dorsal view of apterous ♀; antennae and legs omitted.

qual in length to tibia, with short segment 1 and much longer segments 2 and 3; segment 3 about 1.2x segment 2, split for about 3/4 of its length (Fig. 11.1A); from the bottom of the cleft arises an elaborate swimming fan composed of about 20 plumose, hair-like branches; claws (cl) modified, flattened and blade-like. Hind femur much thicker than middle femur, especially in ♂, ventrally armed with one or more large teeth and numerous smaller teeth and spinules. Abdomen relatively long, sides almost straight or more or less rounded; connexiva broad, obliquely raised or sometimes vertical in posterior parts in ♀. Abdominal venter of male not modified; male genital segments conspicuous but simple; parameres large, falciform with blunt apices. Posterior margin of sternum 7 of female straight, exposing parts of gonocoxae from beneath; proctiger broad, protruding or slightly deflected.

The genus *Rhagovelia* Mayr is distributed worldwide with about 250 species. Only one species is recorded from Australia, *R. australica* Kirkaldy (redescribed by Lansbury 1993), length 4.0-4.2 mm (apterous ♂), 4.0-4.4 mm (apterous ♀). Queensland.

Biology

Riffle bugs have earned their name because they are almost exclusively found on running freshwater such as torrents and streams. *Rhagovelia* species are small but fast moving surface bugs. They are commonly found in loose swarms or "schools" of individuals. Hynes (1955) made the observations that there usually was a segregation of different developmental stages in such schools. When disturbed by a potential predator, these swarms tend to disperse, but reassociate later (Cheng & Fernando 1971; Deshefy 1980). Species are often local and long stretches of a stream may be uninhabited (Bacon 1956). Some species

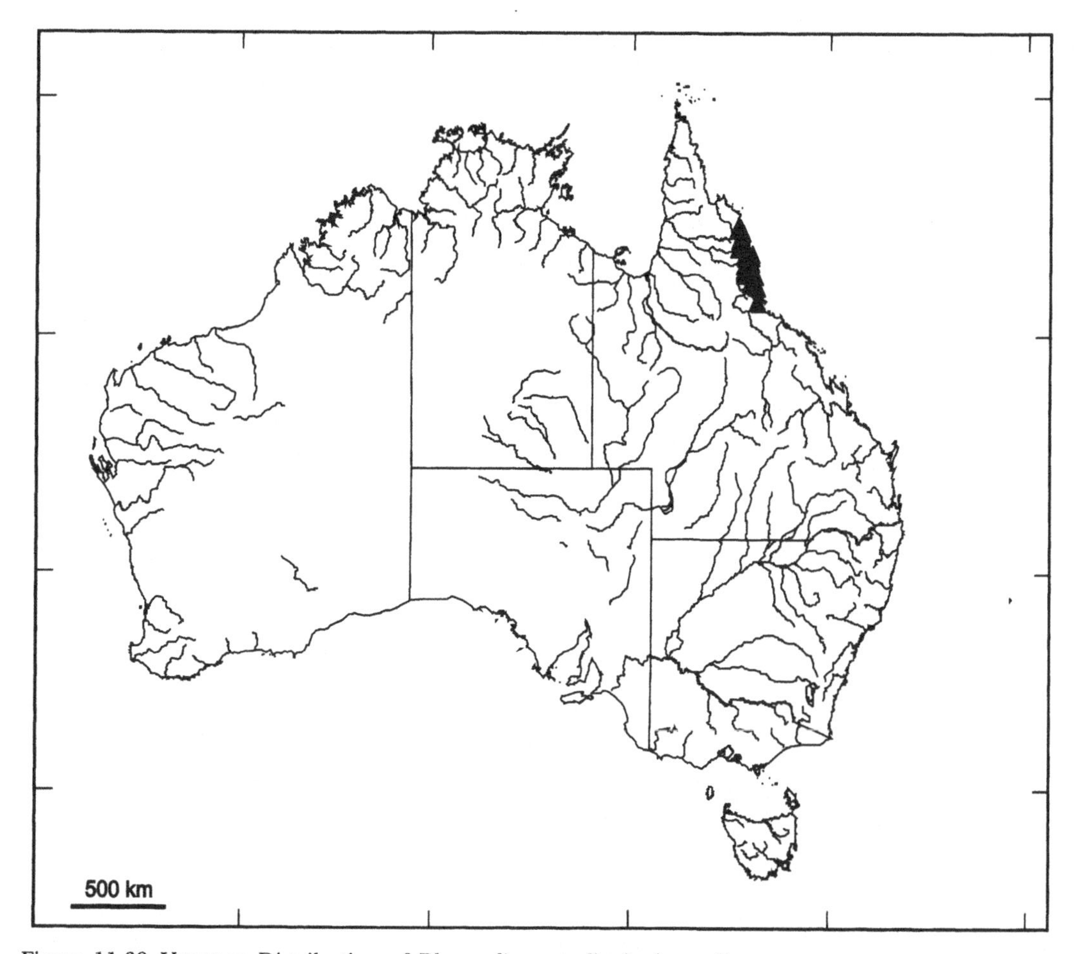

Figure 11.39. VELIIDAE. Distribution of *Rhagovelia australica* in Australia.

aggregate near the shore where the water is shaded by the vegetation or behind stones and elsewhere with moderate water flow.

The remarkable swimming fan of *Rhagovelia* species pierces the surface film and then expands in the water during the thrust stroke of the middle leg. The mechanism of operation of the fan was analysed by Andersen (1976). The spreading of the fan follows the activation of the claw retractor muscle, which pulls on the claw plate and causes a retraction of the anterior claw and the basal part of the fan. The posterior claw remains concealed in the tarsal cleft. When the middle leg is lifted off the water, the fan is folded into the tarsal cleft by protraction of the claws caused by the elasticity of pretarsal structures. Each stroke by the middle legs propels the insect for a longer distance than in veliids without such structures. The very effective way of locomotion in *Rhagovelia* species is undoubtedly an adaptation to life in torrents and fast-flowing streams.

Distribution

The Indo-Malayan Archipelago (J. Polhemus & D. Polhemus 1988) and New Guinea (Lansbury 1993) are among the areas with the highest density of *Rhagovelia* species. It is therefore surprising that only a single species is found in Australia where it is confined to the coastal rainforests of northeastern Queensland, from North of Cairns to Townsville (Fig. 11.39).

12. Family GERRIDAE

Water Striders, Pond Skaters

Identification

The "true" water striders cannot easily be confused with other surface bugs on account of their distinctly prolonged mesothorax and long and slender middle and hind legs which have their coxae inserted close to border between meso- and metathorax, pointing almost directly backwards (Figs 12.1A, F, 12.3D). The head is usually extended forward, with large, globular eyes and 3 or 4 pairs of cephalic trichobothria inserted along the margins of eyes (Figs 12.1B, 12.2B, 12.15A, 12.34F); ocelli absent. Most species are wing dimorphic, with macropterous and apterous adult forms (Figs 12.3A, B). Pronotum of apterous form variable in length; pronotum of macropterous form large, pentagonal (Fig. 12.3A; pn). Fore wings largely membranous, venation variable (Figs 12.1G, H). Metasternal scent gland apparatus present or absent. All tarsi 2-segmented (hind tarsal segments sometimes fused), claws inserted preapically in a cleft on the last tarsal segment. Abdomen variable in length; abdominal scent gland absent in both nymphs and adults. Male genital segments usually large; parameres of variable size, rarely absent; phallic vesica with well-defined sclerites. Female ovipositor plate-shaped, gonapophyses short and non-serrate (with exceptions).

Overview

The Gerridae is the second largest family of semi-aquatic bugs with about 640 described species in 72 genera worldwide. Andersen (1982) classified the Gerridae in 7 subfamilies: Charmatometrinae, Cylindrostethinae, Eotrechinae, Gerrinae, Halobatinae, Ptilomerinae, Rhagadotarsinae, and Trepobatinae. An additional subfamily, Electrobatinae, was described from the Oligocene/Miocene Dominican amber (Andersen & Poinar 1992). Only the subfamilies Gerrinae, Halobatinae, Rhagadotarsinae, and Trepobatinae are represented in the Australian fauna (Cassis & Gross 1995; Andersen & Weir 1994a, 1994b, 1997, 1998).

Gerrid water striders are conspicuously adapted to a life on the water surface upon which they "skate" or "jump-and-slide" in a very characteristic way (Andersen 1976, 1982). The adoption of this type of locomotion is associated with radical changes in the locomotory apparatus. The mesothorax is prolonged to accomodate the powerful legs muscles of the middle legs; the coxae of the two posterior pairs of legs are rotated to an almost horizontal position, pointing directly backwards; and the segments of the middle and hind legs are long and slender. These are adaptations which effectively transform the leg strokes into propulsion of the insect's body. This has opened up the possibility for water striders to increase their body size without loss of efficiency in locomotion (the largest species is the Southeast Asian gerrine water strider *Gigantometra gigas* (China), which reach a body length of 3.4 cm), as well as the possibility for higher reproductive rates, a more diverse exploitation of environmental resources (e.g. food) through size diversification, etc. Gerrid water striders live in a variety of habitats, from stagnant freshwater pools and ponds, streams and rivers, to various marine habitats, including the surface of the open ocean (*Halobates*). Many gerrids assemble in "schools" on the water surface. Such gregarious behaviour may improve the success in finding mates, in capturing prey (emerging aquatic insects, terrestrial insects accidentally caught in the surface film, etc.), but may also improve the chances of escaping predators (J. Polhemus & Chapman 1979e; Andersen 1982; Foster & Treherne 1980).

Gerrid males are usually slightly smaller than the females, but more pronounced sexual size dimorphism may occur (e.g., *Rheumatometra*; Fig. 12.41). Matings are usually initiated by the male simply lunging at the female, rarely associated with any kind of courtship behaviour. A male that successfully contacts a female will grasp her thorax with his fore legs and attempts to insert his genitalia. Females usually respond with some form of resistance which may result in vigorous struggles and "somersaulting" (where the female rolls over on her back). This kind of behaviour has been explained on basis of the fundamental conflict of interest between sexes. Thus, the female only needs few matings to fertilise all of her eggs whereas the male wants to mate with as many females as possible. I addition, repeated harassments by males and the burden of carrying a male on her back may be risky for the female. Ultimately, this conflict between sexes may lead to adaptations and counteradaptations in primary and secondary sexual structures resulting in a kind of "arms-race" between sexes (Andersen 1997). Since they usually are easy to observe, both in the field and in the laboratory, northern temperate pondskaters of the genera *Aquarius*, *Gerris*, and *Limnoporus* are widely used as model organ-

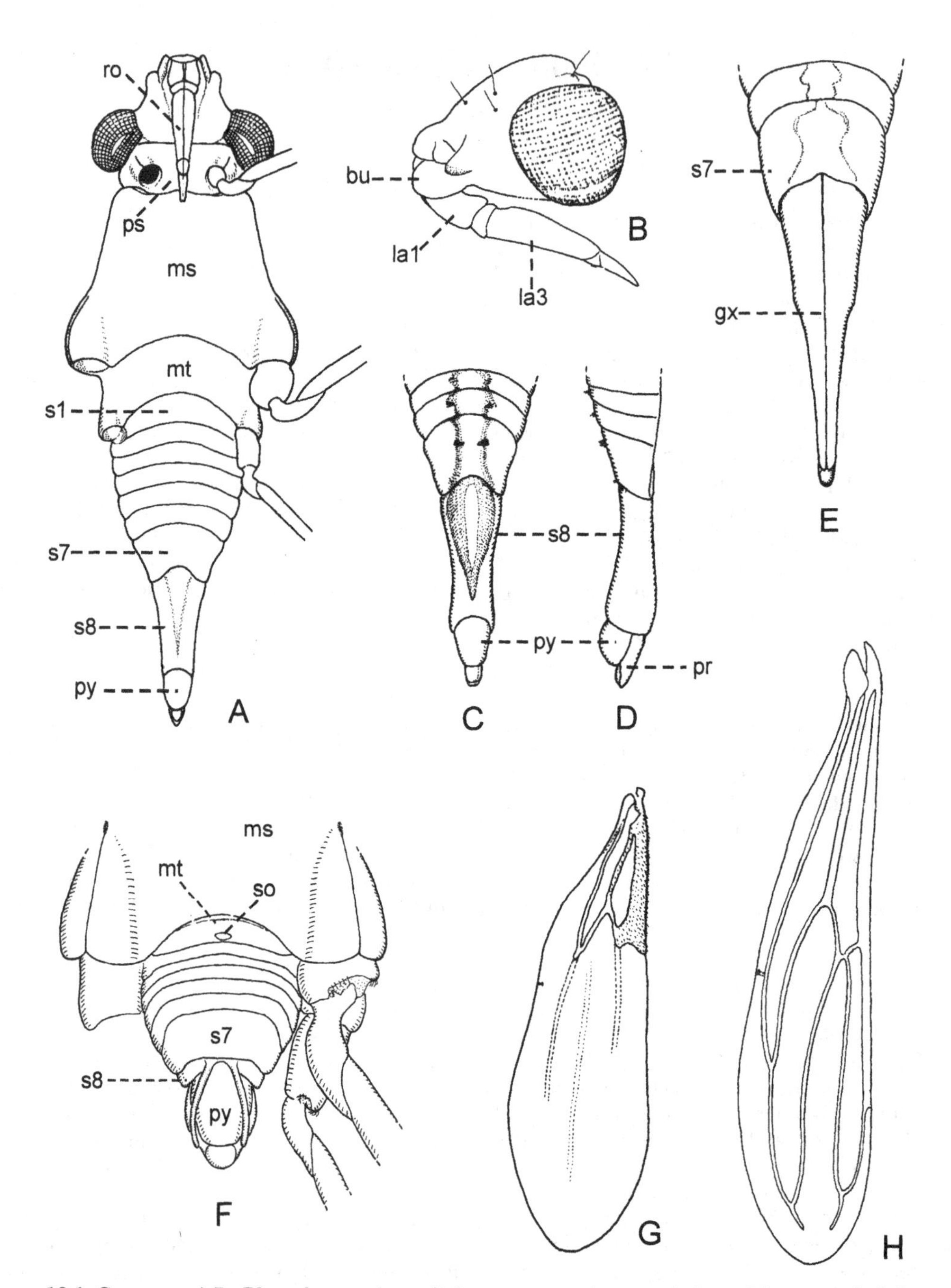

Figure 12.1. GERRIDAE. A-B, *Rhagadotarsus kraepelini*, apterous ♂: A, ventral view of thorax and abdomen (ms, mesosternum; mt, metasternum, ps, prosternum; py, pygophore; ro, rostrum; s1, s7, s8, sterna 1; 7 and 8). B, lateral view of head; antennae removed (bu, bucculum; la1, la3, labial segments 1 and 3). C-D, *R. anomalus*, apterous ♂: C, ventral view of abdominal end (pr, proctiger; py, pygophore; s8, segment 8). D, lateral view of abdominal. E, *R. anomalus*, apterous ♀: ventral view of abdominal end (gx, gonocoxa; s7, sternum 7). F, *Austrobates rivularis*, apterous ♂: ventral view of thorax and abdomen (ms, mesosternum; mt, metasternum, py, pygophore; s7, s8, sterna 7 and 8; so, scent orifice). G, *Rheumatometra philarete*: right fore wing. H, *Tenagogerris euphrosyne*: right fore wing. Reproduced with modification from: A, B, Andersen (1982); C-E, G Andersen & Weir (1998); F, Andersen & Weir (1994a); H, Andersen & Weir (1997).

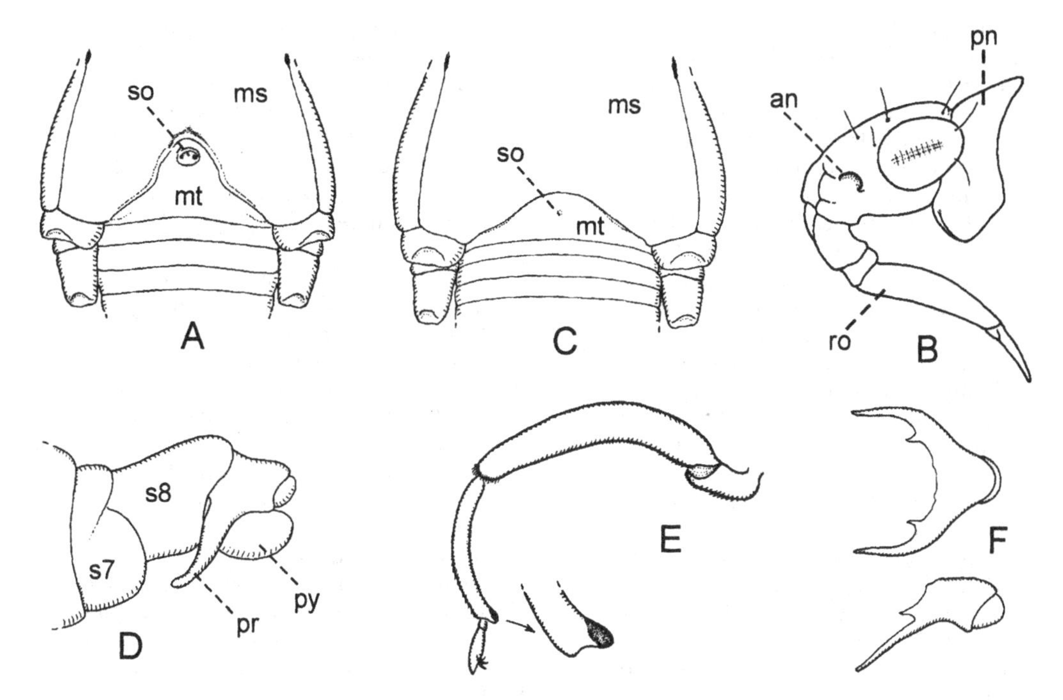

Figure 12.2. GERRIDAE. A, *Stenobates australicus*, apterous ♂: ventral view of thorax and abdomen (ms, mesosternum; mt, metatsernum; so, scent orifice). B, *S. biroi*: lateral view of head, antennae removed (an, antennal tubercle; pn, pronotum; ro, rostrum). C-D, *Rheumatometroides carpentaria*, apterous ♂: C, ventral view of thorax and abdomen (ms, mesosternum; mt, metasternum; so, scent orifice). D, lateral view of genital segments (pr, proctiger; py, pygopohore; s7, s8, segments 7 and 8). E, *Rheumatometra philarete*, apterous ♂; left fore leg (apex of tibia shown at higher magnification). F, *Calyptobates jourama*, apterous ♂: proctiger in dorsal view (top) and lateral view (bottom). Reproduced with modfication from: A, C-F, Andersen & Weir (1998); B, Andersen (1982).

isms in studies of sexual behaviour and sexual selection (Arnqvist 1997). Gerrid eggs are relatively large (0.8-1.2 mm), but tends to become smaller, relatively, in larger species (Andersen 1982). With the exception of the rhagadotarsines (Fig. 12.35B), the eggs are deposited below water or at the water surface, glued lengthwise to emergent or floating water plants. The embryo has a frontal egg-burster which is used to produce a longitudinal split in the egg shell through which the first instar nymph hatch.

Key to the Australian genera of the family Gerridae

1. First abdominal sternum distinctly visible (Fig. 12.1A; s1). Ventral lobes (bucculae) of head distinctly produced anteriorly (Fig. 12.1B; bu). Both male and female terminalia greatly prolonged (Figs 12.1 C-E), female with well developed, serrate ovipositor. (RHAGADOTARSINAE) .... *Rhagadotarsus*
- First abdominal sternum fused with metasternum (Fig. 12.3D). Ventral lobes (bucculae) of head not produced. Male and female terminalia not greatly prolonged, ovipositor not serrate (except second gonapophyses in *Stenobates* and *Rheumatometroides*) 2
2. Middle femur distinctly shorter than middle tibia (Figs 12.38A, C, 12.40A, 12.43B). Macropterous (long-winged) adults: venation of fore wing greatly reduced, distinct veins only present in basal third of wing, forming two closed cells (Fig. 12.1G). (TREPOBATINAE) 3
- Middle femora distinctly longer than middle tibia. Macropterous adults (if present): venation of fore wings not reduced, distinct veins present throughout wing, forming four closed cells (Fig. 12.1H) 6
3. Metasternal scent gland orifice indistinct

or absent in both sexes. Head short, antennal socket closer to eye than width of socket (Figs 12.34E). Fore tibia more or less cylindrical throughout, tarsi cylindrical or slightly flattened. Mesonotum of female medially sclerotised (at most with a median suture). Male proctiger unmodified or with slender antero-laterally directed lateral processes (Figs 12.2F). Freshwater 4

– Metasternal scent gland orifice distinct, especially in males (Figs 12.2A, C). Head long, antennal socket removed from eye by width of socket (Fig. 12.2C; an). Fore tibia distally widened, tarsi broad, distinctly flattened. Mesonotum of female with median, longitudinal membranous region (Fig. 12.43C). Male proctiger with long stout antero-laterally directed lateral processes (Fig. 12.2D; pr). Marine 5

4. Fore tibia with distal spur-like process (Fig. 12.2E). Antennal segments 1-3 with spinous hairs in distal fourth. Male fore femur strongly curved (Fig. 12.2E). Proctiger of ♂ not modified laterally .... *Rheumatometra*

– Fore tibia without distal spur-like process. Antennal segments without spinous hairs. Male fore femur only slightly curved. Proctiger of ♂ with slender antero-laterally directed lateral processes (Fig. 12.2F) .... *Calyptobates*

5. Pale markings on mesonotum consisting of elongate streaks or stripes on either side of the midline (Fig. 12.43B). Metasternum of ♂ strongly extended anteriorly, anterior margin curved; metasternal scent orifice large, situated on a low tubercle (Fig. 12.2A). First antennal segment of ♂ ventrally set with closely packed short to medium length hairs ..... *Stenobates*

– Pale markings on mesonotum large, elongate or quadrate on either side of the midline (Fig. 12.43A). Metasternum of ♂ weakly extended anteriorly (Fig. 12.2B), anterior margin straight, forming a regular triangle; metasternal scent orifice small, not situated on a tubercle. First antennal segment of ♂ without closely packed short to medium length hairs .... *Rheumatometroides*

6. Metasternum well developed, longer than second abdominal sternum, clearly reaching metacetabula laterally (Fig. 12.3D; mt). Middle tibiae without ventral hair-fringe. Claws of hind tarsus (if present) falcate. (GERRINAE) 7

– Metasternum reduced in size, not much longer than second (first visible) abdominal sternum, barely reaching metacetabula laterally (Fig. 12.1F; mt). Middle tibiae with ventral hair-fringe (Figs 12.21, 12.24, 12.28F, 12.29). Claws of hind tarsus modified, straight or S-shaped. (HALOBATINAE) 11

7. First antennal segment distinctly longer than second and third segment together (Fig. 12.5). Metathoracic spiracle located only a little more than its own length from the fore wing base (in apterous individuals from the lateral margin of pronotal lobe) .... *Aquarius*

– First antennal segment distinctly shorter than second and third segment together (Figs 12.8, 12.11A, 12.16). Metathoracic spiracle located much more than its own length from the fore wing base (Fig. 12.3A; sp; macropterous adult), in apterous individuals from the lateral margin of pronotal lobe (Fig. 12.3B; sp) 8

8. Pronotal lobe dark (black or dark brown) and shiny, with a median, longitudinal pale stripe. Anterior pronotum with a pair of elongate, pale spots in middle (Fig. 12.8). Metathoracic scent orifice circular, located on small, conical tubercle lying in a depressed area of metasternum .... *Limnogonus*

– Pronotal lobe chiefly pale and dull with median, longitudinal dark stripe (Figs 12.11B, 12.12A, B, 12.19). If pronotal lobe dark, then anterior pronotum with a pair of elongate, pale spots in middle (Figs 12.11A, 12.16). Metathoracic scent orifice transverse ovate, enclosed between two lip-shaped tubercles in posterior parts of metasternum............................. 9

9. Rostrum relatively short; third labial segment not reaching mesosternum when rostrum is folded back under the body (Fig. 12.3C; la3) .... *Tenagogerris*

– Rostrum relatively long; third segment reaching mesosternum when rostrum is folded back under the body 10

10. Postero-lateral corners of seventh abdominal segments produced into distinct connexival spines (Figs 12.3F, G; co). Male abdomen relatively long; abdominal sternum 7 shorter than preceding two sterna together, usually longer. Hind coxae of ♂ rarely extending beyond posterior margin of second abdominal sternum .... *Limnometra*

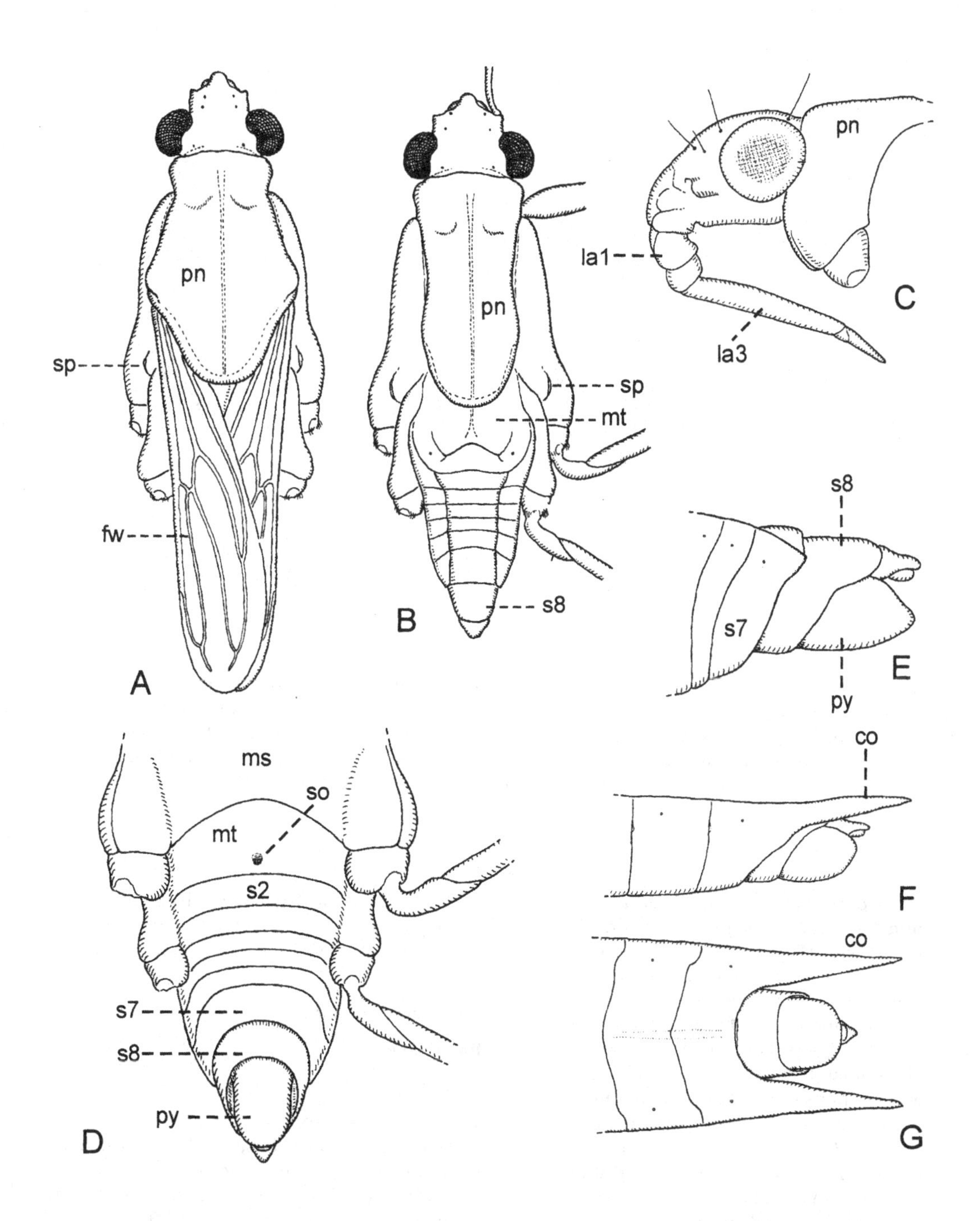

Figure 12.3. GERRIDAE. A, *Tenagogerris euphrosyne*, macropterous ♂ dorsal view, antennae and legs omitted (fw, fore wing; pn, pronotum; sp, metathoracic spiracle). B-E, *T. euphrosyne*, apterous ♂: B, dorsal view, antennae and most of legs omitted (mt, metanotum; pn, pronotum; s8, segment 8; sp, metathoracic spiracle). C, lateral view of head, antennae removed (la1, la2, labial segments 1 and 3). D, ventral view thorax and abdomen, most of legs omitted (ms, mesosternum; mt, metasternum, py, pygophore; so, scent orifice; s2, s7, sterna 2 and 7; s8, segment 8). E, *Tenagogonus australiensis*, apterous ♂: lateral view of abdominal end (py, pygophore; s7, sternum 7; s8, segment 8). F-G, *Limnometra cursitans*, macropterous ♂: F, lateral view of abdominal end. G, ventral view of abdominal end. Reproduced with modification from Andersen & Weir (1997).

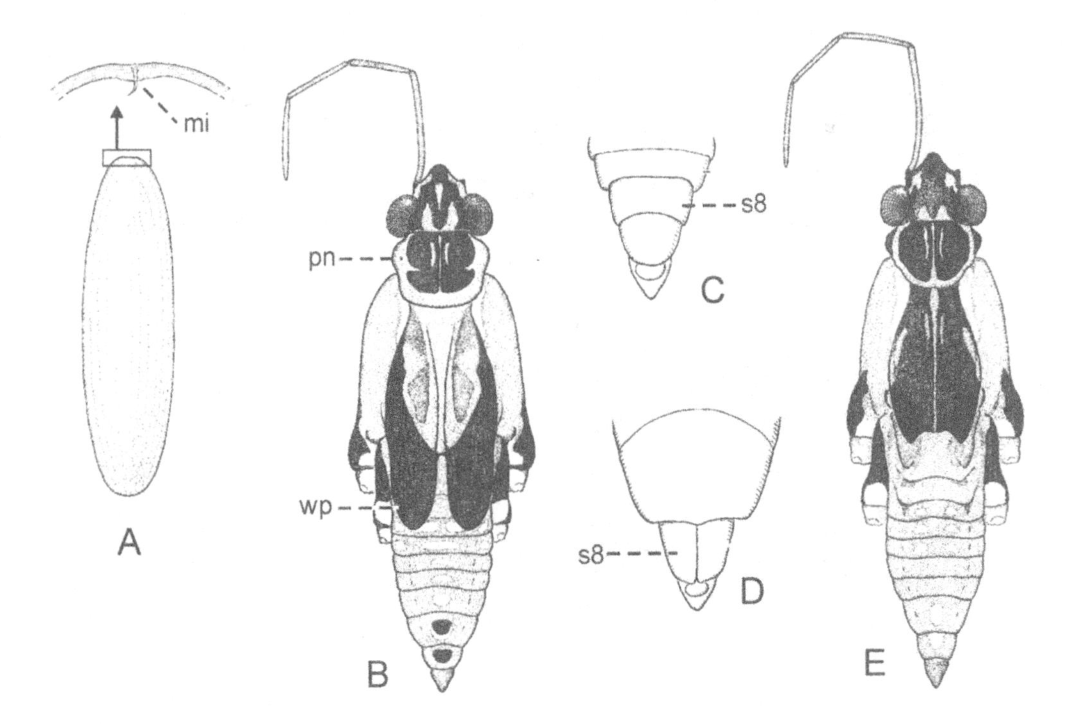

Figure 12.4. GERRIDAE. A, *Tenagogerris euphrosyne*, ripe ovarian egg with anterior end and micropyle (mi) shown at higher magnification. B-D, *T. euphrosyne*, 5th instar nymph: B, dorsal view of macropterous form; legs omitted (pn, pronotum; wp, wing-pad). C, ventral view of abdominal end of ♂ (s8, segment 8). D, ventral view of abdominal end of ♀. E, *T. femoratus*, dorsal view of 5th instar nymph of apterous form; legs omitted. Reproduced with modification from Andersen & Weir (1997).

– Postero-lateral corners of seventh abdominal segments not produced into distinct spines (Fig. 12.3E). Male abdomen relatively short; abdominal sternum 7 at least as long as preceding two sterna together. Hind coxae of ♂ almost reaching or surpassing posterior margin of second abdominal sternum .... *Tenagogonus*

11. Median parts of thoracic and abdominal dorsum with extensive pale markings (Fig. 12.21). Intersegmental suture between meso- and metanotum distinct (Fig. 12.23A). Male fore femora with a tubercle on ventral surface (Fig. 12.21). Proctiger of ♂ suboval in outline (Fig. 12.23C; pr). Freshwater .... *Austrobates*

– Thoracic and abdominal dorsum uniformly dark, usually covered by greyish pubescence (Figs 12.24, 12.29). Intersegmental suture between meso- and metanotum indistinct (Fig.12.23D). Male fore femora not modified on ventral surface as above. Proctiger of ♂ laterally expanded, more or less pentagonal in outline (e.g., Figs 12.26A-C; pr). Marine (with few exceptions) .... *Halobates*

SUBFAMILY GERRINAE

Pond Skaters

Identification

This subfamily comprises medium-sized to large or very large water striders, with the middle femora distinctly longer than the middle tibiae, and with distinct veins throughout the fore wings, forming four closed cells (Fig. 12.1H). The head is elongate, directed forward, with large eyes and four pairs of cephalic trichobothria located along the inner margin of eyes (Fig. 12.3C, 12.15A). The ventral lobes (bucculae) of head small, not produced in front of rostral base (Fig. 12.15C). Pronotum of apterous form longer than head length (with exceptions), covering mesonotum (Fig. 12.3B; pn). Fore femora usually simple, more or less incrassate, rarely modified beneath;

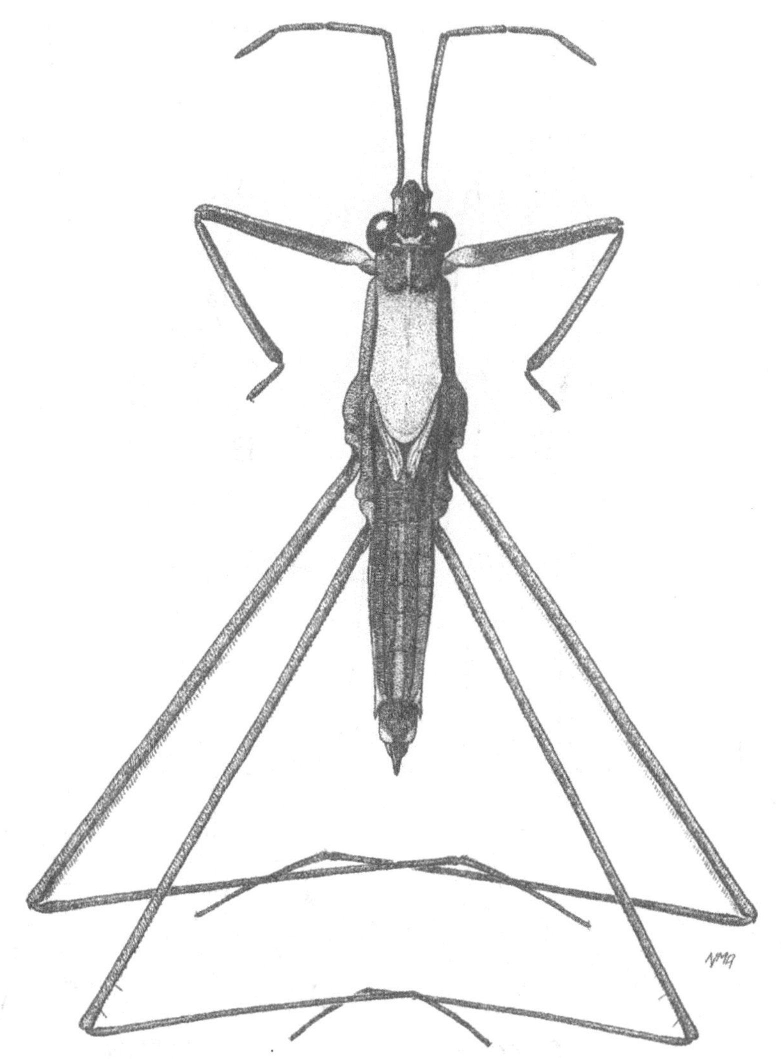

Figure 12.5. GERRIDAE. *Aquarius fabricii*, micropterous ♂. Reproduced from Andersen & Weir (1997).

tibiae cylindrical, without subapical spur; first segment of fore tarsus subequal to or slightly shorter than second segment. Middle femora distinctly longer than tibiae; claws of middle legs small but distinct, those of hind legs sometimes reduced. Metasternal scent apparatus present, scent orifice usually located on a tubercle. Genital segments of male relatively large, distinctly protruding from pregenital abdomen. Male segment 8 long, pygophore elongate, usually simple (Fig. 12.3D; py), parameres small, cone-shaped. Female ovipositor relatively short, first gonapophyses nonserrate.

Overview

Andersen (1995b) gave an overview of the Gerrinae and presented a phylogeny for the genera included in this subfamily. Following the taxonomic revisions by Andersen (1975, 1990, 1993b), Andersen & Spence (1992), Nieser & Chen (1992), Andersen & Weir (1997), and D. Polhemus & J. Polhemus (1997), there are 176 described species placed in two tribes and 14 genera worldwide. Additional genera are known as fossils from the Paleocene-Eocene of northern Europe and North America (Andersen 1998, 2000c). Five genera and 13 species have been recorded from Australia. They all belong to the tribe Gerrini.

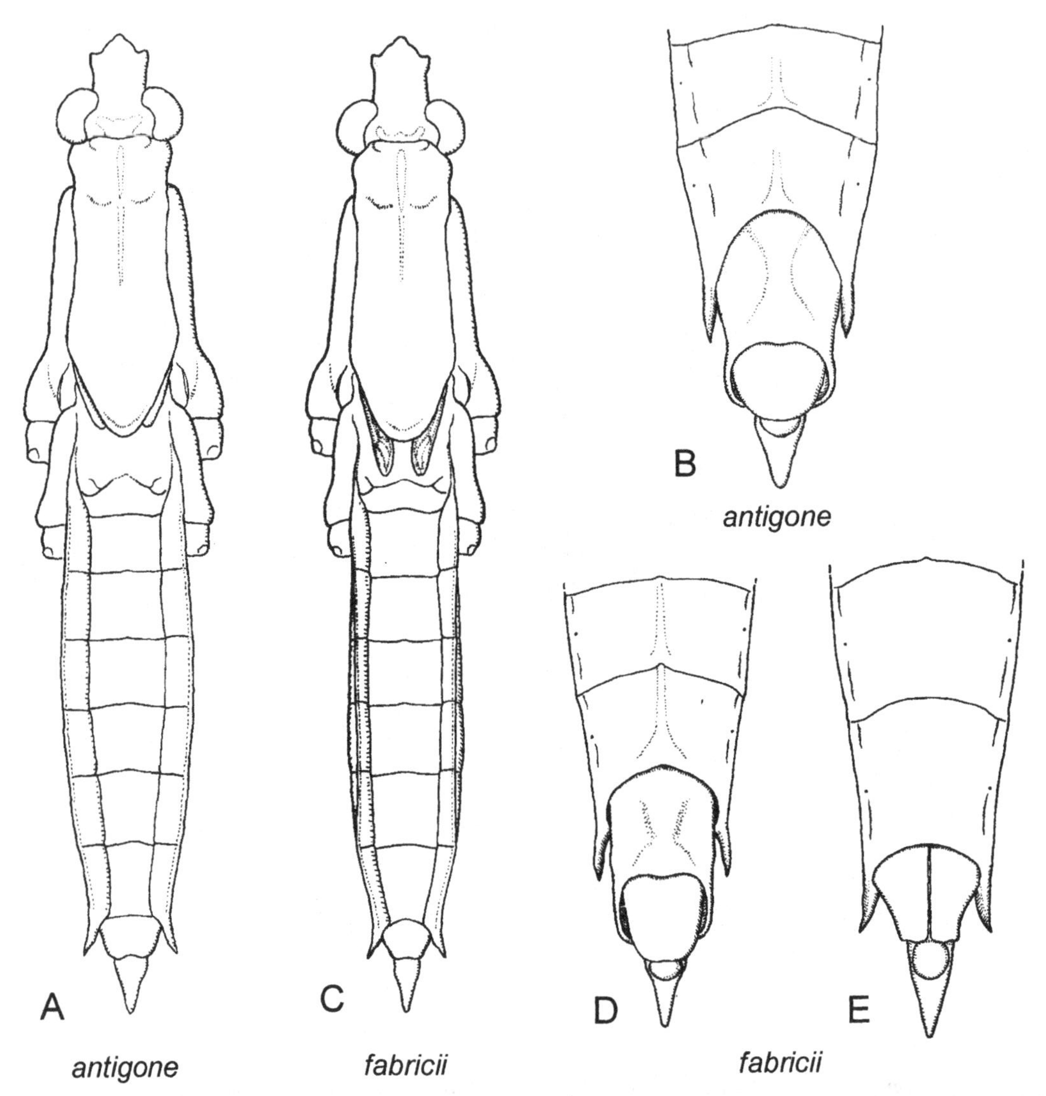

Figure 12.6. GERRIDAE. A-B, *Aquarius antigone*. A, dorsal view of apterous ♀; antennae and legs omitted. B, ventral view of abdominal end of apterous ♂. C-E, *A. fabricii*: C, dorsal view of micropterous ♀; antennae and legs omitted. D, ventral view of abdominal end of micropterous ♂. E, ventral view of abdominal end of micropterous ♀. Reproduced from Andersen & Weir (1997).

AQUARIUS

Identification

Large, chiefly dark coloured water striders, length 10-14 mm, with relatively short rostrum and antenna, antennal segment 1 being longer than segment 2 and 3 together. Most adults encounted are either apterous (*A. antigone*; Fig. 12.6A) or micropterous (*A. fabricii*; Fig. 12.6C), although the macropterous form may be more frequent in the latter. Body chiefly dark coloured; pronotum with pale median longitudinal stripe anteriorly. Antenna short and robust, much shorter than half body length, with first segment distinctly longer than second and third segment together. Pronotum large, always covering mesonotum. Metathoracic spiracle located only a little more than its own length from the fore wing base (in apterous individuals from the lateral margin of pronotal lobe). Fore wings not reaching abdominal end in either sex; second bifurcation of M+Cu distinctly removed from point of origin

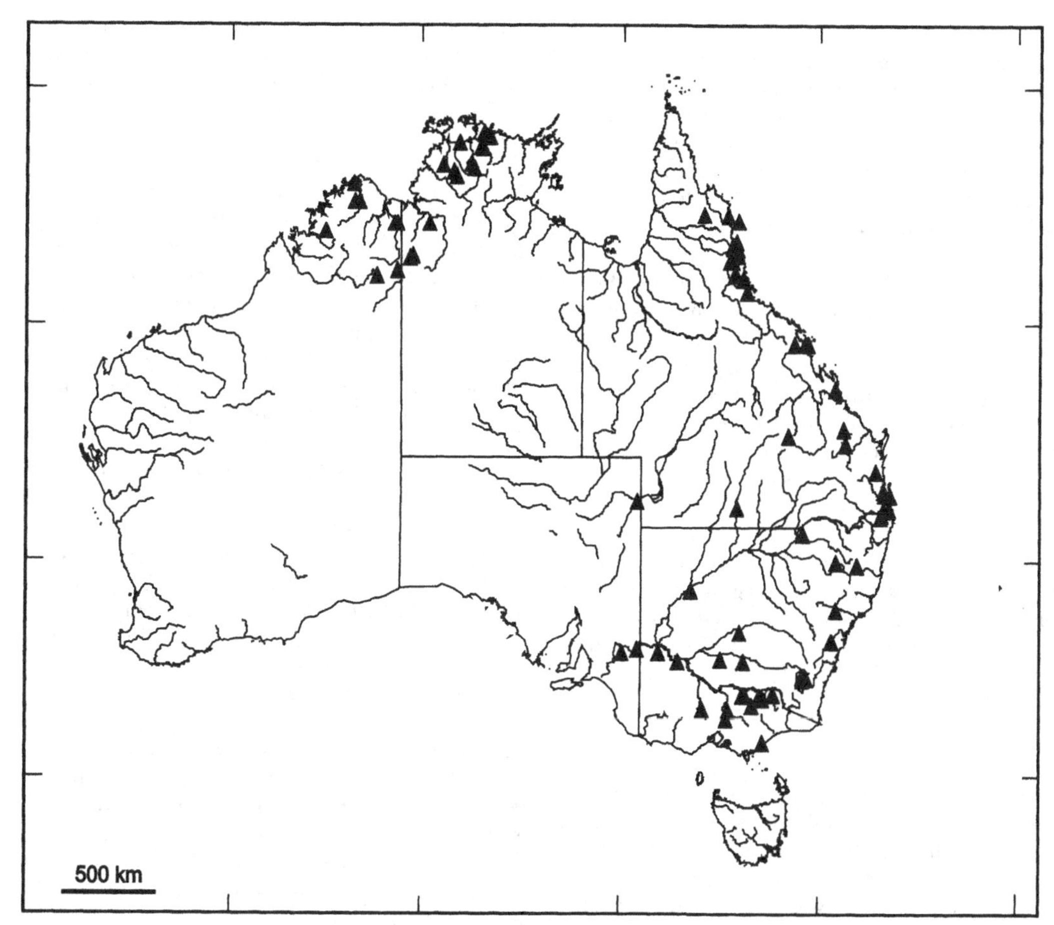

Figure 12.7. GERRIDAE. Distribution of *Aquarius* species in Australia.

of anterior cross-vein. Metasternal scent orifice transverse ovate, enclosed between a pair of low, transverse tubercles situated in posterior part of metasternum. Fore femur more or less incrassate, especially in male. Middle femur slightly shorter than body; tibia distinctly shorter than femur; tarsus less than half as long as tibia. Hind femur subequal in length to middle femur. Claws distinct in all legs. Abdomen long, slender in males, sometimes slightly dilated in females; posterior corners of the connexiva of both sexes spine-like produced (Figs 12.6A-E); hind margin of sternum 7 simple, concave. Male genital segments large; proctiger elongate, more or less acuminate (Figs 12.6B, D).

The genus *Aquarius* Schellenberg was revised by Andersen (1990). At present there are 19 known species and subspecies with an additional fossil species known from the Miocene of Tibet (Andersen 1998). Two species are known from Australia.

Key to Australian species

1. Pronotum uniformly dark. Proctiger not distinctly pointed (Figs 12.6A, B). Macropterous (rare) and apterous (Fig. 12.6A) with wing rudiments not surpassing apex of pronotal lobe. Length 11.0-12.5 mm (♂), 12.8-14.1 mm (♀). Australian Capital Territory, New South Wales, Queensland, South Australia, Victoria..... .... *antigone* (Kirkaldy)
– Pronotum dark anteriorly, largely yellowish brown posteriorly (Fig. 12.5). Proctiger distinctly pointed (Figs 12.6C-E). Macropterous, brachypterous, or micropterous (Figs 12.5, 12.6C) with wing rudiments reaching first or second abdominal segment. Length 10.3-12.4 mm (♂), 11.6-13.8 mm (♀). Northern Territory, Western Australia .... *fabricii* Andersen

Biology

The two Australian species of *Aquarius* inhabit rivers and streams, but *A. fabricii* is also found in riverine pools and billabongs. Their life history and biology is poorly known. In the southeast, nymphs of *A. antigone* are found from March to April and November to December, whereas in the northeast, nymphs are also found from March to August. In the tropical North, nymphs of *A. fabricii* are recorded from all months from May to December. Since the species inhabit flowing, relatively cold waters, there may be one or at most two generations per year. Dispersal between drainage systems may be limited since winged adults are rare, especially in *A. antigone*. The biology, ecology, and behaviour have been extensively studied in several species of *Aquarius*, in particular in Europe and North America. In species like the European *A. najas* (De Geer) and North American *A. remigis* (Say), the male stays with the female for a long time following mating. Females of these species exhibit food-based territoriality. Casual observations of *A. antigone* have failed to reveal similar behavioural features (Andersen, unpublished).

Distribution

In Australia (Fig. 12.7), the two *Aquarius* species are confined to the Kimberley district of Western Australia, the northernmost parts of Northern Territory (*A. fabricii*), the coastal areas of Queensland, large parts of New South Wales, Australian Capital Territory, Victoria, and eastern South Australia (*A. antigone*). The Species of *Aquarius* occur world wide, but are most abundant in the northern temperate and subtropical zones. The two Australian species belong to the *A. paludum* group, otherwise distributed in the Palearctic, Afrotropical, and Oriental Regions (but absent from New Guinea).

LIMNOGONUS

Identification

Medium-sized water striders, length 4.0-11 mm, with pronotal lobe dark (black or dark brown), with a median, longitudinal pale stripe; anterior pronotum with a pair of elongate, pale spots in middle (Fig. 12.8). *Limnogonus* species have a unique structure of the metasternal scent apparatus, which has a circular orifice situated on a cone-shaped tubercle in the middle of a depression (Figs 12.15F, G). Adults are either apterous (Fig. 12.8) or macropterous. Body chiefly dark and shiny above and pale below; meso- and metapleuron with longitudinal pale stripes (Figs 12.9B, E, H). Antenna long and slender, subequal to or slightly shorter than body length; first segment longest, but distinctly shorter than second and third segment together. Rostrum slender, its apex reaching anterior part of mesosternum. Pronotum large, always covering mesonotum. Fore wings clearly surpassing abdominal end in both sexes; second bifurcation of M+Cu approaching point of origin of anterior cross-vein. Metathoracic spiracle located more than its own length from wing bases. Fore femora moderately thickened. Middle femur slightly shorter than body; tibia only slightly shorter than femur; tarsus almost half as long as tibia. Hind femur a little longer than middle femur. Claws distinct in fore and middle legs, reduced in hind legs. Abdomen slightly shortened in male; posterior corners of the connexiva usually obtuse in both sexes, only spine-like produced in few species (not in Australia); hind margin of sternum 7 simple, concave (♂), straight or produced in middle (♀) (Figs 12.9A, C, G). Male genital segments large but simple; hind margin of segment 8 concave or slightly produced in middle.

The Old World species of the genus *Limnogonus* Stål was revised by Andersen (1975). There are two subgenera, *Limnogonus* s. str. and *Limnogonoides* Poisson, the latter being predominantly Afrotropical. The 19 known species and subspecies belonging to the subgenus *Limnogonus* are distributed world wide in the tropics. The four Australian species was revised by Andersen & Weir (1997).

Key to Australian species

1. Pronotal lobe cinnamon brown or brownish yellow (Fig. 12.8). Body robust with length about 2.7x maximum width. Posterior margin of female sternum 7 simple, not produced in middle (Fig. 12.9A). Length 5.2-7.0 mm (♂), 6.0-8.2 mm (♀). Northern Territory, Queensland, Western Australia .... *windi* Hungerford & Matsuda
– Pronotal lobe blackish brown or black (teneral individuals usually paler). Body slender with length 3x maximum width or more. Posterior margin of female sternum 7 produced in middle (Figs 12.9C, D, G) 2
2. Pale stripe on upper part of mesopleuron tapering in width posteriorly, ending in front of and below the metathoracic spiracle (Fig. 12.9B). A broad transverse band of silvery pubescence in middle of abdominal terga (apterous form). Fore-

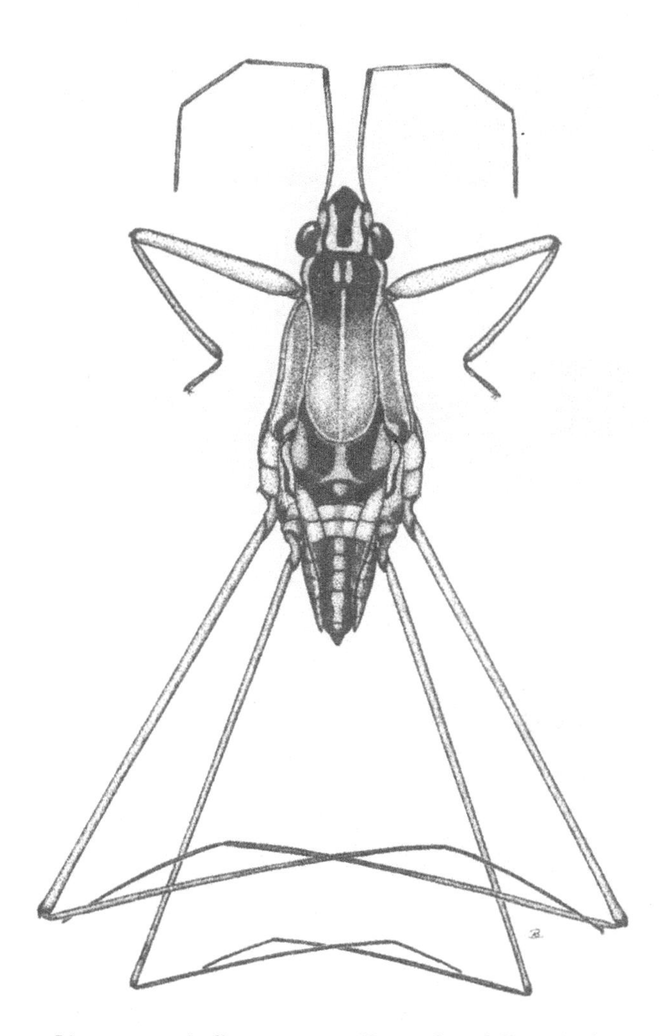

Figure 12.8. GERRIDAE. *Limnogonus* windi, apterous ♂. Reproduced from Andersen & Weir (1997).

wings (macropterous form) greyish brown with dark brownish veins. Posterior margin of female sternum 7 with prominent tooth in middle (Figs 12.9C, D). Length 6.8-8.4 mm (♂), 7.4-11.2 mm (♀). Northern Territory, Queensland, South Australia, Western Australia........... .... *fossarum gilguy* Andersen & Weir

– Pale mesopleural stripe widened posteriorly, ending in front of or above the metathoracic spiracle (Figs 12.9E, H). Abdominal terga (apterous form) with patches of silvery pubescence but no transverse band. Forewings (macropterous form) greyish with contrasting black veins. Posterior margin of female sternum 7 produced (Fig. 12.9G) but without a prominent tooth in middle 3

3. Pale mesopleural stripe posteriorly ending in front of metathoracic spiracle (Fig. 12.9E). Male abdominal venter seen in profile distinctly convex (Fig. 12.9F). Supracoxal lobe of metathorax dorsally dark, ventrally light. Length 7.0-8.4 mm (♂), 7.4-8.7 mm (♀). Northern Territory, Queensland, Western Australia .... *luctuosus* (Montrousier)

– Pale mesopleural stripe posteriorly enclosing metathoracic spiracle (Fig. 12.9H). Male abdominal venter seen in profile not distinctly convex (Fig. 12.9I). Supracoxal lobe of metathorax light with a median dark band. Length 6.2-7.6 mm (♂), 6.8-8.0 mm (♀).Northern Territory, Queensland, Western Australia .... *hungerfordi* Andersen

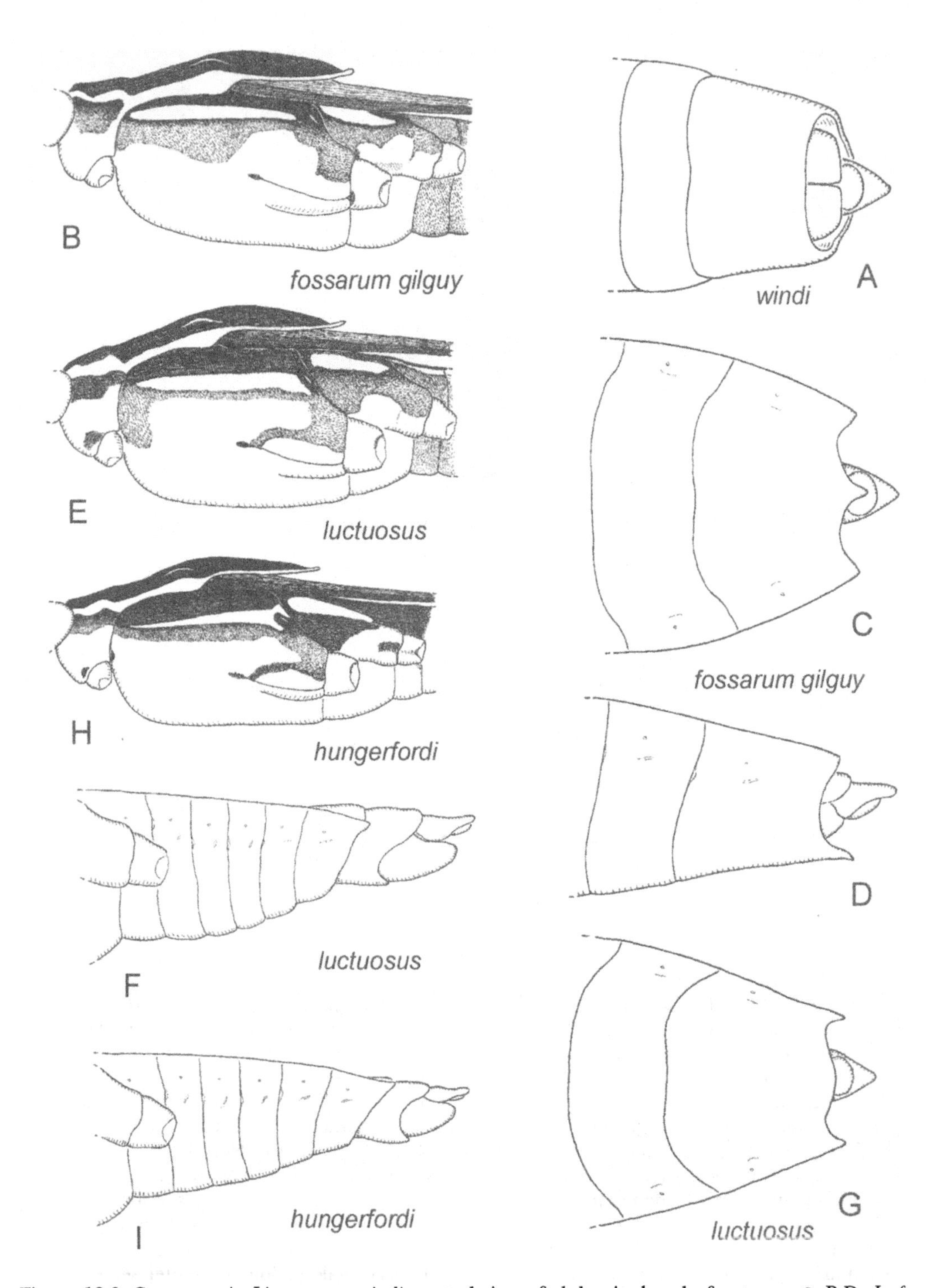

Figure 12.9. GERRIDAE. A, *Limnogonus windi*, ventral view of abdominal end of apterous ♀. B-D, *L. fossarum gilguy*: B, lateral view of thorax of apterous ♂; legs removed. C, ventral view of abdominal end of apterous ♀. D, lateral view of abdominal end of apterous ♀. E-G, *L. luctuosus*: E, lateral view of thorax of apterous ♂. F, lateral view of abdomen of apterous ♂. G, ventral view of abdominal end of apterous ♀. H-I, *L. hungerfordi*: H, lateral view of thorax of apterous ♂. I, lateral view of abdomen of apterous ♂. Reproduced from Andersen & Weir (1997).

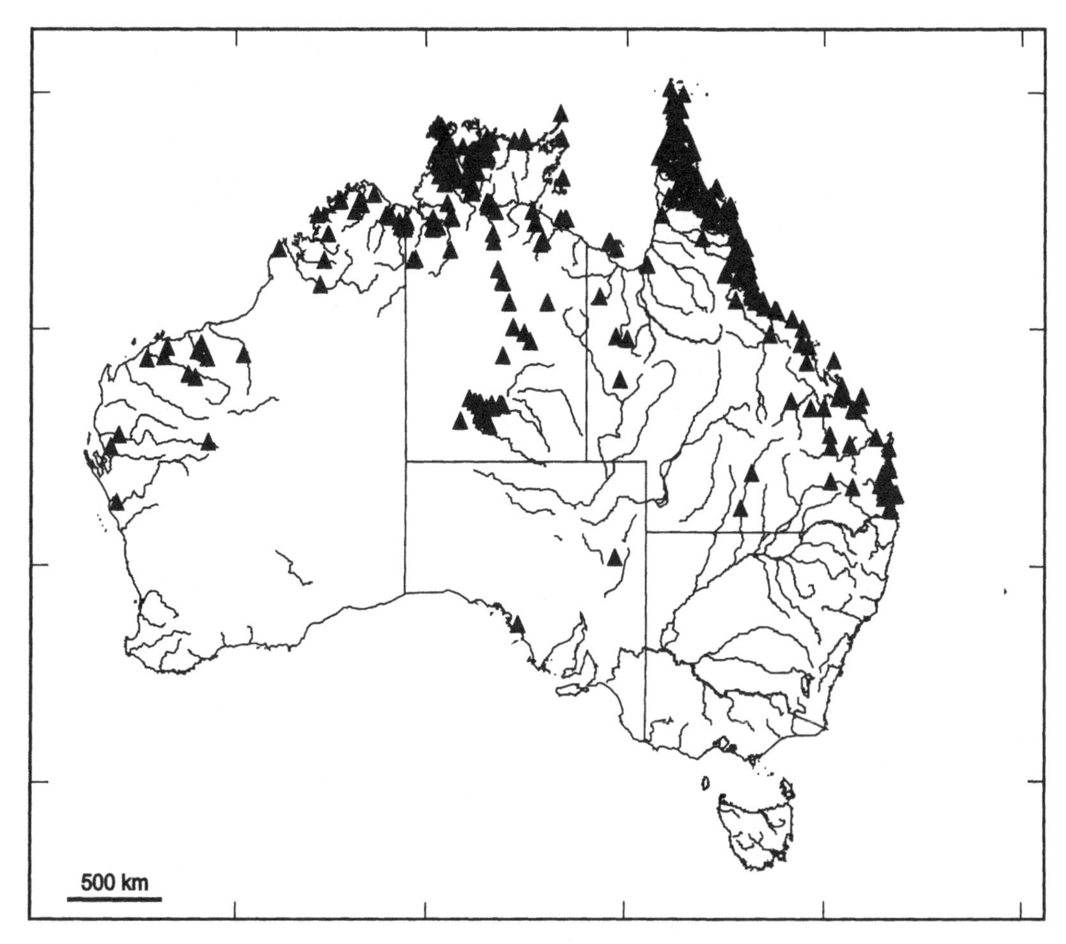

Figure 12.10. GERRIDAE. Distribution of *Limnogonus* species in Australia.

Biology

Limnogonus species are the most eurytopic Australian gerrids with habitats ranging from slow flowing streams, vegetated, still water bodies, to the most temporary waterholes. They may also be found in brackish waters. Observations or experimental data on life history traits have not been previously published for Australian species of *Limnognus*. In the tropical North of Australia, nymphs are found throughout the year with peak numbers from May to October. These species may therefore breed more or less continuously throughout the year, without any detectable reproductive diapause. Life history studies of the common pond skater *L. fossarum fossarum* in tropical and subtropical Asia, point to the same conclusion, although Selvanayagam & Rao (1988) collected more eggs in natural habitats during the rainy season (November-December) than in the following three months. In a laboratory study by Hoffmann (1936a), a total of 141-179 eggs were laid by four females. Egg-laying took place mostly if not entirely at night, and eggs were laid in batches of 10-20 eggs. The development from egg to adult lasted from 46 to 57 days at a temperature of 28-30°.

All four Australian species of *Limnogonus* are wing dimorphic with a high frequency of macropterous adults in populations of *L. fossarum gilguy*, *L. luctuosus*, and *L. hungerfordi*. The flight activity of these species is seemingly high as they often are collected in light traps. This means that they have the capacity to colonise new and temporary water bodies. They are also the most widespread species of water striders in the Indo-West Pacific region where they have colonised even the most remote islands of the Pacific Ocean (Andersen 1971, 1975).

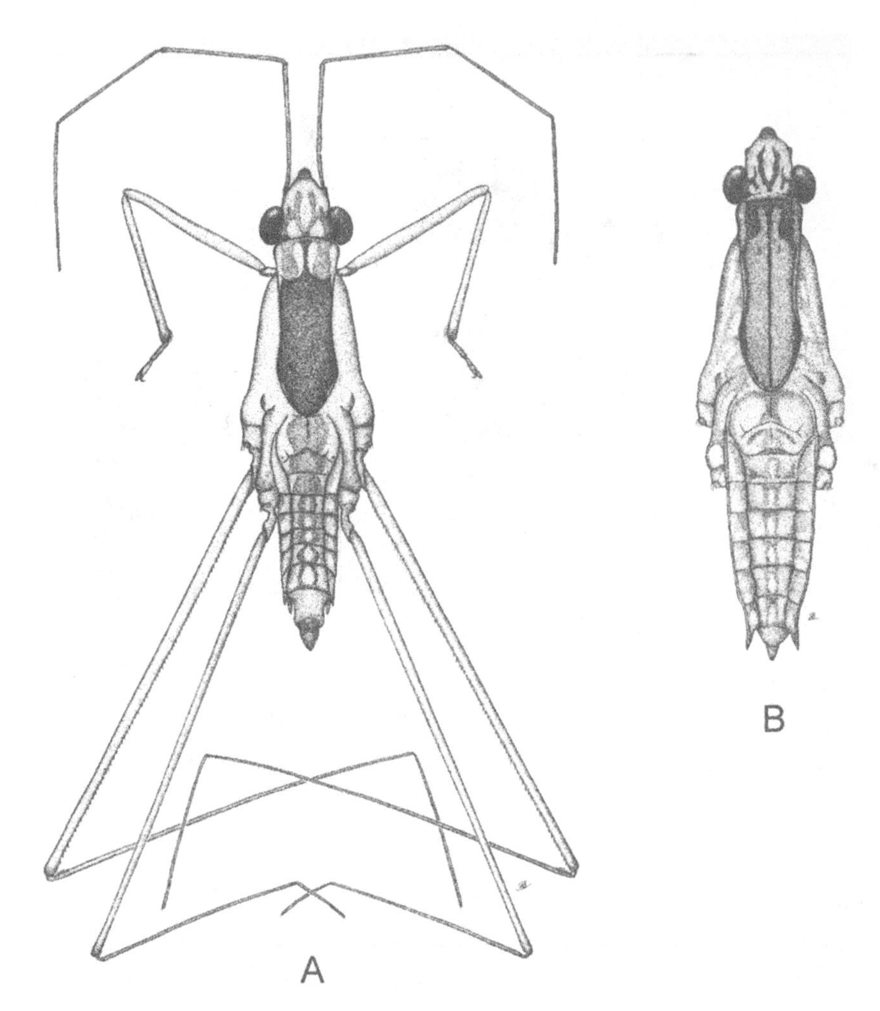

Figure 12.11. GERRIDAE. A-B, *Limnometra lipovskyi*: A, dorsal view of apterous ♂. B, dorsal view of apterous ♀; antennae and legs omitted. Reproduced from Andersen & Weir (1997).

Distribution

Three of the four known Australian species of *Limnogonus* are widely distributed in the Oriental and Australasian Region, extending eastwards into the Pacific Islands. Only *L. windi* is endemic to Australia. In Australia (Fig. 12.10), species of *Limnogonus* are found in Western Australia, in Northern Territory southwards to Alice Springs, throughout Queensland, and in the Flinders Ranges, South Australia (*L. fossarum gilguy*).

LIMNOMETRA

Identification

This genus includes large or very large water striders, length 6.9-21 mm long (the largest Australian water striders known so far). The pronotal lobe is chiefly pale and with a median, longitudinal dark stripe (Figs 12.11B, 12.12A, B). The metathoracic scent orifice is transverse ovate, enclosed between two lip-shaped tubercles in the posterior part of metasternum. The postero-lateral corners of seventh abdominal segments are produced into distinct connexival spines (Figs 12.13A-D, H, J). Adults are either macropterous (Figs 12.12A-C) or apterous (Figs 12.11A, B), rarely micropterous or brachypterous. The body is elongate, chiefly brownish above and yellowish below; pronotum usually with dark median, longitudinal stripe throughout its length (Figs 12.11B, 12.12A, B), rarely with more extensive darker areas (Figs 12.11A, 12.12C). Antennae long and

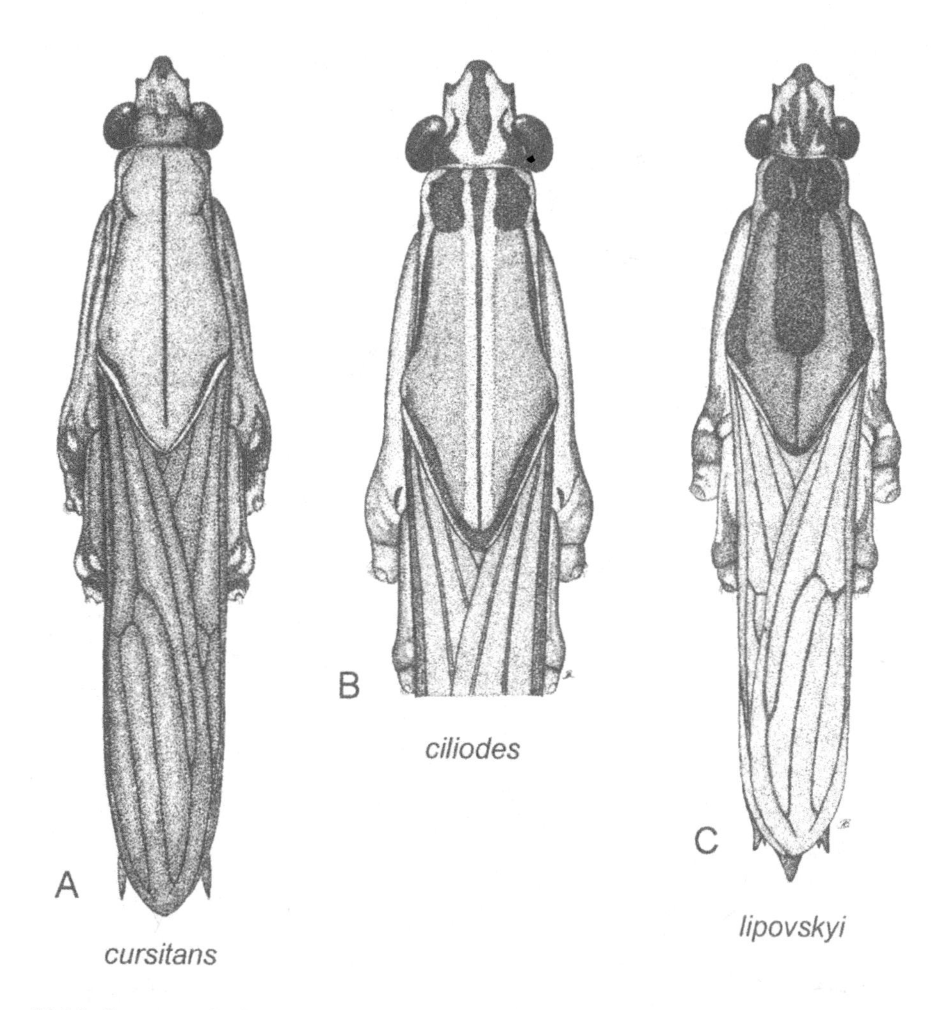

Figure 12.12. GERRIDAE. A, *Limnometra cursitans*, macropterous ♂; antennae and legs omitted. B, *L. ciliodes*, macropterous ♂; antennae and legs omitted. C, *L. lipovskyi*, macropterous ♂; antennae and legs omitted. Reproduced from Andersen & Weir (1997).

very slender, more than two thirds of the total length total length; segment 1 longest, but much shorter than segment 2 and 3 together. Rostrum slender and long, its apex reaching anterior part of mesosternum. Pronotum large, always covering mesonotum. Fore wings reaching abdominal end; second bifurcation of M+Cu originating from point of origin of anterior cross-vein. Metathoracic spiracle located more than its own length from wing bases. Male fore femora slender, a little thicker than middle femur; first tarsal segment subequal to or longer than second segment. Middle femur as long as (♀) or slightly longer than body length (♂); tibia only slightly shorter than femur. Hind femora subequal to or longer than middle femur. Claws distinct in fore and middle legs, reduced in hind legs. Abdomen long, almost parallel-sided; hind margin of sternum 7 simple, concave (Figs 12.13A-D, H, I). Male genital segments large, but usually simple (exception *L. lipovskyi*; Figs 12.13A, B).

Previously treated as a subgenus of *Tenagogonus* (see below). Species of the genus *Limnometra* Mayr are generally larger and more elongate-bodied than species of *Tenagogonus*, with postero-lateral corners of abdominal segment 7 produced, forming distinct connexival spines. *Limnometra* was revised by Hungerford & Matsuda (1958) and Nieser & Chen (1992), and at present contains 27 species distributed throughout the Oriental and Australian regions, as far eastward as the Fiji Islands. The three Australian species were revised

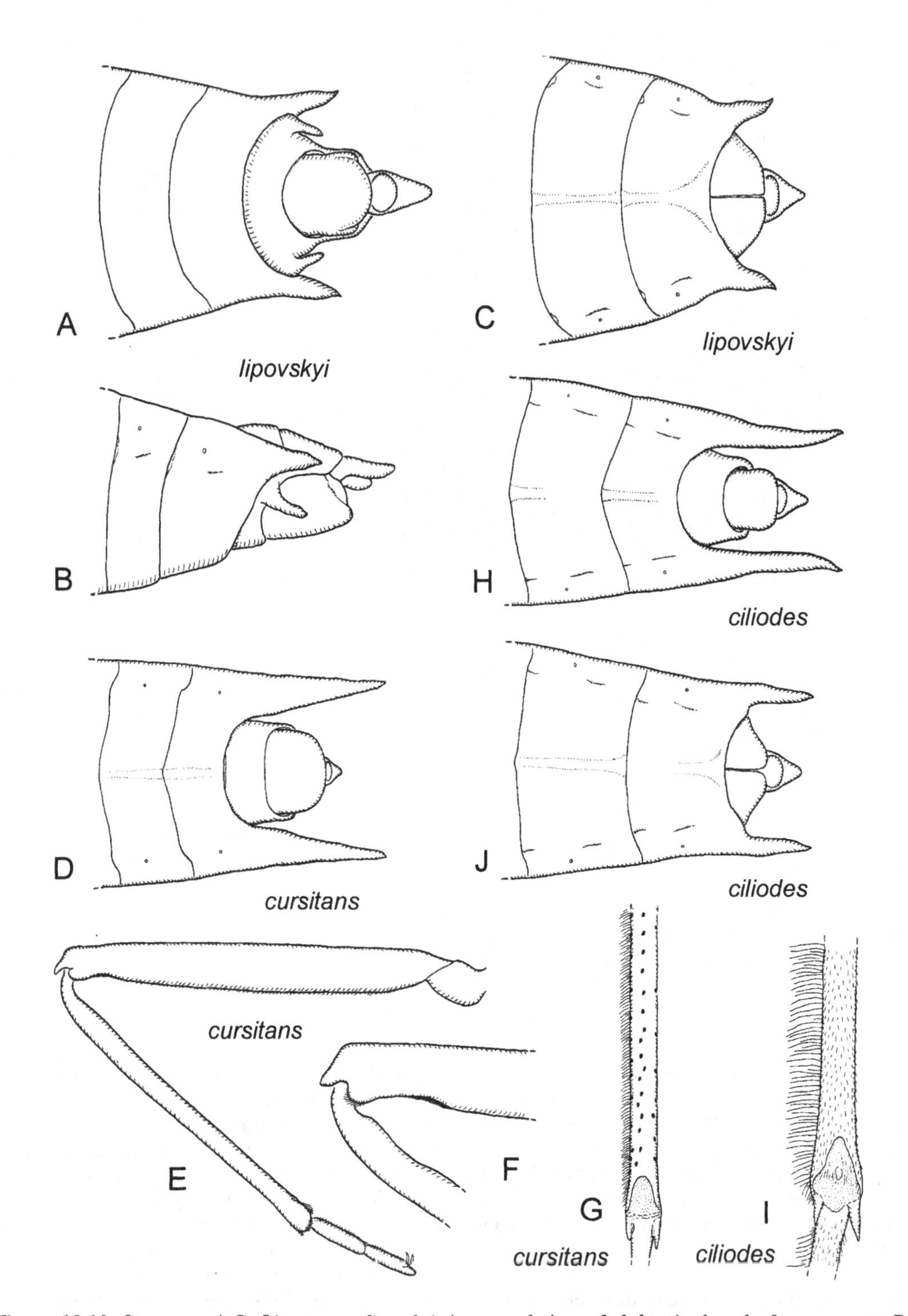

Figure 12.13. GERRIDAE. A-C, *Limnometra lipovskyi*: A, ventral view of abdominal end of apterous ♂. B, lateral view of abdominal end of apterous ♂. C, ventral view of abdominal end of apterous ♀. D-G, *L. cursitans*, macropterous ♂: D, ventral view of abdominal end. E, left fore leg. F, apex of left fore femur. G, apex of middle tibia. H-I, *L. ciliodes*, macropterous ♂: H, ventral view of abdominal end. I, apex of middle tibia. J, *L. ciliodes*: ventral view of abdominal end of macropterous ♀. Reproduced from Andersen & Weir (1997).

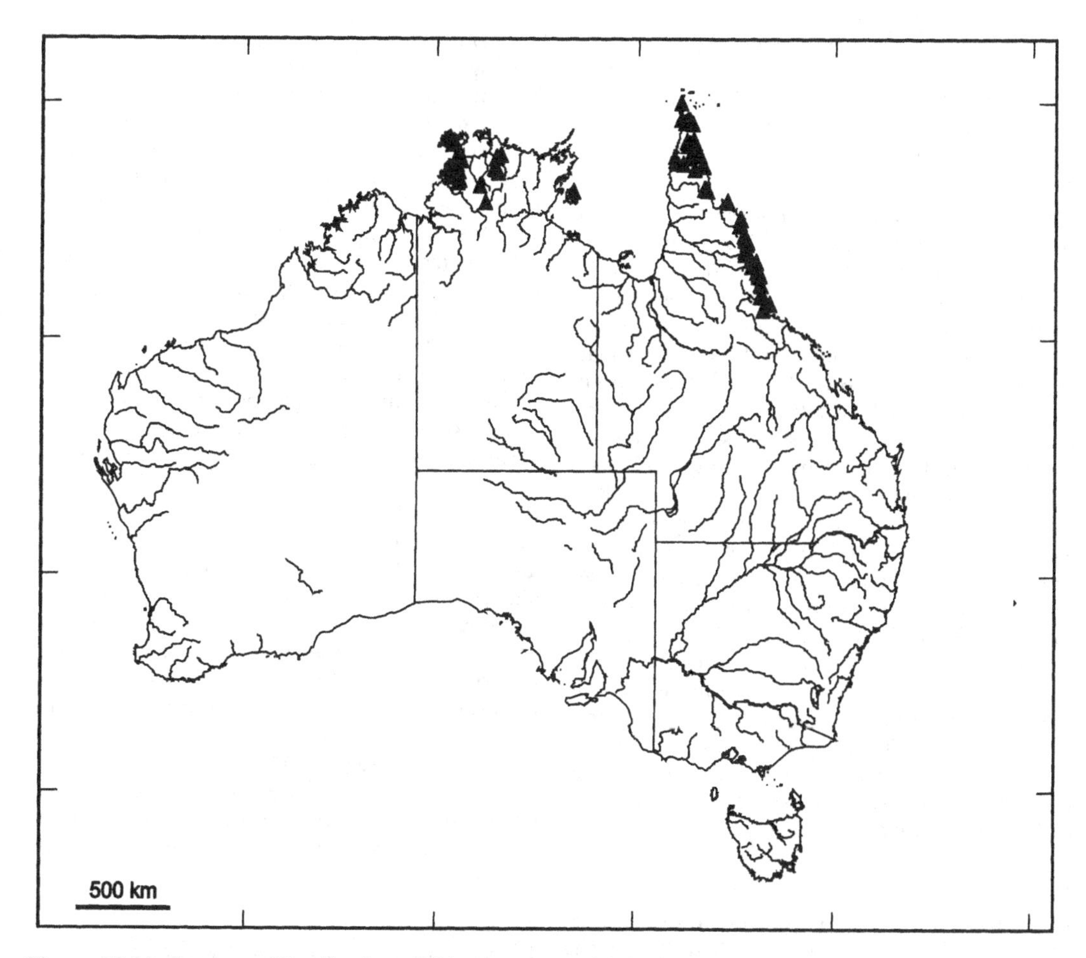

Figure 12.14. GERRIDAE. Distribution of *Limnometra* species in Australia.

by Andersen & Weir (1997) who clarified the identity of *L. cursitans* described as early as in 1775 by the famous Danish entomologist I. C. Fabricius.

Key to Australian species

1. Connexival spines not reaching beyond apex of genital segments (Figs 12.13A-C). Male abdominal segment 8 with a backward pointing tooth-like projection on either side (Figs 12.13A, B). Macropterous (Fig. 12.12C) or apterous (Figs 12.11A, B). Length 7.1-15.5 mm (♂), 9.7-12.8 mm (♀). Northern Territory, Queensland ..*lipovskyi* Hungerford & Matsuda
- Connexival spines reaching beyond apex of genital segments (Figs 12.13D, H, J). Male abdominal segment 8 not modified as above 2

2. Head and pronotum almost uniformly brownish (Fig. 12.12A). Middle and hind femora with pale apical annulations. Fore femur of ♂ thickened with subapical constriction (Figs 12.13E, F). Middle femora of ♂ with 1-3 rows of small blackish pegs along its entire length (Fig. 12.13G) and a hair fringe which is not as wide as femur. Always macropterous (Fig. 12.12A). Length 13.4-18.7 mm (♂) or 11.8-15.0 mm (♀). Northern Territory, Queensland .... *cursitans* (Fabricius)
- Head and pronotum with extensive dark markings (Fig. 12.12B). Middle and hind femora without pale annulations. Fore femur of ♂ slender throughout. Middle femur of ♂ without black pegs but with a hair fringe which is as wide as or wider than femur and continues onto tibiae

(Fig. 12.13J). Macropterous, rarely micropterous. Length 15.7-20.6 mm (♂), 14.4-17.3 mm (♀). Queensland................. ....*ciliodes* Andersen & Weir

Biology

Most Australian samples of *Limnometra* are from streams and stream pools in closed rainforests; occasionally from shallow pools in dry riverbeds. Very little is known about the biology of species. In the tropical north of Australia, nymphs are recorded from most months and the species may well breed more or less continuously throughout the year. Males of *L. cursitans* and *L. ciliodes* are among the largest water striders, not only in Australia, reaching 19-21 mm in length. *L. lipovskyi* males are extremely variable in size, with a size range of 7.1 mm to 15.5 mm, compared to a size range of 9.7 mm to 12.8 mm in females. Thus in this species some males are significantly larger than females, unlike most other water striders where the female is the larger sex. This pattern of sexual size dimorphism may be associated with male territoriality as known from some species of the holarctic genus *Limnoporus* (Wilcox & Spence 1986; Spence & Wilcox 1986; Andersen & Spence 1992; Andersen 1994). Unlike most members of the Gerrinae, several species of *Limnometra*, such as *L. cursitans*, are always macropterous.

Distribution

In Australia (Fig. 12.14), species of *Limnometra* are confined to the tropical north of Northern Territory and Queensland.

TENAGOGERRIS

Identification

Medium sized water striders, length 6.3-10 mm, with chiefly dark pronotal lobe and anterior pronotum with a pair of elongate, pale spots in middle (Figs 12.16, 12.17A, E, I). The rostrum is relatively short, third segment not reaching mesosternum when rostrum is folded back under the body (Figs 12.3C, 12.15C, D). Adults are either macropterous (Fig. 12.3A, Plate 6, G) or apterous (Fig. 12.3B, Plate 6, E). Body elongate oval, chiefly yellowish brown with dark markings above, yellowish below; pronotum either pale with dark median, longitudinal stripe throughout (Figs 12.16, 12.17E, I), or uniformly dark except for a pair of pale spots anteriorly (Fig. 12.17A). Antenna long and slender, 0.6-0.8x total length; segment 1 longest, slightly shorter than segment 2 and 3 together. Pronotum of apterous form narrowed but prolonged, reaching a little beyond posterior margin of mesonotum. Fore wings surpassing (♂) or reaching (♀) abdominal end. Second bifurcation of M+Cu either approaching point of origin of anterior cross-vein (Fig. 12.1H)or more or less distinctly removed from that point. Metathoracic spiracle located more than its own length from wing bases (Fig. 12.3A; sp). Mesosternum 3.8-4.2x length of metasternum; metasternum with transverse oval scent orifice located near posterior margin of metasternum, enclosed between pair of low, transverse tubercles (Fig. 12.3D; so). Male fore femora moderately thickened, a little less than twice as thick as middle femur; first segment of fore tarsus distinctly shorter than second segment. Middle femora about as long as body (♂), or slightly shorter (♀); middle tibia only slightly shorter than femur. Hind femora subequal to or little longer than middle femora. Claws distinct in fore and middle legs, reduced in hind legs. Male abdomen shortened and distinctly tapering in width posteriorly; posterior corners of connexiva obtuse. Male genital segments relatively large, but otherwise simple (Fig. 12.3D). Female abdomen relatively longer than in male; posterior corners of connexiva obtuse or slightly pointed but never forming spines (Figs 12.17D, H).

The genus *Tenagogerris* Hungerford & Matsuda was revised by Andersen & Weir (1997) and now contains three species which are all endemic to Australia.

Key to Australian Species

1. Pronotum uniformly dark except for a pair of pale, elongate spots anteriorly (Fig. 12.17A). Ventral surface of male fore femur with distinct, setose tubercle basally (Figs 12.17B, C). Sternum 7 of female abdomen distinctly produced in middle (Fig. 12.17D). Length 6.9-8.2 mm (♂), 9.0-10.0 mm (♀). Northern Territory, Western Australia............................. ....*femoratus* Andersen & Weir
- Pronotum pale, yellowish brown or ferrugineous, with more or less distinct dark markings (Figs 12.16, 12.17E, I). Ventral surface of male fore femur not modified (Figs 12.17F, G). Sternum 7 of female abdomen simple, not produced in middle (Fig. 12.17H)... 2
2. Larger species, length 6.8-8.9 mm (♂) or 8.4-10.0 mm (♀). Head and pronotum yellowish brown or ferrugineous with distinct dark markings (Figs 12.16, 12.17E). Australian Capital Territory, New South Wales, Queensland, South Australia, Victoria................................ *euphrosyne* (Kirkaldy)

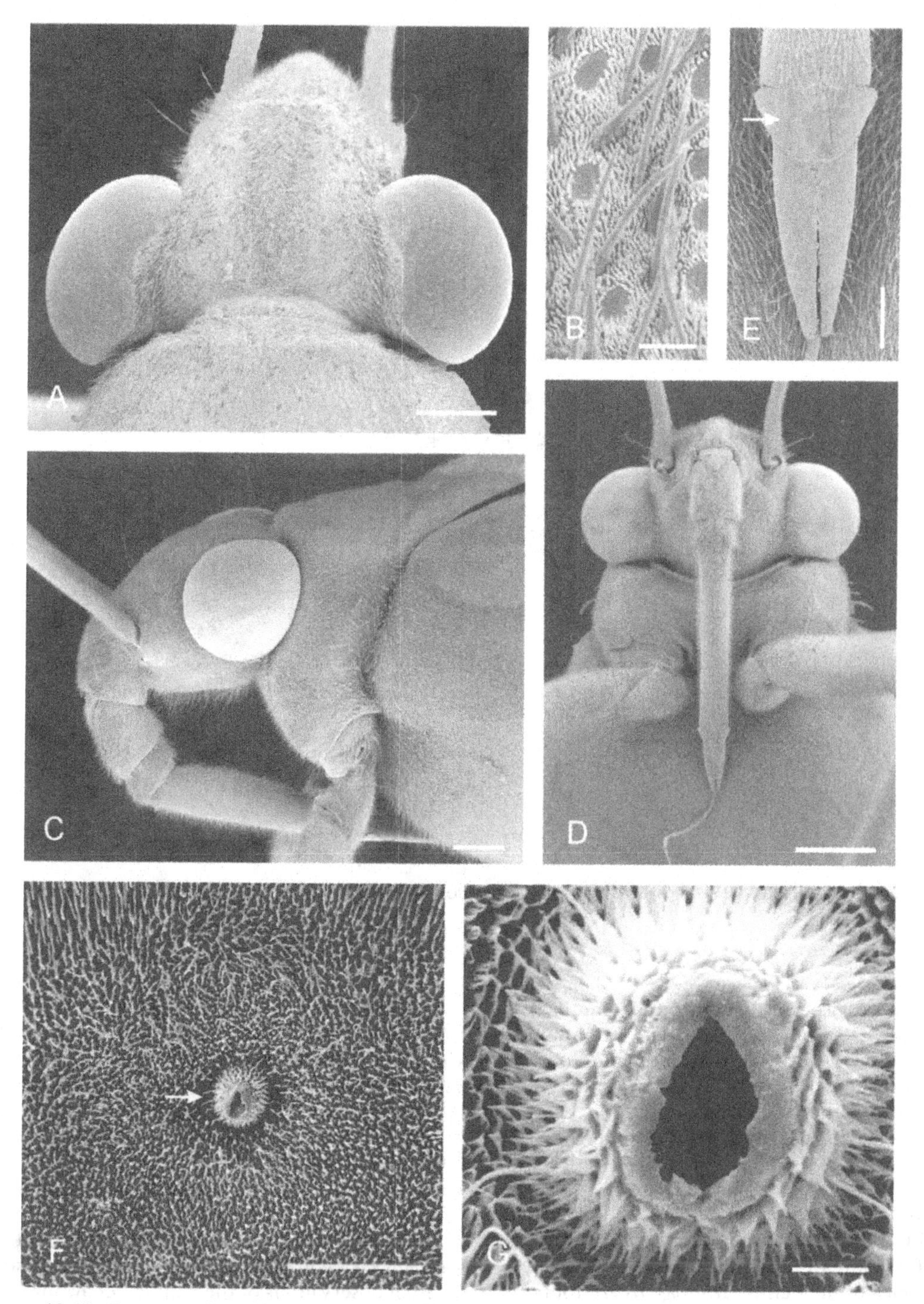

Figure 12.15. GERRIDAE. A-B, *Tenagogerris euphrosyne*, apterous ♂: A, dorsal view of head. B, structure of dorsal head surface. C-E, *Tenagogerris euphrosyne*, apterous ♀: C, lateral view of head and anterior thorax. D, ventral view of head and anterior thorax. E, apex of rostrum showing intercalary sclerites (arrow). F-G, *Limnogonus luctuosus*, apterous ♂: F, metasternum with scent orifice (arrow). G, scent gland orifice and its surroundings. Scales: A, C, 0.25 mm; B, 0.01 mm; D, 0.5 mm; E, 0.1 mm; F, 0.1 mm; G, 0.02 mm. A-E, Scanning Electron Micrographs (CSIRO Entomology, Canberra); F-G, reproduced from Andersen (1975).

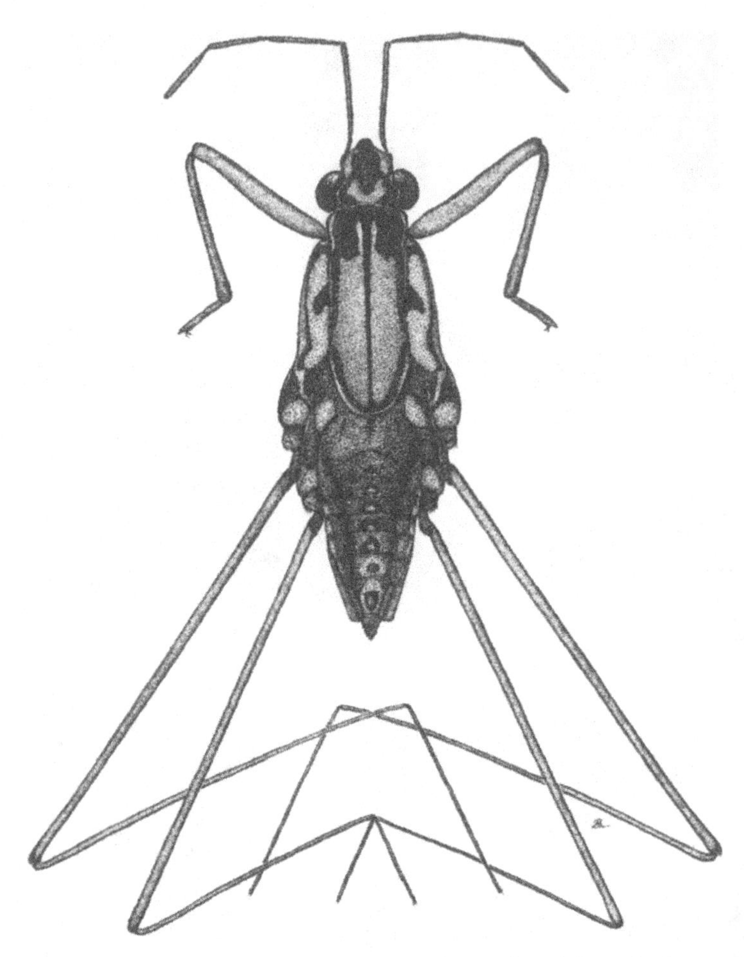

Figure 12.16. GERRIDAE. *Tenagogerris euphrosyne*, apterous ♀. Reproduced from Andersen & Weir (1997).

- Smaller species, length 5.6-7.8 mm (♂) or 7.8-9.3 mm (♀). Head and pronotum yellowish brown with indistinct dark markings (Fig. 12.17I). Northern Territory, Western Australia .... *pallidus* Andersen & Weir

Biology

Although *Tenagogerris euphrosyne* is the most common water strider in southeastern Australia, very little is known about the biology of this species. Species of the genus *Tenagogerris* are found in a wide variety of freshwater habitats, such as lakes, ponds, temporary pools, slowly running streams and small rivers, as well as pools and billabongs in river beds. In the Australian Capital Territory, New South Wales, and Victoria, adults are scarce during the months from May to October, and most samples with nymphs are from the period November-April. This may indicate that adults of this species have a reproductive "diapause" during the coldest part of the year, where they retreat to hiding places on land. The flightless adults, at least, may hide in the vegetation close to the breeding habitat as known from the northern temperate regions. In Queensland, adults and nymphs are found throughout the year, indicating a more or less continuous breeding activity. The male is smaller than the female in *T. euphrosyne* (female/male size ratio of 1.3) and casual observations of the mating behaviour indicate that the male remains in his position on the females back for some time after copulation (Plate 6, F). Ripe ovarian eggs (Fig. 12.4A) are 1.3-1.4 mm long, with a single micropyle (mi) transversing the anterior end. 5th and last instar nymphs have species-specific dark markings on the dorsal surface (Figs 12.4B, E).

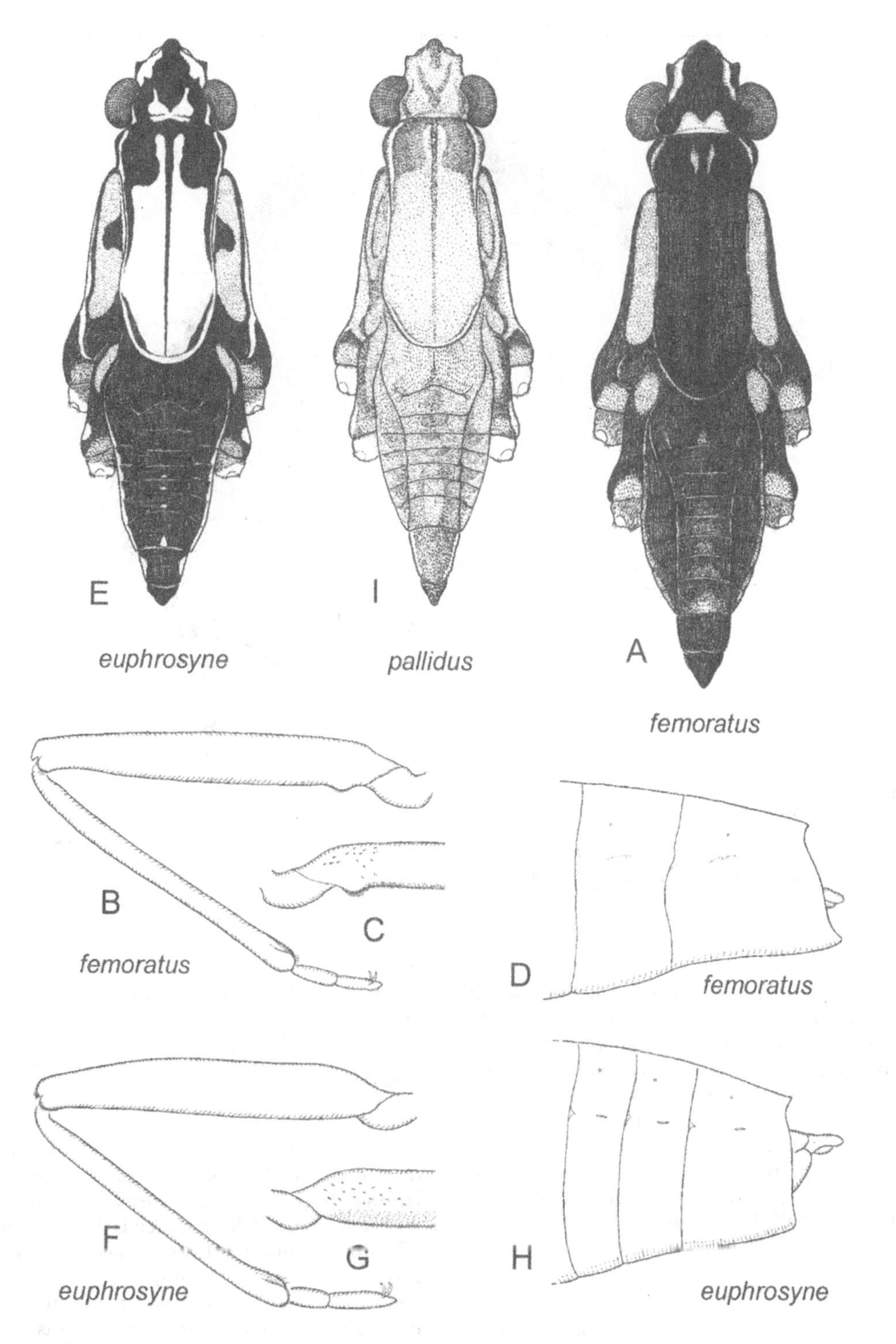

Figure 12.17. GERRIDAE. A-C, *Tenagogerris femoratus*, apterous ♂: dorsal view; antennae and legs omitted. B, left fore leg. C, apex of left fore femur. D, *T. femoratus*: lateral view of abdominal end of apterous ♀. E-G, *T. euphrosyne*, apterous ♂: E, dorsal view; antennae and legs omitted. F, left fore leg. G, apex of left fore femur. H, *T. euphrosyne*: lateral view of abdominal end of apterous ♀. I, *T. pallidus*, apterous ♂: dorsal view; antennae and legs omitted. Reproduced from Andersen & Weir (1997).

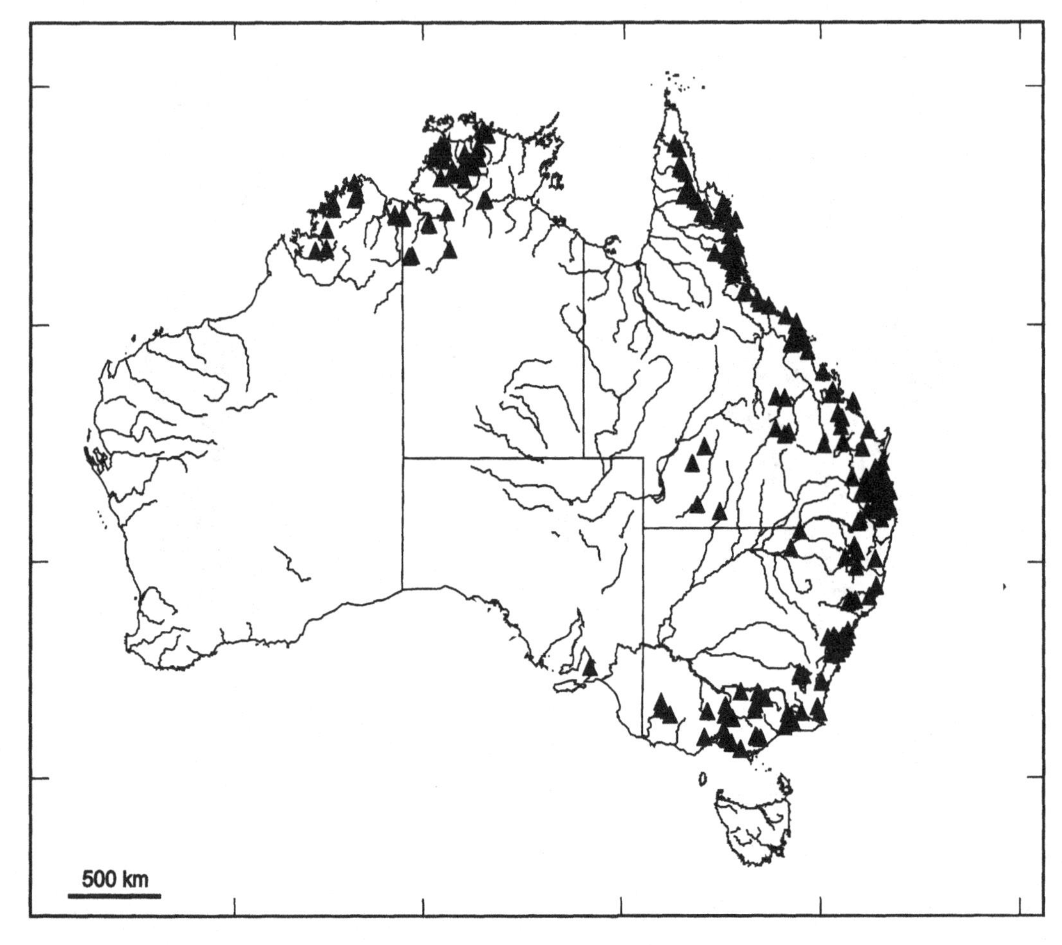

Figure 12.18. GERRIDAE. Distribution of *Tenagogerris* species in Australia.

Tenagogerris species are wing dimorphic with macropterous and apterous adult forms (Fig. 12.3A, B). The frequency of wingless adults is relatively high in *T. euphrosyne*, whereas the two adult forms are equally common in *T. femoratus* and *T. pallidus* (Andersen & Weir 1997). Fifth instar nymphs moulting into macropterous adults have distinct wing pads (Fig. 12.4B, wp), which separates them from 5th instar nymphs moulting into apterous adults (Fig. 12.4E).

Distribution

Tenagogerris is only found in Australia. *T. euphrosyne* is widely distributed along the coasts of Queensland and New South Wales, the Australian Capital Territory, and Victoria, and a few inland localities in central and southern Queensland (Fig. 12.18). *T. femoratus* and *T. pallidus* occur in the Kimberleys of Western Australia and northern parts of Northern Territory.

TENAGOGONUS

Identification

Medium sized water striders, 5.2-9.9 mm long, quite similar to *Limnometra* in colour-pattern, but lacking the connexival spines characteristic of that genus. The male abdomen is relatively shorter; with sternum 7 at least as long as preceding two sterna together. The hind coxae of the male almost reaching or surpassing posterior margin of second abdominal sternum. Adults either macropterous or apterous (Figs 12.19A, B). Body elongate oval, chiefly yellowish brown to reddish brown above, yellowish below; pronotum with dark median, longitudinal stripe throughout its length (Plate 6, H). Antenna long and very slen-

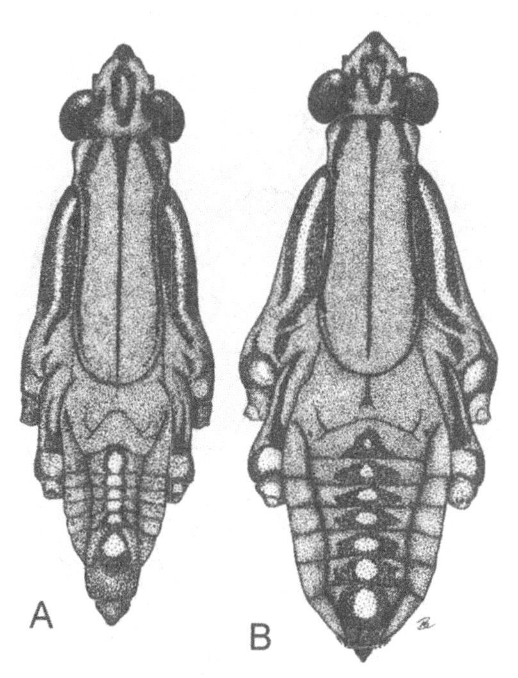

Figure 12.19. GERRIDAE. A, *Tenagogonus australiensis*, apterous ♂: dorsal view; antennae and legs omitted. B, *T. australiensis*, apterous ♀: dorsal view; antennae and legs omitted. Reproduced from Andersen & Weir (1997).

der, segment 1 longest, but much shorter than segments 2 and 3 together. Rostrum slender, its apex reaching anterior part of mesosternum.. Fore wings clearly surpassing abdominal end; second bifurcation of M+Cu usually approaching point of origin of anterior cross-vein. Metathoracic spiracle located more than its own length from wing bases. Metasternal scent orifice transverse ovate, located near posterior margin of metasternum, enclosed between a pair of low, transverse tubercles. Male fore femora moderately thickened, first segment of fore tarsus shorter than second segment. Middle femur about as long as body; tibia only slightly shorter than femur. Hind femur subequal to or a little longer than middle femur. Claws of middle legs small but distinct, those of the hind legs reduced. Male abdomen relatively short with posterior corners of connexiva obtuse and never spinously produced (Fig. 12.3E); posterior margin of sternum 7 simple, concave. Female abdomen relatively longer than in male, with posterior corners of connexiva produced into short spines in some species; posterior margin of sternum 7 straight or slightly produced.

The genus *Tenagogonus* Stål was revised by H ungerford & Matsuda (1958) and Chen & Nieser (1992). So far, 14 species are described, distributed throughout the Afrotropical, Oriental, and Australian regions, extending eastward to the Fiji Islands. Only one species in Australia, *Tenagogonus australiensis* Andersen & Weir, length 7.0-8.7 mm (♂) or 8.4-9.0 mm (♀). Queensland.

Biology

Almost nothing is known about the biology of *Tenagogonus* species. *T. australiensis* seems to be restricted to streams in closed forests. Nymphs of have been recorded from the period July-December. Like in other species of this genus, most samples consist of flightless, apterous specimens. So far, the only macropterous specimens available come from a large sample from Mossman Gorge, composed of 19 apterous and 3 macropterous adults and 10 nymphs, indicating a breeding population.

Distribution

In Australia, the single species of *Tenagogonus* is restricted to northern Queensland (Fig. 12.20). This or a closely related species also occurs in Papua New Guinea (J. Polhemus, personal communication).

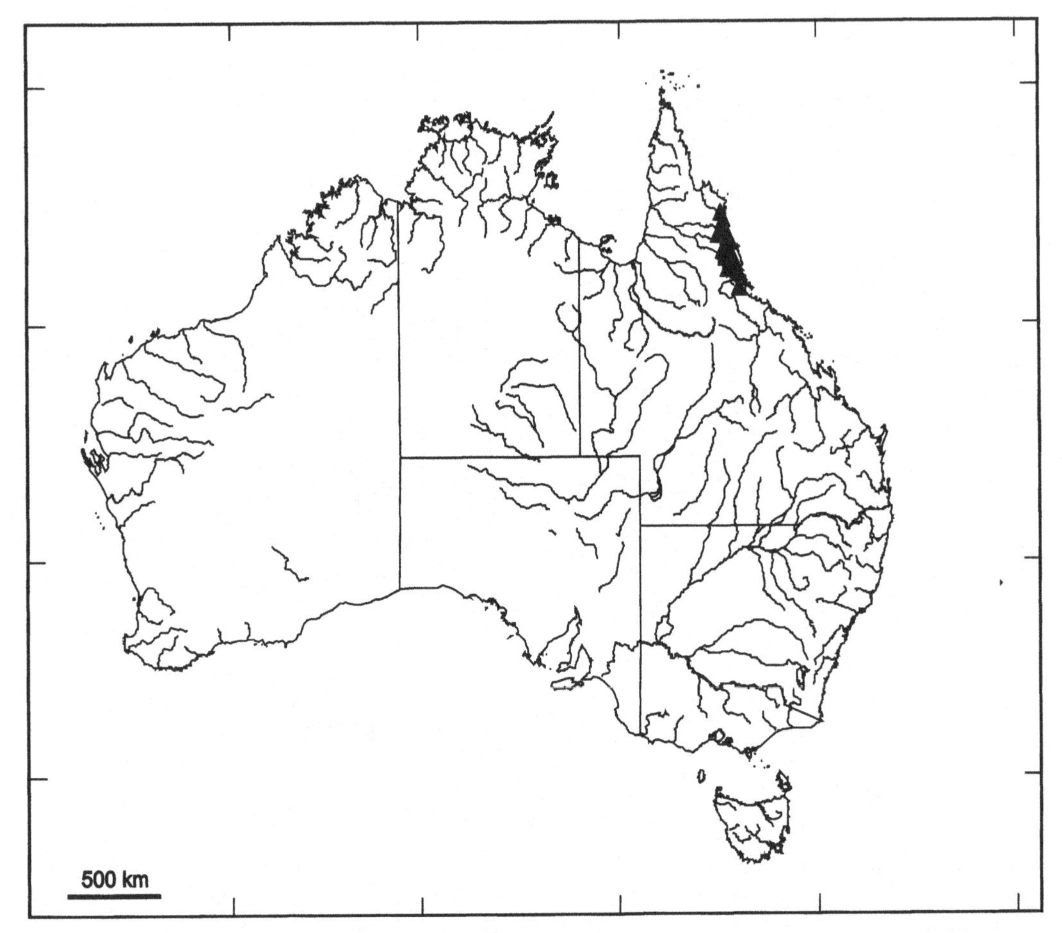

Figure 12.20. GERRIDAE. Distribution of *Tenagogonus australiensis* in Australia.

SUBFAMILY HALOBATINAE

Identification

The halobatines are small or medium-sized water striders with a relatively broad body and short abdomen (Figs 12.23A, B). The head is relatively short and broad with large eyes, three pairs of cephalic trichobothria, and small ventral lobes (bucculae). Antennae relatively long, segment 1 distinctly longer than segment 2 and 3 together. Rostrum short and stout, barely reaching hind margin of prosternum (Fig. 12.23C). The pronotum is shorter than the head length in apterous adults (Figs 12.23A, B; pn), large, subpentagonal in macropterous adults (unknown in Australian species). The mesonotum and metanotum usually confluent. The metasternum is very short, at most barely reaching the metacetabula laterally (Fig. 12.1F; mt); metasternal scent apparatus including scent orifice (so) always present. The fore femora are usually thickened, rarely modified ventrally; tibiae sometimes curved, always with a subapical spur; first segment of fore tarsus variable in length. Basal abdominal laterotergites and sternum not separated. Genital segments of male large, distinctly protruding from pregenital abdomen (Figs 12.23A, B, 12.28A); segment 8 relatively short, sometimes strongly modified, with styliform processes; proctiger large, plate-shaped (Fig. 12.23A, B, 12.26A; pr), parameres present (but small) or absent. Ovipositor of female relatively short, first gonapophyses non-serrated.

Overview

Species of Halobatinae are restricted to the Old World and most numerous in tropical regions. The family is classified in two tribes. Species

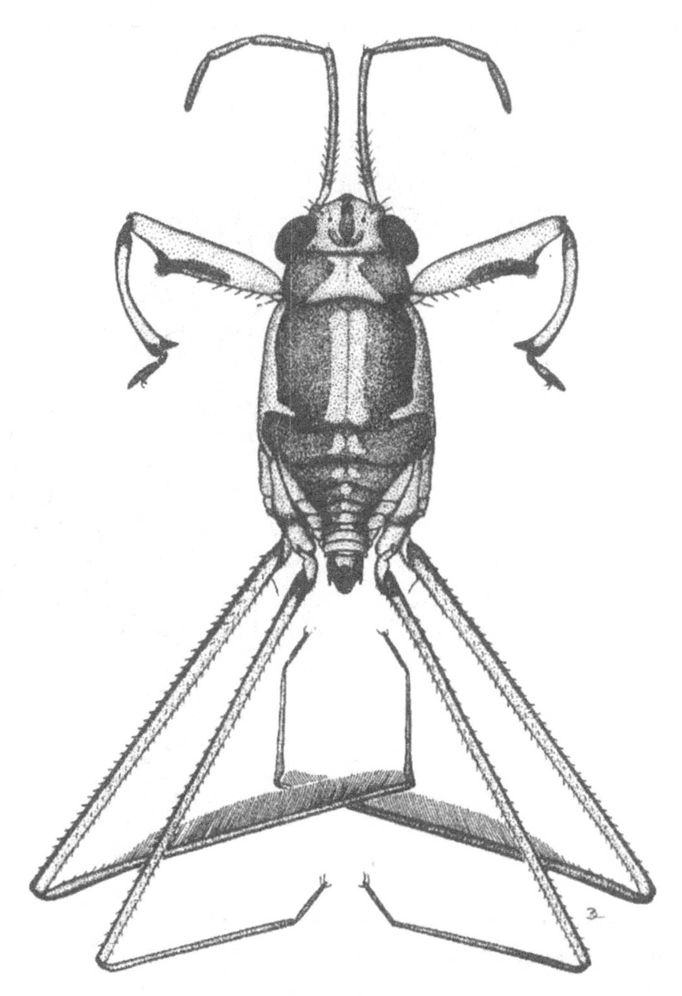

Figure 12.21. GERRIDAE. *Austrobates rivularis*, apterous ♂. Reproduced from Andersen & Weir (1994a).

belonging to the Metrocorini are typical inhabitants of flowing freshwater throughout the Afrotropical and Oriental regions (as far East as Sulawesi), with about 100 species in 7 genera of which the genus *Metrocoris* is the most speciose. Most species belonging to the tribe Halobatini live in various types of marine habitats, with about 50 species in 3 genera of which *Halobates* is the most species-rich. Only the last mentioned tribe has been recorded from Australia, with two genera and 15 species.

AUSTROBATES

Identification

Small water striders with oval body, greyish black above with extensive orange yellow markings on head, thoracic notum, abdomen, and legs (Fig. 12.21). Antennal segment 1 slightly shorter than the remaining three segments together; segments 3-4 subequal in length, each distinctly shorter than segment 2. Mesonotum prolonged, distinctly widened posteriorly. Intersegmental suture between meso-and metanotum distinct (Fig. 12.23A). Metasternum very short, barely reaching the metacetabula laterally (Fig. 12.1F, mt); metasternal scent orifice (so) oval, situated on a tubercle. Fore femora thickened, with a tubercle on ventral margin; fore tibiae distinctly curved (♂, Fig. 12.21), with prominent apical spur; first segment of fore tarsus very short. Middle femora distinctly longer than middle tibiae and hind femora; middle tibia ventrally with a conspicuous fringe of long hairs in distal four-fifths (Fig.

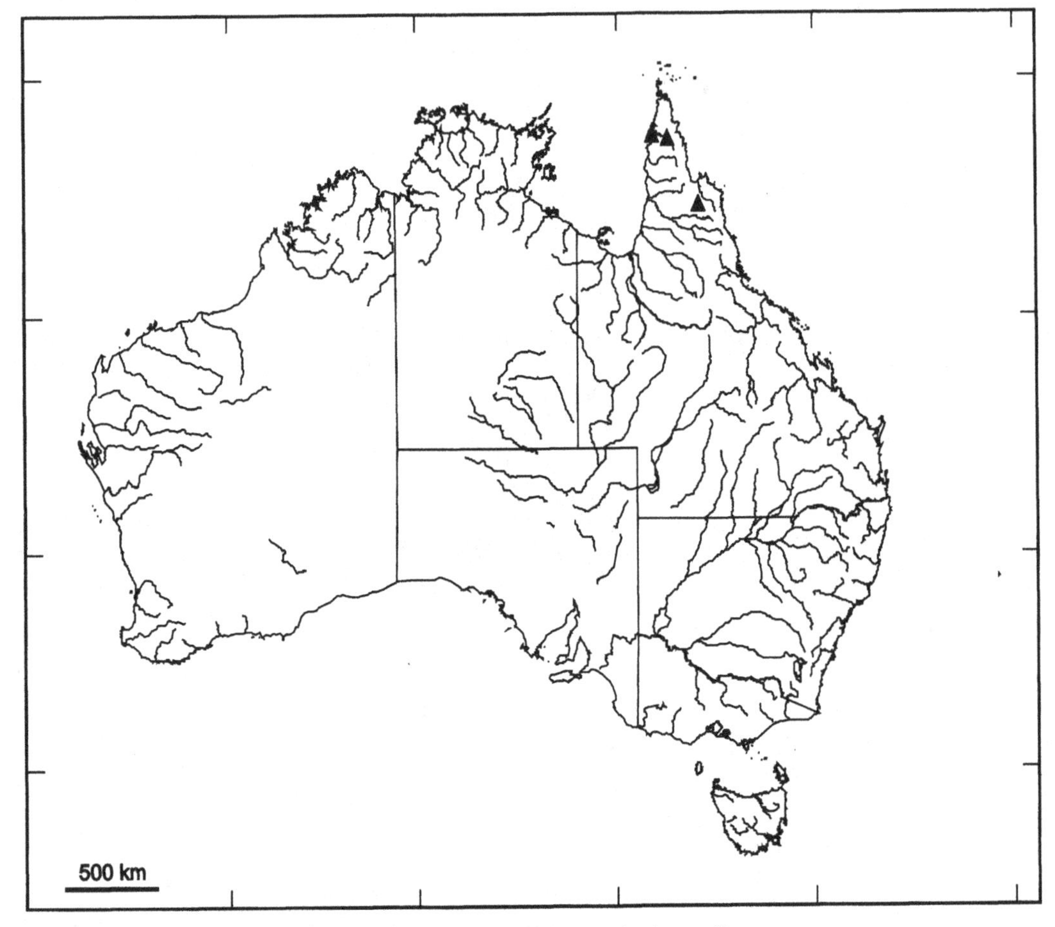

Figure 12.22. GERRIDAE. Distribution of *Austrobates rivularis* in Australia.

12.21). Abdomen shortened, especially in the male (Fig. 12.1F). Genital segments of male relatively small; segment 8 cylindrical, slightly wider than long, with a pair of low spiracular processes; ventral hind margin of segment 8 concave, with a pair of long, slender styliform processes (Fig. 12.1F). The pygophore (py) is suboval in ventral outline; proctiger is plate-like, suboval, slightly dilated at middle (Fig. 12.23D; pr); parameres absent. The genital segments of female are large with well developed gonocoxae; proctiger button-shaped, protruding.

Only one species: *Austrobates rivularis* Andersen & Weir (1994a), length 3.1-3.3 mm (♂) or 4.1-4.4 mm (♀). Queensland.

Biology

The first record of *Austrobates rivularis* was from a freshwater creek shaded by fringing vegetation of a riverine evergreen, notophyl, closed forest type. Both adults and nymphs were collected in January where the creek was 20-30 metres wide at the collecting point and the water was flowing at a reasonable speed. It was clear over a sandy/pebbly bottom with no vegetation growing in the creek itself. Eggs dissected from a female are relatively large, about one fourth the length of the females body, with a single micropyle at the anterior end (Andersen & Weir 1994a).

Distribution

Austrobates rivularis is only known from the northern parts of Cape York Peninsula, Queensland (Fig. 12.22). The type locality is Lydia Creek at Batavia Downs, in the middle of Cape York Peninsula. The species have since then been collected from two other streams on cape York Peninsula: Andoom Creek (in February) and Kennedy River (in April).

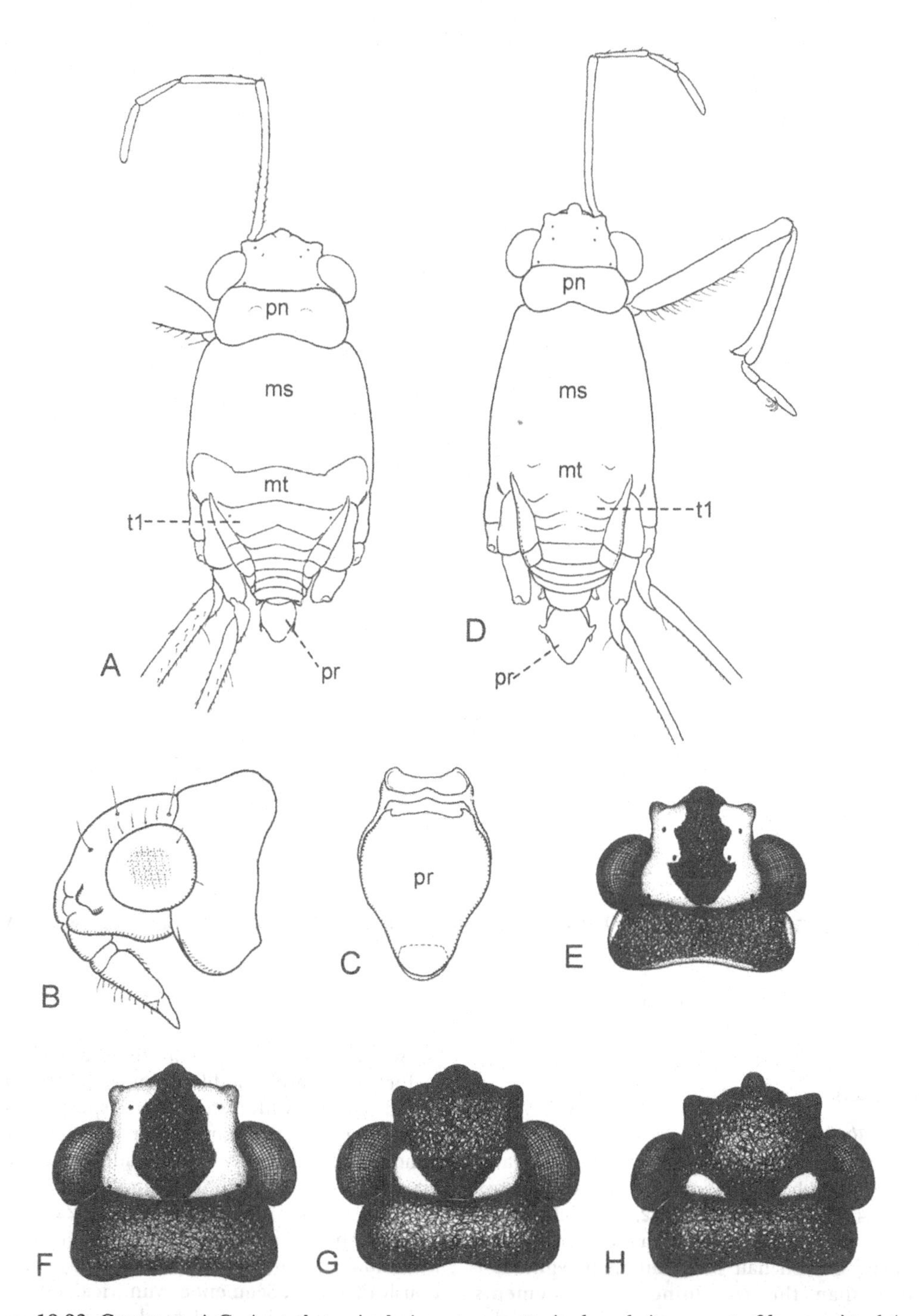

Figure 12.23. GERRIDAE. A-C, *Austrobates rivularis*, apterous ♂: A, dorsal view; most of legs omitted (ms, mesonotum; mt, metanotum; pn, pronotum; pr, proctiger; t1, abdominal tergum 1). B, lateral view of head. C, dorsal view of proctiger (pr). D, *Halobates* (s.str.) *darwini*, apterous ♂: dorsal view, most of legs omitted (ms, mesonotum; mt, metanotum; pn, pronotum; pr, proctiger; t1, abdominal tergum 1). E, *H.* (*Hilliella*) *mjobergi*, dorsal colour pattern of head, antennae omitted. F, *H.* (s.str.) *zephyrus*, dorsal colour pattern of head. G, *H.* (s.str.) *hayanus*, dorsal colour pattern of head. H, *H.* (s.str.) *germanus*, dorsal colour pattern of head. Reproduced with modification from : A-C, Andersen & Weir (1994a); D-H, Andersen & Weir (1994b).

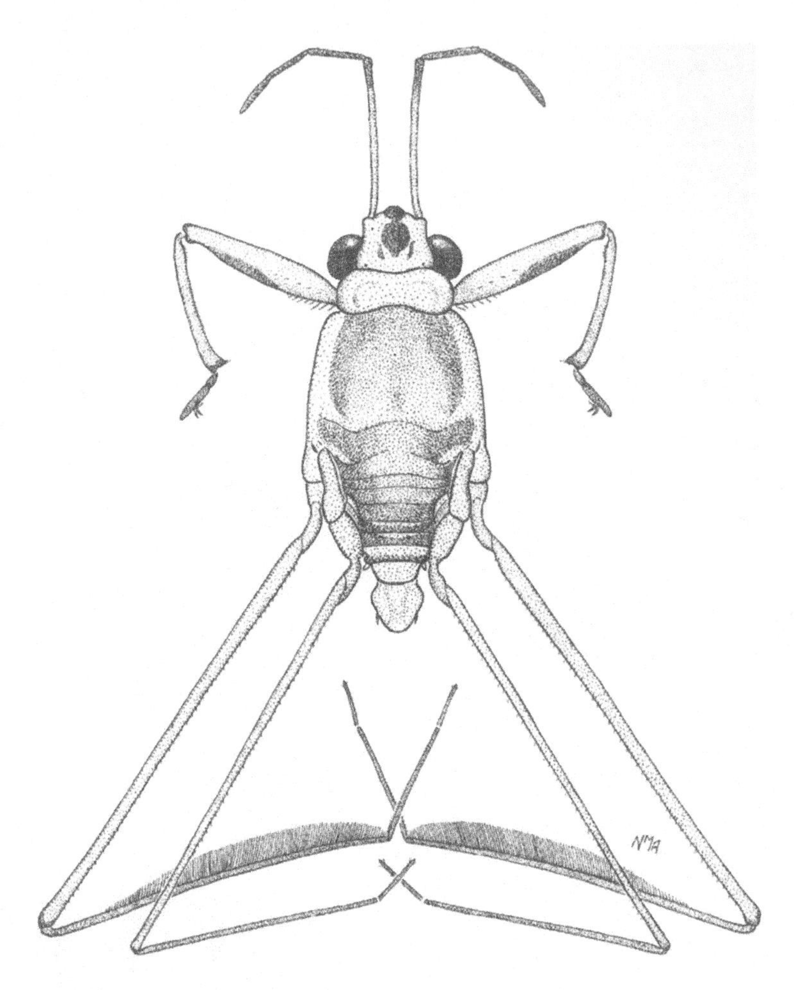

Figure 12.24. GERRIDAE. *Halobates* (*Hilliella*) *robinsoni*, apterous ♂. Reproduced from Andersen & Weir (2003a).

HALOBATES

Sea Skaters

Identification

The sea skaters are small to medium-sized water striders, length 3.2-6.6 mm, which are always apterous (Figs 12.23B, 12.24, 12.29). Body oval to broadly oval, usually dark with a more or less pronounced greyish hair pile. Antennal segment 1 longer than the remaining three segments together; segments 3-4 subequal in length, slightly shorter than segment 2. Mesonotum prolonged, distinctly widened posteriorly, in particular in females. Intersegmental suture between meso-and metanotum incomplete, usually reduced to a pair of lateral V-shaped pits (Fig. 12.23B). Metasternum very short, barely reaching the metacetabula laterally (Fig. 12.28A); metasternal scent orifice oval. Fore femora moderately thickened, rarely modified ventrally; fore tibiae slightly curved, with distinct apical spur; first segment of fore tarsus variable in length. Middle femora distinctly longer than middle tibiae and hind femora; middle tibia ventrally with a conspicuous fringe of long hairs (Figs 12.24, 12.28F, 12.29). Abdomen shortened, especially in the male. Male abdominal end (terminalia) conspicuously modified. Segment 8 cylindrical, usually as wide as long, with dorsal hind margin roundly produced in most species; each of the spiracles of the segment 8 typically placed upon a spiracular process (Fig. 12.26C; sp); ventral hind margin of segment 8 concave, with a pair of slender styliform processes (Fig. 12.26C; st) varying in relative length, degree of asymmetry, orientation, shape of apices, etc. The pygophore is suboval in

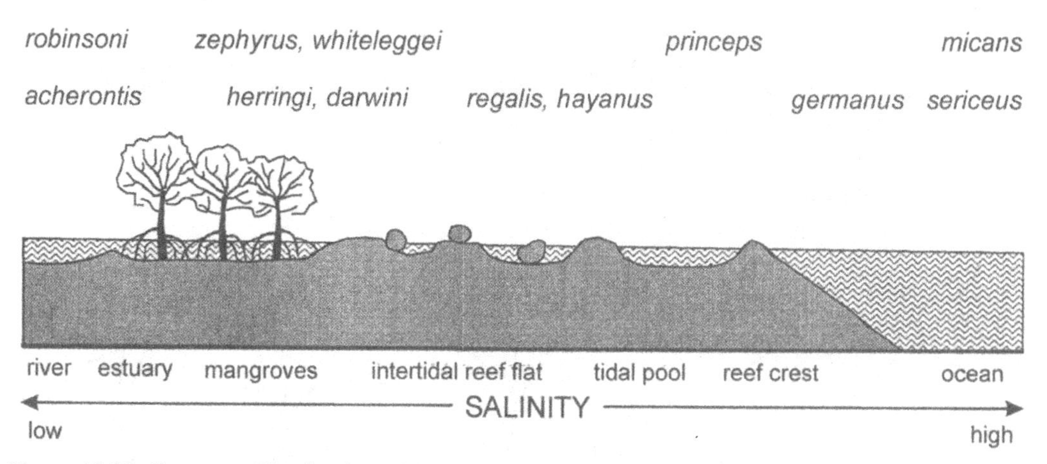

Figure 12.25. GERRIDAE. Distribution of Australian species of *Halobates* in marine habitats. Reproduced with modification from Andersen & Weir (1994b).

ventral outline (Fig. 12.26C; py). The proctiger is plate-like, usually pentagonal in outline (Fig. 12.26A; pr), with its lateral margins more or less produced in many species. The genital segments of female are usually small and inconspicuous.

The genus *Halobates* Eschscholtz has 46 described species which can be arranged in a number of species-groups. The Australian fauna includes species belonging to 5 of the 14 species-groups so far recognized (Andersen 1991b; Andersen & Weir 1994b): the *H. mjobergi* group (subgenus *Hilliella* China), and the *H. hayanus, micans, princeps, regalis,* and *zephyrus* groups (subgenus *Halobates* s. str.). These groups only partly conform with the species-groups delimited by Herring (1961). Most recently, the phylogeny of sea skaters has been reconstructed based on morphological as well as molecular characters (Andersen 1991b; Damgaard et al. 2000). In addition, the discovery of a fossil *Halobates* from the Middle Eocene of northern Italy indicates that sea skaters are at least 45 million years old (Andersen et al. 1994).

The 14 Australian species of *Halobates* are classified in two subgenera:

(1) Subgenus *Halobates* Eschscholtz: Small to medium-sized water striders, length 3.2-6.6 mm, chiefly dark-coloured or silver-grey without extensive pale markings except on venter (immature specimens have more extensive pale body areas). Yellow markings on head usually restricted to a basal, crescent-shaped mark (Figs 12.23G, H). First segment of fore tarsus variable in length. Middle tarsal segment with a narrow hair fringe (Fig. 12.29). Segments of hind tarsus fused in some species. Styliform processes of male segment 8 and proctiger often more or less asymmetrical; parameres absent. Female gonocoxae relatively small, partly concealed; proctiger button-shaped, sometimes reflexed.

(2) Subgenus *Hilliella* China: Small water striders, length 3.3-4.7 mm, chiefly dark brown with extensive yellow markings on head (Figs 12.23E, 12.24), pronotum, pleura, legs, and venter. First segment of fore tarsus much shorter than second segment. Middle tarsal segments without hair fringe (Fig. 12.24). Segments of hind tarsus always fused. Styliform processes of male segment 8 and proctiger symmetrical; parameres present. Female gonocoxae relatively large, partly exposed; proctiger button-shaped, protruding.

Key to Australian species

1. First segment of fore tarsus shorter than 0.3x second segment. Middle femur at most 1.05x longer than hind femur 2
- First segment of fore tarsus longer than 0.3x second segment. Middle femur 1.1x or more longer than hind femur (subgenus *Halobates*) 5
2. Brown and yellow species usually with most of head (Fig. 12.23E), posterior

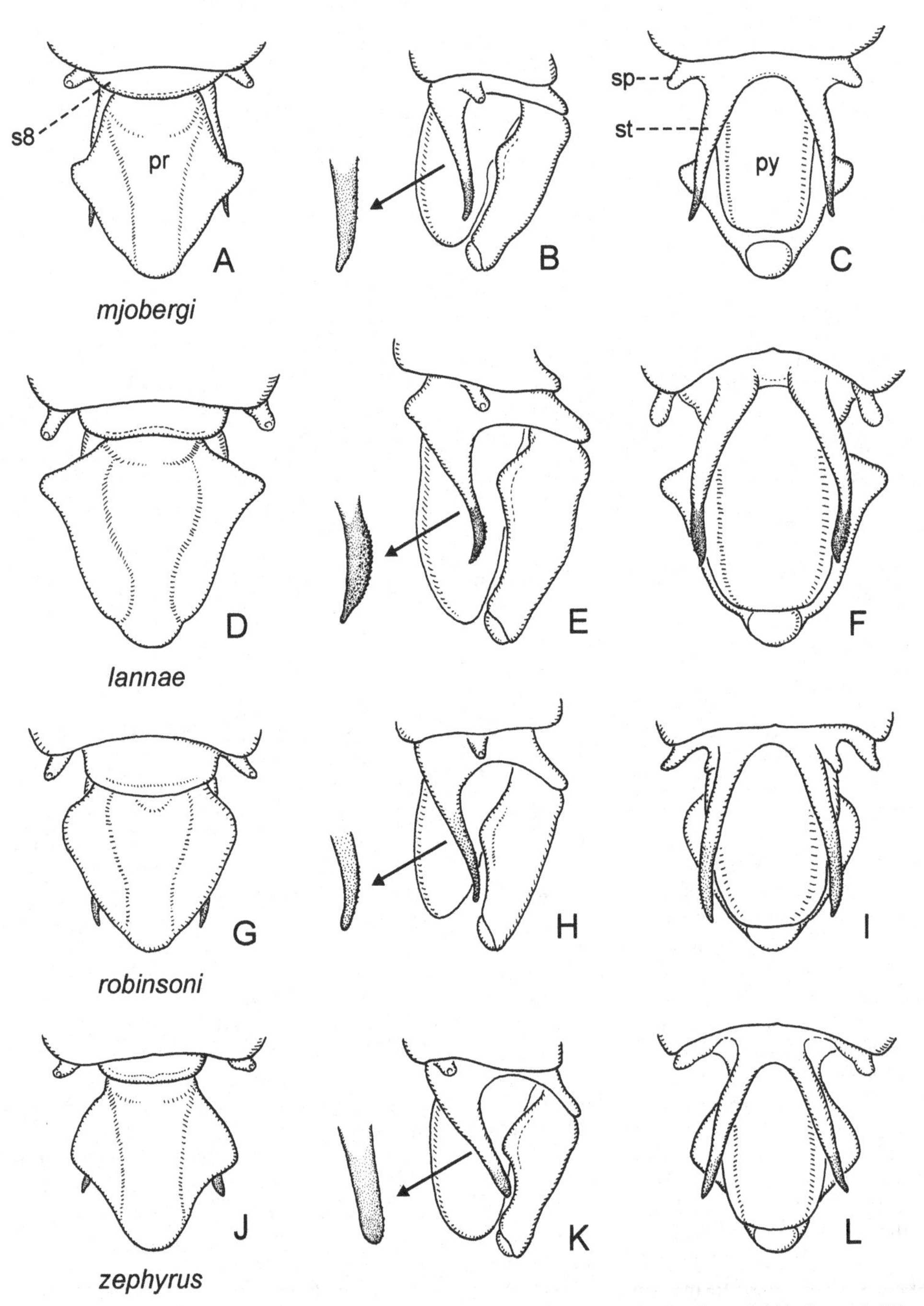

Figure 12.26. GERRIDAE. A-C, *Halobates* (*Hilliella*) *mjobergi*: male genitalia in dorsal (A), lateral (B), and ventral view (C); apex of left styliform process shown at higher magnification (pr, proctiger; py, pygophore; sp, spiracular process; st, styliform process; s8, segment 8). D-F, *Halobates* (*Hilliella*) *lannae*: male genitalia in dorsal (D), lateral (E), and ventral view (F); apex of left styliform process shown at higher magnification. G-I, *Halobates* (*Hilliella*) *robinsoni*: male genitalia in dorsal (G), lateral (H), and ventral view (I); apex of left styliform process shown at higher magnification. J-L, *Halobates* (s. str.) *zephyrus*: male genitalia in dorsal (J), lateral (K), and ventral view (L); apex of left styliform process shown at higher magnification. Reproduced with modification from: A-F, J-L, Andersen & Weir (1994b); G-I, Andersen & Weir (2003a).

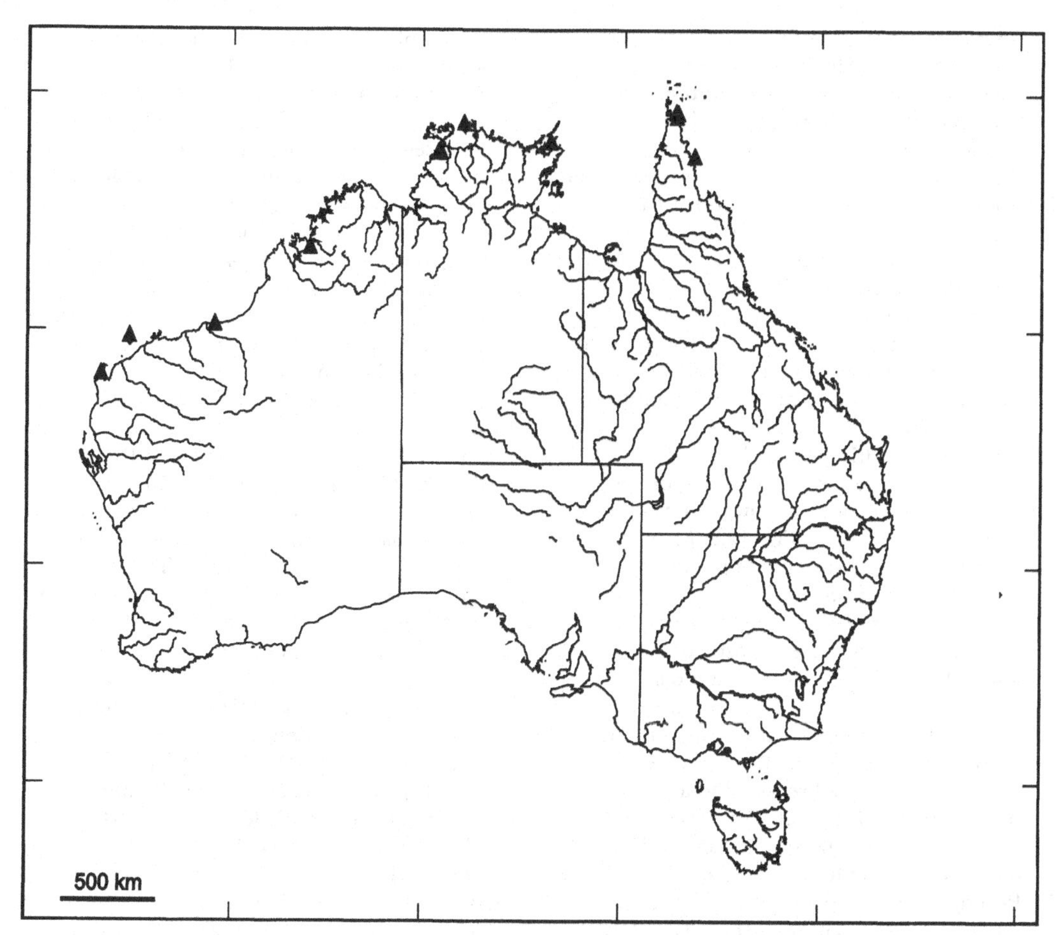

Figure 12.27. GERRIDAE. Distribution of *Halobates* (*Hilliella*) species in Australia.

margin of pronotum, most of thoracic pleura, and prominent stripes on femora, yellow (Fig. 12.24). (subgenus *Hilliella*) 3

– Dark brown to black species with most of thoracic pleura dark and only bases of anterior femora, yellow. Apices of ♂ styliform processes slender and straight (Figs 12.26J-L). Head with prominent pale markings along inner margins of eyes (Fig. 12.23F). Fore femur of ♂ distinctly constricted in distal third. Length 3.6-4.0 mm (♂), 4.2-4.5 mm (♀). New South Wales, Queensland .... *zephyrus* Herring

3. Fore femur of ♂ depressed ventrally in distal fourth with an elongate patch of stiff hairs. Proctiger of ♂ widest across middle (Fig. 12.26A, pr); apices of styliform processes simple (Fig. 12.26B). Female meso-metanotum with scattered, long dark bristles. Body length 3.3-3.8 mm (♂) or 3.4-4.2 mm (♀). Northern Territory, Queensland, Western Australia .... *mjobergi* Hale

– Fore femur of ♂ not modified ventrally, more or less tapering in width toward apex. Proctiger of ♂ widest in basal third (Figs 12.26D, G) 4

4. Proctiger of ♂ distinctly produced laterally in basal third (Fig. 12.26D); apices of styliform processes slightly widened, pointed (Fig. 12.26E). Thorax and abdomen above chiefly dark brown. Female meso-metanotum without dark bristles. Body length 4.1-4.5 mm (♂) or 4.3-4.7 mm (♀). Northern Territory, Western Australia .... *lannae* Andersen & Weir

– Proctiger of ♂ widened but not distinctly produced laterally in basal third (Fig. 12.26H); apices of styliform processes simple, blunt (Fig. 12.26 K). Thorax and abdomen above chiefly brown or yellowish brown. Female meso-metanotum with scattered, long dark bristles. Body length 3.6 mm (♂) or 4.4-4.5 mm (♀). Western Australia .... *robinsoni* Andersen & Weir

5. Body length 6.0-6.3 mm (♂), 6.1-6.7 mm (♀). Male proctiger pentagonal, with a recurved prominence on lateral angle (Figs 12.30A-C). Fore femur of ♂ with a tubercle beneath. Northern Territory, Western Australia .... *princeps* White
– Body length at most 5.5 mm. Structure of male proctiger not as above. Fore femur of ♂ unarmed 6
6. Interocular width of head less than 3.6x width of an eye. Yellow colouration on head extensive, with prominent yellow stripes along eyes (as Fig. 12.23F) or reduced to a crescent-shaped mark extending forward toward the eyes (Fig. 12.23G), occasionally with a yellow spot at the base of antennae. Conspicuous yellow or brown markings on some parts of venter 7
– Interocular width of head subequal to or more than 3.6x width of an eye.Yellow colouration on head reduced to a pair of triangular markings at the base (Fig. 12.23H). Body including thoracic and abdominal venter, uniformly dark 12
7. Proctiger of ♂ with a patch of spinous dark hairs on each side (Fig. 12.30D), styliform processes almost symmetrical, apices slender and diverging (Fig. 12.30E). Female meso-metanotum without dark bristles. Length 3.7-4.2 mm (♂), 4.1-4.6 mm (♀).Northern Territory, Western Australia .... *hayanus* White
– Proctiger of ♂ without a patch of spinous dark hairs on each side, styliform processes neither symmetrical nor diverging apically, apices stout. Female meso–metanotum with scattered, dark bristles 8
8. Both male and female completely yellow beneath. Apices of styliform processes of male stout, but not boot-shaped (Figs 12.30H, K). Hind trochanters of female pilose, but without a tuft of long hairs 9
– Male not completely yellow beneath, thoracic venter darkened. Apices of styliform processes of male boot-shaped (Figs 12.31B, E, I). Hind trochanters of ♀ with a tuft of long hairs 10
9. First segment of fore tarsus 0.5x (♂) or 0.6x (♀) longer than second segment. Male terminalia (Figs 12.30G-I). Hind coxae of ♀ shorter, only about 2.0x as long as wide. Length 4.3-4.7 mm (♂), 4.8-5.2 mm (♀). Queensland, Western Australia .... *regalis* Carpenter
– First segment of fore tarsus 0.4x (♂) or 0.5x (♀) longer than second segment. Male terminalia (Figs 12.30J-L). Hind coxae of ♀ very long, about 3.0x times as long as wide. Length 3.7-4.0 mm (♂), 4.2-4.9 mm (♀). New South Wales, Queensland .... *whiteleggei* Skuse
10. Lateral projections of ♂ proctiger finger-like but relatively short (Figs 12.31D-F, H-J). Male meso-metanotum with scattered stiff, black bristles 11
– Lateral projections of ♂ proctiger finger-like and long (Figs 12.31A-C). Male meso-metanotum without scattered stiff, black bristles. Length 4.0-4.4 mm (♂), 4.1-5.1 mm (♀). Northern Territory, Queensland ...... *herringi* J. Polhemus & Cheng
11. Body length 4.5-4.8 mm (♂) or 4.3-4.8 mm (♀). Meso-metanotum with few short, dark bristles (Fig. 12.31G). Male terminalia (Figs 12.31D-F). Northern Territory, Queensland) .... *darwini* Herring
– Body length 3.8-4.2 mm (♂) or 3.8-3.9 mm (♀). Meso-metanotum with numerous long, dark bristles (Fig. 12.31K). Male terminalia (Figs 12.31H-J). Northern Territory .... *acherontis* J. Polhemus
12. Body length 4.1-4.6 mm (♂) or 3.5-4.6 mm (♀). Styliform processes of ♂ strongly asymmetrical, left process bent abruptly upwards at a right angle, apices modified (Figs 12.32A-D). New South Wales, Northern Territory, Queensland, Western Australia .... *micans* Eschscholtz
– Smaller species, length less than 4.0 mm (♂) or 3.8 mm (♀). Styliform processes of ♂ almost symmetrical, apices simple 13
13. Second antennal segment 0.5x as long as fourth segment. First segment of fore tarsus about 0.4x longer than second segment. Male terminalia (Figs 12.32E-G) . Length 3.2-3.6 mm (♂), 3.0-3.4 mm (♀). New South Wales (Lord Howe Island), Queensland .... *sericeus* Eschscholtz
– Second antennal segment much more than 0.5x as long as fourth segment. First segment of fore tarsus about 0.6x longer than second segment. Male terminalia (Figs 12.32H-J). Length 3.3-3.9 mm (♂), 3.5-4.0 mm (♀).Northern Territory, Queensland, Western Australia .... *germanus* White

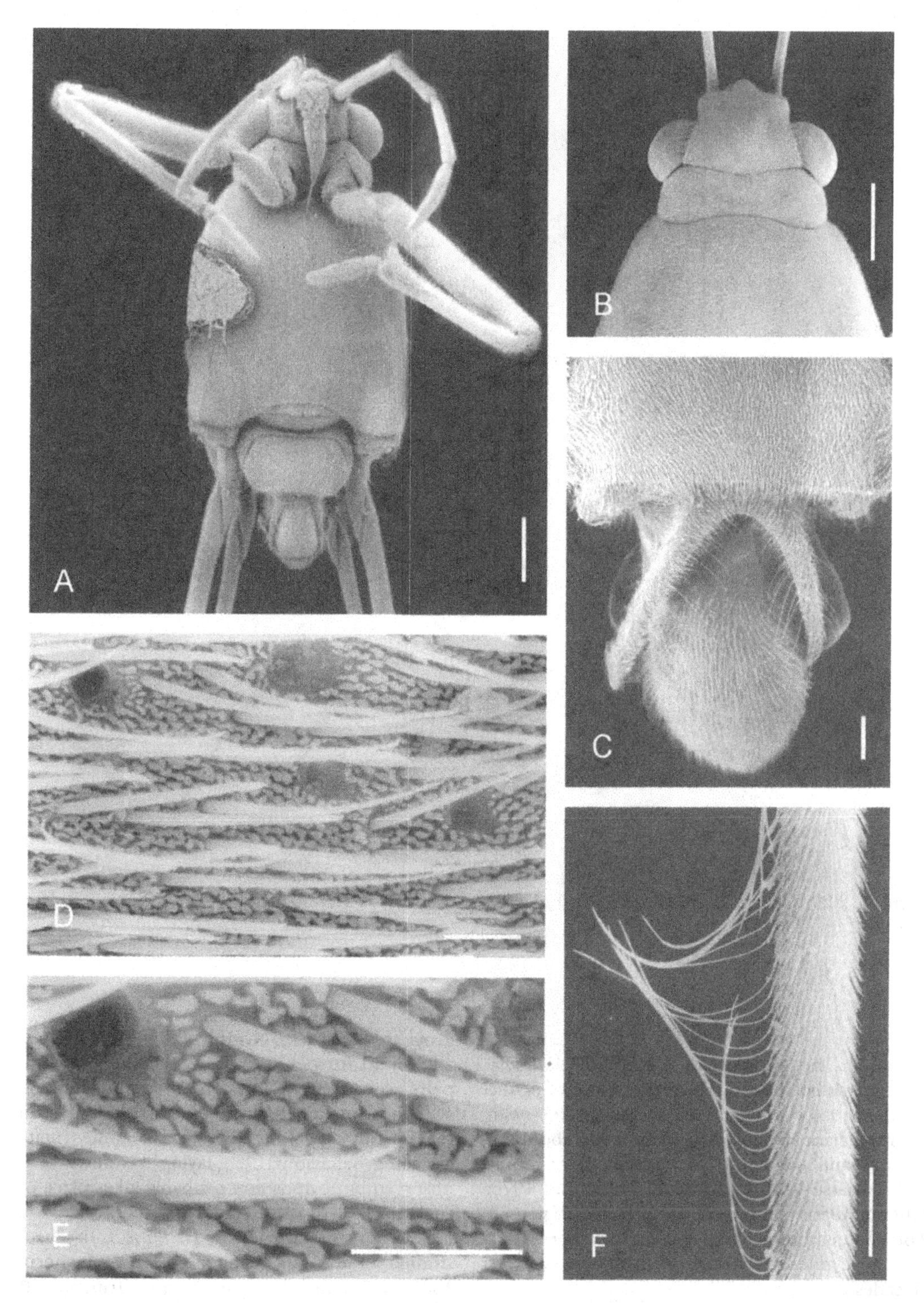

Figure 12.28. GERRIDAE. A, *Halobates* (*Hilliella*) *mjobergi*: ventral view of apterous ♂. B, *H.* (*Hilliella*) *mjobergi*: dorsal view of head and pronotum of apterous ♀. C-E, *H.* (s. str.) *zephyrus*: C, ventral view of abdominal end of apterous ♂. D-E, surface of thoracic venter with micro-hairs and macro-hairs. F, middle tibia of apterous ♀ with fringe of hairs. Scales: A-B, 0.5 mm; D-E, 0.01 mm; F, 0.1 mm. Scanning Electron Micrographs (CSIRO Entomlogy, Canberra).

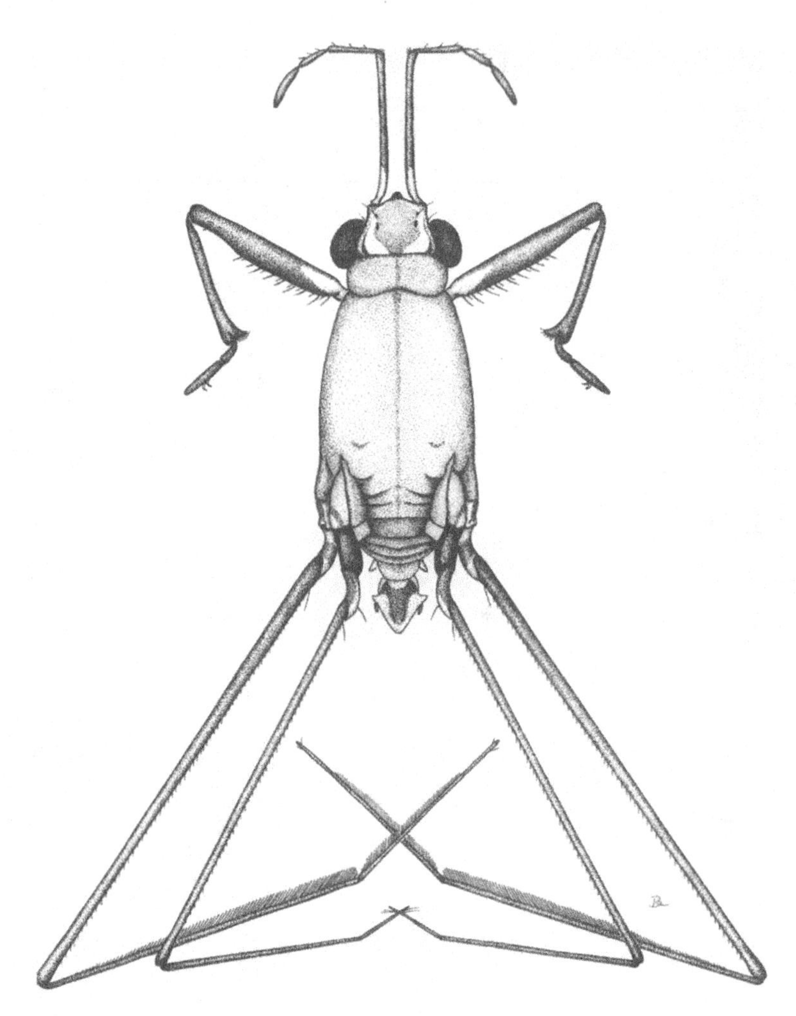

Figure 12.29. GERRIDAE. *Halobates* (s.str.) *darwini*, apterous ♂. Reproduced from Andersen & Weir (1994b).

Biology
Although the open-ocean species of *Halobates* have attracted most interest, the majority of the 45 described species of sea skaters prefer near-shore, marine habitats. Our knowledge about the biology and ecology of sea skaters is generally quite sparse (Andersen & J. Polhemus 1976; Cheng 1985; Andersen & Cheng in press). Fortunately, a few coastal species have been studied more intensively during the past couple of decades.

Halobates robustus has been studied in the Galapagos Islands (Birch et al. 1979; Foster & Treherne 1980, 1982). This species inhabits protected, rocky coasts with mangroves. Adults tend to aggregate in large "flotillas" very close to mangrove trees or rocks. Nymphs (all instars) are usually found further away from the shore. Mating pairs are very frequently observed and the male (which is smaller than the female) stays with the female for a prolonged period of time (mate-guarding behaviour). Egg-laying has never been observed but oviposition probably takes place on rocks and/or roots of mangrove trees.

Another coastal species, *H. fijiensis*, has been studied in the Fiji Islands (Foster & Treherne 1986). It inhabits bays and lagoons fringed with mangroves. Younger nymphs are always found in sheltered waters amongst mangroves. Older nymphs and adults are found in more open water, sometimes several hundred meters from the mangroves. Mating pairs are infrequently observed,

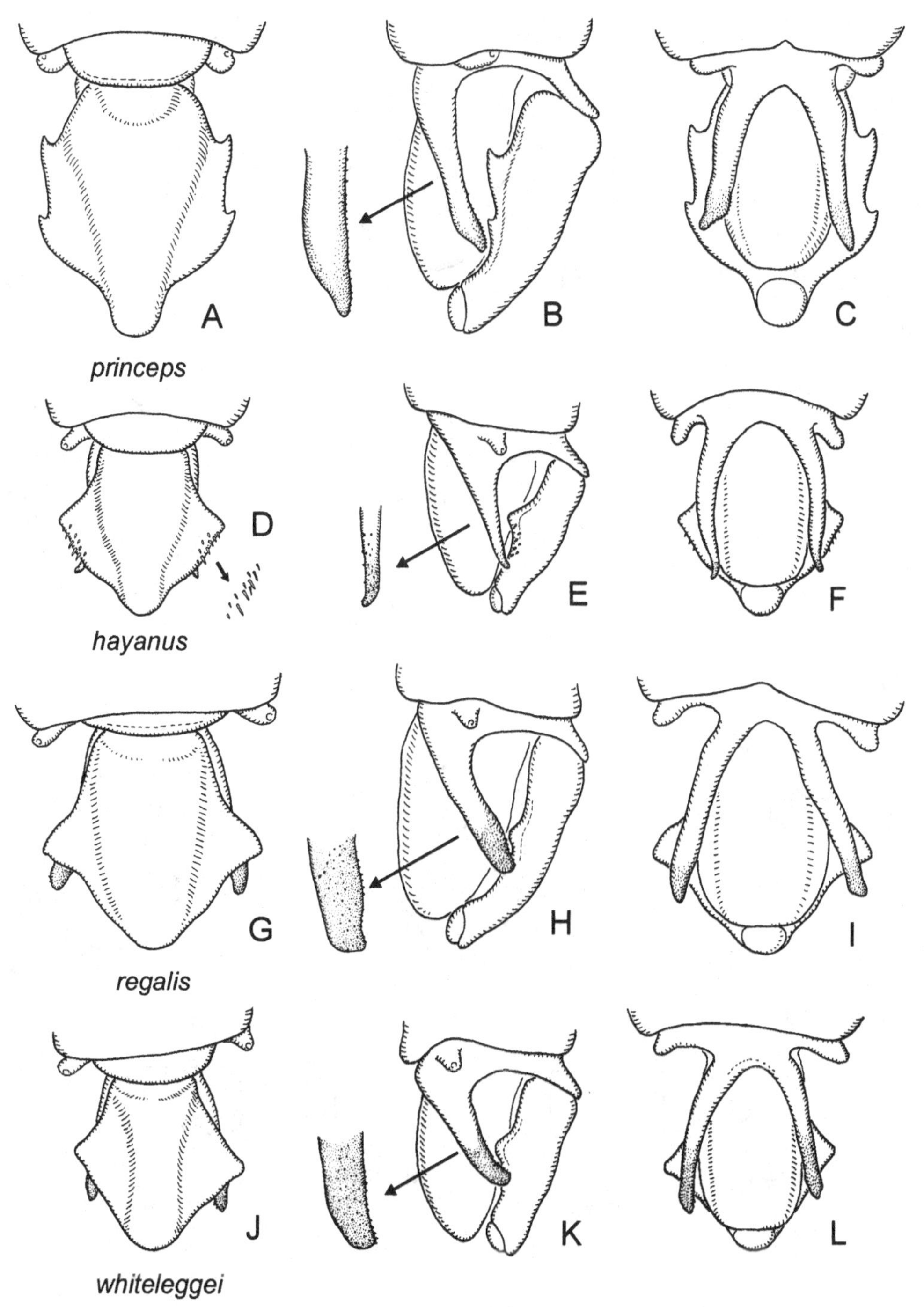

Figure 12.30. GERRIDAE. A-C, *Halobates* (s. str.) *princeps*: male genitalia in dorsal (A), lateral (B), and ventral view (C); apex of left styliform process shown at higher magnification. D-F, *Halobates* (s. str.) *hayanus*: male genitalia in dorsal (D), lateral (E), and ventral view (F); apex of left styliform process shown at higher magnification. G-I, *Halobates* (s. str.) *regalis*: male genitalia in dorsal (G), lateral (H), and ventral view (I); apex of left styliform process shown at higher magnification. J-L, *Halobates* (s. str.) *whiteleggei*: male genitalia in dorsal (J), lateral (K), and ventral view (L); apex of left styliform process shown at higher magnification. Reproduced with modification from Andersen & Weir (1994b).

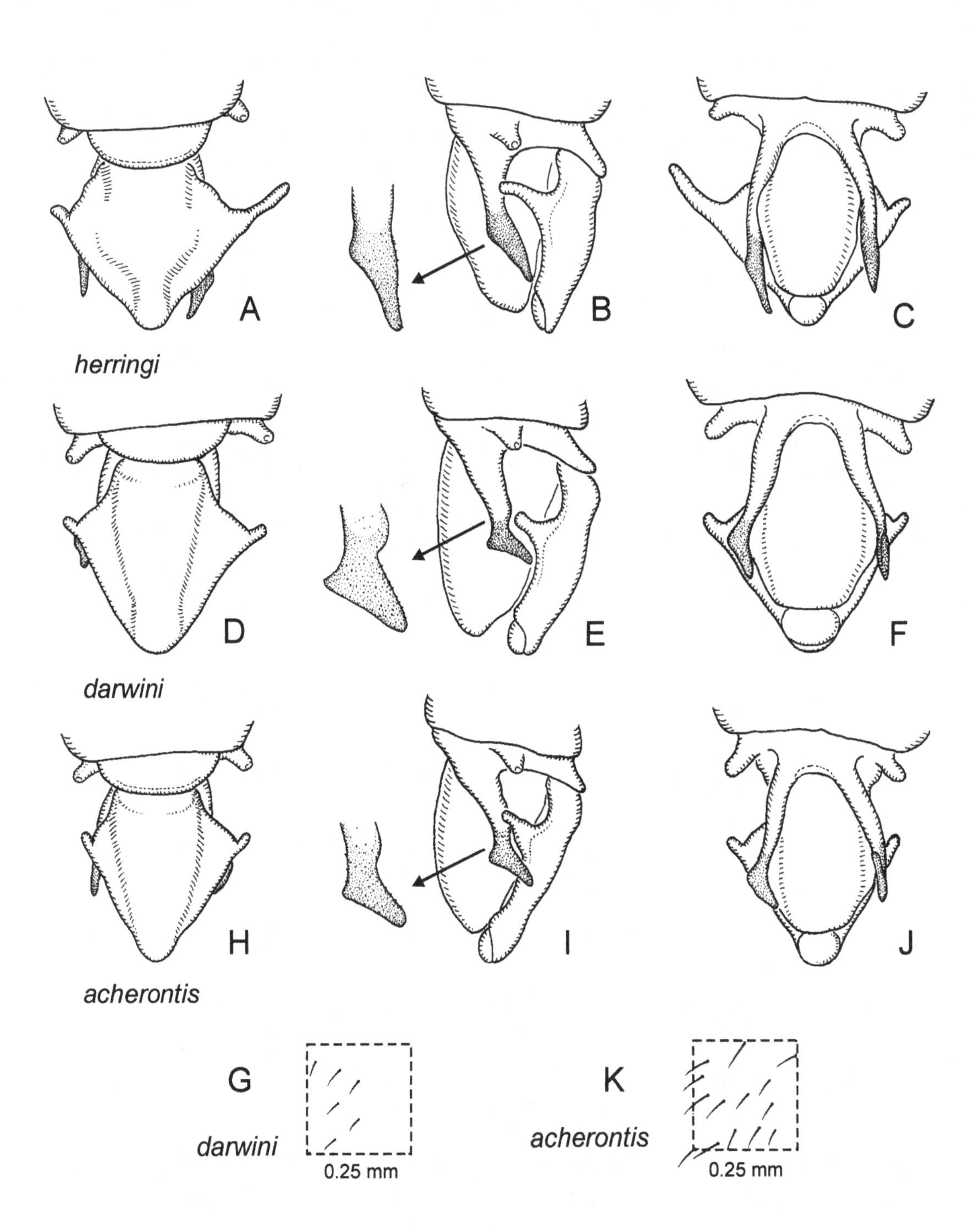

Figure 12.31. GERRIDAE. A-C, *Halobates* (s. str.) *herringi*: male genitalia in dorsal (A), lateral (B), and ventral view (C); apex of left styliform process shown at higher magnification. D-F, *Halobates* (s. str.) *darwini*: male genitalia in dorsal (D), lateral (E), and ventral view (F); apex of left styliform process shown at higher magnification; G, *Halobates* (s. str.) *darwini*: density of hairs on body surface. H-J, *Halobates* (s. str.) *acherontis*: male genitalia in dorsal (H), lateral (I), and ventral view (J); apex of left styliform process shown at higher magnification. K, *Halobates* (s. str.) *acherontis*: density of hairs on body surface. Reproduced with modification from Andersen & Weir (1994b).

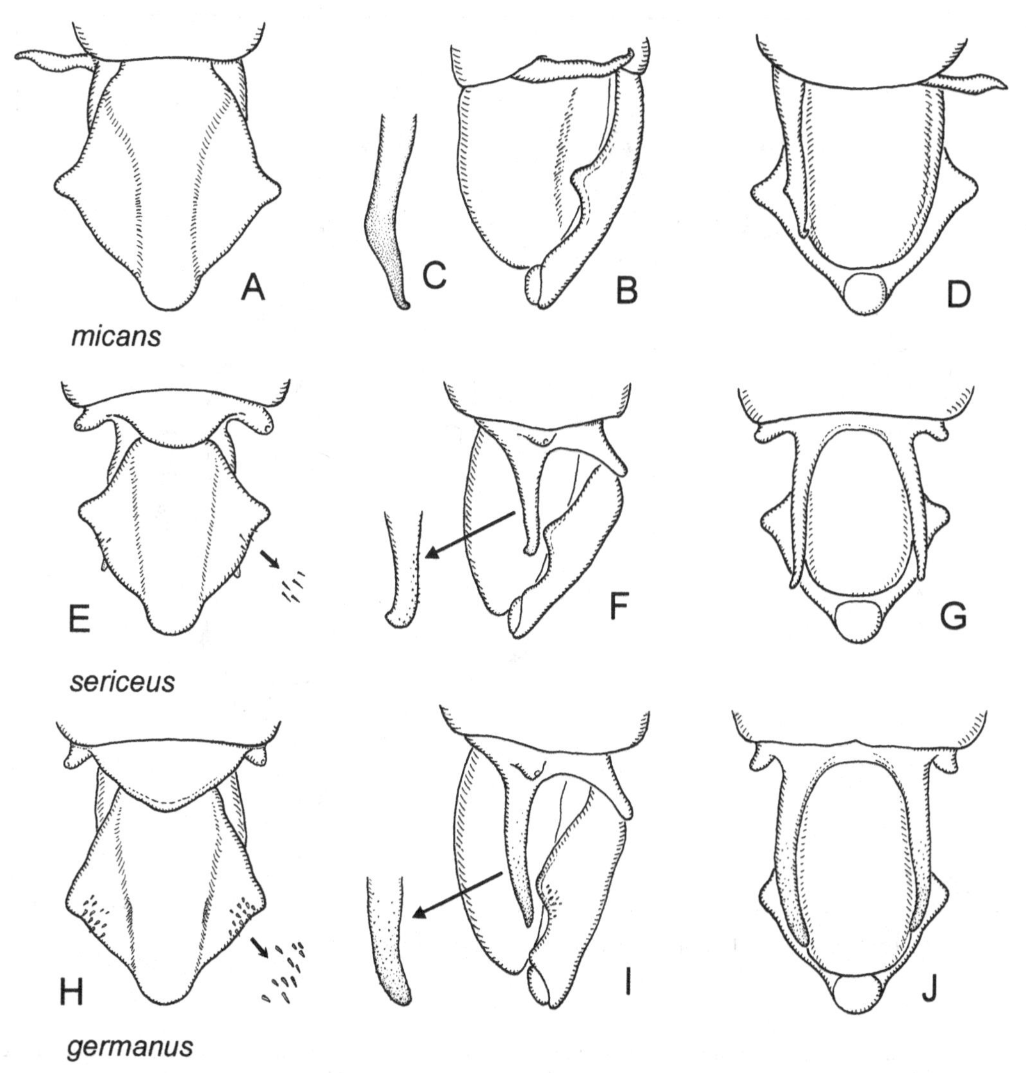

Figure 12.32. GERRIDAE. A-D, *Halobates* (s. str.) *micans*: male genitalia in dorsal (A), lateral (B), and ventral view (D); apex of right styliform process shown at higher magnification (C). E-G, *Halobates* (s. str.) *sericeus*: male genitalia in dorsal (E), lateral (F), and ventral view (G); apex of left styliform process shown at higher magnification. H-J, *Halobates* (s. str.) *germanus*: male genitalia in dorsal (H), lateral (I), and ventral view (J); apex of left styliform process shown at higher magnification. Reproduced with modification from Andersen & Weir (1994b).

and the encounter between male and female (male slightly larger than female) is brief. Egg-laying was observed to take place on stands of seagrass near the low-water mark, at extreme low spring tides. The newly hatched nymphs then have to make their way for several hundred meters to the protecting mangroves.

Five species of sea skaters have successfully colonized the open ocean (Cheng 1989): *H. germanus* (Indian and West Pacific Ocean), *H. sericeus* (Pacific Ocean), *H. sobrinus* (tropical eastern Pacific), *H. micans* (Atlantic, Indian, and Pacific Ocean), and *H. splendens* (tropical southeastern Pacific). Both adults and nymphs of these species live permanently upon the sea surface, always at some distance from nearest land. Eggs are deposited on various floating objects (Lundbeck 1914; Andersen & J. Polhemus 1976). They probably feed on other animals living at the sea-air interface (called *pleuston*) such as planctonic

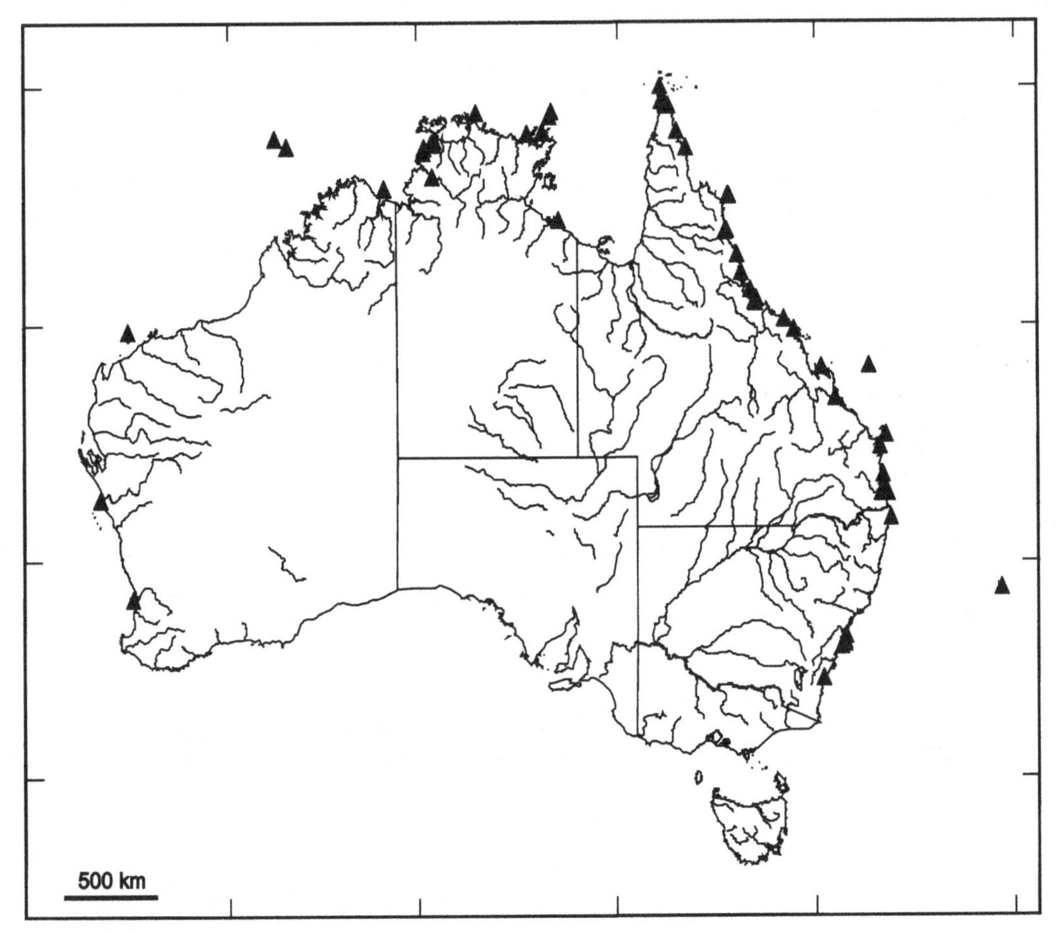

Figure 12.33. GERRIDAE. Distribution of *Halobates* (*Halobates*) species in Australia.

crustaceans and siphonophoran jellyfish, and are themselves fed upon by pelagic fish and seabirds (Andersen & J. Polhemus 1976; Cheng 1985). A reconstructed phylogeny for *Halobates* suggests that the oceanic habitat was invaded twice (Damgaard et al. 2000), once by a lineage comprising *H. germanus* and *sericeus*, and once by a lineage comprising *H. micans* and two other oceanic species, *H. sobrinus* and *splendens*, confined to the eastern Pacific Ocean.

Although observations of preferred habitats are limited, the species of sea skaters found along the coasts of Australia are probably distributed among different marine habitats as shown on the idealized transect through various aquatic habitats (Fig. 12.25). Only two *Halobates* species have been recorded from freshwater. *H. acherontis* was collected more than 100 km above the mouth of Daly River, Northern Territory (J. Polhemus 1982) and *H. robinsoni* about 50 km upstream Robinson River, Western Australia (Andersen & Weir 2003a). Most Australian species inhabit mangroves in river estuaries, tidal creeks, or protected bays of coastal lagoons and are probably close to *H. robustus* in their way of living (see above). Here belong *H. darwini, herringi, lannae, mjobergi, whiteleggei*, and *zephyrus*. Species inhabiting tidal pools and lagoons along coral coasts are *H. regalis, hayanus*, and *princeps* which may be ecologically close to *H. fijiensis* (see above). *H. micans* and *sericeus* live in the open ocean and are only found close to the beach after storms. A transitional stage between the coastal and oceanic way of life in sea skaters is represented by *H. germanus* which usually is found closer to land than the truly open-ocean species, *H. micans* and *sericeus* (Cheng 1989; Andersen 1991b).

Coastal sea skaters may be collected by ordinary aquatic nets whereas the capture of the oceanic species requires a specialized neuston-

Plate 5. Collecting and preserving water bugs. A, The second author collecting in a small flowing creek with falls and pools; Limestone Creek, Gregory National Park, Northern Territory (May 2001). Habitat of 20 species of Gerromorpha representing 8 genera and 5 families. B, picking specimens from soup strainer. C, labelling and preserving specimens in alcohol collecting tubes. D, specimens in alcohol collecting tubes. E, dorsal view of point-mounted *Microvelia* (*Austromicrovelia*) *childi*, ♀ (Veliidae). F, dorsal view of point-mounted *Hydrometra strigosa*, ♂ (Hydrometridae). G, lateral view of point-mounted *Ochterus australicus* (Ochteridae). H-I, pinned *Anisops stali* (Notonectidae): H, dorsal view. I, lateral view. Photographs by: A-D, David McClenaghan; E-I, Ben Boyd.

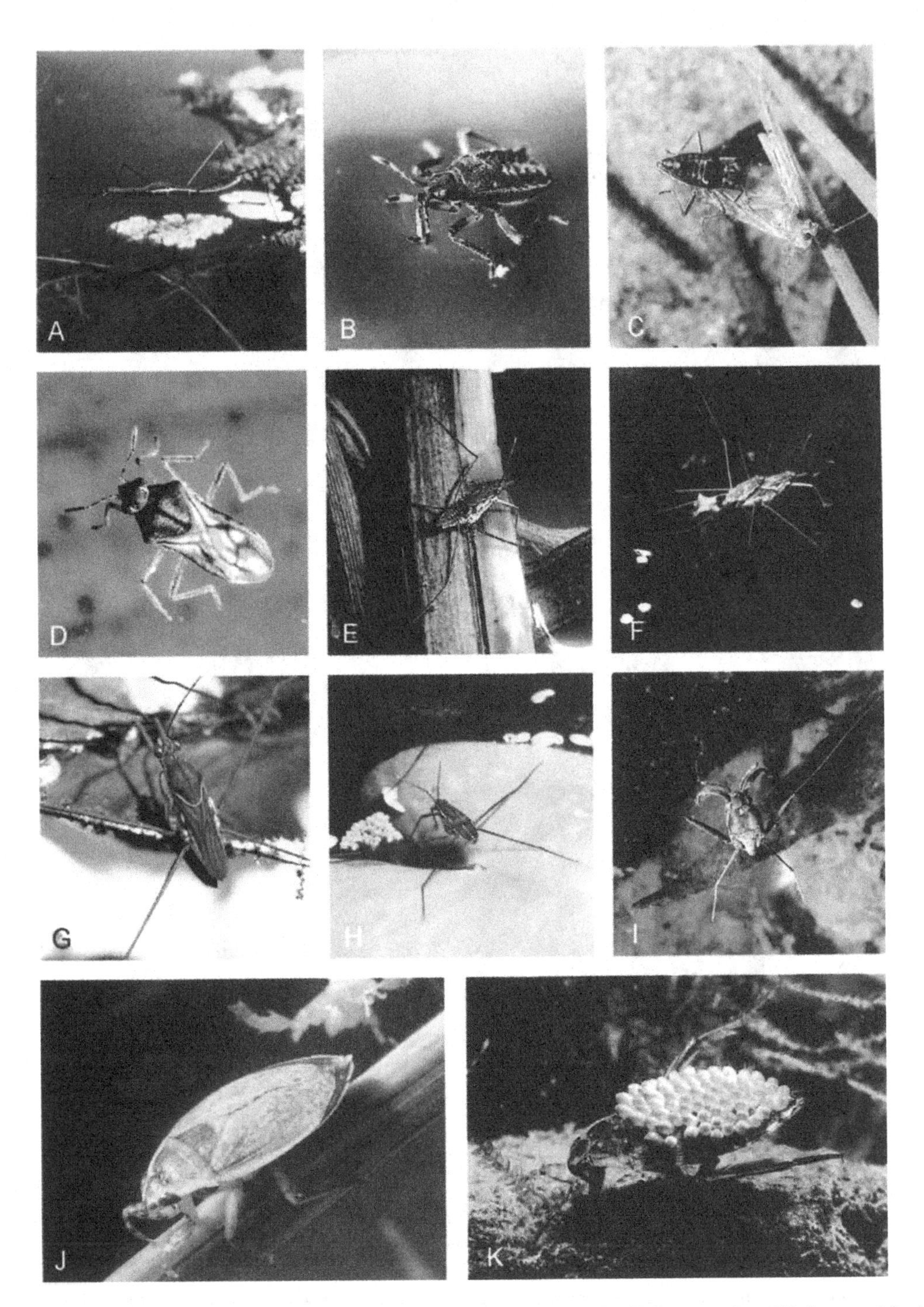

Plate 6. Australian water bugs (Gerromorpha & Nepomorpha). A, *Hydrometra feta* (Hydrometridae), macropterous ♀. B, *Drepanovelia dubia* (Veliidae), apterous ♂; C, *Microvelia (Austromicrovelia) peramoena* (Veliidae), apterous. D, *Microvelia (Pacificovelia) oceanica* (Veliidae), macropterous. E, *Tenagogerris euphrosyne* (Gerridae), apterous ♀. F, *T. euphrosyne*, in copula. G, *T. euphrosyne*, macropterous. H, *Tenagogonus australiensis* (Gerridae), apterous ♀. I, *Rheumatometra philarete* (Gerridae), apterous ♂. J, *Diplonychus* sp. (Belostomatidae). K, *Diplonychus* sp., ♂ with eggs on its back. Photographs by: A, F-H, J, Paul Zborowski; B-E, I, K, John Gooderham & Edward Tsyrlin.

Plate 7. Australian water bugs (Nepomorpha). A, *Lethocerus* sp. (Belostomatidae), waiting for prey. B, *L. distinctifemur*, raptorial fore legs and rostrum. C, *Laccotrephes tristis* (Nepidae), adult waiting for prey (small fish). D, *L. tristis*, large nymph, feeding on *Ranatra*. E, *Ranatra* sp. (Nepidae), feeding on *Tenagogonus australiensis* (Gerridae). F-G, nymphs of water boatmen (Corixidae): F, early instar. G, last instar with wing-pads. H, *Agraptocorixa* sp. (Corixidae), resting on bottom. I, *Agroptocorixa* sp., head and fore legs. J, *Agraptocorixa* sp., resting in vegetation. Photographs by: A, F, H-I, Paul Zborowski; B, David McClenaghan; F-H, John Gooderham & Edward Tsyrlin.

Plate 8. Australian water bugs (Nepomorpha). A, *Sigara* sp. (Corixidae), with abdominal air bubble. B, *Sigara* sp., resting on bottom. C, *Micronecta* sp. (Corixidae), resting on bottom. D, *Nerthra* sp. (Gelastocoridae), on land. E, *Ochterus* sp. (Ochteridae), on land. F, *Naucoris* sp. (Naucoridae), renewing its air-store at the water surface. G, *Enithares* sp. (Notonectidae), renewing its air-store at the water surface. H, *Enithares* sp., dark form. I, *Enithares* sp., pale form. J, *Paraplea* sp. (Pleidae). Photographs by: A-E, G, J, John Gooderham & Edward Tsyrlin; F, H-I, Paul Zborowski.

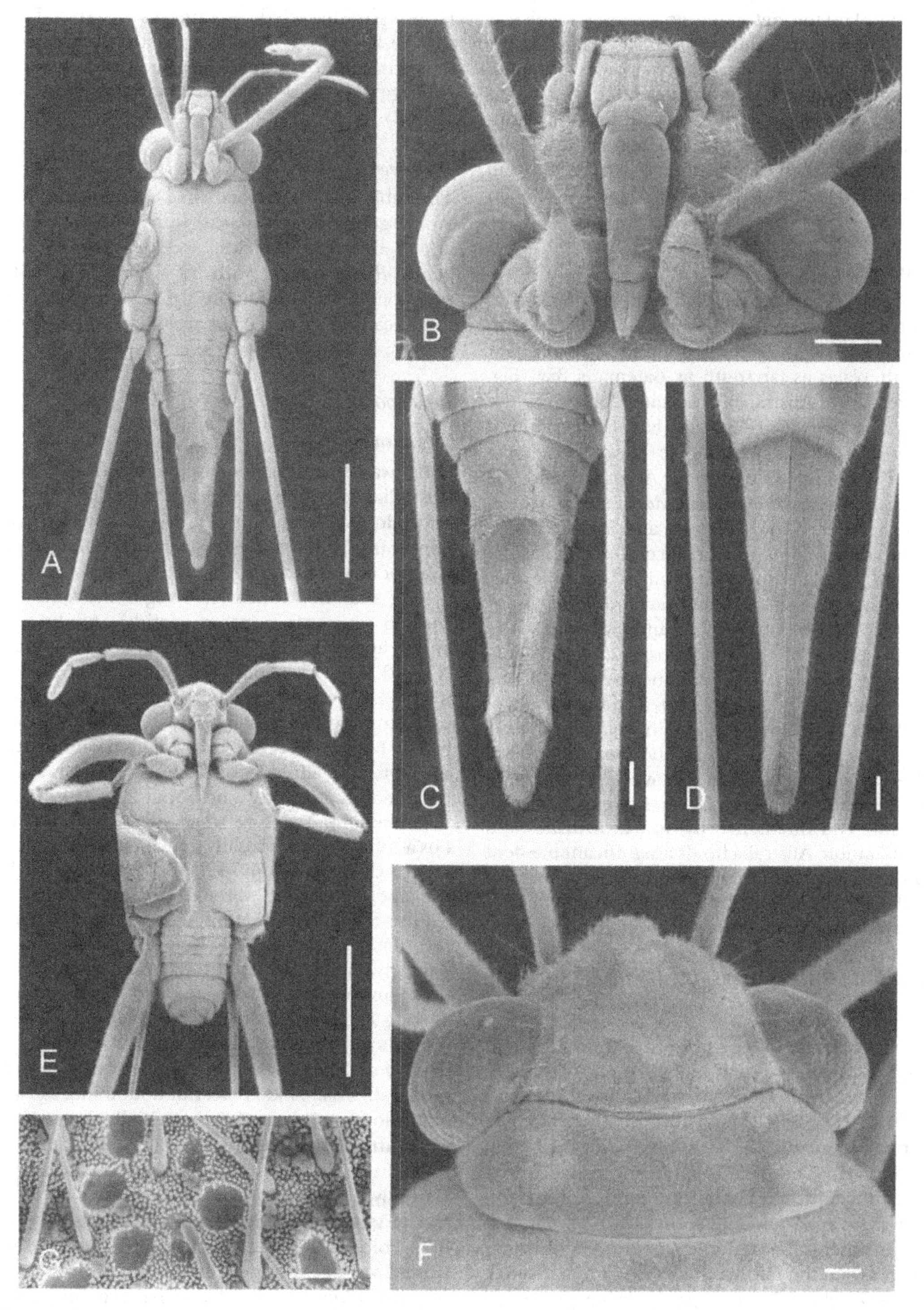

Figure 12.34. GERRIDAE. A-C, *Rhagadotarsus anomalus*, apterous ♂: A, ventral view. B, ventral view of head. C, ventral view of abdominal end. D, *R. anomalus*, apterous ♀: ventral view of abdominal end. E, *Rheumatometra philarete*, apterous ♂: ventral view. F-G, *R. philarete*: apterous ♀: F, dorsal view of head and pronotum. G, surface of dorsal head. Scales: A, E, 1 mm; B-D, F, 0.1 mm; g, 0.01 mm. Scanning Electron Micrographs.

net (a plankton net mounted on floats) which is towed along the side of a boat (Plate 4, F). However, since sea skaters are strongly attracted to light at night, the most profitable way of collecting these insects is to suspend a strong light-source over the side of a boat. (Additional information about sea skaters is available from: http://www.zmuc.dk/EntoWeb/Halobates/Halobat1.htm).

Distribution

Most Australian sea skaters occur in the tropical North, but two Australian endemics, *H. zephyrus* and *H. whiteleggei*, extend along the coast of New South Wales as far south as Batemans Bay (Fig. 12.33). Most Australian sea skaters belong to the *H.* (*s. str.*) *regalis*-group, and have a distribution limited to Australia and adjacent areas. *H. whiteleggei* is confined to eastern Australia, and the species-pair *H. acherontis* and *darwini* to northern Northern Territory. *H. regalis* and *herringi* have a wider distribution in northern Australia. Widespread *Halobates* species found in Australia may be either coastal (*H. hayanus* and *princeps*) or oceanic (*H. micans, sericeus,* and *germanus*). The three species of the subgenus *Hilliella* are confined to northern Australia and southern Papua New Guinea (Fig. 12.27). The Australian fauna of *Halobates* is ecologically and historically diverse and in several respects unique. Sea skaters are above all inhabitants of mangroves and coral reefs, marine communities recognized as extremely vulnerable and threatened by human activities and exploitation. Australia holds a significant element of the biological diversity of marine water striders and is capable of making a worthy contribution to its protection and conservation.

SUBFAMILY RHAGADOTARSINAE

Identification

Rhagadotarsine water striders are unique in having a distinct first abdominal sternum (Fig. 12.1A; s1), distinctly produced ventral lobes (bucculae) of head (Fig. 12.1B; bu), and greatly prolonged genital segments in both sexes (Figs 12.1C-E, 12.34C, D). The head is short and broad with 4 pairs of cephalic trichobothria on dorsal surface. The scent apparatus and scent orifice of metathorax is absent. The middle femora are distinctly longer than the middle tibiae. Abdomen distinctly longer than thorax, with 7 visible pregenital segments on both dorsum and venter (Fig. Fig. 12.1A). The genital segments of both sexes are very large, elongate, distinctly protruding from pregenital abdomen (Figs 12.1C-E); the pygophore of ♂ is elongate oval, the proctiger is large, parallel-sided; parameres absent. The ovipositor of ♀ well developed, first gonapophyses sclerotised and serrated. The shell of the elongate egg is differentiated at the anterior end (Fig. 12.36B).

Overview

The subfamily Rhagadotarsinae contains the Old World genus *Rhagadotarsus*, with two subgenera and 5 described species, and the New World genus *Rheumatobates*, with 32 species. The last-mentioned genus is famous for having elaborate modifications of the male antennae and legs in many species (Andersen 1982).

RHAGADOTARSUS

Identification

Small water striders easily separated from other Australian gerrids by the greatly prolonged terminal abdominal segments of both sexes, the prominent ventral lobes of head (buccula) produced in front of rostral base (Fig. 12.1B, bu), and relatively long abdomen which has 7 segments ventrally (only 6 segments in other gerrids). Adults are either apterous (Fig. 12.35A) or macropterous (Fig. 12.35B). The body chiefly blackish, covered with grey or bluish grey pubescence on dorsum; pronotum with a median yellowish mark anteriorly. The head is short and broad. The pronotum of the apterous adult form is much shorter than head length. Mesosternum about 2.5x as long as metasternum. The fore wings of the macropterous adult form have basally thickened veins which form 3 closed cells (Fig. 12.35B); the wings reach the abdominal end when folded (♂) or only expose the tip of the genital segments (♀). The fore femora are relatively long, slender in basal part, gradually thickened towards apex; the fore tibiae are broad and flattened and less than half as long as femur; first segment of fore tarsus is very short; claws relatively long, inserted in a deep cleft. The claws of the middle and hind legs are small, bristle-like. Posterior abdominal sterna of ♂ more or less modified, medially depressed, with tufts of hairs, etc. (Figs 12.1C, D, 12.34C); segment 8 dorso-ventrally depressed, modified ventrally.

The genus *Rhagadotarsus* Breddin was revised by J. Polhemus & Karunaratne (1993) and comprises five species widely distributed in the Afrotropical, Oriental, and Australian regions. Only one Australian species, *Rhagadotarsus* (*Rhagadotarsus*) *anomalus* J. Polhemus & Karunaratne, length 3.4-4.2 mm (♂) or 4.2-5.8 mm (♀). New South Wales, Northern Territory, Queensland, Western Australia.

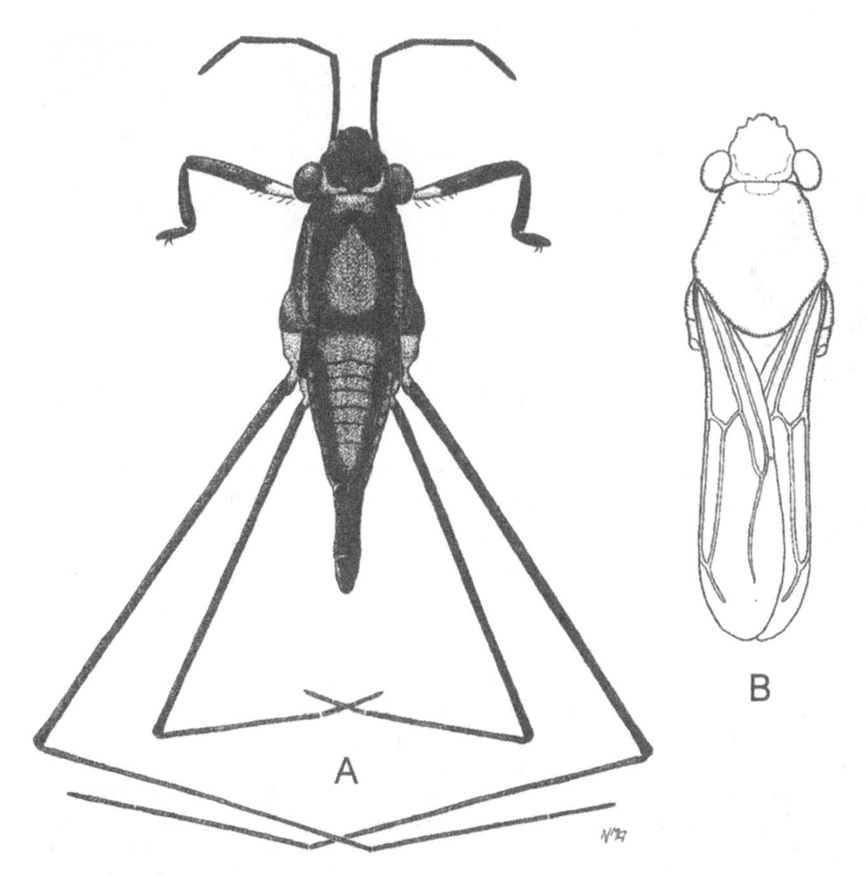

Figure 12.35. GERRIDAE. A-B, *Rhagadotarsus anomalus*: A, apterous ♂. B, macropterous ♂; antennae and legs omitted. Reproduced from Andersen & Weir (1998).

Biology

Rhagadotarsus anomalus is found in a wide variety of freshwater habitats, including lakes, ponds, pools, and slowly running streams and small rivers, often in company with species of *Tenagogerris* and *Limnogonus* (Gerrinae). The mating behaviour of *Rhagadotarsus* species involves communication by patterned surface waves, produced and perceived by both males and females (Wilcox 1972). These signals are generated by leg movements while the insect is free on the surface or while it grasps floating or fixed objects; these objects then become copulation and oviposition sites (Fig. 12.36A). Several different signals (waves with different amplitudes and frequency patterns) have been recorded in *R. anomalus*. In the precopulatory phase, males produce calling signals of different kinds (by movements of middle and hind legs). When a female approaches within 5-10 cm, the male switches to courtship signals (by movements of the fore legs and a floating object). When within 2-3 cm the female responds with courtship signals. The male then releases the object, mounts the female and copulates. The copulation proper is relatively brief. After dismounting, the male produces sporadic postcopulatory signals, while the female embeds its eggs into the floating object. After oviposition the female moves away and the male usually returns to the object and begins signalling again. Surface wave communication also includes aggressive signals towards other males. Wilcox (1972) observed that females use their long and serrated ovipositor to insert eggs into floating objects just used by the male for signalling. The egg of *Rhagadotarsus* is elongate (Fig. 12.36B), densely sculptured at the anterior end which also carries a small micropylar tubercle (mi).

Populations of *Rhagadotarsus* species are predominantly apterous, but macropterous individuals are known in most species including *R. anomalus*. The fore wings of the macropterous form

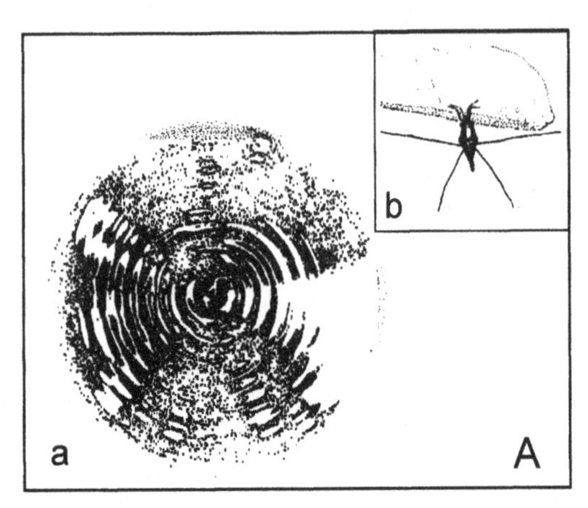

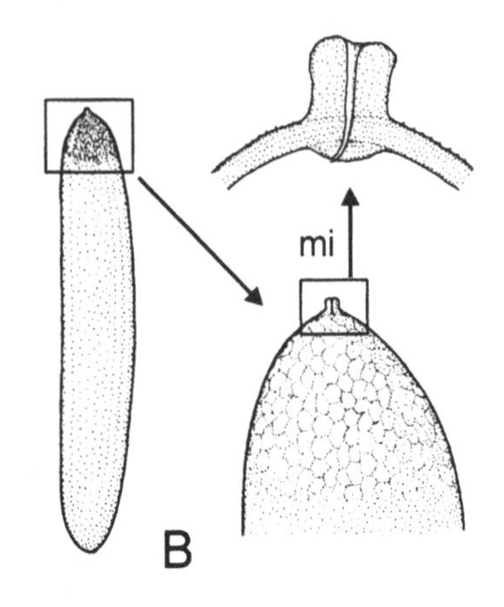

Figure 12.36. GERRIDAE. A, *Rhagadotarsus anomalus*: pattern of surface ripples (a) produced by male (b). B, *Rhagadotarsus kraepelini*, ripe ovarian egg with anterior end and micropylar process (mi) shown at higher magnification. Reproduced with modification from Andersen (1982).

reach the tip of the abdomen and self-inflicted damage to the fore wings is limited to tearing off small pieces between veins and along the margins. The life history of *R. kraepelini* Breddin was reported by Hoffmann (1936b) at Canton, China, where these insects breed over most of the year and overwinter as adults, producing 5 or 6 generations per year. In Australia, nymphs of *R. anomalus* are recorded from most months of the year, with peak numbers during the periods May-August, and October-November.

Distribution

The Australian species, *Rhagadotarsus anomalus*, was previously confused with the Oriental species *R. kraepelini* Breddin (Wilcox 1972). It is widely distributed along the east coast of Queensland, in the northern parts of Northern Territory, the Kimberleys, Pilbarra and Carnavon regions of Western Australia, and also recorded from a few inland localities in New South Wales and Queensland (Fig. 12.37). It is also recorded from southern Papua New Guinea (J. Polhemus & Karunaratne 1993).

SUBFAMILY TREPOBATINAE

Identification

The trepobatines are small water striders with their middle femora distinctly shorter than their middle tibiae (Figs 12.38A, C, 12.40A). The head relatively short and broad with large eyes, four pairs of cephalic trichobothria, and relatively small ventral lobes (bucculae). The pronotum is shorter than the head length in the apterous adults (Figs 12.38A, C, 12.40A, 12.43A-C), large, subpentagonal in the rare macropterous adults (Figs 12.38B, 12.40B). The fore wings are chiefly membranous, with three distinct veins forming two closed cells in basal part and two weak, distal veins (Fig. 12.1G). The fore femora are usually simple, rarely modified (curved); tibiae flattened or with subapical spur (Fig. 12.2E); first segment of fore tarsus very short. Metasternal scent apparatus including scent orifice present or absent. Basal abdominal laterotergites and sternum not separated. Genital segments of male large, distinctly protruding from pregenital abdomen; segment 8 long, pygophore elongate, either simple or provided with a pair of antero-laterally directed processes (Figs 12.2D, F), parameres small. Ovipositor of female relatively short, first gonapophyses non-serrated.

Overview

Species of Trepobatinae can be found in most regions of the world and in all major types of aquatic habitats, including marine. Following the taxonomic revisions by J. Polhemus & D. Polhemus (1993, 1994b, 1995b, 1996, 2000a, 2002), there are 118 described species placed in 4 tribes and 25 genera worldwide. Four genera and 7 species have been recorded from Australia. They belong to the tribes Metrobatini (*Rheumatometra*), Naboandelini (*Calyptobates*), and Stenobatini (*Rheumatometroides* and *Stenobates*).

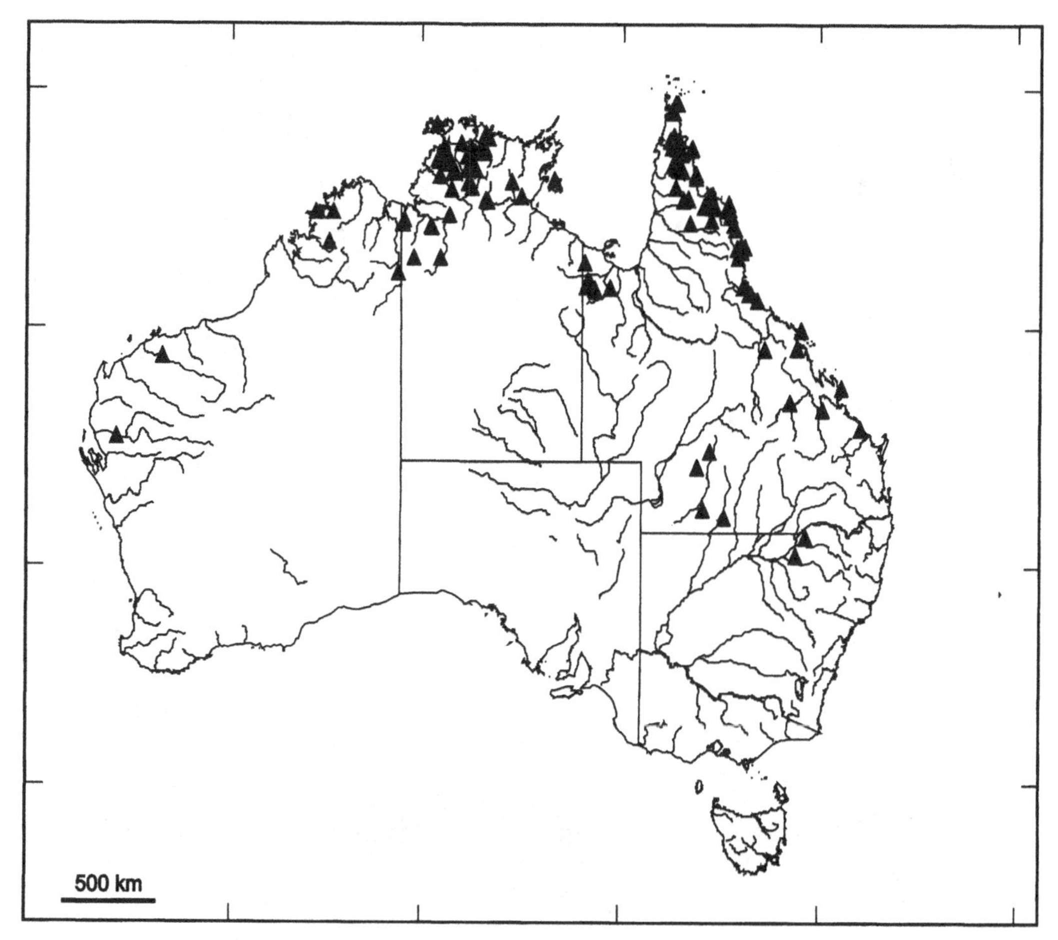

Figure 12.37. GERRIDAE. Distribution of *Rhagadotarsus anomalus* in Australia.

CALYPTOBATES

Identification

Here belongs some of the smallest Australian water striders, length 1.8-3.7 mm, easily overlooked or confused with nymphs of larger gerrids. Adults are either apterous (Figs 12.38A, C) or macropterous (Fig. 12.38B). The body is suboval, brownish to blackish above, without grey pruinose pubescence except occasionally on metanotum, abdominal dorsum and pleuron; pronotum with a median yellowish mark anteriorly; venter usually pale with dark markings (Figs 12.38F-H). Antennae 0.6-0.75x body length; segment 1 longest, slightly thicker and distinctly longer in male than in female; segments 2-4 subequal in length; segment 4 flattened in male. Fore wings of macropterous form dark brownish, extending far beyond abdominal end when folded. The metacetabula of female with a dorsal tuft of stiff erect black hairs. Mesosternum about 10x as long as metasternum; metasternal scent apparatus present, but scent orifice very small, located in anterior half of metasternum. Fore femur of male slightly thickened but only weakly arched; fore tibia cylindrical, without apical spur. Abdomen relatively short, tergites 1 and 7 longest. Genital segments of male relatively large, proctiger with slender, antero-laterally directed lateral processes (Fig. 12.2F).

The genus *Calyptobates* J. Polhemus & D. Polhemus belongs to the tribe Naboandelini. This tribe also includes the genera *Naboandelus* Distant, with 14 species distributed throughout the paleotropics, reaching its greatest diversity in the sub-Saharan Africa, and *Hynesionella* Poisson, with 3 species in the last mentioned region. In these genera the fore tibiae are much shorter than the femora, lacking apical processes or spurs in the male, the metasternal scent apparatus is

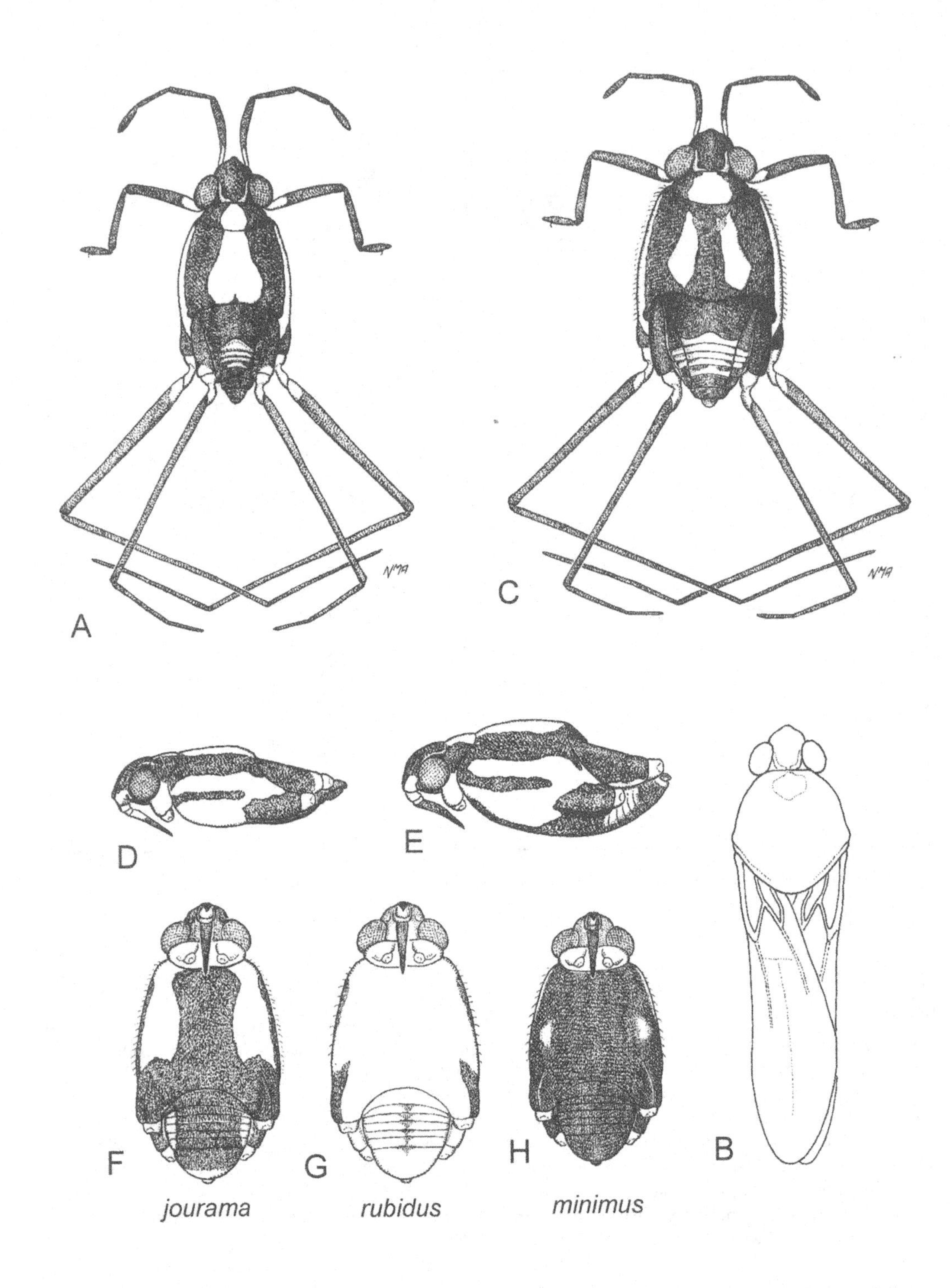

Figure 12.38. GERRIDAE. A-F, *Calyptobates jourama*: A, apterous ♂. B, dorsal view of macropterous ♂; antennae and legs removed. C, apterous ♀. D, lateral view of apterous ♂; antennae and legs removed. E, lateral view of apterous ♀; antennae and legs removed. F, ventral view of apterous ♀; antennae and legs removed. G, *C. rubidus*, apterous ♀: ventral view; antennae and legs removed. H, *C. minimus*, apterous ♀: ventral view; antennae and legs removed. Reproduced from Andersen & Weir (1998).

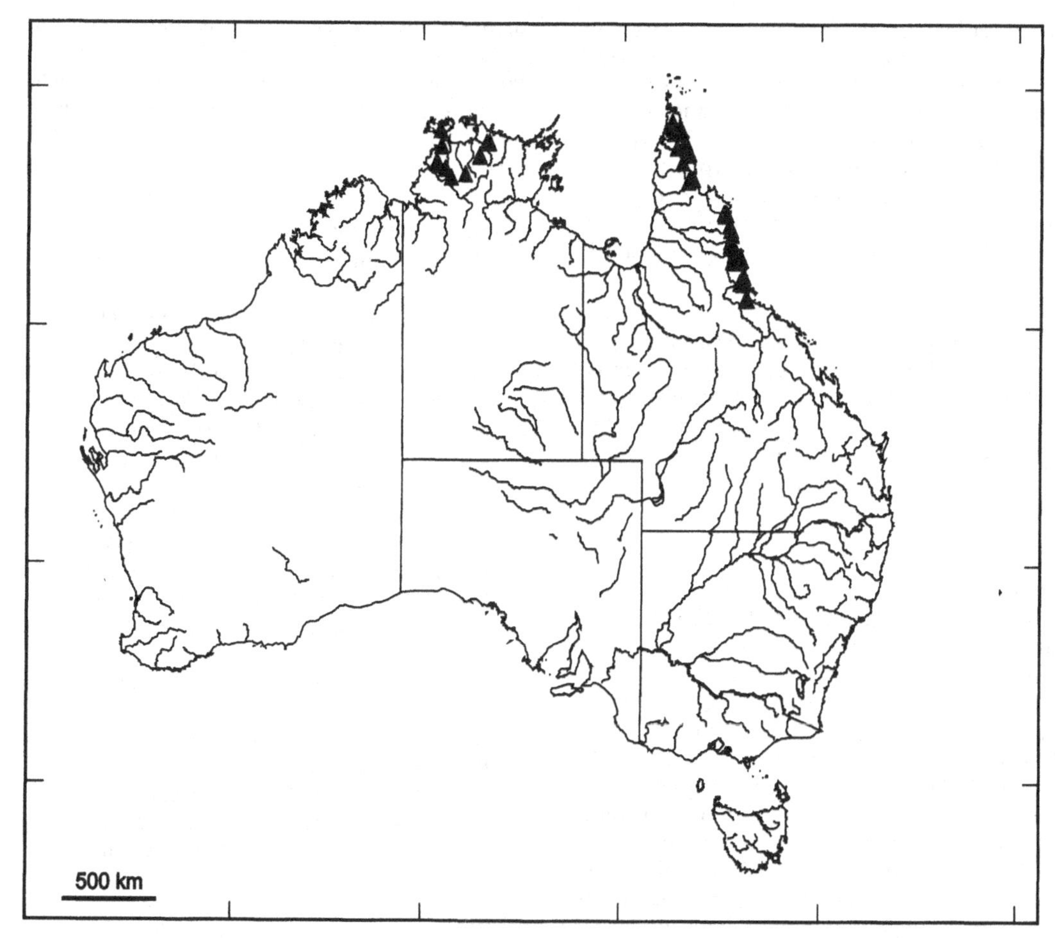

Figure 12.39. GERRIDAE. Distribution of *Calyptobates* species in Australia.

present, although the scent orifice is small and inconspicuous, and the male proctiger has long lateral processes (shared with the tribe Stenobatini, see below).

Calyptobates was revised by J. Polhemus & D. Polhemus (1994) and contains 7 species sparsely distributed in the Oriental and Australian regions. Three species are recorded from Australia.

Key to Australian species

1. Venter except prosternum completely dark (♂); pale markings (if present) poorly defined, on lower pleural region of mesothorax (♀; Fig. 12.38H). Length 1.8-3.4 mm (♂), 2.0-3.0 mm (♀). Northern Territory.. .... *minimus* J. Polhemus & D. Polhemus
– Venter with more or less extensive pale markings on mesosternum (Figs 12.38F-G) .. 2
2. Mesosternum and often abdominal venter completely yellowish to orange, without any medial darkening (♂) or rarely with infuscated region medially, poorly defined posteriorly (♀; Fig. 12.38G)). Length 1.8-3.1 mm (♂), 2.1-2.3 mm (♀). Queensland.. .... *rubidus* J. Polhemus & D. Polhemus
– Mesosternum and abdominal venter of both sexes with extensive median dark markings (Fig. 12.38D-F), reaching posterior margin of mesosternum. Length 2.0-3.5 mm (♂), 2.1-3.7 mm (♀). Queensland........ *jourama* J. Polhemus & D. Polhemus

Biology

Almost nothing is known about the biology of *Calyptobates* species. These tiny freshwater gerrids are weak skaters, occupying sheltered habitats in ponds and along the margin of streams. They are easily overlooked by the casual collector or mistaken for being early instar nymphs of larger gerrids. In Australia, nymphs have been recorded from most months of the year, with peak numbers during the period June-August and October-December. They are dimorphic with respect to wing development, but macropterous specimens are generally rare. Macropters tear the distal parts of their wings off in a way similar to that described below for *Rheumatometra* (see below).

Distribution

Calyptobates species are sparsely distributed in the Oriental and Australian regions (J. Polhemus & D. Polhemus 1994). The three Australian species are confined to coastal areas of northern Queensland and the northernmost part of the Northern Territory (Fig. 12.39).

RHEUMATOMETRA

Identification

Small water striders, length 2.3-4.1 mm, with a striking sexual size dimorphism, the male being much smaller than the female (Fig. 12.41). Adults are either apterous (Fig. 12.40A) or macropterous (Fig. 12.40B), suboval (♂) or broadly oval (♀), moderately dorso-ventrally flattened, blackish, partly covered with grey pruinose pubescence; pronotum dark with median yellowish mark (Plate 6, I). Head short and broad (Fig. 12.34F). Antennae only 0.4-0.5x body length, with a preapical spinous hair on segments 1-3; segment 1 longest; segments 2-4 subequal in length; segment 4 oval. Fore wings of macropterous form dark brownish and largely membranous, extending far beyond abdominal end when folded. Mesonotum about 2x (♂), or more than 3x pronotum length, with a longitudinal depression in middle (♀). Mesosternum 6-7x as long as metasternum; metasternal scent apparatus (including orifice) absent. Structure of fore legs sexually dimorphic; male femora incrassate and strongly curved (Fig. 12.2E, 12.34E), tibiae curved, with elongate dark pad on apical spur; female femora thickened but not curved, tibiae straight. Abdomen relatively short, subequal in length to thorax, tergites 1 and 7 longest. Genital segments of male small, at most slightly protruding from pregenital abdomen; proctiger small, parallel-sided.

The genus *Rheumatometra* Kirkaldy belongs to the tribe Metrobatini revised by J. Polhemus & D. Polhemus (1993) which also comprises the genus *Metrobates* Uhler from North and South America, and a number of genera distributed in New Guinea, the Bismarck Archipelago, and the Solomon Islands. The striking sexual size dimorphism of *Rheumatometra* (Fig. 12.41) is shared with the monotypic genus *Andersenella* J. Polhemus and D. Polhemus (1993) from Papua New Guinea which, however, does not have as strongly modified male fore legs. *Rheumatometra* has only two described species, both Australian.

Key to Australian species

1. Middle femur of ♂ distinctly thickened, almost as thick as an eye width (Fig. 12.40D). Hind femur of ♀ distinctly widened proximally, with a conspicuous pilosity of long, suberect hairs (Fig. 12.40E). Male abdominal venter flattened and medially depressed; venter of segment 8 medially depressed (Fig. 12.40F). Length 2.3-4.6 mm (♂), 3.4- 5.7 mm (♀).Australian Capital Territory, New South Wales, Queensland, Tasmania, Victoria .... *philarete* Kirkaldy
- Middle femur of ♂ slender, only half of an eye width. Hind femur of ♀ only slightly widened proximally, without conspicuous pilosity (Fig. 12.40A). Male abdominal venter flattened but not depressed; venter of segment 8 not modified (Fig. 12.40C). Length 2.3-4.0 mm (♂), 3.4-4.6 mm (♀). New South Wales, Queensland, Victoria............................... .... *dimorpha* Andersen & Weir

Biology

Very little is known about the biology of *Rheumatometra* species. They inhabit stony streams and rivers where they are usually found on the quiet surface of pools around rocks or logs or along the bank. Adults are recorded from January to March in Tasmania, November-May in Victoria and New South Wales, and throughout most of the year in Queensland. Nymphs are recorded from November to March in southeastern Australia, including Tasmania, and also from May to September in Queensland indicating a possibility of several generations per year. Both species of *Rheumatometra* exhibit a very pronounced sexual dimorphism (Fig. 12.41), the male being a little more than half the size of the female. The male fore legs are modified to fit the curve of the female mesothorax (Fig. 12.34E). The grip of the male during matings is so persistent that he usually remains seated on the back of the female after being killed in alcohol. Pairs caught in tandem are never in genital contact which suggests that this behaviour may be an extreme type of postcopulatory mate-guarding, as known from other water strider species (Spence & Andersen 1994).

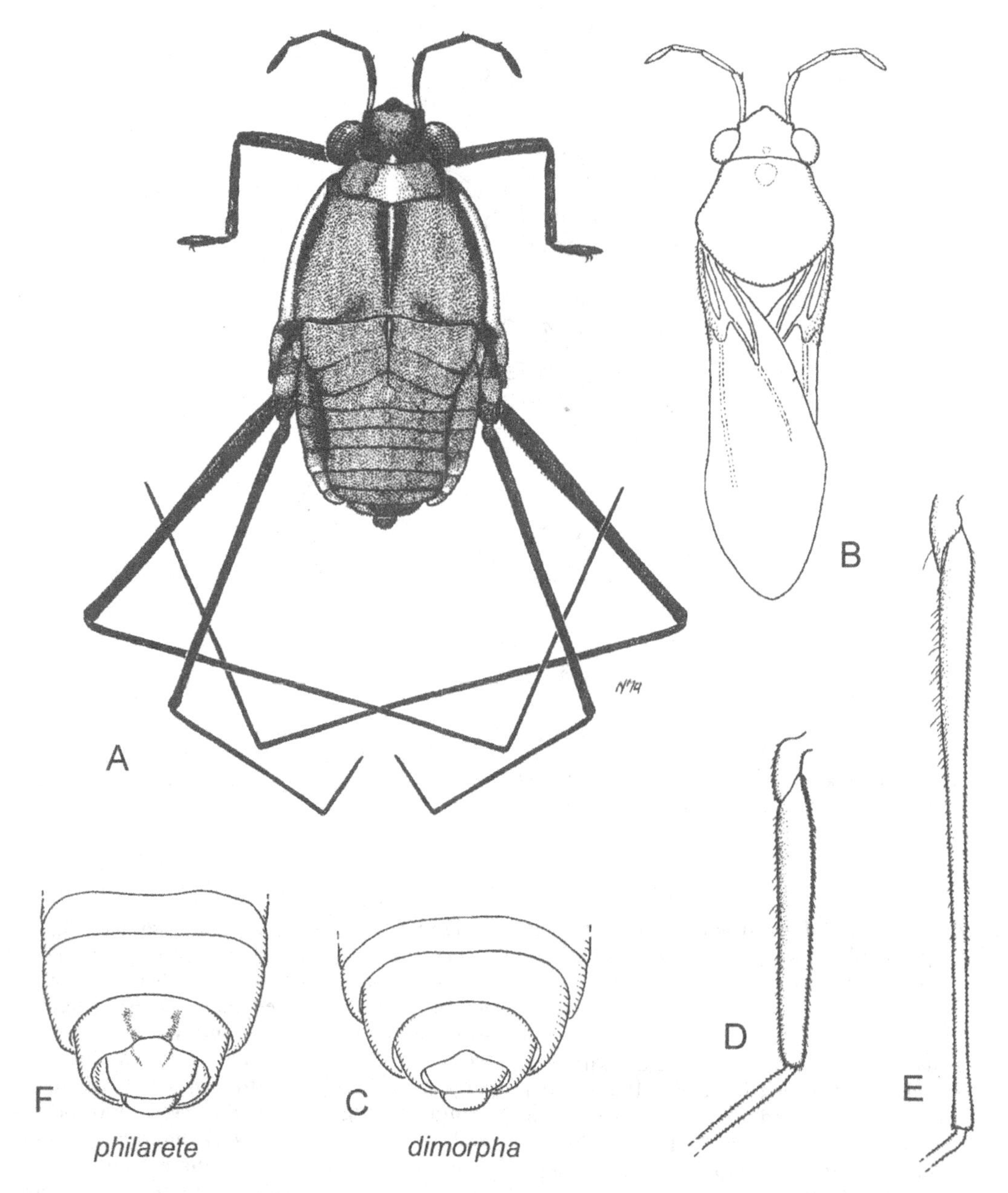

Figure 12.40. GERRIDAE. A-C, *Rheumatometra dimorpha*: A, apterous ♀. B, dorsal view of macropterous ♀; legs omitted. C, ventral view of abdominal end of apterous ♂. D-F, *R. philarete*: D, basal segments of middle leg of apterous ♂. E, basal segments of hind leg of apterous ♀. F, ventral view of abdominal end of apterous ♂. Reproduced from Andersen & Weir (1998).

Rheumatometra species are polymorphic with respect to wing development. Although the apterous form is usually the most abundant, macropterous individuals may be quite frequent in some populations. A sample of *R. dimorpha* from Little Mulgrave River, northern Queensland, contains 47 macropterous and 33 apterous adults; 97 out of 127 last instar nymphs had distinct wing-pads. After maturation (as indicated by a relatively hard, darkly pigmented cuticula), macropterous

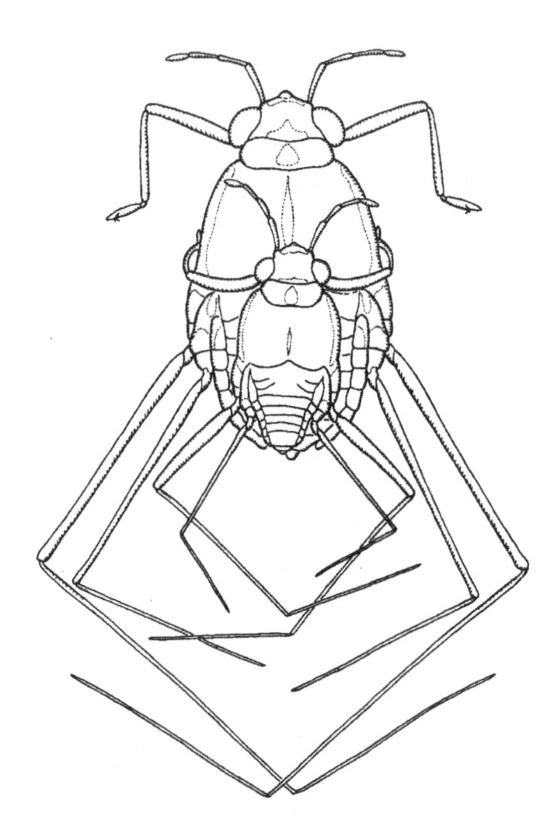

Figure 12.41. GERRIDAE. Pair of *Rheumatometra philarete* with ♂ in postcopulatory guarding position. Reproduced from Andersen & Weir (1998).

adults usually tear off the distal two-thirds of their wings, just behind the apices of the basal, thickened veins of the fore wings (Fig. 12.40B).

Distribution

Rheumatometra is endemic to Australia with two described species (Andersen & Weir 1998) which are widely distributed in coastal areas of Queensland and New South Wales, and in the Australian Capital Territory, Victoria, and Tasmania (Fig. 12.42).

RHEUMATOMETROIDES

Identification

Small water striders, 2.7-4.0 mm long, which are always apterous (Fig. 12.43A). Body ovate, thorax relatively high as viewed in profile; dorsum usually blackish brown with grey pruinose pubescence and with contrasting brown or yellowish markings; pale markings on mesonotum large, elongate or quadrate on either side of the midline (Fig. 12.43A). Head with large, protruding eyes. Antennae relatively long and slender, slightly thicker in male; segment 1 longest, almost twice as long as segment 2; segments 3-4 subequal in length, slightly shorter than second segment. Rostrum relatively stout and ventrally curved, with segment 3 swollen in male. Mesonotum large and widening posteriorly, length about 3x pronotum length, in female divided into two lateral sclerites connected by a membranous region which diverge through metanotum and continues along the sides of abdominal tergites. Metasternum of male weakly extended anteriorly (Fig. 12.CB; mt), anterior margin straight, forming a regular triangle; metasternal scent orifice (so) small, not situated on a tubercle. Male fore femora often thickened basally; fore tibiae broadened distally, without apical spur. Abdomen relatively short, tergite 5 longest, tergite 7 very short, concealed; tergites 3-6 of female abdomen connected with membranes that allow longitudinal expansion of abdomen. Genital segments of male relatively large, distinctly protruding from pregenital abdomen; proctiger with long lateral,

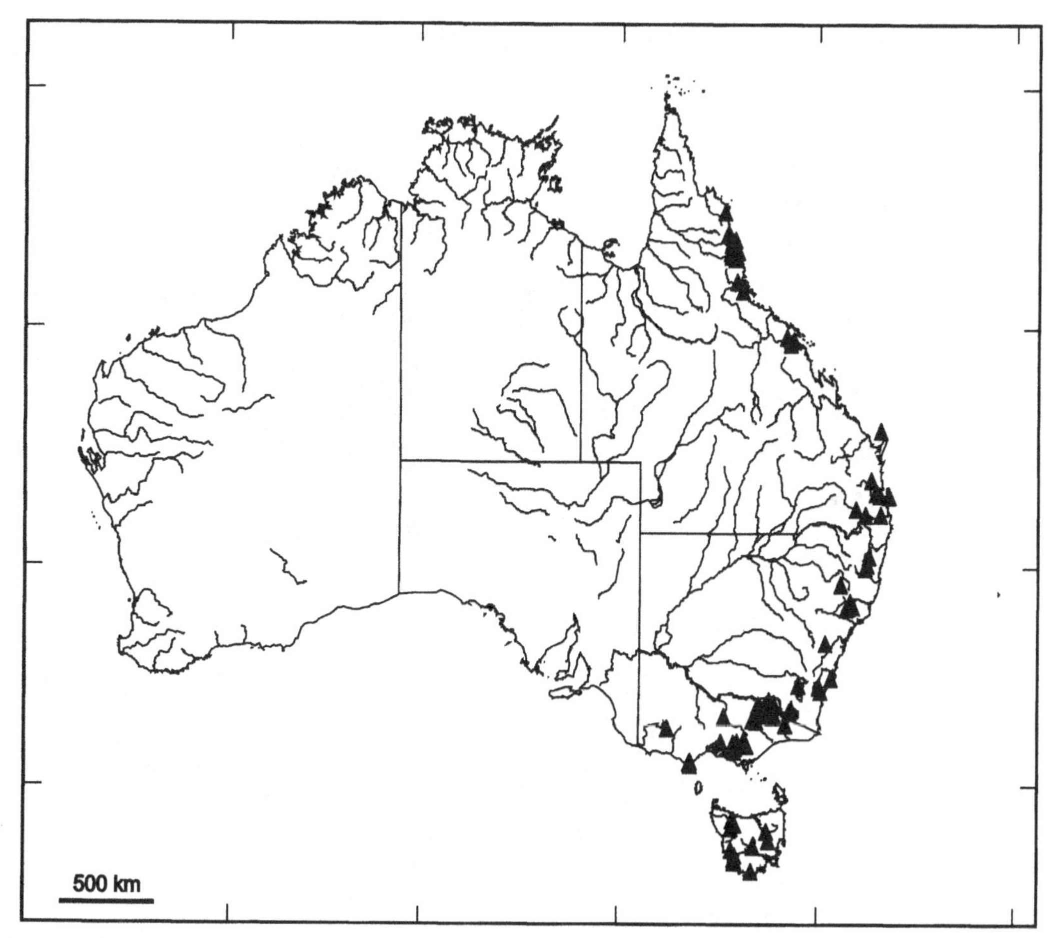

Figure 12.42. GERRIDAE. Distribution of *Rheumatometra* species in Australia.

antero-ventrally directed processes (Fig. 12.2D). Ovipositor of female well developed; first gonapophyses with membranous apices; second gonapophyses serrated.

The genus *Rheumatometroides* Hungerford & Matsuda belongs to the tribe Stenobatini which also includes three other genera of marine water striders. *Rheumatometroides* was revised by J. Polhemus & D. Polhemus (1996) and includes seven species. The single Australian species, *Rheumatometroides carpentaria* was originally described by J. Polhemus & D. Polhemus (1991) in the following genus (*Stenobates*), length 3.7-3.9 mm (♂), 3.6-3.9 mm (♀). Northern Territory.

Biology

Species of *Rheumatometroides* are marine, inhabiting mangrove swamps. Like other marine water striders, adults are always wingless.

Distribution

As presently known, species belonging to *Rheumatometroides* are distributed from Singapore to the Solomon Islands. The single Australian species is only known from Groote Eilandt in the Gulf of Carpentaria (Fig. 12.44).

STENOBATES

Identification

Quite similar to *Rheumatometroides*, but the pale markings on mesonotum consist of elongate streaks or stripes on either side of the midline (Fig. 12.43B), the male metasternum is strongly extended anteriorly, with curved anterior margin, and the metasternal scent orifice is large, situated on a low tubercle (Fig. 12.2A; so). Small water striders, length 3.1-5.1 mm long, which are always apterous (Figs 12.43B, C). Body elongate oval, thorax relatively high as viewed in profile,

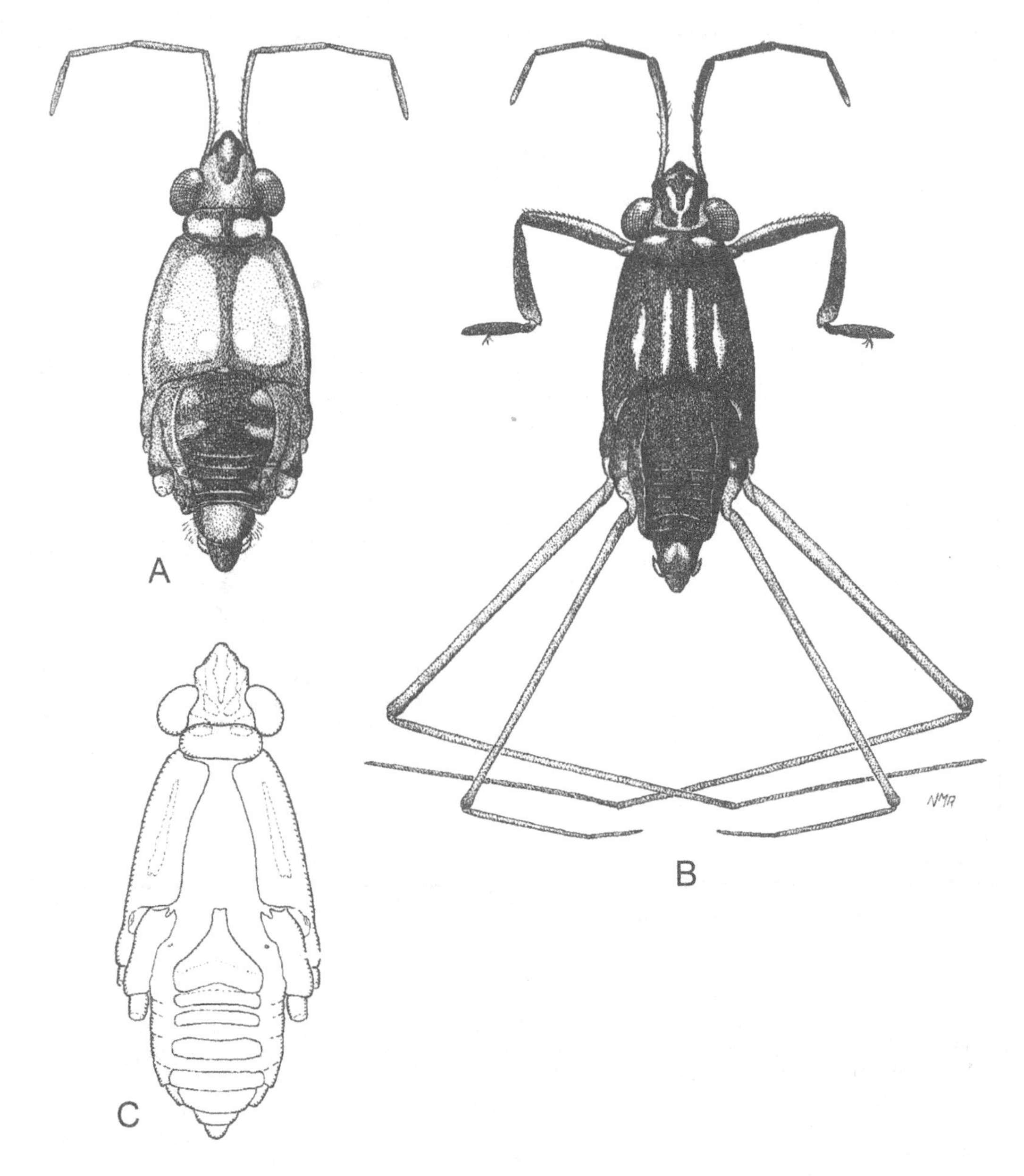

Figure 12.43. GERRIDAE. A, *Rheumatometroides carpentaria*, apterous ♂: legs omitted. B-C, *Stenobates australicus*: B, apterous ♂. C, dorsal view of apterous ♀; antennae and legs omitted. Reproduced from Andersen & Weir (1998).

dorsum usually blackish brown with grey pruinose pubescence, always with contrasting brown or yellowish markings (Fig. 12.43B). Antennal segment 1 longest, almost twice as long as segment 2, ventrally set with closely packed short to medium length hairs (♂); segments 3-4 subequal in length. Mesonotum large and widening posteriorly, length about 3-3.5x pronotum length. Male metasternum large, with a prominent scent orifice located on a low tubercle in anterior part (Fig. 12.2A; so); female metasternum much shorter, with small scent orifice. Male fore femora slightly thickened basally; fore tibiae broadened distally, with at least a modest apical spur.

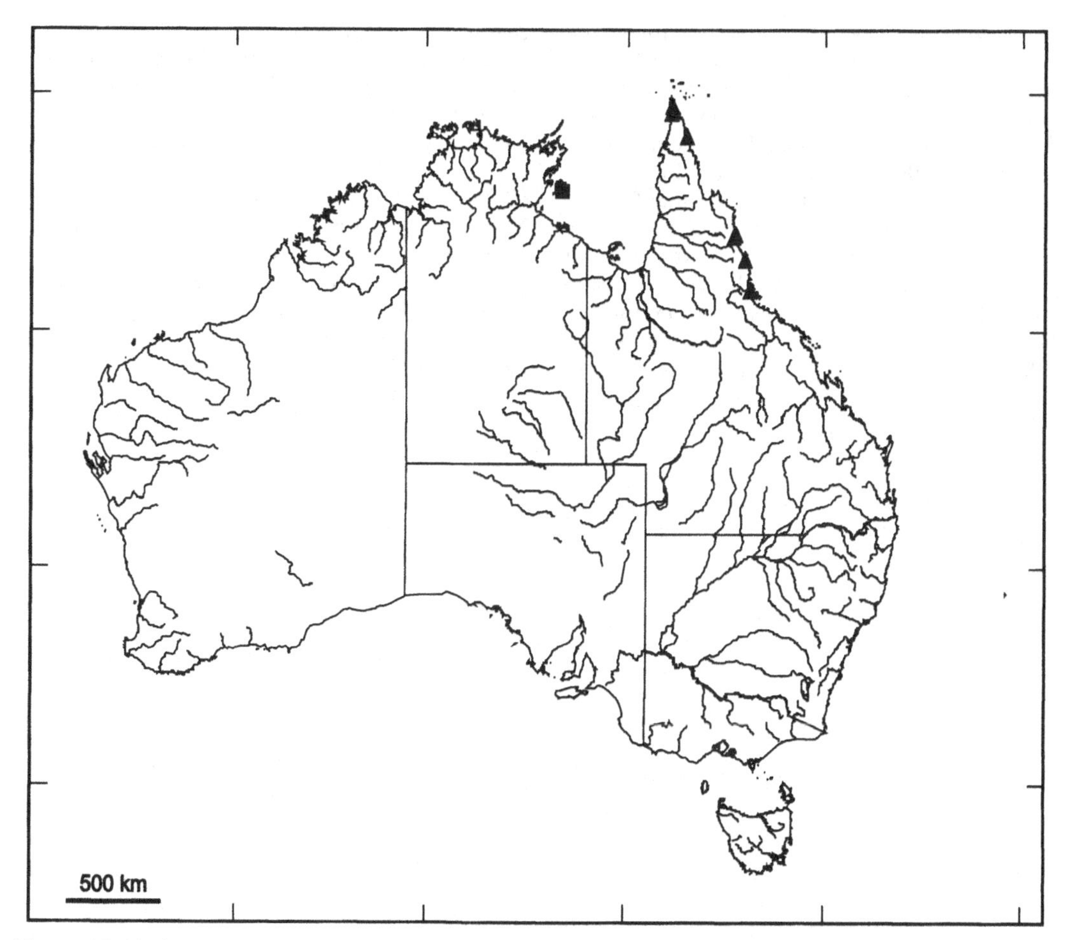

Figure 12.44. GERRIDAE. Distribution of *Rheumatometroides carpentaria* (square) and *Stenobates australicus* (triangles) in Australia.

Abdomen relatively short, tergites 6 and 7 longest. Genital segments of male relatively large; proctiger with long lateral, antero-ventrally directed processes. Ovipositor of female as in *Rheumatometroides*.

The genus *Stenobates* Esaki belongs to the tribe Stenobatini, revised by J. Polhemus & D. Polhemus (1996), which also includes three other genera of marine water striders.
The genus comprises a total of 11 species. Only one species in Australia: *Stenobates australicus* J. Polhemus & D. Polhemus, length 3.6-3.8 mm (♂), 3.5-4.2 mm (♀). Queensland.

Biology
Species belonging to the tribe Stenobatini are marine, occurring in mangrove estuaries which have a pronounced salinity gradient from euhaline to fresh as one goes inland and upstream (J. Polhemus & D. Polhemus 1996). In southern Papua New Guinea, *Stenobates australicus* was abundant not only near the mouth of the Kikori Delta (salinity 30-35 ppt.), but also in great numbers on the creeks of the interior delta where salinities were as low as 1.5 ppt. as well as in completely fresh water at the upper limit of tidal influence. In the same area, *Rheumatometroides kikori* J. Polhemus & D. Polhemus was found in low numbers along the heavily shaded mixohaline creeks, but not seen on the open channels of the tidally influenced freshwater rivers. In northern Queensland, *S. australis* is rather common at the shady margins of mangrove estuaries. It frequently occurs together with nymphs of *Halobates herringi* Polhemus & Cheng, the adults of which are much stronger skaters and usually inhabit more open water further from shore. Like other marine water striders, adult stenobatines are

always wingless. The life history is largely unknown. In northern Queensland, nymphs of *Stenobates australicus* have been collected in August, but the records are too few to give any clue about breeding season. J. Polhemus & D. Polhemus (1996) characterized the female ovipositor as "serrate" and suggested that females emplace their eggs in hard substrates, most likely mangrove roots. However, only the second gonapophyses have distinct teeth and it is unlikely that the stenobatine ovipositor can be used in the way suggested by J. Polhemus & D. Polhemus (1996). Furthermore, ripe ovarian eggs of *S. australicus* (Andersen & Weir 1998: fig. 40) are not differentiated in a way suggesting that they are deposited other than superficially on the substrate. Observations of egg-laying in the field or laboratory are needed to settle this question.

Distribution

Stenobates comprises 11 species, distributed from Singapore to Papua New Guinea and northeastern Australia. Only one species, *S. australicus*, occurs along the coast of northern Queensland (Fig. 12.44). It is also recorded from southern Papua New Guinea.

Infraorder NEPOMORPHA

Aquatic Bugs

Identification

Small, medium-sized to large or very large insects inhabiting various types of freshwater. Also including littoral, non-aquatic species (Gelastocoridae, Ochteridae). Nepomorphans are characterised by: (1) antennae very short, usually concealed in grooves below the eyes (Figs 4.2D, G; an; Figs 13.10B, 14.3B, 15.4A, 20.4A; arrows); (2) rostrum usually short and stout, often apparently 3-segmented (except in Corixidae); (3) fore legs often modified (raptorial or otherwise), pretarsus (distal part of last tarsal segment) usually with dorsal and ventral arolium; (4) hind legs often flattened and fringed with swimming hairs (natatorial; Fig. 4.4F); and (5) fore wings modified as hemelytra, with a coriaceous basal part and a membranous distal part (Fig. 4.5B, C), with well developed wing-to-body coupling mechanism.

Overview

The true aquatic bugs are formally classified as an infraorder, Nepomorpha, of heteropterous bugs. This infraorder was formally proposed by Popov (1971a), but previously known by the name Hydrocorisae proposed by Dufour (1833) as part of his classification of true bugs. Based on the concealed antenna, Fieber (1851) used the name Cryptocerata (revived by Mahner 1993) for this group (as opposed to Gymnocerata comprising all other bugs which have their antenna exposed). In its current taxonomical state, the Nepomorpha comprises about 2,000 described species classified in 11 families (Stys & Jansson 1988).

13. Family NEPIDAE

Water Scorpions

Identification

Brownish, either dorso-ventrally flattened, suboval or subcylindrical water bugs which are medium-sized to very large, body length from 12 to 60 mm, excluding the slender respiratory siphon arising from the end of abdomen; the siphon varies in length from being rather short (Figs 13.4, 13.6A, 13.8A) to being as long as or even longer than the body (Figs 13.2, 13.11A). The fore legs are raptorial, femur widened and with a ventral groove to receive tibia and tarsus (Figs 13.6B, 13.8D, H, 13.10F, G); all tarsi 1-segmented. Hind coxae short, free, not united with metapleuron. Head relatively short, pointed anteriorly; eyes relatively small. Antennae shorter than head and hidden in a groove beneath the eyes (Fig. 13.10B), usually 3-segmented, segment 2 and sometimes segment 3 provided with a finger-like projection (Fig. 13.11E); ocelli absent. Rostrum 4-segmented, short (Fig. 13.10B), the first segment represented by a short ventral piece. Metathoracic scent glands absent. Fore tarsus with a single blunt claw, middle and hind tarsi with two equal-sized claws. Membrane of hemelytra well developed, in most species with numerous veins (Fig. 13.8E). Ventral laterotergites 4-6 with static sense organs near the spiracles (Figs 13.1A, D; so). Male genitalia symmetrical. Eggs uniquely with from 2 to 26 respiratory horns at the anterior end (Figs 13.1H, I, 13.8F).

Overview

Water scorpions are quite common and very distinctive water bugs throughout the World. They earn this name from the tail-like, non-retractile respiratory siphon arising from the end of their abdomen (Menke 1979a). In the adult, the siphon is a tube derived from the abdominal tergum 8 and composed by two channel-like parts held tightly together; the siphon has two spiracles at its base. Water scorpions breathe air from an air-store concealed beneath the wings. To replenish this air-store, the bug breaks through the water surface film with its siphon. In adults, the air is transported through the siphon to the eighth abdominal spiracles, into the tracheal system, and thence to and through the first abdominal spiracles which are situated dorsally and open into the subhemelytral air-store (Parsons 1972a, 1972b, 1973). In nepid nymphs the two parts of the much shorter respiratory siphon do not form a tube; instead the siphon has a ventral

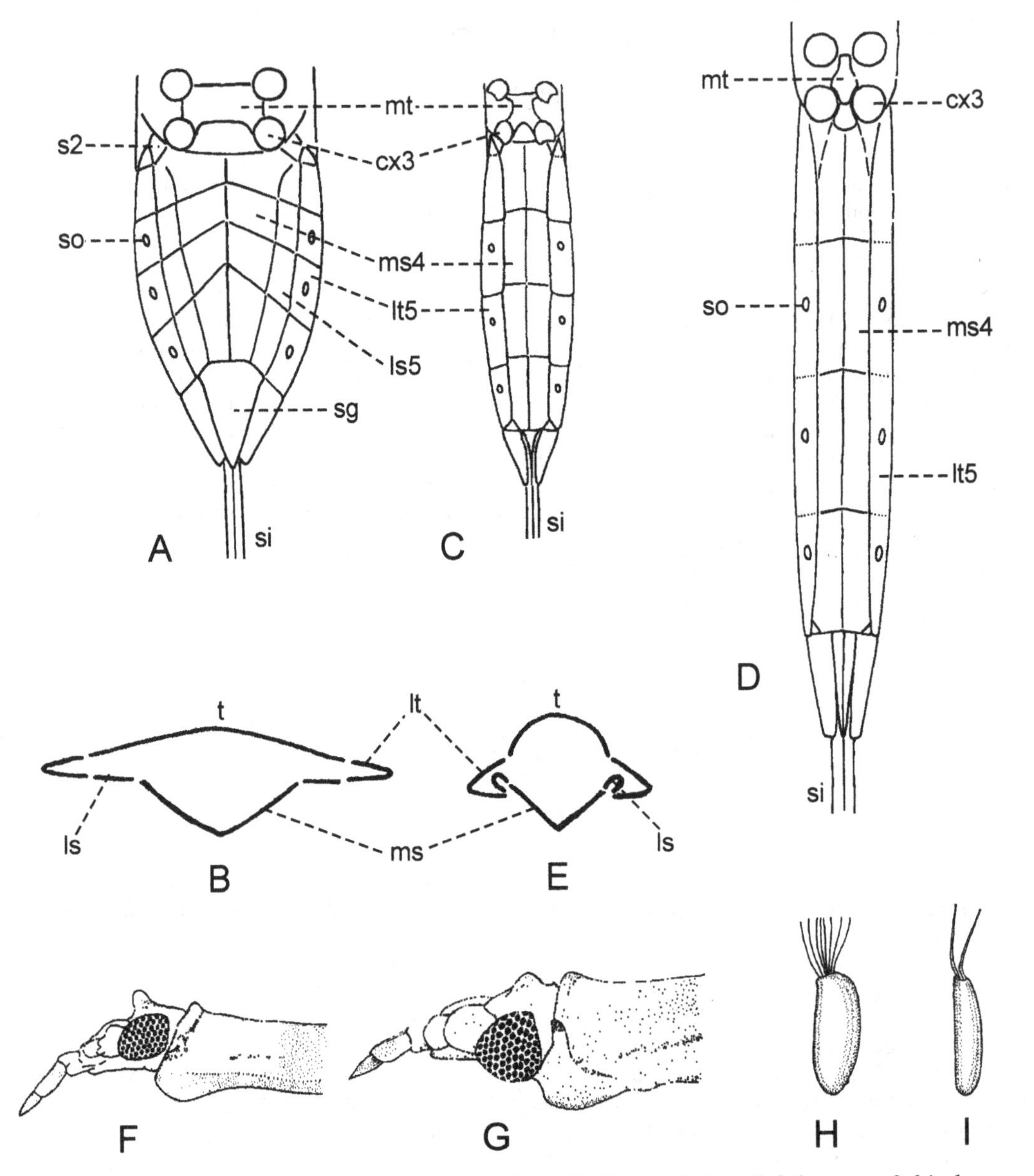

Figure 13.1. NEPIDAE. A-B, *Nepa* sp. (Nepinae): A, schematised ventral view of abdomen cx3, hind coxa; ls5, laterosternite 5; lt5, laterotergite 5; ms4, mediosternite 4; mt, metasternum; s2, sternum 2; si, respiratory siphon; sg, subgenital plate; so, sense organ). B, schematic cross section of abdomen (ls, laterosternite; lt, laterotergite; ms, mediosternite; t, tergum). C, *Austronepa angusta* (Ranatrinae): schematised ventral view of abdomen. D-E, *Ranatra* sp. (Ranatrinae): D, schematised ventral view of abdomen. E, schematised cross section of abdomen. F, *Ranatra* sp.: lateral view of head. G, *Cercotmetus* sp.: lateral view of head. H, *Laccotrephes tristis*: egg. I, *Ranatra dispar*: egg. Reproduced with modification from: A-E, Menke & Stange (1964); F, Lansbury (1972); G, Lansbury (1973); H-I, Hale (1924a).

longitudinal channel; the air-store is located on the abdominal venter and all the spiracles are ventral in position. Other unique structures of nepids are the static sense organs which are distinctive, oval structure closely associated with the fourth, fifth, and sixth abdominal spiracles (Baunacke 1912). These structures have been shown experimentally to function to keep the bug correctly oriented in the water by detecting differences in water pressure among the three pairs of spiracles (Thorpe & Crisp 1947a); these experiments proved that the static sense organs are not depth perceptors as found in *Aphelocheirus* (Aphelocheiridae).

Nepid eggs are unique in possessing two or more slender respiratory horns on their anterior end (Figs 13.1H, I; 13.8F). Female water scorpions normally deposit their eggs so that the respiratory horns are exposed to the air. These horns have a peripheral plastron meshwork which is connected through an inner gas-containing meshwork with the air-filled meshwork of the inner shell wall of the egg (Hinton 1961; Cobben 1968). The plastron of the horn thus ensures respiration even when an egg is completely submerged, as is frequently the case because of flooding.

Most nepids prefer stagnant or slowly moving waters. Only species of the endemic Australian genus *Goondnomdanepa* are restricted to flowing water (see below). Being poor to moderately good swimmers, nepids prefer to hide in the mud or to perch in the submerged vegetation, waiting for the prey to come within reach of their raptorial fore legs. Nepids are predacious bugs reported to feed on a variety of other aquatic animals, but mosquito larvae and tadpoles are among the most frequently mentioned. Although most species have well-developed fore and hind wings, flight is rarely observed in nepids and both hind wing brachyptery and flight muscle degeneration occurs in the family (Larsén 1949).

The family occur in all regions of the world but the number of species is highest in the tropics. Most authors recognise two subfamilies, Nepinae and Ranatrinae, with 10 and 4 genera, respectively, and the number of described species is approaching 250. In the Indo-Australian region there are eight genera of which five occur in Australia. The following key is adapted from Lansbury (1972, 1974a).

Key to the Australian genera of the family Nepidae

1. Body broad and flat, laterosternites visible, abdominal venter divided into six more or less distinct longitudinal zones (Figs 13.1A, B) (NEPINAE) .... *Laccotrephes*
- Body usually cylindrical, laterosternites concealed by the ventral parts of the laterotergites, abdominal venter only divided into four longitudinal zones (Figs 13.1C-E); if body flattened then distance between middle and hind coxae broader than the width of a coxa. (RANATRINAE) 2
2. Body subcylindrical, distance between middle and hind coxae narrower than the width of a coxa (Fig. 13.1D) 3
- Body flattened, distance between middle and hind coxae broader than the width of a coxa (Fig. 13.1C) 4
3. External margin of eyes not reaching downward below ventral margin of head (Fig. 13.1F). Fore femur longer than pronotum. Respiratory siphon one third or more of the body length (Fig. 13.11A) .... *Ranatra*
- External margin of eyes reaching downward below ventral margin of head (Fig. 13.1G). Fore femur shorter than pronotum. Respiratory siphon only about one fourth of the body length (Fig. 13.6A) .... *Cercotmetus*
4. Humeral width of pronotum less than its median length (Fig. 13.4). Fore femur with longitudinal groove shorter, extending about halfway along its inner margin .... *Austronepa*
- Humeral width of pronotum larger than its median length (Figs 13.8A, B). Fore femur with longitudinal groove extending about four fifths along its inner margin .... *Goondnomdanepa*

SUBFAMILY NEPINAE

Identification

These are the true "water scorpions" characterised by their large, flattened body, raptorial fore legs with distinctly widened femora, and long respiratory siphon arising from the tip of abdomen (Figs. 13.2). The abdominal laterosternites are not concealed by folding of abdomen (Fig. 13.1B; ls); abdominal venter appears to be divided in six longitudinal zones (Fig. 13.1A). The head is narrower than the anterior width of pronotum, anterior angles of pronotum embracing posterior part of head up to half length of eyes. The fore coxae are short and stout, less than twice as long as wide. The subgenital plate of the female is broad and flattened, never extending beyond end of abdomen (Fig. 13.1A; sg). The eggs carry a variable number of respiratory horns, but always more than two (Fig. 13.1H).

Overview

The subfamily Nepinae contains 10 genera of which four occur in the Indo-Australian region, but only *Laccotrephes* Stål is represented in Australia. Important genera outside this region are the Neotropical genus *Curicta* Stål with 16-18 species (Keffer 1996) and the Holarctic genus *Nepa* Linnaeus with five species among which is

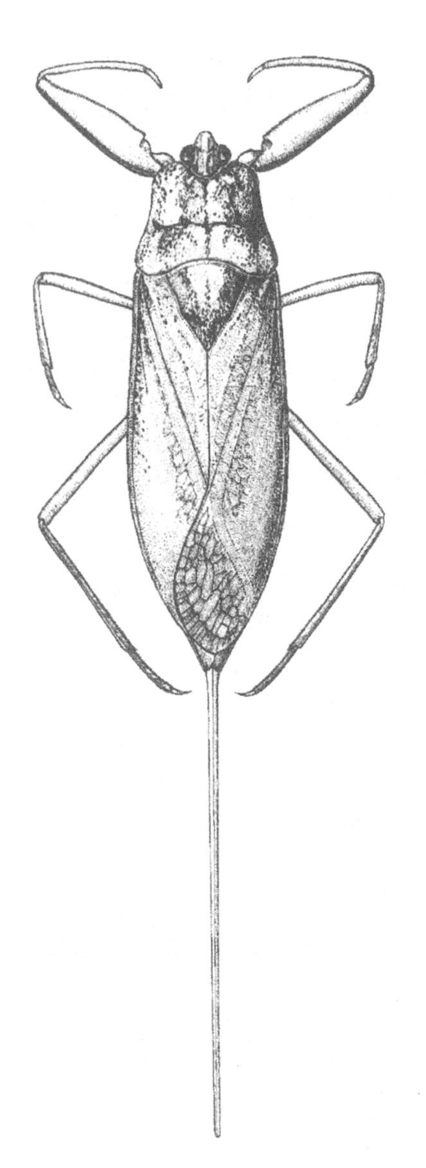

Figure 13.2. NEPIDAE. *Laccotrephes tristis*: dorsal habitus. Reproduced from CSIRO (1991).

an eyeless, cave-dwelling (troglobitic) species recently described from Romania (Decu et al. 1994). Four monotypic genera have been described from Africa (Poisson 1965a).

Water scorpions are poor swimmers and tend to "crawl" through the water. Most species inhabit quiet water in pools, ponds, and streams. The abdominal dorsum is usually bright red, in strong contrast with the dull brownish colour of the remaining body surface. When the bugs are in flight the red colouration is striking.

LACCOTREPHES

Identification

Large, broad and rather flattened water bugs (Fig. 13.2), body length from 11 to 45 mm, with respiratory siphon at least two thirds as long as body length, but usually as long as or somewhat longer than body length. Antennae 3-segmented, segment 2 with a finger-like projection. Lateral margins of pronotum slightly concave but not constricted at the transverse groove. Prosternum without sulci posterior of fore coxae. Hemelytra smooth, membrane thin with distinct venation.

The genus *Laccotrephes* Stål is only represented by one Australian species, *L.* (*Laccotrephes*) *tristis* (Stål), body length 26-32 mm (♂) or 31-35.5 mm (♀), respiratory siphon 30-35 mm (♂) or 26-33 mm (♀). New South Wales, Northern Territory, Queensland, South Australia, Victoria, Western Australia.

Biology

Species of *Laccotrephes* live at the edge of the water usually in very shallow muddy places. Hale (1924a) found *L. tristis* in slowly moving or stagnant waters with weeds. However, Lansbury (personal communication in Chen et al. in press) observed this species commonly in rock pools up to 5 m deep with very clear water. They often cover their back with mud and wait concealed with their head directed towards the deeper water. Some species can walk on land, in day time on cloudy days after rain. *Laccotrephes* species are often the last predators to leave drying pools and large numbers can sometimes be found in the sludge in the bottom of these pools, even when there is very little or no free water left. They feed on various kinds of invertebrates, tadpoles, and small fish (Plate 7, C). Mosquito larvae appear to be a favourite prey and the nymphs could be used for mosquito control in small garden ponds and pots with marsh plants. Hoffmann (1927) reports success in mosquito control in Guandong, China using *Laccotrephes*. Unlike most nepines, *Laccotrephes* species seem to be strong fliers.

During copulation the male grasps the female with a fore leg over the prothorax and with a middle leg just behind the corresponding middle leg of the female. In the observed cases the other legs were used to secure footing in the vegetation. When they have a more stable support possibly both fore legs will grasp the female thorax as in *Nepa*. When secured the male curves the tip of its abdomen beneath the abdominal tip of the female and the genital capsule is pushed upward between the bases of the respiratory siphon in order to make connection. The eggs of *Lacco-*

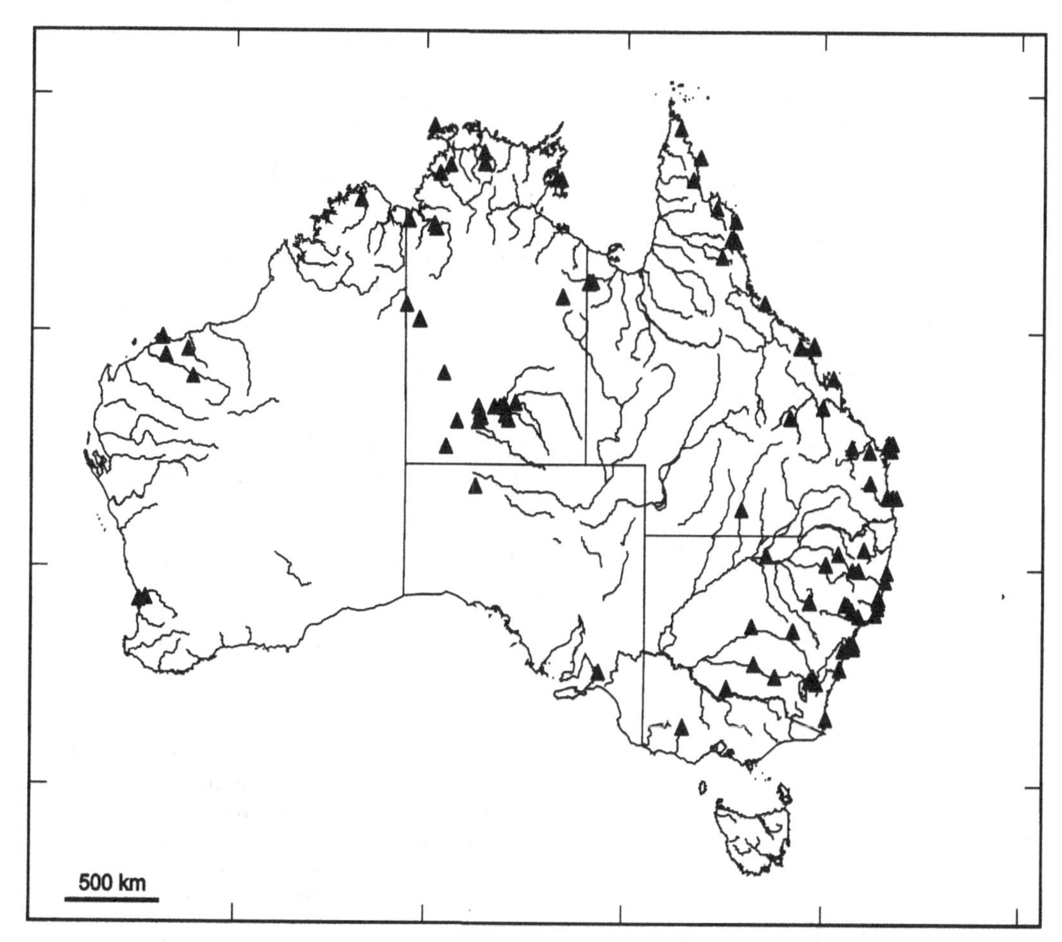

Figure 13.3. NEPIDAE. Distribution of *Laccotrephes tristis* in Australia.

trephes tristis are 3.3 mm long (Fig. 13.1H), with 8-10 respiratory horns arranged in a circular formation at the anterior end, each horn being 1.4-1.5 mm long (Hale 1924a). The eggs are deposited at the edge of the habitat where they are covered by mud by the female until only the respiratory horns are exposed. Oviposition seems to take place at night only. After hatching the young nymphs move down the bank and quickly enter the water (Hoffmann 1927, 1933). The larger nymphs feed on various kinds of aquatic invertebrates, including other water bugs (Plate 7, D).

Distribution

Laccotrephes species are widely distributed in the Old World tropics with about 80 described species, but the genus needs revision (Polhemus & Keffer 1999). The only Australian species, *L. tristis*, is distributed throughout the Indo-Australian region. In Australia it is recorded from all states except Tasmania (Fig. 13.3).

SUBFAMILY RANATRINAE

Identification

These bugs are commonly referred to as "water stick insects" because of their large, elongate, subcylindrical body, long respiratory siphon, and long and slender legs. This does not apply, however, to the Australian species of *Austronepa* (Fig. 13.4) and *Goodnomdanepa* (Fig. 13.8A) which have a more flattened and broader body resembling species of Nepinae. The respiratory siphon is very short in *Cercotmetus* species (Fig. 13.6A). Abdominal laterosternites (Fig. 13.1E; ls) concealed by infolded ventral parts of laterotergites (lt), so that the abdominal venter appears to be divided into four longitudinal zones (Figs 13.1C, D). Head usually wider than the anterior width of prono-

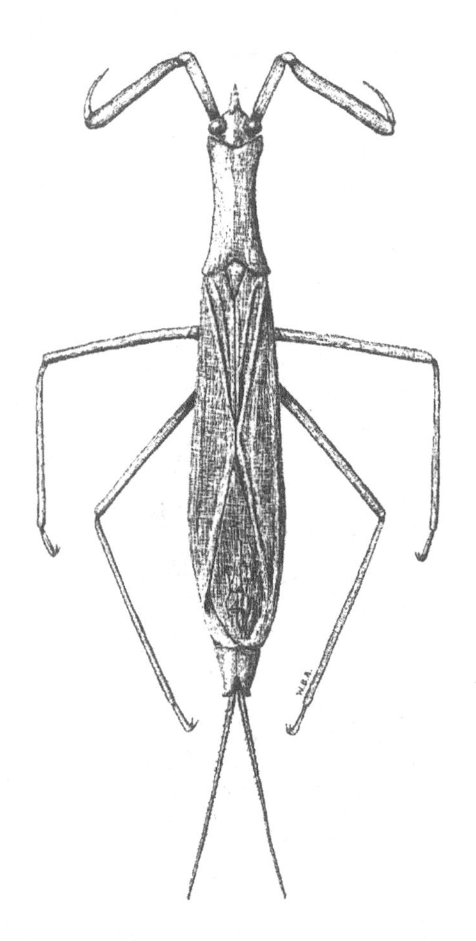

Figure 13.4. NEPIDAE. *Austronepa angusta*: dorsal habitus of female. Reproduced from Lansbury (1967).

tum, if narrower then the anterior angles of pronotum not embracing the posterior half of the head. Fore coxae elongate, about twice as long as wide in *Goondnomdanepa*, much longer relatively in other genera. Subgenital plate of female laterally compressed, keeled, and often extending beyond the abdominal end. Eggs with two respiratory horns (Figs 13.1I, 13.8F).

Overview

The subfamily Ranatrinae contains four genera all of which occur in the Indo-Australian region. The only genus occurring also outside this region is the large cosmopolitan genus *Ranatra* Fabricius with about 100 species. Its greatest species density is in South America with some 45 species. The genus *Cercotmetus* Amyot & Serville is restricted to tropical Asia with nine species. The genera *Austronepa* and *Goondnomdanepa* with one and three species, respectively, are so far only known from Australia.

Most ranatrines are better swimmers than the nepines. They are sit-and-wait predators which perch or "hang" in the vegetation with their respiratory siphon protruding through the water surface and the fore legs outstretched, ready to capture a prey. The fore tibia has an apical sense organ, which apparently senses vibrations generated by the prey, but visual responses are also involved in prey capture (Cloarec 1976b).

AUSTRONEPA

Identification

Large water bugs, body length from 17 to 17.3 mm; body elongate ovate, flattened, with a comparatively short respiratory siphon which is about one third of the body length (Fig. 13.4). Head about as wide as anterior pronotum. Antennae 3-segmented, segment 2 with a short finger-like projection. Pronotum narrow, with humeral width less than its median length, and all margins concave. Fore femur about as long as prothorax, with an acute thorn-like spine in middle, and with a longitudinal groove extending about halfway along its inner margin. Distance between middle and hind coxae broader than the width of a coxa. Fore wings heavily coriaceous, membrane opaque with numerous cells and brachiating veins.

The genus *Austronepa* Menke & Stange contains only one species, *A. angusta* (Hale), body length 17-17.3 mm, respiratory siphon 5-5.5 mm. This species was initially placed in the New World genus *Curicta* Stål which belongs to the subfamily Nepinae (Hale 1924a). However, whereas species of the latter genus have eggs with 5 or more respiratory horns, *Austronepa* has only two horns like other Ranatrinae. This and other characters lead Menke & Stange (1964) to erect a new genus for the Australian species.

Biology

The biology of the single included species is virtually unknown. Ovarian eggs have two respiratory horns at the anterior pole (Menke & Stange 1964; Lansbury 1967). Breeding sites vary but habitats are lentic or slowly lotic. They include road-side ponds and ditches, with or without dense growths of water plants, small and shallow grassy pools, and large billabongs where the species was found in deep water amongst *Pandanus* roots (Lansbury 1984).

Distribution

This genus is endemic to Australia. The single described species, *Austronepa angusta*, is widely

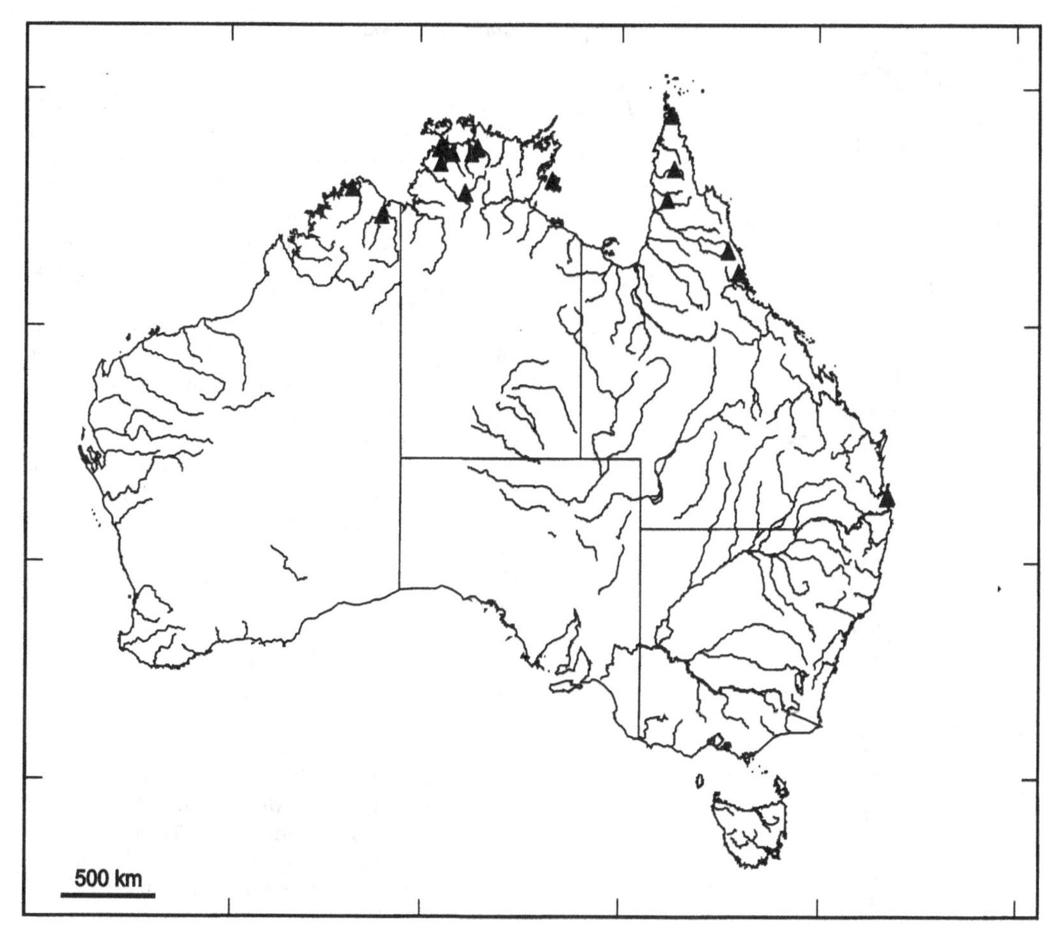

Figure 13.5. NEPIDAE. Distribution of *Austronepa angusta* in Australia.

distributed through the tropics of Western Australia, Northern Territory, and Queensland down to North Stradbroke Island (Fig. 13.5).

CERCOTMETUS

Identification

Very large bugs, body length from 32 to 60 mm, subcylindrical, with a relatively short respiratory siphon which is only about one fourth of the body length (Fig. 13.6A). The ventral margin of eyes reaching below the venter of head, obscuring the ventral margin of head in lateral view (Fig. 13.1G); antennae three-segmented, segment 2 with or without a finger-like process. Fore femur is shorter than the prothorax with a ventral tooth midway, but without a sharp constriction at the level of the ventral tooth (Fig. 13.6B). Middle and hind tibiae with well developed fringes of swimming hairs.

The genus *Cercotmetus* Amyot & Serville was revised by Lansbury (1973) who recognised 9 species in the Indo-Australian region. Only one species, *C. brevipes australis* Lansbury (1975), is found in Australia; body length 36.5-38 mm, respiratory siphon 8.5 mm. Northern Territory, Queensland.

Biology

The biology of species belonging to this genus is poorly known. They swim much better than *Ranatra* species owing to their well-developed fringes of swimming hairs on the middle and hind tibiae. Although they are often found in streams they usually live among plant debris in virtually stagnant water along the banks (Chen et al. in press). Hinton (1961) reports that the two respiratory horns of the eggs are much longer than in *Ranatra* but since then *Cercotmetus* eggs with short and *Ranatra* eggs with long respiratory

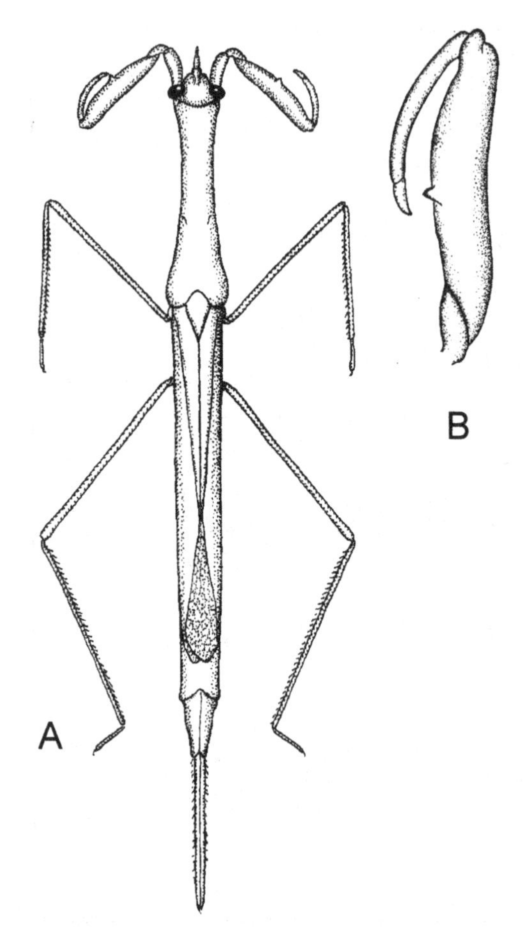

Figure 13.6. NEPIDAE. A-B, *Cercotmetus brevipes*: dorsal habitus of male (original). B, fore leg; reproduced from Lundblad (1933).

horns have become known (Lansbury 1973). The species *Cercotmetus asiaticus* is reported to feed almost exclusively on mosquito larvae in West Malaysia (Williamson 1949) and Laird (1956) suggested to introduce it into permanent water bodies on some Pacific Islands in order to help in mosquito control.

Distribution

Cercotmetus with nine described species is restricted to tropical Asia (Lansbury 1973). The single Australian species, *C. brevipes australis* Lansbury (1975) is sparsely recorded from rivers, creeks and springs in Queensland and Northern Territory (Fig. 13.7).

GOONDNOMDANEPA

Identification

Medium-size to large water bugs, body length from 10 to 17.5 mm; body elongate ovate, flattened, with a comparatively short respiratory siphon which is about one fourth of the body length (Fig. 13.8A). Head slightly narrower than anterior pronotum. Antennae 2- or 3-segmented, without finger-like projections on any segment (Figs 13.8C, I). Pronotum relatively broad, with humeral width larger than its median length, and all margins concave (Fig. 13.8B). Fore femur slightly longer than prothorax, with a prominent tubercle in basal part (Figs 13.8D, H) and a longitudinal groove extending about four fifths along its inner margin. Distance between middle and hind coxae broader than the width of a coxa. Fore wings coriaceous, membrane with numerous veins forming a clearly reticulate pattern (Fig. 13.8E). Ovarian eggs with two very long respiratory horns (Figs 13.8F, G).

The genus *Goodnomdanepa* Lansbury is endemic to Australia with three described species. The following key is adapted from Lansbury (1974a, 1978).

Key to Australian species

1. Larger species, body length 14.5-17.5 mm, respiratory siphon 7.5-9.5 mm. Fore femur basally with very large tubercle (Fig. 13.8H). Hemelytra with raised reticulate pattern. Respiratory siphon longer than claval commissure. Northern Territory .... *prominens* Lansbury
- Smaller species, body length 10-14.8 mm, respiratory siphon 1.9-3.75 mm. Fore femur basally with smaller and stouter tubercle (Fig. 13.8D). Hemelytra without raised reticulate pattern. Respiratory siphon shorter than claval commissure 2
2. Antennae 3-segmented (Fig. 13.8C). Body length 12.4-14.8 mm, respiratory siphon 2.6-3.75 mm. Northern Territory, Western Australia .... *weiri* Lansbury
- Antennae 2-segmented (Fig. 13.8I). Body length 10 mm, respiratory siphon 1.9 mm. Northern Territory....... *brittoni* Lansbury

Biology

Very little is known about the biology of this genus. The first described species, *Goondnomdanepa weiri* Lansbury, was collected by Tom Weir beneath rocks and stones in a shallow running creek. They were plentiful in June, but had become rather scarce in November owing to the drying up of the creek (Lansbury 1974a). Sub-

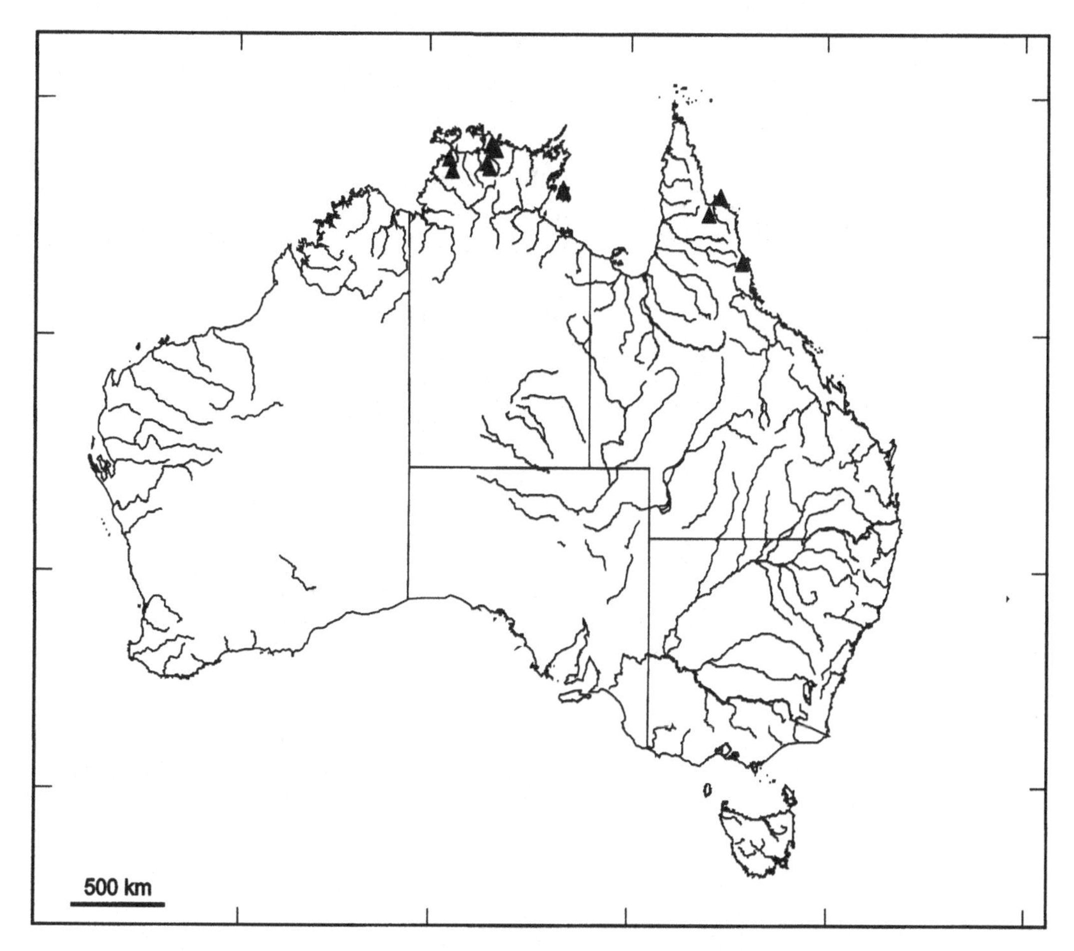

Figure 13.7. NEPIDAE. Distribution of *Cercotmetus brevipes australis* in Australia.

sequently, another species, *G. prominens* Lansbury, was found together with *G. weiri*, seemingly living under the same stone.

Distribution

Endemic to Australia with three species sparsely distributed in the Mt. Cahill area, Northern Territory (*Goondnomdanepa brittoni*, *prominens*, and *weiri*) and in the Kimberleys, Western Australia (*G. weiri*) (Fig. 13.9).

RANATRA

Identification

Large to very large water bugs, body length from 20 to 53 mm, with a subcylindrical body and a respiratory siphon of varying length, but usually one third or more of the body length (Fig. 13.11A). The ventral margin of the eyes not obscuring the ventral margin of the head in lateral view (Fig. 13.1F); antennae 3-segmented, segment 2 with a finger-like projection in most species (Fig. 13.11E). Prothorax narrow anteriorly, distinctly widened posteriorly (Figs 13.11D, G). In some species (e.g., *Ranatra diminuta* and *occidentalis*) there are two forms, "normal" and "reduced", with respect to the development of the posterior lobe of the prothorax. Fore femur longer than prothorax with one or two teeth at or distally of the middle of the ventral margin and with a sharp constriction at that point (Figs 13.10A, F, 13.11B, C). Apart from the characteristics mentioned in the key, *Ranatra* differs from *Cercotmetus* by the less developed fringes of swimming hairs on the middle and hind tibiae.

The genus *Ranatra* Fabricius is distributed world-wide with about 100 species of which three are recorded from Australia. The following key is adapted from Lansbury (1972).

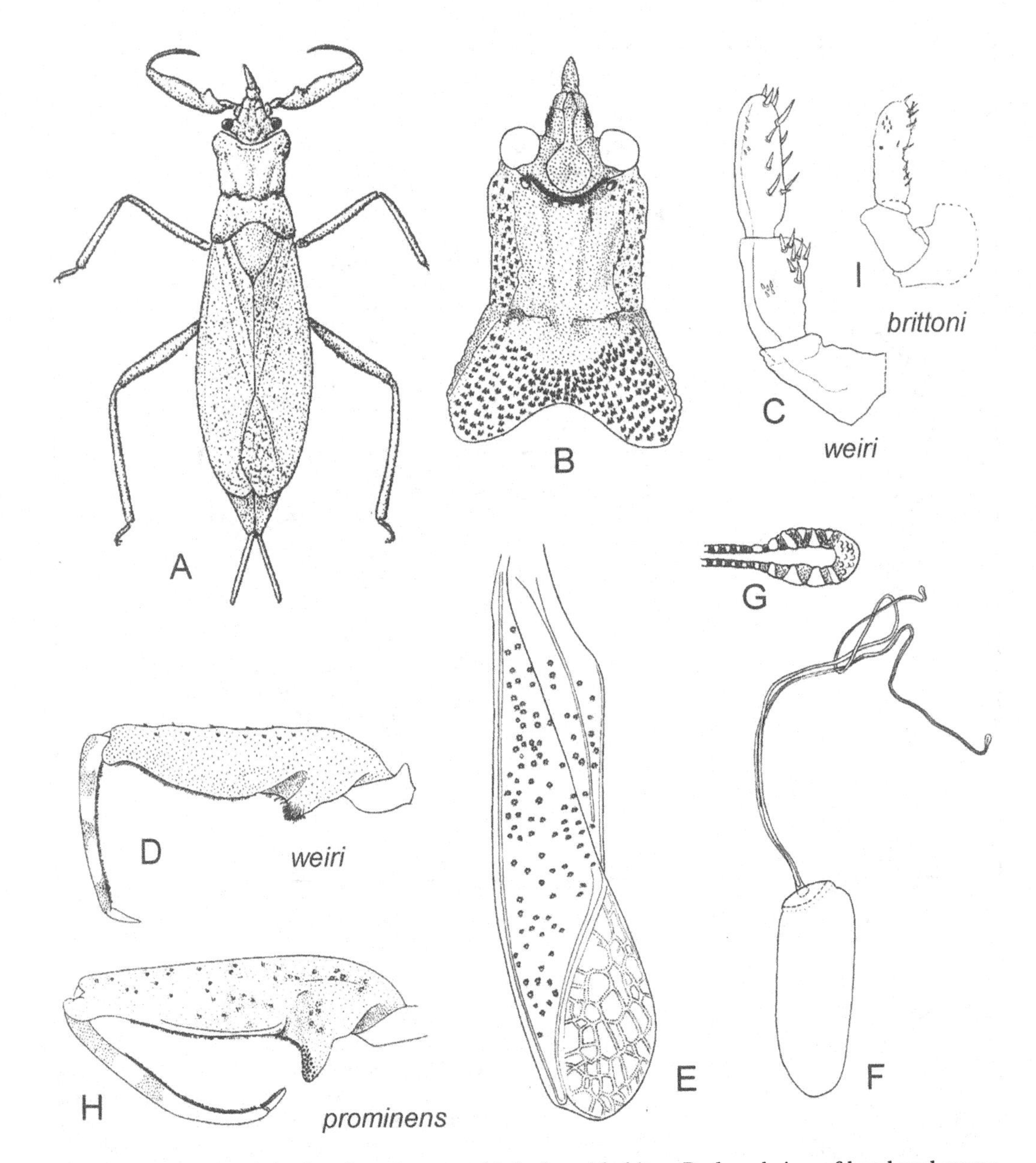

Figure 13.8. NEPIDAE. A-G, *Goondnomdanepa weiri*: A, dorsal habitus. B, dorsal view of head and pronotum. C, antenna. D, fore leg. E, left fore wing. F, egg. G, apex of respiratory horn of egg. H, *G. prominens*, fore leg. I, *G. brittoni*: antenna. Reproduced from: A, H, I, Lansbury (1978); B-G, Lansbury (1974a).

Key to Australian species

1. Fore femur with a single prominent tooth centrally on ventral margin (Fig. 13.11B). Body length 37-50 mm, respiratory siphon 34-49 mm. Australian Capital Territory, New South Wales, Queensland, South Australia, Tasmania, Victoria, Western Australia................ *dispar* Montandon
- Fore femur with two teeth centrally on ventral margin (Figs 13.10F, G, 13.11C, F)... 2
2. Posterior lobe of prothorax of normal form with sparsely or irregularly spaced punctures (Fig. 13.11D). Fore femur rel-

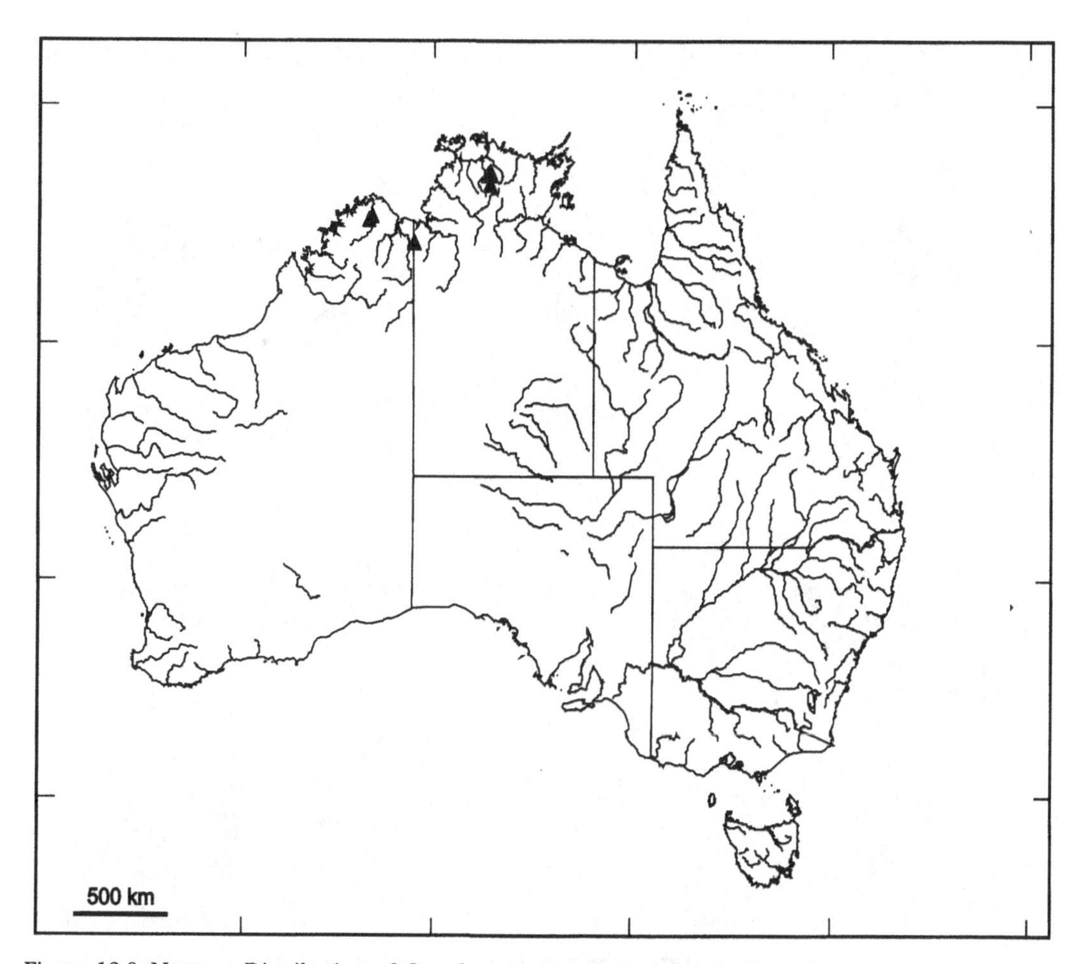

Figure 13.9. NEPIDAE. Distribution of *Goondnomdanepa* species in Australia.

atively narrow (Figs 13.10A, 13.11C). Body length 23.5-28.5 mm, respiratory siphon 18-25 mm. Queensland, South Australia, Victoria, Western Australia .... *diminuta* Montandon

– Posterior lobe of prothorax of normal form with densely and evenly spaced punctures (Fig. 13.11G). Fore femur relatively broad (Fig. 13.11F). Body length 22.5-25 mm, respiratory siphon 17-18 mm. Western Australia .... *occidentalis* Lansbury

Biology

These are the true "water stick insects" with elongate, subcylindrical body, long respiratory "tail", and long and slender legs. In the water they commonly hide between vegetation with the head and fore legs pointing downward, waiting for suitable prey to come into reach (Plate 7, E). Due to long fore coxae and femora the reach of the raptorial fore legs is considerably longer than in most other nepids. The preferred perch sites varies with developmental stage. First instar nymphs tend to stay in the upper centimetres of the water perching on *Lemna* and *Elodea*. During development the preferred perch sites become gradually deeper and plant preference shifts to reeds like *Calamagrostis* and *Phragmites* (Cloarec 1976a). In addition younger nymphs prefer green plants, older nymphs and adults brown plants, to perch on. These preferences probably aid in keeping microhabitats of different instars separated and thus avoid cannibalism to some extent. First instar nymphs catch small prey such as copepods, but although the maximum size of the prey increases with the size of the growing nymph, even adults will still catch small prey when available (Cloarec 1972; 1976a; Cloarec & Joly 1988). Bailey (1986a, 1986b, 1987, 1989) published exten-

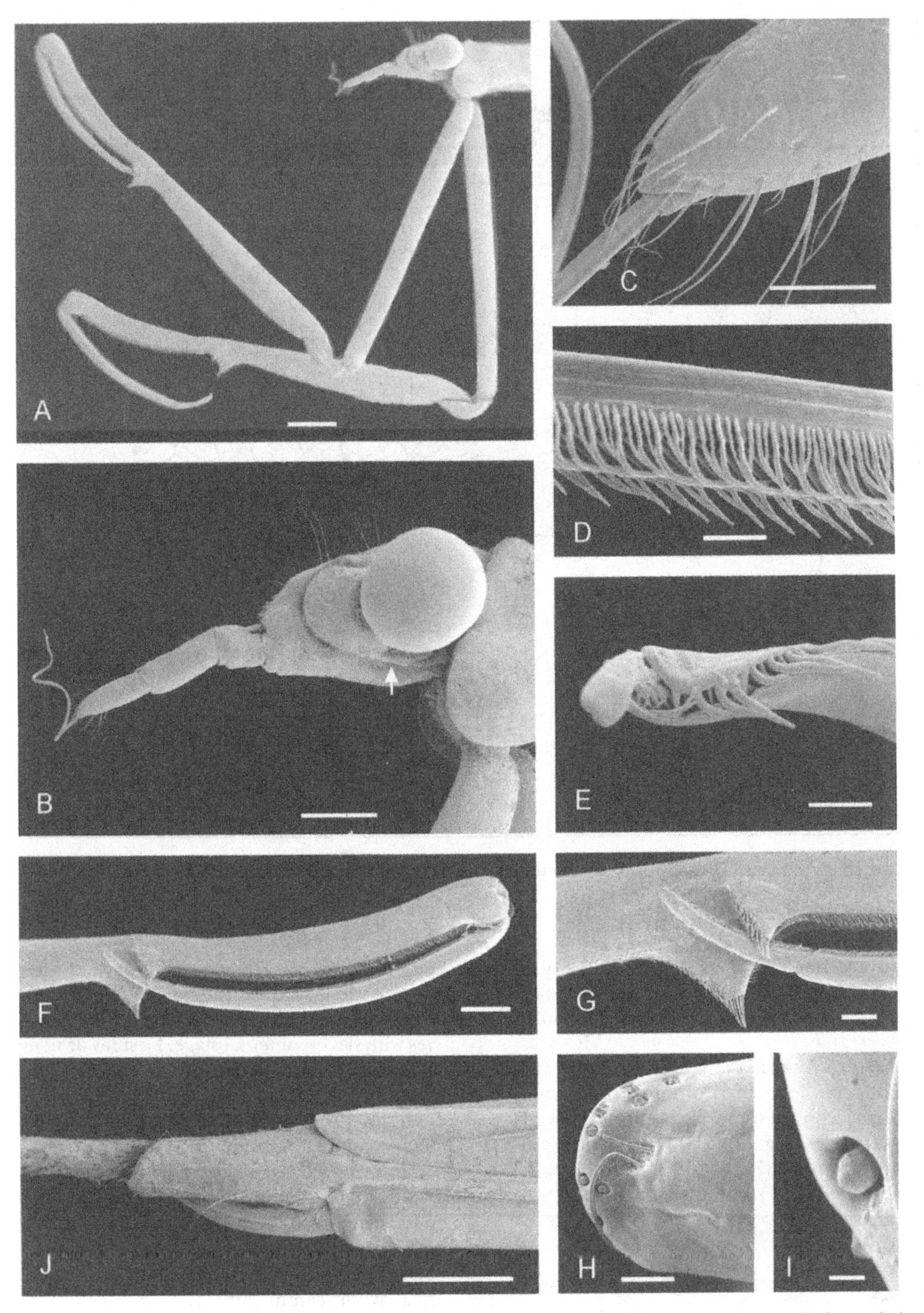

Figure 13.10. NEPIDAE. A-J, *Ranatra diminuta*: A, lateral view of head and anterior thorax. B, lateral view of head showing concealed antenna (arrow). C, apex of rostrum. D, detail of maxillary stylets. E, apex of maxillary stylets. F, fore leg. G, medial teeth of fore leg. H, apex of fore tarsus with sensilla. I, sensillus at apex of fore tarsus. J, lateral view of abdominal end of ♀. Scales: A, J, 1.0 mm; B, F, 0.5 mm; C, G, =.1 mm; D, E, H, 0.01 mm; I, 0.002 mm. Scanning Electron Micrographs (CSIRO Entomology, Canberra)

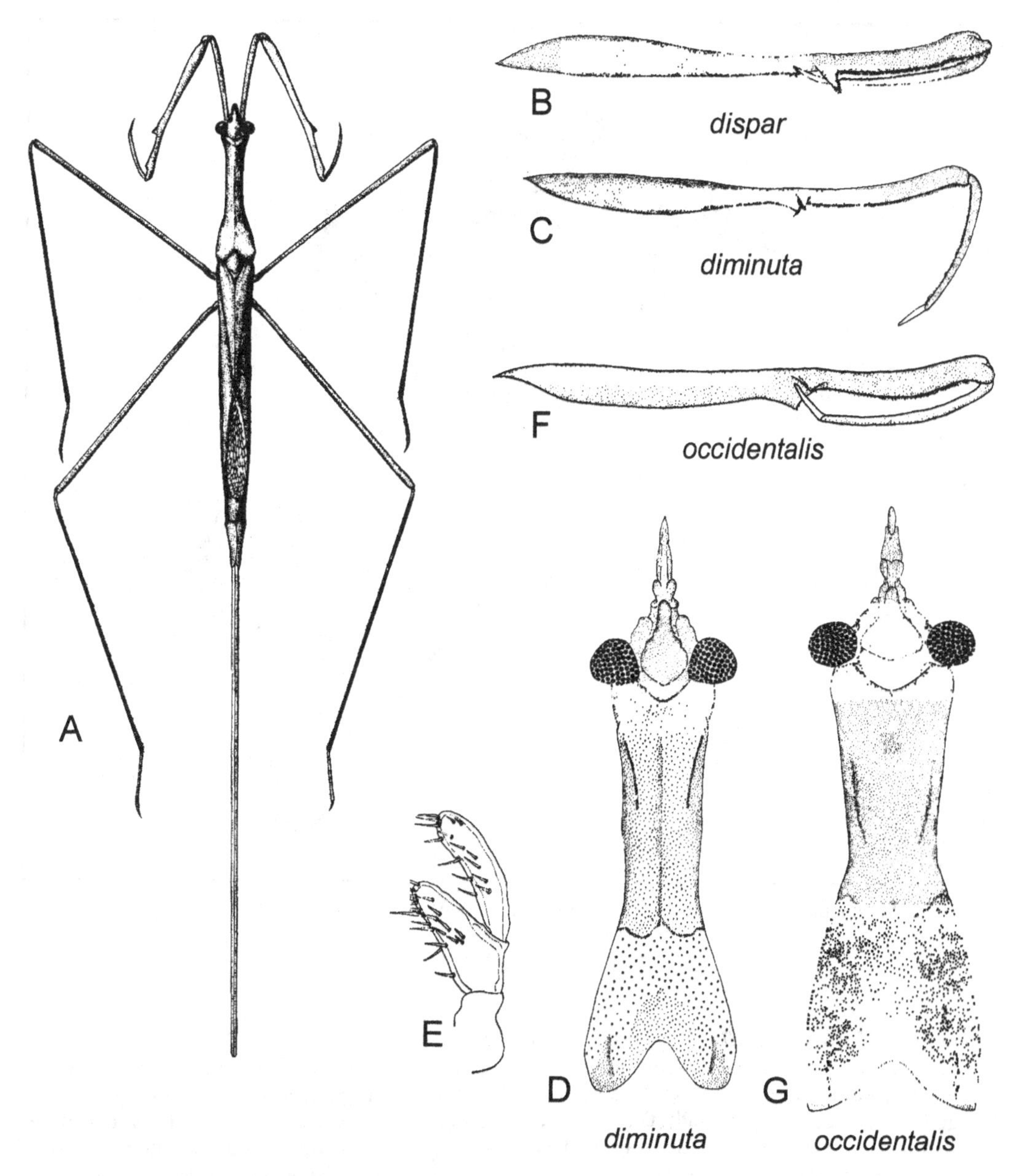

Figure 13.11. NEPIDAE. A-B, *Ranatra dispar*. A, dorsal habitus. B, fore leg. C-E, *R. diminuta*: C, fore leg. D, dorsal view of head and pronotum. E, antenna. F-G, *R. occidentalis*: F, fore leg. G, dorsal view of head and pronotum (normal form). Reproduced with modification from: A, CSIRO (1991); B-G, Lansbury (1972).

sive studies of the predatory behaviour of the Australian species *Ranatra dispar*.

Stridulatory structures have been recorded in species of *Ranatra*, composed of a roughened elevated area on the outer surface of the fore coxa which is rubbed against the striate inner surface of the fore acetabula (Schuh & Slater 1995). Hale (1924a) described and illustrated the egg and all nymphal instars of *Ranatra dispar* (as *R. australiensis*). The eggs of *Ranatra* are usually inserted into slits in floating vegetation although the Asia species *R. chinensis* Mayr was reported to deposit

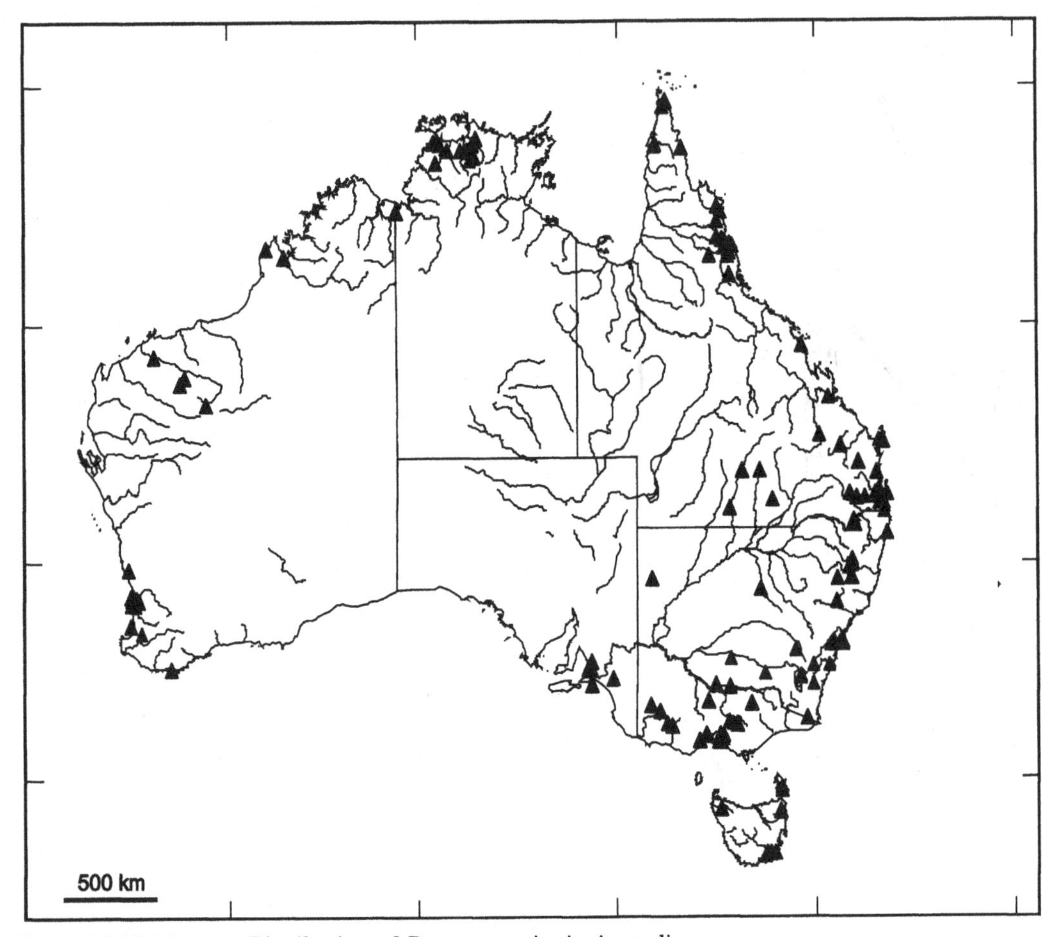

Figure 13.12. NEPIDAE. Distribution of *Ranatra* species in Australia.

eggs in mud (Hoffmann 1927). The eggs of *R. dispar* are long and slender, 3.4 3.6 mm long, with the two respiratory horns slightly shorter than the egg proper (Fig. 13.1I). Usually, the egg is oriented so that the egg itself is in the water whereas the respiratory horns are exposed to the air.

Distribution

This genus is cosmopolitan with about 100 species with greatest species density in South America (about 45 species). It is distributed throughout the Indo-Australian Region with 27 species and subspecies (revised by Lansbury 1972). Three species occur in Australia and are recorded from all states (Fig. 13.12).

14. Family BELOSTOMATIDAE

Giant Water Bugs

Identification

Body oval to elongate oval, dorso-ventrally flattened (Figs 14.1A, 14.4A). Includes the largest aquatic bugs as well as the largest true bugs (Hemiptera-Heteroptera), the Australian species reaching a total length of 75 mm. The antennae are short, concealed in grooves on the ventral side of head (Fig. 14.1B; an); usually 4-segmented, segments 2 and 3 with lateral processes (Figs 11.4F, 14.3C; 14.4D). Fore legs raptorial (except in the snail-eating, African genus *Limnogeton* Mayr), fore femora enlarged with tibia apposed (Figs 14.1G, 14.3E, 14.4C). Fore tarsi with 1 to 3 segments (Figs 14.3F, 14.4C; ta), with one or two claws (cl). Middle and hind legs usually flattened and fringed with swimming hairs. Membrane of hemelytra with reticulate venation. Abdominal tergum 8 is modified into a pair of retractable respiratory air-straps (Fig. 14.1C; as); static sense organs associated with abdominal spiracles 2-7. Male genital segments symmetrical, with subgenital plate (operculum) more or less pointed (Fig. 14.1I). The subgenital plate (operculum) of the female is truncate and carries a pair of small tufts of bristles sublaterally (Figs 14.1D, 14.3D). The nymphs have their metepisterna extending posteriorly to cover abdominal sterna 2 or 3.

Overview

Belostomatids are remarkable water insects. The gigantic size reached by species belonging to the genus *Lethocerus* Mayr enables them to subdue vertebrate prey such as tadpoles, small frogs and fish (Plate 7, A-B). Many belostomatids, including Australian species of the genus *Diplonychus* Laporte, have a unique reproductive biology in which the female lays its eggs on the back of the male (Fig. 14.1A, Plate 6, K) which "brood" the eggs until they hatch. The Belostomatidae are currently subdivided into three subfamilies with 9 genera and about 140 species. The subfamily Horvathiinae with one genus, *Horvathinia* Montandon, is only found in South America. The subfamily Lethocerinae contains the cosmopolitan genus *Lethocerus* with approximately 25 species of which 16 species are confined to the New World. The remaining seven genera belong in the subfamily Belostomatinae of which the largest genera are *Belostoma* Latreille, with about 65 New World species, *Appasus* Amyot & Serville, with 2 species in Asia and 10 in Africa, and *Diplonychus* with five species in Asia, New Guinea, and Australia. The family is rather poorly represented in the Indo-Australian region with a total of 11 species. Only the genera *Diplonychus* and *Lethocerus* are represented in Australia, each with two species.

Instead of a respiratory siphon as in the water scorpions, Nepidae (see above), belostomatids have two retractable air-straps which originate from tergum 8 of the adult (Fig. 14.1C; as). These are used in leading air to the air-store beneath the wings, but because they are flat and lie side by side, the air-straps do not form a tube. Instead, a channel is formed by long, marginal hairs through which the air passes directly into the subhemelytral air-store rather than into the tracheae as in water scorpions. Gas exchange between the air-store and the tracheal system occurs primarily through the first pair of abdominal spiracles which are large and facing dorsal. The large eighth pair of spiracles are primarily exhalant. The smaller, ventrally located second to seventh pairs of spiracles are adjacent to a broad mat of hairs holding a thin film of air. This ventral air-store functions in leading air from the subhemelytral air-store to the spiracles and, since it is exposed to the water, may also act as a physical gill. In nymphs, the air-store is confined to the abdominal venter, and because air-straps are not developed, the abdominal end is used to break the surface film (Parsons 1972a, 1972b, 1973; Menke 1979b). Presumed static sense organs (or pressure receptors) are associated with the spiracles of abdominal sterna 2-7. These are similar in structure to, but smaller and simpler than those found in water scorpions (see above).

Giant water bugs like *Diplonychus* and *Lethocerus* species prefer standing water habitats. Although they are strong swimmers, most are sedentary hunters which prefer to perch on submerged water plants and wait for prey to pass by. They hold their raptorial fore legs ready to seize any moving object that comes near, seemingly triggered by visual stimuli although prey detection by water movements cannot be ruled out (Cullen 1969). The diet of belostomatids is varied, and these bugs will probably prey upon anything they can subdue. *Lethocerus* species have been reported to feed upon a variety of aquatic organisms (dytiscid beetles, tadpoles, small frogs and fish, etc.) and giant water bugs are sometimes important pests in fish hatcheries. The salivary secretions of *Lethocerus* contain a potent mixture of hydrolytic enzymes (proteases) which may also have a paralysing or toxic effect on the prey. Bites by giant water bugs are not harmful to humans except that they produce a burning sensation which may last for hours. However, since the belostomatid rostrum is short (Fig. 14.3A) it is easy to avoid being bitten.

Females of *Lethocerus* lay their eggs on emergent vegetation above water. The eggs are very large, with dark brown stripes extending downward from the free end. They are glued in adjoining rows forming batches which average several scores of tens. The male stays in the water at the base of the egg batch and may from time to time emerge to "water" the eggs. Unattended eggs will perish both when submerged and when kept above water. Apart from watering, males also defend the egg batch from predators.

Females of the New World genera *Abedus* and *Belostoma*, as well as of the Old World genus *Diplonychus* (Fig. 14.1A, Plate 6, K), glue their eggs to the dorsum of males, who carry and care for them until they hatch. In insects, paternal care for the offspring is rather unique. A general overview of this behaviour is given by Smith (1997). Males carrying eggs engage in a variety of "brooding" activities that include keeping the eggs wet, frequently exposing them to atmospheric air, and maintaining an irregular flow of water over them by stroking them with the hind legs. Eggs fail to develop if kept submerged (inadequate exchange of oxygen) or are left out of water (desiccation). The number of eggs carried by males vary from scores of tens to more than one hundred. All of the eggs on the back of one male are normally laid by a single female

There has been much speculation about the adaptive significance of the brooding behaviour in belostomatids. The eggs are too big to obtain sufficient oxygen for development when simply laid under water. On the other hand, if the eggs are deposited above water they will dry out since they lack the wax layer and complicated shell structure which prevent water loss in terrestrial bugs. Brooding in the way described above ensures that the eggs have a sufficient supply of oxygen without facing the risk of drying out. Since the female needs to feed continuously to produce eggs, brooding her own eggs would be less advantageous for the species than leaving this to the male. Finally, when the female is depositing her eggs, the male ensures by repeated copulations that the eggs he broods are fertilised by his sperm alone (Ichikawa 1989).

Belostomatids are recorded from the Upper Triassic and Lower Jurassic (Popov 1971a, 1996; Rasnitsyn & Quicke 2002), thus being some of the oldest water bugs known.

Key to the genera of the family Belostomatidae

1. Medium-sized species, length 15-20 mm. Fore tarsus 1-segmented, with two small claws (Fig. 14.3F). Abdominal sternites 4-6 not subdivided by a suture-like fold- (Fig. 14.1D) (BELOSTOMATINAE) ..... *Diplonychus*
– Very large species, length 50-75 mm. Fore tarsus 3-segmented, with one long claw (Fig. 14.4C). Abdominal sternites 4-6 subdivided into mediosternites and parasternites by a weak suture-like fold. (LETHOCERINAE) .... *Lethocerus*

SUBFAMILY BELOSTOMATINAE

Identification

Medium-sized to large, oval or elliptical water bugs. Antennal segments 2 and 3 each with a finger-like projection (except in some species of the American genus *Abedus* Stål), segment 4 long (Fig. 14.1F). Metathorax without scent glands. Tibia and tarsus of middle and hind legs are similar in structure, flattened but nor broadly dilated. Abdominal sternites without fold dividing them in mediosternites and laterosternites; abdominal spiracles situated in the centre of the ventral laterotergites. Abdominal air-straps variable. Parameres of ♂ almost straight, with pointed and recurved apex (Figs 14.1E, J).

Overview

Species belonging to this subfamily have a unique reproductive biology. After mating, the female glues its eggs to the dorsum of the male, who carries and cares for them until they hatch. Included genera are *Abedus* (10 species in southwestern U.S.A. to Central America), *Appasus* (c. 15 species in Africa and Asia), *Belostoma* (c. 60 species, North and South America), *Diplonychus* (5 species in Asia, New Guinea, and Australia), *Hydrocyrius* Spinola (five species in Africa and Madagascar), and *Limnogeton* (four Afrotropical species).

DIPLONYCHUS

Identification

Medium-sized water bugs, length from 15 to 23 mm, with broadly oval body shape (Plate 6, J). Eyes not protruding, their lateral margins continuous with the lateral margins of head; first antennal segment equal to or longer than segment 4 and the lateral projections of segments 2 and 3. Fore tarsi 1-segmented, with 2 small claws (Fig. 14.3F); middle and hind tibiae flattened. Lateral band of short pilosity on ventral part of laterotergites 4 not reaching the lateral margin of the segment (Figs 14.1D, H). Subgenital plate (opercu-

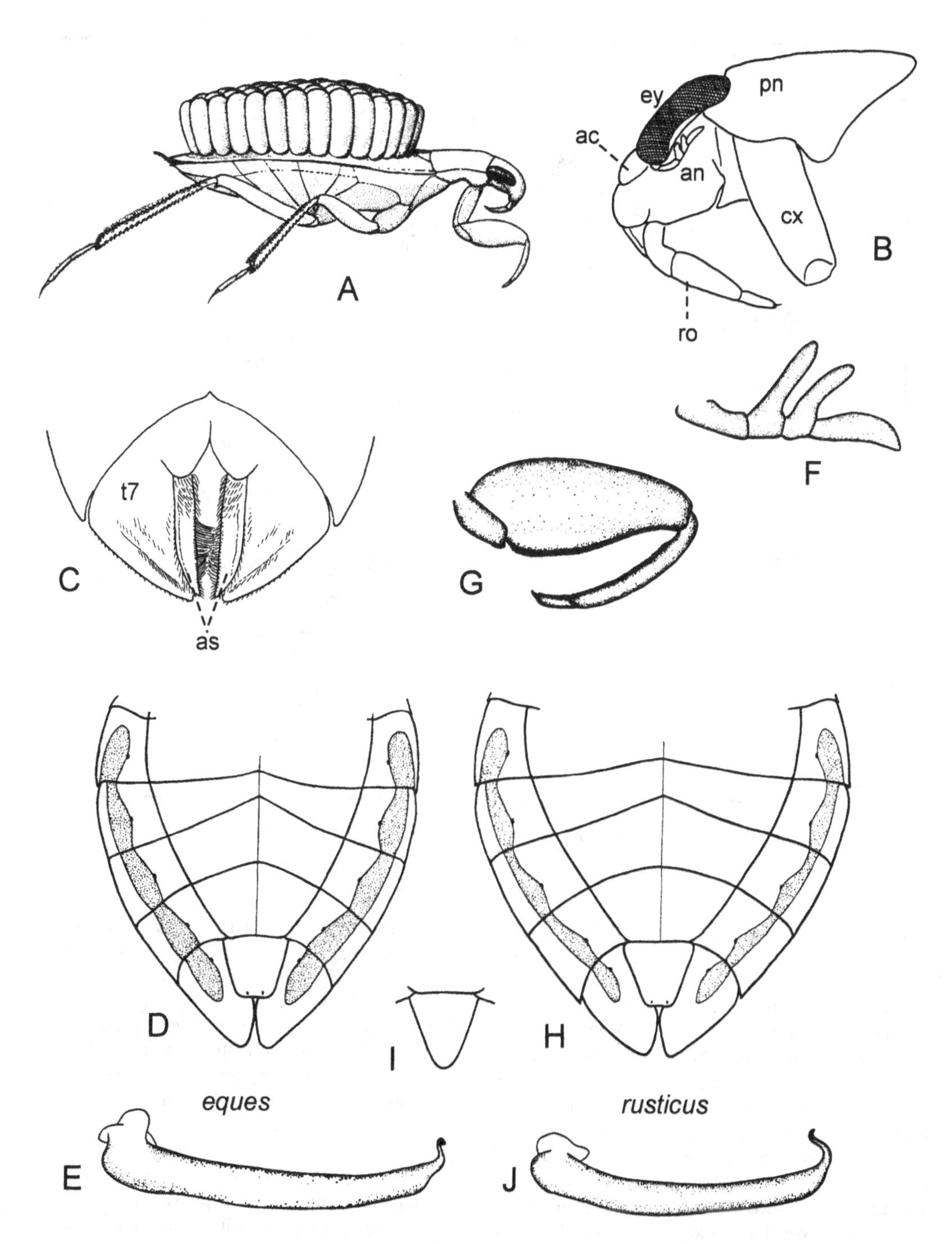

Figure 14.1. BELOSTOMATIDAE. A-E, *Diplonychus eques*: A, lateral view of ♂ with batch of eggs on its back. B, lateral view of head and prothorax (ac, anteclypeus; an, antenna; cx, fore coxa; ey, eye; pn, pronotum; ro, rostrum). C, dorsal view of abdominal end showing air-straps (as) in middle of tergum 7 (t7). D, ventral view of abdomen of ♀; E, paramere of ♂. F-J, *D. rusticus*: F, antenna; G, fore leg; H, ventral view of abdomen of ♀. I, ventral view of subgenital plate (operculum) of ♂. J, paramere of ♂. Reproduced with modification from: A, CSIRO (1991); D-J, Lundblad (1933). B-C, original.

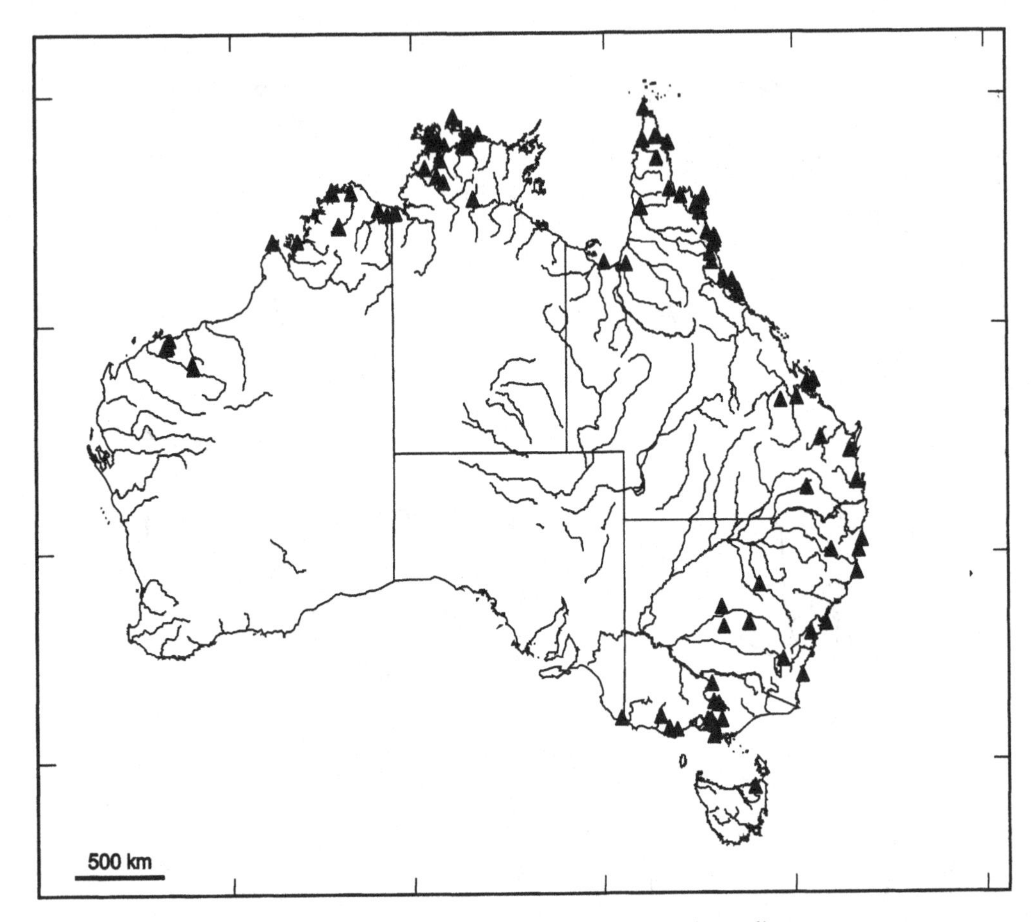

Figure 14.2. BELOSTOMATIDAE. Distribution of *Diplonychus* species in Australia.

lum) more or less pointed in ♂ (Fig. 14.1I), truncated and carrying a pair of small tufts of bristles sublaterally in ♀ (Figs 14.1D, 14.3D).

The genus *Diplonychus* Laporte (formerly *Sphaerodema* Laporte) comprises 5 species from Asia, New Guinea, and Australia. The following key is adapted from Lundblad (1933) who separated the two Australian species, *D. eques* (Dufour) and *D. rusticus* (Fabricius) (J. Polhemus 1994b; *D. planus* (Sulzer) of Cassis & Gross 1995) on the relative width and shape of the lateral, longitudinal hair-bands on abdominal venter (Figs 14.1D, H). We have examined hundreds of Australian *Diplonychus* and are unable to find constant differences in this and any other character to distinguish two species.

Key to Australian species

1. Lateral bands of fine hairs on ventral abdominal laterotergites broad and only weakly constricted, not shaped like a pearl-string (Fig. 14.1D); posterior half of hair-band on ventral laterotergite 6 about as wide as innermost part of segment. Male paramere (Fig. 14.1E). Length 15-23 mm. Australian Capital Territory, New South Wales, Northern Territory, Queensland, South Australia, Tasmania, Victoria, Western Australia.. *eques* (Dufour)
– Lateral bands of fine hairs on ventral abdominal laterotergites distinctly constricted, shaped like a pearl-string (Fig. 14.1H); posterior half of hair-band on ventral laterotergite 6 distinctly narrower than innermost part of segment. Male paramere (Fig. 14.1J). Length 15-23 mm. Northern Territory, Queensland.. *rusticus* (Fabricius)

Biology

Little is known about the biology of Australian

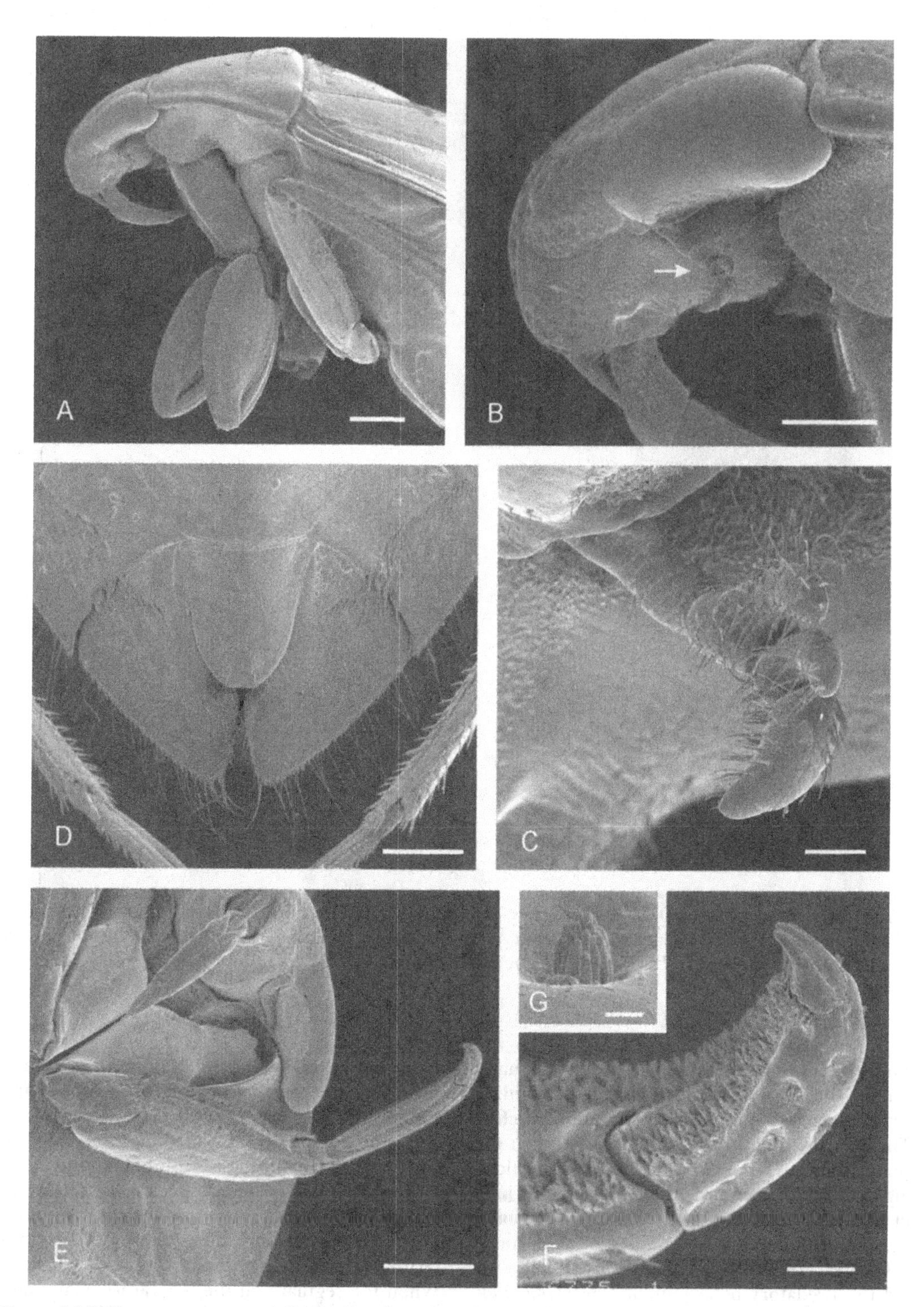

Figure 14.3. BELOSTOMATIDAE. A-G, *Diplonychus eques*: A, lateral view of head and thorax. B, lateral view of head with concealed antenna (arrow). C, left antenna. D, ventral view of abdominal end of ♂. E, ventral view of head and anterior thorax with extended fore leg. F, fore tarsus. G, sensory organ on fore tarsus. Scales: A, D, E 1 mm; B, 0.5 mm; F, 0.1 mm; G, 0.01 mm. Scanning Electron Micrographs (CSIRO Entomology, Canberra).

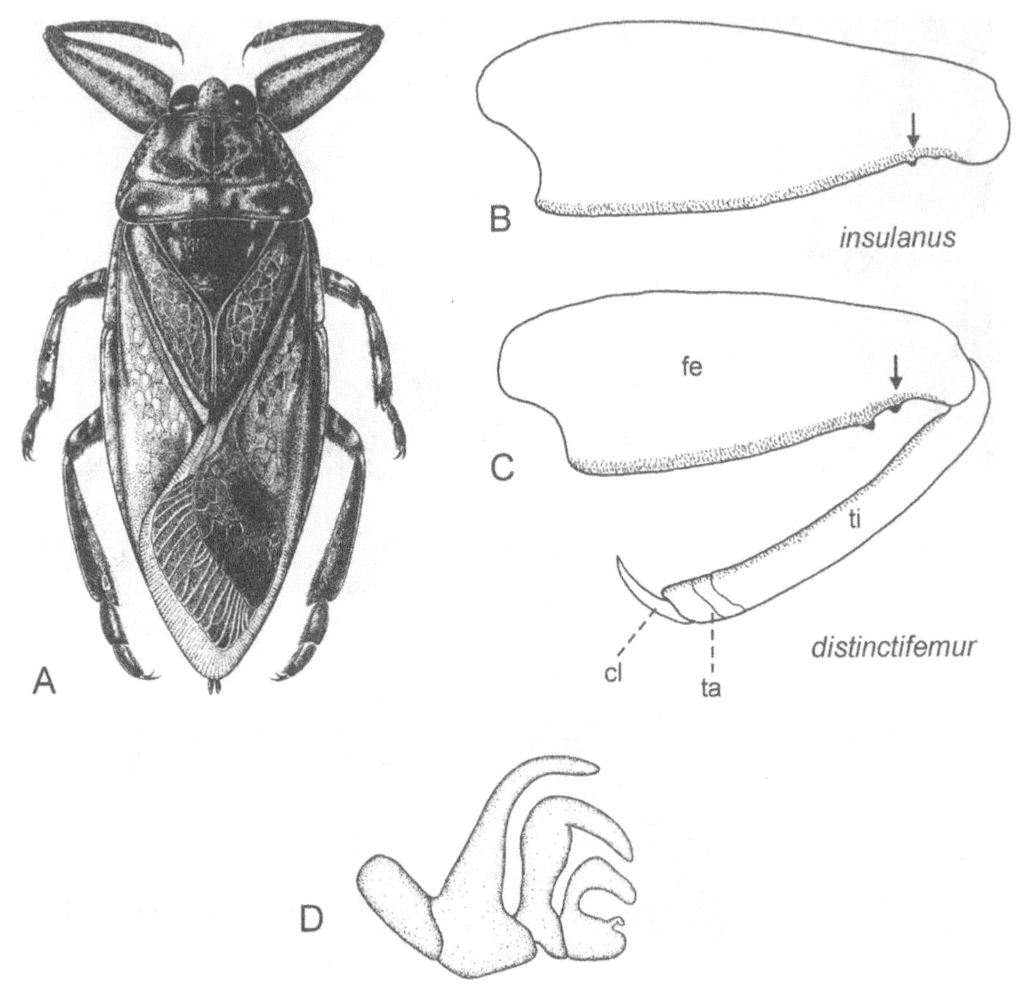

Figure 14.4. BELOSTOMATIDAE. A-B, *Lethocerus insulanus*: A, dorsal habitus. B, fore femur. C, *Lethocerus distinctifemur*: fore leg (cl, claw; fe, femur; ta, tarsus; ti, tibia). D, *Lethocerus* sp.: antenna. Reproduced from: A, CSIRO (1991); D, Lundblad (1933). B-C, original.

Diplonychus species. Most species live in stagnant, usually shallow, waters with vegetation including artificial concrete tanks and irrigation ditches. They usually hide near the surface between the vegetation, and seem to prefer to stay under floating plants. However, in South Australia they hide in decaying vegetation at the bottom (Lansbury personal communication in Chen et al., in press). Cloarec (1989, 1990a, 1990b, 1991,1992) studied the predatory behaviour of *Diplonychus indicus*, an Asian species. In the laboratory they alternate between actively foraging and ambushing for prey. When sucking on a prey they will strike at others coming near enough. Up to three preys can be held at one time. This means that with high prey densities, they will kill more prey than they actually consume (Dudgeon 1990). Early instars will only catch invertebrate prey of suitable size, adults will take anything of a size they can master including amphibians and fish. Hoffmann (1927) used a species of *Diplonychus* successfully in mosquito control on the campus of Lignan University in China. When fed mosquito larvae, adults clearly prefer the older instars. When fed regularly in the laboratory, males and females forage at different times. Females are most active in the daytime from 8.00 to 17.00h, where after there is a steep drop in feeding activity. Males are less active between 8.00h to 14.00h, but has their highest feeding activity around

20.00h, when female activity is low. It is supposed that this pattern restricts intraspecific competition for food.

In *Diplonychus* species, mating is initiated by the male making "display-pumping", that is vertical movements near the water surface causing ripples which attract females (Wilcox 1995). When a female reaches the male she attempts to deposit eggs on the back of the male which resists and start copulating with the female. Only after, usually several, copulations the male allows the female to start laying the eggs on his back, starting at end of the hemelytra and proceeding forward until the eggs cover the entire surface of the hemelytra (Fig. 14.1A). After every few eggs laid the male copulates with the female again. Egg laying may go on for over 24 hours. When deposited, eggs are about 1.6 mm long, 1.1 mm wide, and greenish. During the embryonic development they increase six fold in volume and become greenish brown with the anterior end darker. The male cares for the eggs chiefly by "surface brooding" where he positions himself close to the water surface so that the eggs are exposed to the air. In addition submerged males may display "brood-pumping", similar to "display-pumping" (see above) but slower, and rhythmically brush the eggs with their legs in order to circulate water over the egg batch. Apart from aerating the eggs this "grooming" of the eggs keeps the batch clean and may prevent infections by mould or bacteria.

Distribution

The genus *Diplonychus* is distributed throughout Asia and the Indo-Australian region. *D. rusticus*, one of the supposedly two Australian species, is very widespread, ranging from India through Southeast Asia, to the Philippines and New Guinea. In Australia, this species has been recorded from Northern Territory (Groote Eylandt) and Queensland. The other species, *D. eques*, is endemic to Australia and widespread, recorded from all states (Fig. 14.2).

SUBFAMILY LETHOCERINAE

Identification

Large to very large, elongate oval water bugs, reaching 115 mm in body length. Antennal segments 2 and 3 each with one finger-like projection, segment 4 with two projections (Fig. 14.4D). Hind tibia and tarsus very flat and broadly dilated, much broader than middle tibia and tarsus. Fore tarsus 3-segmented (Fig. 14.4C; ta), although often appearing 2-segmented externally; anterior claw of fore leg large, posterior claw vestigial or absent. Abdominal sternites 4-6 subdivided into mediosternites and parasternites by a weak suture-like fold which originates at the basal angle of the subgenital plate (operculum); abdominal spiracles situated on the mesal margin of the ventral laterotergites. Abdominal air-straps long, mesal margins nearly contiguous.

Overview

These giant water bugs are the largest of all true bugs (Hemiptera-Heteroptera). The subfamily contains only the cosmopolitan genus *Lethocerus* Mayr, with approximately 25 species. The Old World species were keyed by Menke (1960).

LETHOCERUS

Identification

These are the largest Australian water bugs, reaching a length of 50-70 mm (Fig. 14.4A, Plate 7, A-B). Other characters as given for the subfamily (see above). The two Australian species can be separated by the following key (adapted from Menke 1960).

Key to Australian species

1. Ridge separating sulci of fore femur bearing a single subapical tubercle or one large and one small tubercle (Fig. 14.4 B; arrow). Mesosternum with two prominent raised areas. Length 54-69 mm. Northern Territory, New South Wales, Queensland .... *insulanus* (Montandon)
- Ridge separating sulci of fore femur bearing 2 subapical tubercles of about equal size (Fig. 14.4C; arrow). Mesosternum without two prominent raised areas. Length 52-63 mm. Northern Territory, Queensland .... *distinctifemur* Menke

Biology

Lethocerus species are frequently attracted to light in large numbers which in North America has earned them the name "Electric Light Bugs". Giant water bugs thrive in ponds and lakes and are occasionally found in quiet parts of streams and several Asian species occur regularly in paddy fields. Despite the fact that they are powerful swimmers they usually ambush their prey which consists of any animal they can master, including vertebrates. In some localities *Lethocerus* is the main predator on tadpoles.

Brooding in *Lethocerus* is very different from that of the Belostomatinae (Ichikawa 1989, 1990, 1991a, 1991b; Smith 1997). The eggs are deposited on emergent vegetation or other objects above the water. Usually the male takes a position at the

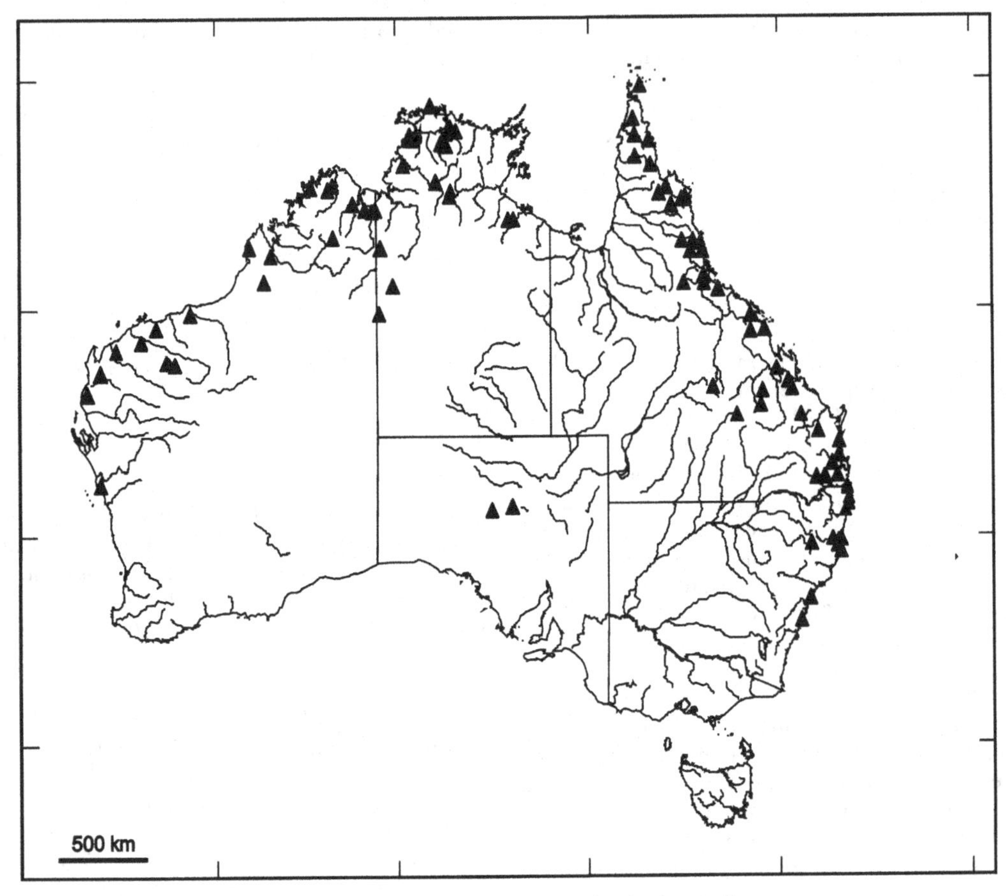

Figure 14.5. BELOSTOMATIDAE. Distribution of *Lethocerus* species in Australia.

base of a suitable object for egg-laying and starts "display-pumping" producing ripples in the surface film in order to attract females. After a receptive female has arrived and several copulations the female will climb out of the water to deposit her eggs. After the female has deposited all her eggs she leaves the male never to return to the egg batch. The male stays in the water at the base of the egg batch and from time to time climbs towards the batch, inserting its rostrum between the eggs of the upper part of the batch and lets water flow into the batch of eggs. Apart from "watering" the eggs, the male also defends the egg batch against predators. Curiously, the most important predators of the eggs seem to be other females of giant water bugs. As males are, on the average, smaller than the females, it is usually difficult for a male to defend a batch of eggs against a female. After the eggs have been sucked out by a female the male usually mates with this female and starts brooding the new egg batch. Since males spend more time engaged in reproductive activities there are usually a scarcity of free males. It may therefore be profitable for a female to take over an already brooding male.

Lethocerus species are the only members of the superfamily Nepoidea which have metathoracic scent glands which in the male are 10-25 times larger than in the female (Staddon 1971). The glands produce an aromatic substance which is supposed to be used to mark a trail to the egg batch. This would explain why the female gland is much smaller (Smith 1997). In large parts of tropical Asia *Lethocerus* species are used as food.

Distribution

The genus *Lethocerus* is found throughout the World with 5 species in the Indo-Australian region. In Australia, the two species are widely distributed in the coastal parts of eastern and northeastern Australia, upper half of Northern Territory, and northwestern parts of Western Australia (Fig. 14.5). A few records from the interior of South Australia.

15. Family CORIXIDAE

Water Boatmen

Identification

Small to medium-sized water bugs, dorsum flattened. Head broad, strongly deflected (hypognathous). Eyes very large, kidney-shaped. Ocelli absent except in Diaprepocorinae. Antennae 3- or 4-segmented, concealed between eyes and prothorax (Figs 15.4A-B). Rostrum modified, immovably fused to head, broad at base, tapering distally; labium without distinct segmentation (Figs 15.1B, 15.4D), usually with transverse grooves and a longitudinal channel; labrum reduced, covered by labium. Meso-scutellum exposed (Figs 15.1A, G; sc) or hidden by pronotum. Fore wings (hemelytra) of uniform texture, membrane without veins. Fore legs relatively short; fore tarsus with a single segment, usually modified into a "pala" which has the form of a spoon or scoop, fringed with long hairs (Fig. 15.1D; pa). Middle legs very long and slender, tarsi 1- or 2-segmented, paired claws very long. Hind legs flattened, oar-like, fringed with hairs, tarsi 2-segmented. Stridulatory structures formed by a field of pegs on the basal surface of the fore femora apposed to the lateral edges of the head (Fig. 15.1D; st) (except Micronectinae; see below). Metathoracic scent glands present in adults; nymphal abdominal scent glands present between terga 3/4, 4/5 and 5/6. Male abdomen usually with a "strigil" laterally on tergum 6 (Fig. 15.1E; sg). Male distal abdominal segments and genitalia usually strongly asymmetrical. Genital capsule (pygophore; Fig. 15.1F; s9) boat-like; parameres asymmetrically developed.

Overview

One of the most common and species-rich families of Nepomorpha in Australia, with 5 genera and 35 species. Hungerford (1948) recognized six subfamilies, of which four subfamilies are represented in the Australian fauna; two of these, Diaprepocorinae and Micronectinae, are treated as separate families by some authors (e.g., Chen et al. in press). Worldwide, the family Corixidae contains c. 35 genera and about 450 species. Mahner (1993) discussed the phylogeny of the Corixidae in great detail, referring to an unpublished doctoral thesis by G. Zimmermann (1986). Mahner (1993) places the subfamily Diaprepocorinae as the most basal group followed by the Micronectinae which is sister group to the subfamilies Cymatiainae and Corixinae together.

Water boatmen are common throughout most of the world. They usually live in stagnant freshwater, such as ponds, lakes, and water reservoirs, but have successfully invaded a wide range of habitats, including frigid subarctic waters beneath ice, hot spring water, and saline waters, both inland and along the sea coast (Scudder 1976, 1983; Lauck 1979). They are strong swimmers and with their relatively large eyes and long hind legs with strongly developed fringes of swimming hairs resemble the backswimmers (Notonectidae). However, unlike notonectids, corixids swim with their back facing upwards. When submerged, water boatmen carry considerable amounts of air on their bodies, kept in place by their hydrofuge pilosity. This air-store acts as a physical gill (see Chapter 2) which extracts dissolved oxygen from the water. In corixids atmospheric air enters dorsally through spaces between the head and prothorax when the bug breaks the surface film. Inhalation is primarily through the first pair of abdominal spiracles and, perhaps, through the mesothoracic spiracles (Fig. 2.2). The exposed air-store is found primarily on the abdominal venter, but there is also a partial air-store in the space between the abdomen and the wings (Popham 1960; Parsons 1976). The high respiratory efficiency in water boatmen is enhanced by frequent rowing movements of the hind legs over the body. This causes fresh currents of water to flow over the air-store, thereby presumably increasing the rate of oxygen uptake. Corixids also practise "secretion-grooming" by rubbing secretions from the metathoracic scent glands over the body parts covered with hydrofuge hairs (Kovac & Maschwitz 1991). The secretions have strong antibacterial properties and prevent the contamination of the hair piles by killing attached microorganisms. No secretion-grooming occurs, however, in *Micronecta* species and in all corixid nymphs, although abdominal scent glands are present in the latter.

The unique, triangular rostrum of corixids combined with stylets enables them to ingest particulate matter as well as fluids (Fig. 15.4E, F). To break up small particles they have evolved tooth-like epipharyngeal grinders in their powerful, complex bipartite food pump. A strainer in the pump blocks passage of larger particles (Parsons 1966). Understanding the feeding biology of the Corixidae has been enigmatic for decades and corixids have been described both as omnivores, detritivores, carnivores, and algal feeders by a variety of authors (reviews by Jansson & Scudder 1972; Bakonyi 1978; Popham et al. 1984). Hungerford (1920) observed water boatmen feeding on the cell contents of multicellular algae such as *Spirogyra* by piercing the cell walls with their stylets. Many water boatmen feed by whirling bottom material up with their front legs and picking out organic particles from the ooze. Other species would eat both animal and plant matter which they select from the bottom ooze or detritus. Thus, Griffith (1945) found mainly unicellular Algae, Protozoa, remnants of Rotifera and loose chlorophyll in stomach contents of the North American species *Rhamphocorixa acuminata* (Uhler). Water boatmen commonly stay close to the bottom of the habitat, clinging to the substrate with their long and slender middle legs. However, some hunt for animal prey in open water. Hale (1922) reports to have kept several species of Australian corixids for months on a diet of mosquito larvae only. Even newly hatched nymphs were observed to capture tiny mosquito larvae, increasingly larger prey being taken by the older instar nymphs. Mosquito and chaoborid larvae are apparently preferred by some Alaskan *Callicorixa* species which may play an important role in keeping down populations of mosquitoes (Sailer & Lienk 1954). As pointed out by Jansson & Scudder (1972), water boatmen should clearly not be regarded as predominantly plant feeders. The present view is that most species prefer animal food which seemingly increases the longevity and is necessary for reproduction in adults.

Water boatmen are fed upon by other water insects and by many fish. There are associations between the presence of fish and the species composition and spatial distribution of corixids which may be attributed to predation (Macan 1976; Oscarson 1987; Savage 1989). The acidification of many lakes in industrialized countries, owing partly to atmospheric pollution, has had some surprising consequences. For example, the predominantly carnivore species *Glaenocorisa propingua* (Fieber) greatly increased its range in acidified lakes in Sweden where fish populations had become depleted or extinct (Henrikson & Oscarson, 1978, 1981). Cannibalism and interspecific predation have also been recorded for corixids and may be important factors in limiting population size under crowded conditions (Pajunen & Ukkonen 1987). Finally, all stages of certain corixids are used as human food in Mexico. The eggs, which are often laid in huge numbers, are gathered by placing reeds in the water and returning later to harvest them (Lauck 1979).

Key to the Australian genera of the family Corixidae

1. Two ocelli present on dorsum of head, close to compound eyes (Figs 15.1A, B; oc). Pronotum short, meso-scutellum well developed (Fig. 15.1A; sc) and exposed behind pronotum (DIAPREPOCORINAE) .... *Diaprepocoris*
- Ocelli absent. Pronotum length variable, meso-scutellum either exposed or covered by pronotum 2
2. Meso-scutellum exposed behind pronotum; pronotum short, covering only the anterior area of meso-scutellum (Fig. 15.1G; sc). Length 4.6 mm or less. (MICRONECTINAE) .... *Micronecta*
- Meso-scutellum completely covered by pronotum. Length 5 mm or more 3
3. Rostrum without transverse grooves (Fig. 15.1H; la). Fore tarsus elongate, almost cylindrical, without pegs in male (Fig.-15.1I; ta). (CYMATIAINAE) .... *Cnethocymatia*
- Rostrum with transverse grooves (Figs 15.4D, 15.7I). Fore tarsus more or less spoon-shaped, with rows of pegs in males (Fig. 15.1D; pa). (CORIXINAE) 4
4. Pronotum and hemelytra brownish, with pale transverse bands which are interrupted on hemelytra (Fig. 15.6). Body slender, length 5-7 mm .... *Sigara*

Figure 15.1. CORIXIDAE. A-B, *Diaprepocoris zealandiae*: A, dorsal view; legs omitted (cf, claval furrow; cl, clavus; co, corium; nf, nodal furrow; pn, pronotum; sc, scutellum). B, frontal view of head (ac, anteclypeus; an, antenna; la, labium; oc, ocellus). C, *Agraptocorixa eurynome*: right fore wing (hemelytron) (cf, claval furrow; cl, clavus; co, corium; eb, embolium; nf, nodal furrow). D, *Sigara* (*Tropocorixa*) *australis*: left fore leg of ♂ (fe, femur; pa, pala (fore tarsus); st, stridulatory field; ti, tibia). E-F, *Sigara* sp.: E, dorsal view of male abdomen (g9, genital segment 9 (genital capsule); sg, strigil; t5-t8, abdominal tergites 5-8). F, genital capsule (segment 9) (ae, aedeagus; lp, left paramere; rp, right paramere; s9, segment 9). G, *Micronecta* sp.: dorsal view; legs omitted (cl, clavus; co, corium; me, membrane; pn, pronotum; sc, scutellum). H-I, *Cnetocymatia nigra*: A, frontal view of head (la, labium). I, left fore leg of ♂ (fe, femur; ta, tarsus; ti, tibia). Reproduced with modification from: A, G, Parsons (1976); B, Martin (1969); C, Zimmermann (1986); D, Lansbury (1970); E-F, Savage (1989); H-I, Lansbury (1983).

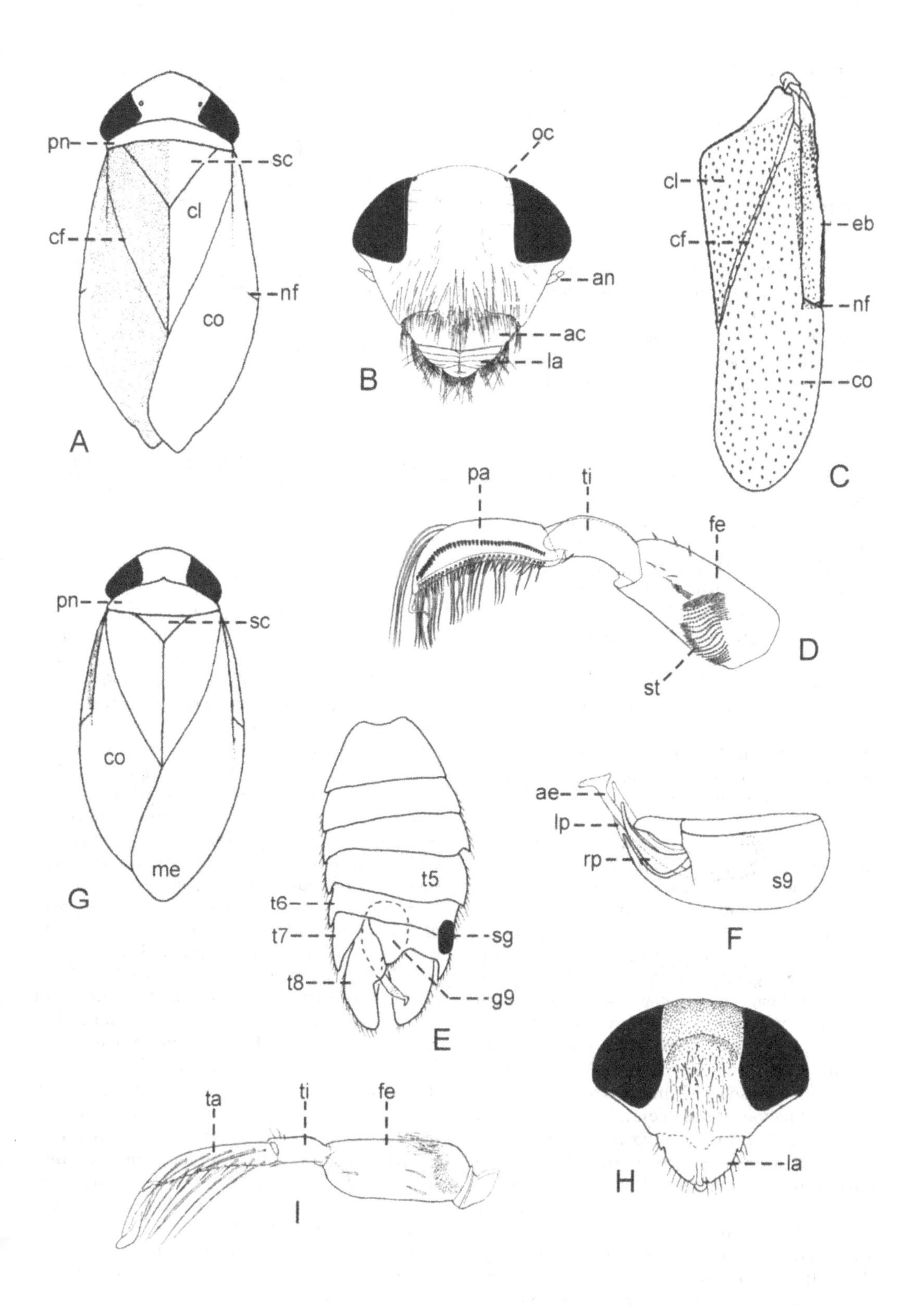

pn
sc
cl
cf
nf
co
A
oc
an
ac
la
B
cl
cf
eb
nf
co
C
pa
ti
fe
st
D
pn
sc
co
me
G
t5
t6
t7
t8
sg
g9
E
ae
lp
rp
s9
F
ta
ti
fe
I
la
H

- Pronotum and hemelytra uniform in colour or with numerous small dark dots (Fig. 15.2). Body stout, length 6-10 mm .... *Agraptocorixa*

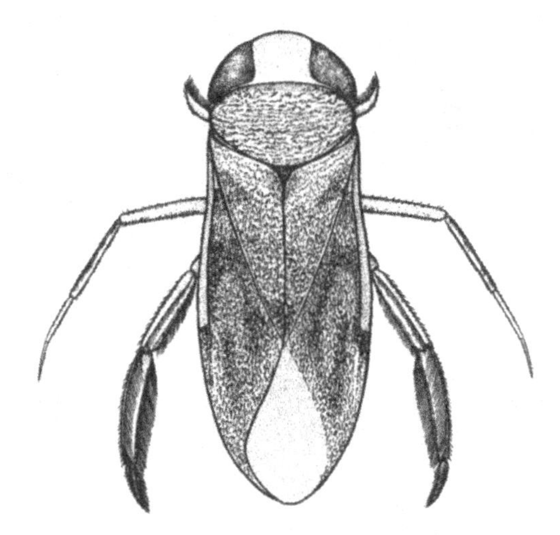

Figure 15.2. Corixidae. *Agraptocorixa eurynome*, dorsal habitus. Reproduced from CSIRO (1991).

Subfamily Corixinae

Identification

Small to medium-sized water boatmen with body lengths from 4 to15 mm. Rostrum with transverse grooves (Fig. 15.4D). Hemelytra with embolar groove (Fig. 15.1C; eb) and nodal furrow (nf). Female pala spoon-shaped, male pala spoon-shaped or of a different form and with a palm-like concavity in both sexes. Hind tarsal claw inserted subapically. Male abdomen usually with a strigil (Fig. 15.1E; sg). Eggs ovoid, affixed to the substrate by a stalk (pedicel) of varying length (Figs 2.8J-K; Cobben 1968).

By far the largest subfamily containing the majority of corixid species placed in four tribes and 26 genera (Hungerford 1948). Species belonging to the Corixinae are distributed worldwide, but predominantly found in temperate and subtropical areas. Two genera occur in Australia.

Overview

Water boatmen are well known for their capability to produce sounds that even can be perceived by the human ear. The mechanism for sound production or stridulation consists of a group of pegs on the base of the fore femur (functioning as the *plectrum*; Fig. 15.1D; st) which is rubbed against the margin of the sharp lateral edge of the head (functioning as the *stridulitrum*) (Hungerford 1948; Jansson 1972). Sound is usually produced by males only, but females may also stridulate (Jansson 1973, 1976). Apparently most corixids only produce one type of "song", but a few species have two. Songs are both species- and sex-specific, and stridulation is associated with sexual maturity. The songs are used to form aggregates, to attract females, and to space out males within a population. The so-called "strigil" on the male abdomen is misnamed and has nothing to do with stridulation. It has been implicated in holding the female during mating (Larsén 1938) and in helping to maintain the air-store beneath the wings (Popham et al. 1984). Water boatmen of the subfamily Corixinae have a mesothoracic scolopophorous organ ("Hagemann's organ") which is believed to be sensitive to sound (Prager 1973). It has a sensory membrane which bears a flask-shaped body and a larger club-shaped structure (Parsons 1976).

The life history of water boatmen is chiefly know from studies of northern temperate species (Southwood & Leston 1959; Lauck 1979; Savage 1989). However, Fernando & Leong (1963) studied the habitats and nymphal instars of the Oriental species *Agraptacorixa hyalinipennis* (Fabricius). The nymphs are similar to the adults except for size and wing development (Plate 7F-H). Corixids lay their eggs on a wide variety of underwater substrates, usually plant material, but stones may be selected in habitats with sparse vegetation. The numbers of eggs laid by individual females varies between 20 and 1000 and eggs may be laid at different times of the year in different species, in relation to the temperature of a particular habitat and the availability of food. The rates of development of eggs and nymphs are primarily determined by temperature and are remarkably consistent between species (Savage 1989). Factors governing the ecological distributions of water boatmen have been extensively studied in northern Europe (Macan 1976; Savage 1989). The species succession of corixids in relation to the presence of vegetation and the accumulation of organic matter was first demonstrated by Macan (1938). Some species clearly tolerate a higher concentration of organic matter (thus more eutrophied conditions) than others and species tend to replace each other in a definite succession (Savage 1989).

Many species of water boatmen migrate by flight and are often attracted to light during night. In species living in temperate climates there are two types of migratory flights (Young 1966): (1) an obligatory early spring migration

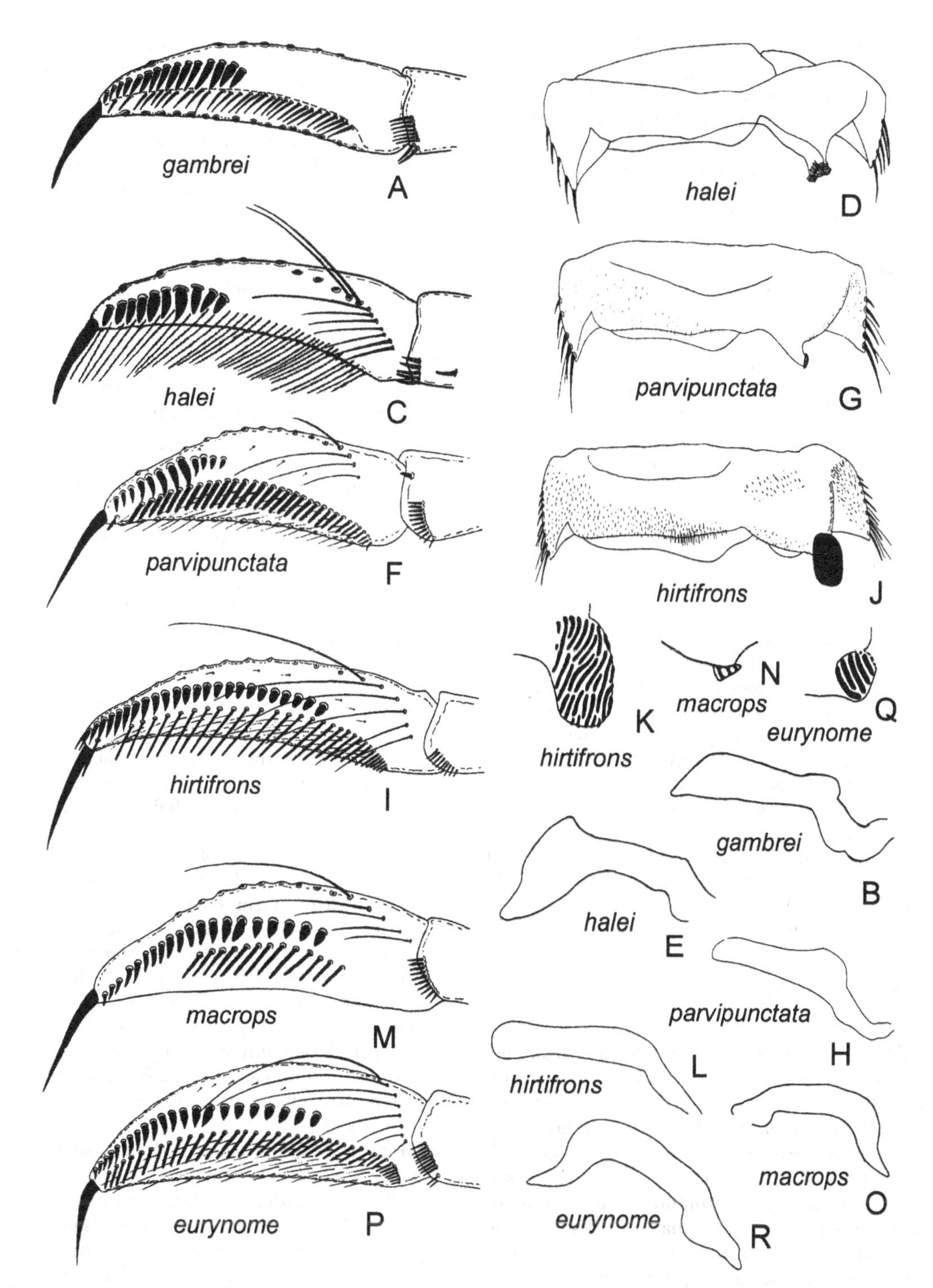

Figure 15.3. CORIXIDAE. A-B, *Agraptocorixa gambrei*, ♂: A, pala. B, right paramere. C-E, *A. halei*, ♂: C, pala. D, tergite 6 with strigil. E, right paramere. F-H, *A. parvipunctata*, ♂: F, pala. G, tergite 6 with strigil. H, right paramere. I-L, *A. hirtifrons*, ♂: I, pala. J, tergite 6 with strigil. K, strigil. L, right paramere. M-O, *A. macrops*, ♂: M, pala. N, strigil. O, right paramere. P-R, *A. eurynome*, ♂: P, pala. Q, strigil. R, right paramere. Reproduced with modification from: A-E, Lansbury (1984); F-L, P-R, Lundblad (1928); M-O, Knowles (1974).

which occurs just before the ovarian development in females; (2) secondary, facultative flights in early summer and fall, triggered by environmental stimuli, such as oxygen deficiency brought on by shrinking of the habitat and increasing temperature (Popham & Lansbury 1960).

Although the corixid hemelytra cover all of the abdomen when folded over the back, many corixines (in particular western Palaearctic species) are polymorphic with respect to the development of flight musculature and/or hind wings, although the latter (concealed brachyptery) is not as common as flight muscle polymorphism (Young 1965a, 1965b). During the so-called "teneral development" of the adult, the flight muscles continue to grow and expand to their functional size along with the hardening and darkening of the thoracic skeleton. This process normally takes from one to four weeks. However, in a percentage of a population the flight muscles do not enlarge and the bugs are therefore incapable of flight. According to Young (1965b), flightless individuals are most common in late spring populations and he hypothesized that the loss of flight capacity confers an advantage to the species through a greater efficiency of flightless bugs in the habitat. Young (1965a) showed experimentally that flightless bugs live longer when starved and at the same time started ovarian development earlier and laid more eggs than normal, flying bugs. Seemingly, there is a tendency for all species to be flightless, but flightlessness is expressed only in populations living in stable, more permanent habitats. Obviously, in temporary habitats survival depends on the emergence of flying adults, probably triggered by environmental factors such as food, temperature and/or day length.

AGRAPTOCORIXA

Identification

Medium-sized water boatmen, length from 6.5 to10 mm. Pronotum and hemelytra concolorous, without transverse dark lines, or densely punctate (Fig. 15.2); both uniformly pilose with short hairs which have dark spots at their bases. Male with facial impression. Hemelytra with media fused with the cubitus before the nodal furrow at a distance equal to or greater than the length of the nodal furrow (Fig. 15.1C). Male pala with a single row of strong pegs varying in number from 8 to 30 (Fig. 15.3) Strigil present on right side of abdomen (except in *A. gambrei*). Eggs with a long stalk, about as long as the egg itself (Fig. 2.8L; Hale 1922; Cobben 1968).

The genus *Agraptocorixa* Kirkaldy includes 11 species in the Afrotropical, Oriental, and Australian regions of which six species are found in Australia; two of these occur also in New Guinea. Hale (1922) described this genus as *Corixa* (*Porocorixa*) to emphasize the punctate dorsal colour-pattern. The Australian species was revised by Lundblad (1928) and Knowles (1974). The following key (adapted from Knowles 1974 and Lansbury 1984) applies only to males which have diagnostic characters in their fore tarsus (pala) and genitalia. Females are difficult to identify because they lack such characters.

Key to Australian Species (males only)

1. Strigil absent. Pala with a row of 15 pegs (Fig. 15.3A). Right paramere (Fig. 15.3B). Length 7.8 mm. Western Australia.......... .... *gambrei* Lansbury
- Strigil present 2
2. Strigil situated on an extension of abdominal tergite 6 (Figs 15.3D, G). Peg rows of pala short, never with more than 12 pegs (Figs 15.3C, F) 3
- Strigil not situated on an extension of abdominal tergite 6 (Fig. 15.3J). Peg rows of pala long, having at least 20 pegs (Figs 15.3I, M, P) 4
3. Right paramere widely expanded distally (Fig. 15.3E). Pala with a row of 8-12 pegs (Fig. 15.3C). Length 6.6-7.2 mm. New South Wales, Northern Territory, Queensland, Western Australia........ *halei* Hungerford
- Right paramere in distal half of equal width along its entire length (Fig. 15.3H). Pala with row of 9-12 pegs (Fig. 15.3F). Length 6.5-7.5 mm. New South Wales, Northern Territory, Queensland, South Australia, Tasmania, Victoria, Western Australia............ *parvipunctata* (Hale)
4. Strigil large, oval in shape (Figs 15.3J, K). Pala with a row of 24-30 pegs (Fig. 15.3I). Right paramere (Fig. 15.3L). Length 7.2-9.2 mm. New South Wales, Queensland, South Australia, Victoria........ *hirtifrons* (Hale)
- Strigil small, either triangular or circular-in shape (Figs 15.3N, Q) 5
5. Strigil acutely triangular in shape, basal edge shorter than other two sides (Fig. 15.3N). Pala with a row of about 20 pegs (Fig. 15.3M). Right paramere (Fig. 15.3O). Length 8.4 mm. Western Australia.................................... *macrops* Hungerford
- Strigil circular in shape (Fig. 15.3Q). Pala with a row of 20-24 pegs (Fig. 15.3P). Right paramere (Fig. 15.3R). Length 8.5-10 mm. New South Wales, Queensland, South Australia, Tasmania, Victoria, Western Australia .... *eurynome* (Kirkaldy)

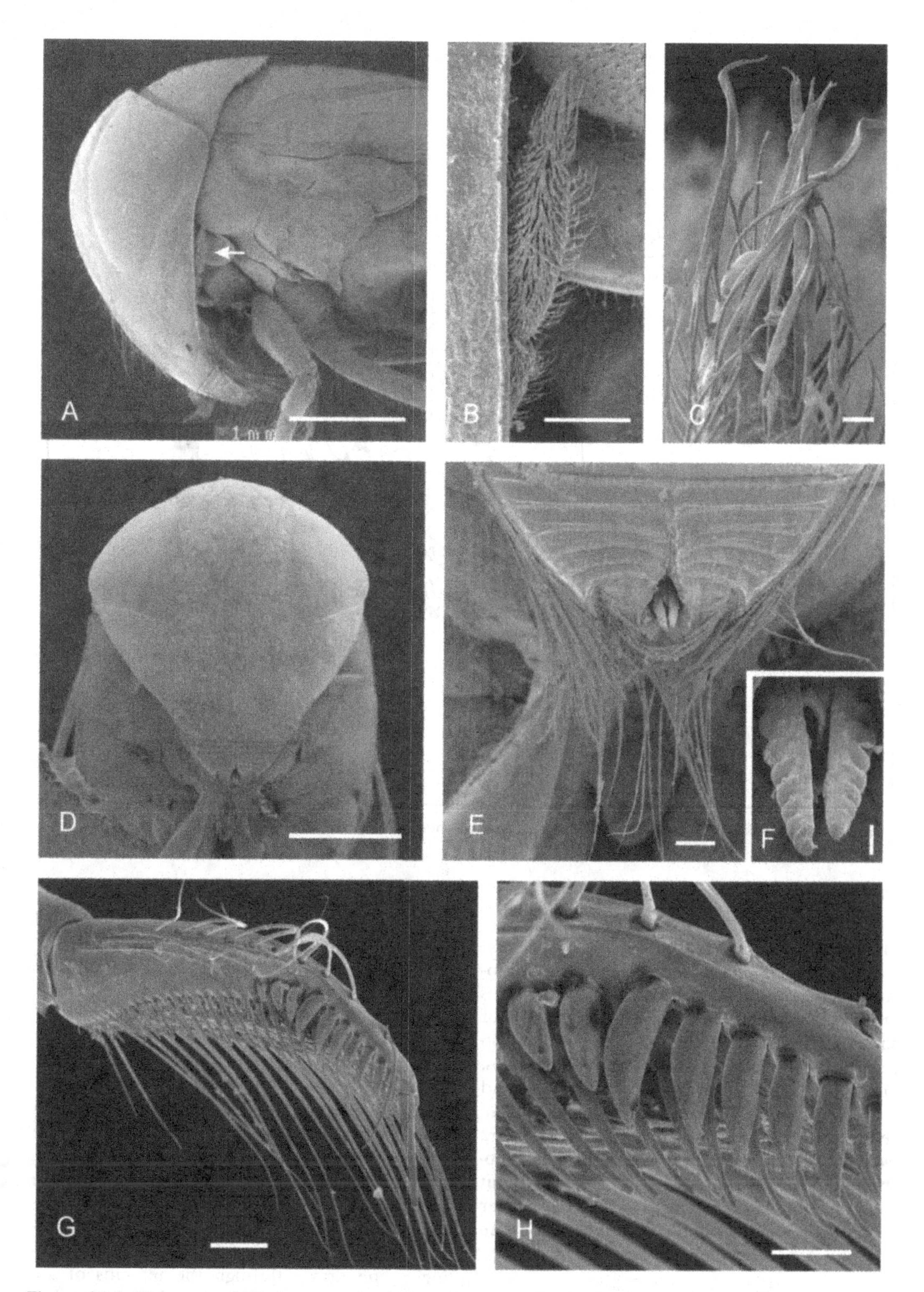

Figure 15.4. CORIXIDAE. A-H, *Agraptocorixa halei*: A, lateral view of head and thorax of ♀; arrow points at antenna. B, antenna. C, distal end of antenna. D, frontal view of head of ♂. E, apex of labium. F, distal end of maxillary stylets. G, pala (fore tarsus) of ♂. H, pegs on pala. Scales: A, D, 1 mm; B, E, G, 0.1 mm; C, F, 0.01 mm; H, 0.05 mm. Scanning Electron Micrographs (CSIRO Entomology, Canberra).

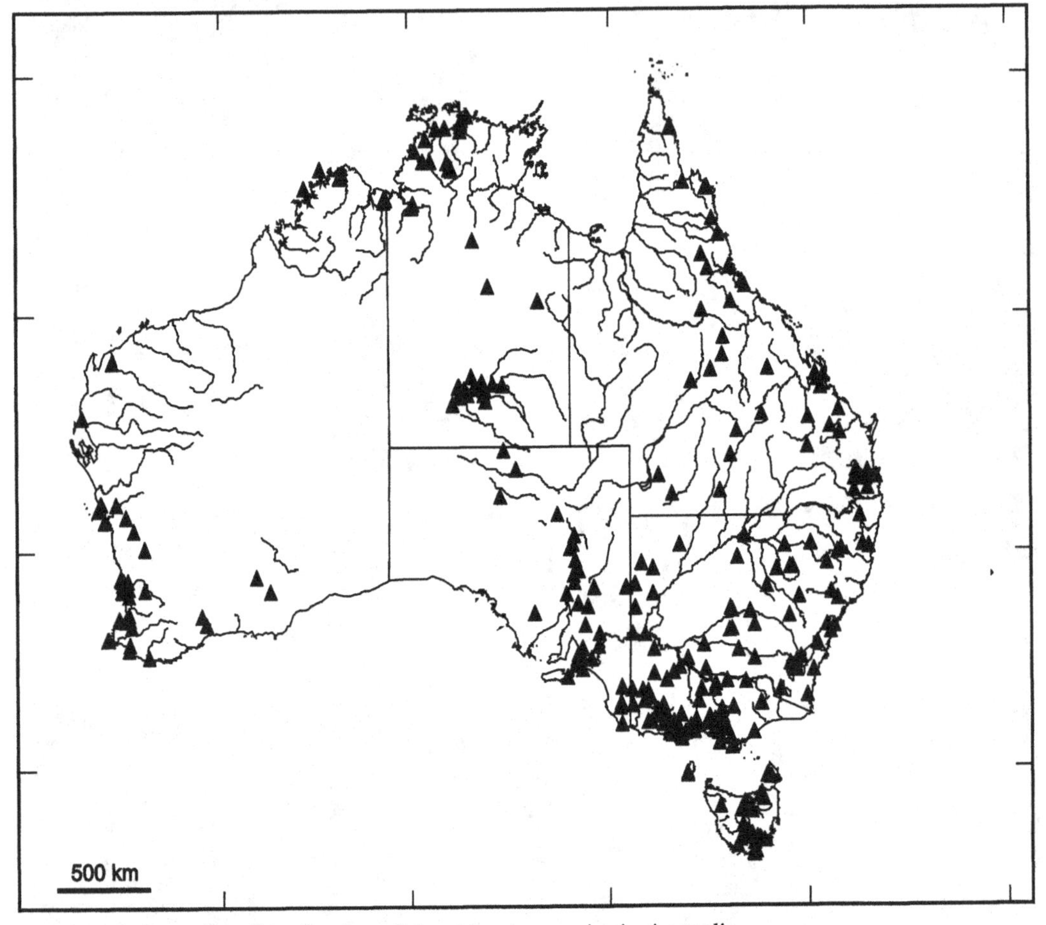

Figure 15.5. CORIXIDAE. Distribution of *Agraptocorixa* species in Australia.

Biology

Only little is known about the biology of the Australian *Agraptocorixa* species. The Indo-Australian species *Agraptocorixa hyalinipennis* (Fabricius) lives in ponds and sluggish parts of streams (Fernando & Leong 1963) which also seem to be the habitats of the Australian species. Hale (1922, 1923) presented a photograph and drawing showing a group of eggs which he referred to "*Porocorixa eurynome?*" (= *Agraptocorixa eurynome*); each egg is oval, attached to the substrate (a plant stem) by a slender pedicel of about the same length as the egg (Fig. 2.8J). A similar structure was described by Cobben for the Oriental *A. hyalinipennis.*

Distribution

Agraptocorixa species are widely distributed in Australia (Fig. 15.5), including many records from southwestern Western Australia, the interior of New South Wales, South Australia, Northern Territory, and Queensland, as well as the top end of Northern Territory. More sparsely recorded from northern Queensland.

SIGARA (TROPOCORIXA)

Identification

Small to medium sized water boatmen, length from 5 to 8 mm. Pronotum brown with transverse yellow stripes (Fig. 15.6, Plate 8, A-B); hemelytra brown with transverse yellow stripes which are interrupted in middle. In the hemelytra, the media is fused with the cubitus at the nodal furrow; apices of the clavi of the female not surpassing a line drawn through the margins of the hemelytra at the nodal furrows. Males with dextral abdominal asymmetry. Egg with short pedicel (stalk) (as in Fig. 2.8K)

The genus *Sigara* Fabricius is a very large and

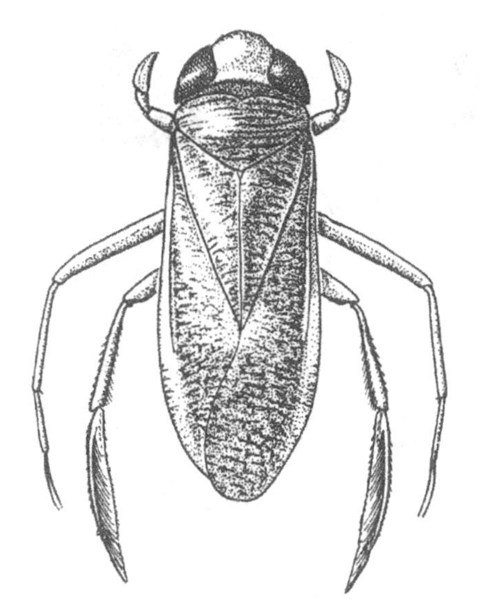

Figure 15.6. CORIXIDAE. *Sigara (Tropocorixa) australis.* Reproduced from Lansbury & Lake (2002).

variable genus with about 150 species worldwide. The characters given above do not necessarily apply to all species groups. Currently, the genus is classified into 16 subgenera. Almost all *Sigara* species occurring in the tropical parts of the World are placed in the subgenus *Tropocorixa* Hutchinson which contains about 50 species of which 7 species occur in Australia. The Australian species can be identified by the following key (adapted from Lansbury 1970, 1975) which only applies to males which have diagnostic characters in their fore tarsus (pala) and genitalia, in particular the shape of the right paramere. On order to examine the parameres, the genital capsule has to be removed from the abdomen which is easily done in alcohol-preserved specimens.

Key to Australian species (males only)

1. Right paramere not spine-like distally- (Figs 15.7D, F, H) 2
- Right paramere spine-like distally (Figs 15.7L, N, P, S) 4
2. Pala elongate, 3x as long as wide (Fig. 15.7A). Pilose area of hind femora not reaching half way along upper margin (Fig. 15.7C). Length 6.9-7.3 mm. New South Wales, South Australia, Victoria .... *australis* (Fieber)
- Pala never more than 2.75x as long as-wide (Figs 15.7E, G) 3
3. Distal pegs on pala much longer than remainder (Fig. 15.7E). Pilose area of hind femora reaching more than half way along upper margin. Length 6-6.2 - mm. Western Australia .... *mullaka* Lansbury
- Distal pegs on pala not significantly longer than remainder (Fig. 15.7G). Pilose area of hind femora not reaching half way along upper margin (as in Fig. 15.7C). Tasmania .... *tasmaniae* (Jaczewski)
4. Frontal impression or fovea of head obsolete (Figs 15.7I, J). Length 5.6-6.2 mm. New South Wales, South Australia, Victoria .... *sublaevifrons* (Hale)
- Frontal impression or fovea of head-clearly defined 5
5. Length at most 5.6 mm, usually just over 5 mm. New South Wales, Queensland, South Australia .... *tadeuszi* (Lundblad)
- Length at least 5.9 mm 6
6. Distal margin of pala concave, not produced along lower margin (Fig. 15.7O). Pilose area of hind femora reaching about half way along upper margin. Length 5.9-6.6 mm. New South Wales, Queensland, South Australia, Tasmania, Victoria .... *truncatipala* (Hale)
- Distal margin of pala not concave, clearly produced along lower margin (Fig. 15.7Q). Pilose area of hind femora not reaching half way along upper margin (Fig. 15.7R). Length 6.1-6.6 mm. Tasmania .... *neboissi* Lansbury

Biology

Almost nothing is known about the biology of Australian *Sigara* species. Lansbury & Lake (2002) recorded *S. australis* and *S. neboissi* from stagnant (lentic) localities and slow-flowing streams in Tasmania. The former also including newly created habitats. Waters usually with neutral pH and low to moderate conductivities (indicating amount of dissolved organic matter). Habitat records indicate that some species prefer ponds and rice-fields, other species live in sluggish parts of streams. The pale and dark patterned upper surface of many corixids, including *Sigara*, is believed to have an effect as camouflage against predators. Popham (1942, 1966) demonstrated that corixids prefer to stay on a background with which their own colour-pattern harmonizes and that fish easily pick out and eat those bugs that contrast with the bottom, in particular when they move. The egg of most *Sigara* species is ovoid with a short stalk at base (as in Fig. 2.8J). The numbers of eggs laid by individual corixids varies from 10 to 1000 (Young 1965b; Peters & Spurgeon 1971) and eggs may be laid at different times depending on the temperature of a particular habitat (Jansson & Scudder 1974).

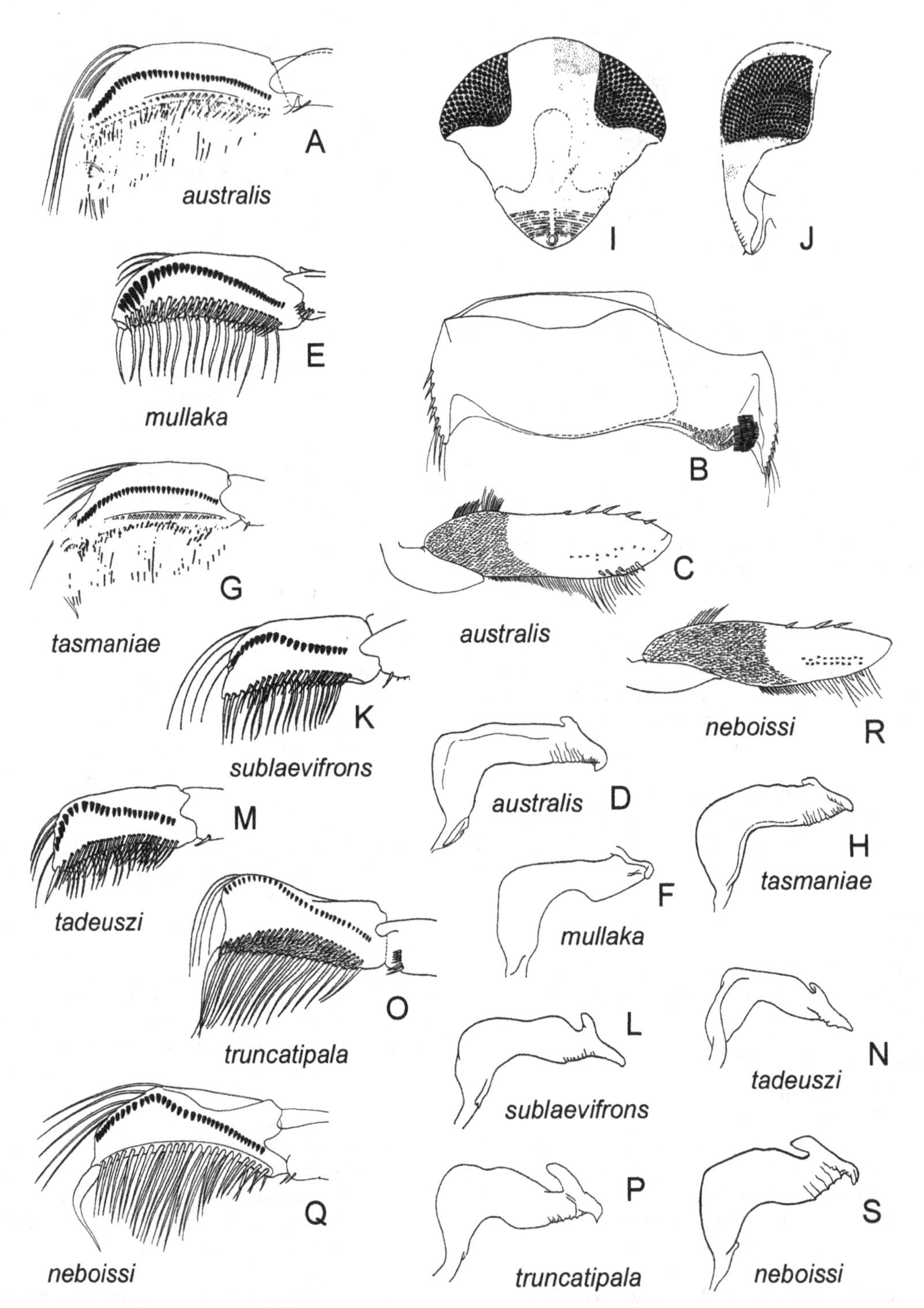

Figure 15.7. CORIXIDAE. A-D, *Sigara* (*Tropocorixa*) *australis*, ♂: A, pala. B, tergite 6 with strigil. C, hind femur. D, right paramere. E-F, *S. mullaka*, ♂: E, pala. F, right paramere. G-H, *S. tasmaniae*, ♂: G, pala. H, right paramere. I-L, *S. sublaevifrons*, ♂: I, frontal view of head. J, lateral view of head. K, pala. L, right paramere. M-N, *S. tadeuszi*, ♂: M, pala. N, right paramere. O-P, *S. truncatipala*, ♂: O, pala. P, right paramere. Q-S, *S. neboissi*, ♂: Q, pala. R, hind femur. S, right paramere. Reproduced with modification from: A, D-Q, S, Lansbury (1970); B-C, R, Lansbury (1975).

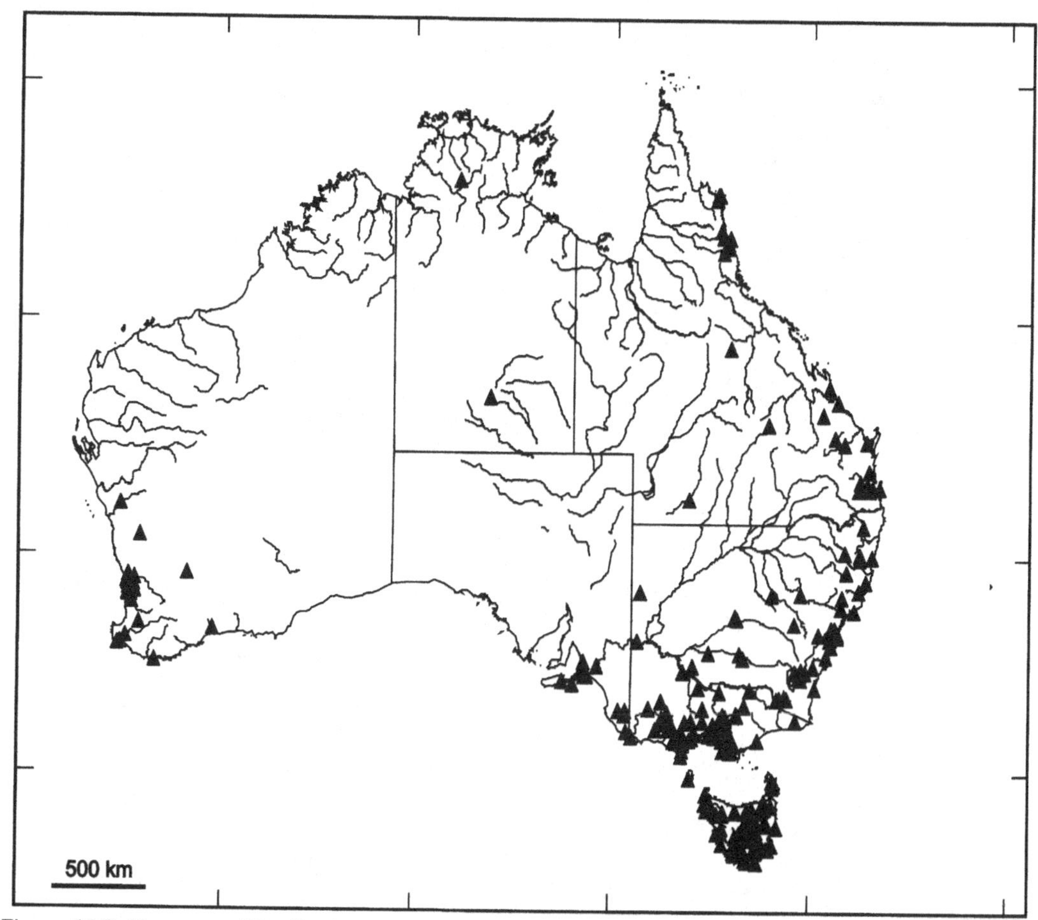

Figure 15.8. CORIXIDAE. Distribution of *Sigara* (*Tropocorixa*) species in Australia.

Distribution

Sigara species are widely distributed in southeastern Australia (Fig. 15.8), in particular in southern Victoria and Tasmania, but also in southwestern Western Australia. A few records from northern Queensland and Northern Territory.

SUBFAMILY CYMATIAINAE

Identification

Small to medium-sized water bugs. Rostrum without transverse grooves (Fig. 15.1H). Scutellum covered by pronotum. Pala (fore tarsus) elongate, almost cylindrical, without pegs in males (Fig. 15.1I). Claw of hind tarsus inserted apically. Hemelytra with embolar grooves and nodal furrow. Male abdomen without a strigil.

Overview

This subfamily contains two genera and 6 species worldwide of which one genus with a single species occurs in Australia.

CNETHOCYMATIA

Identification

Separated from all other water boatmen in Australia by the characters mentioned for the subfamily Cymatiainae above. Pronotum without a median longitudinal carina. Hemelytra with the pruinose area of clavus half a long as the pruinose area of the embolar groove posterior of the point where the media curves to the costal margin. Male abdominal tergite 7 with a finger-like projection anteriorly. Described as a separate genus by Jansson (1982) whereas Lansbury (1983) did not consider it to be sufficiently different from *Cymatia* s.str. and reduced it to subgeneric status. Only one Australian species, *Cnethocymatia nigra* (Hungerford), with prono-

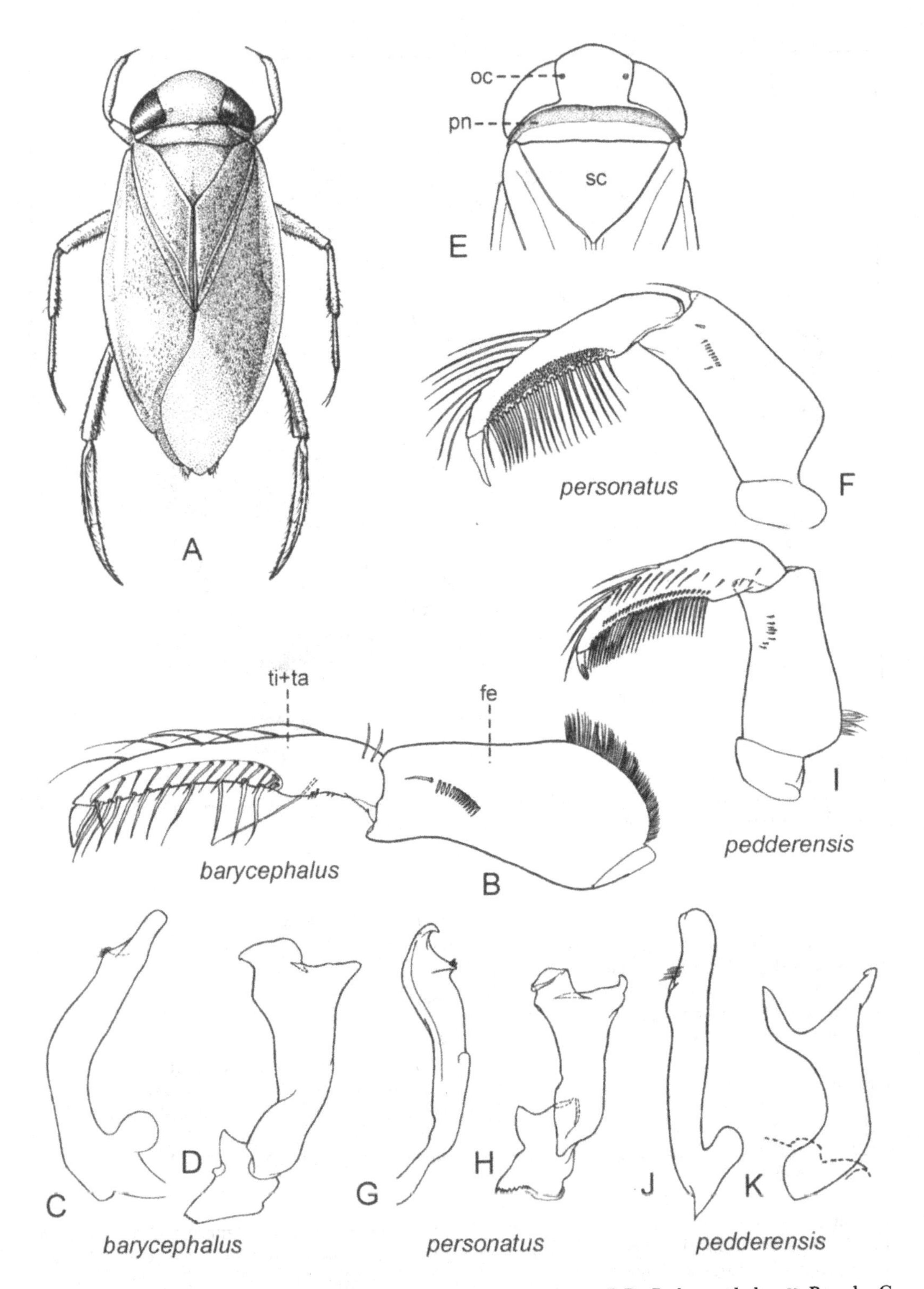

Figure 15.9. CORIXIDAE. A, *Diaprepocoris barycephalus*, dorsal habitus. B-D, *D. barycephalus*, ♂: B, pala. C, right paramere. D, left paramere. E, *D. personatus*, dorsal view of head and thorax. F-H, *D. personatus*, ♂: F, pala. G, right paramere. H, left paramere. I-K, *D. pedderensis*, ♂: I, pala. J, right paramere. K, left paramere. Reproduced with modification from Lansbury & Lake (2002).

tum and hemelytra dark brownish, except for a curved, yellowish spot at the distal apex of the corium. Length 4.6-5.0 mm. Queensland.

Biology

Most of the few specimens known in collections are from light traps and the others do not bear any habitat data. Only Lansbury (1983) studied some specimens collected in the field but as these were from one pond with and one without macrophytes, these data give few clues about habitat selection. Representatives of the other genus in the subfamily, *Cymatia* Flor, which occurs in the Northern Hemisphere are reported to actively catching mosquito larvae and any small invertebrates they can master. The prey is grasped with the elongate cylindrical fore tarsi (Lansbury 1983). The egg of the Palaearctic *Cymatia coleoptrata* (Fabricius) has a long stalk, as long as the egg itself (Cobben 1968).

Distribution

In Australia, *Cnethocymatia nigra* is known from a few records from northern Queensland (Fig. 15.10). Further known from Papua New Guinea and Indonesia (Irian Jaya).

SUBFAMILY DIAPREPOCORINAE

Identification

Separated from other corixid subfamilies by the presence of two ocelli dorsally on the head, close to compound eyes (Fig. 15.9E; oc). A well-developed, exposed meso-scutellum (sc) is only shared with the Micronectinae (below). The morphology of thorax and abdomen was treated by Parsons (1976) who also described the structure of the mesothoracic scolopophorous organ which does not have the club-shaped structure found in the subfamily Corixinae. The male and female genital segments were described by Dunn (1976).

Overview

Fossils placed in the subfamily Diaprepocorinae have been recorded from the Upper Jurassic of Europe and Central Asia (Popov 1971). The four living species are now restricted to New Zealand and the southern half of Australia.

DIAPREPOCORIS

Identification

Small to medium-sized, suboval water boatmen (Fig. 15.9A), length from 3.6 to 8.4 mm. Head more or less produced anteriorly, yellowish. Two dark ocelli present on dorsum of head, close to compound eyes (Fig. 15.9E; oc). Pronotum short, rugose, usually brownish without transverse stripes. Meso-scutellum exposed and well developed (Fig. 15.9E; sc), usually dark brownish. Fore tibia and tarsus fused forming an elongate pala fringed with long hairs (Fig. 15.9B; ti+ta); pala of ♂ without distinct rows of pegs (Figs 15.9B, F, I). Cuticular blades present on hind tarsi (Lansbury 1991a). Hemelytra yellowish brown with darker patches; scutellum and hemelytra covered with numerous small black spinelets. Male abdomen and genital segments strongly asymmetrical; asymmetry usually dextral but sinistral asymmetry may occur (*D. pedderensis* Knowles). A strigil is present but small. Eggs are elongate with one flattened side which is glued to the substrate (Fig. 12.8I; Cobben 1968).

The genus *Diaprepocoris* was described by Kirkaldy (1897) for the species *D. barycephalus* from Tasmania and Victoria. Hale (1924b) added *D. personatus* from Western Australia and *D. zealandiae*, endemic to New Zealand (see also Young 1962). Finally, Knowles (1974) described *D. pedderensis* from Tasmania. Thus, three species of this unique genus of water boatmen are recorded from Australia. The following key (adapted from Knowles 1974 and Lansbury & Lake 2002) will separate the males. Females do not have the useful characters of the male pala and genitalia, but the small size distinguishes females of *D. pedderensis*.

Key to Australian species (males only)

1. Synthlipsis (width of head between eyes) wider than head length, vertex rounded (Fig. 15.9E) 2
- Synthlipsis less than head length; vertex pointed, produced beyond eyes (Fig. 15.9A). Fore leg (Fig. 15.9B). Parameres (Figs 15.9C, D). Length 6.6-8.4 mm. New South Wales, Queensland, South Australia, Tasmania, Victoria, Western Australia .... *barycephalus* Kirkaldy
2. Length 5 mm or more. Pala falcate in dorso-lateral view (Fig. 15.9F). Parameres (Figs 15.9G, H). Queensland, Tasmania, Western Australia .... *personatus* Hale
- Length 3.6-4.3 mm. Pala not falcate in dorso-lateral view (Fig. 15.9I). Parameres (Figs 15.9J, K). Tasmania... *pedderensis* Knowles

Biology

Very little is known about the habitats and life history of this genus. Lansbury & Lake (2002) recorded *Diaprepocoris barycephala* from small stagnant waters or slow-flowing streams in coastal lowland localities in Tasmania. *D. pedderensis* was first

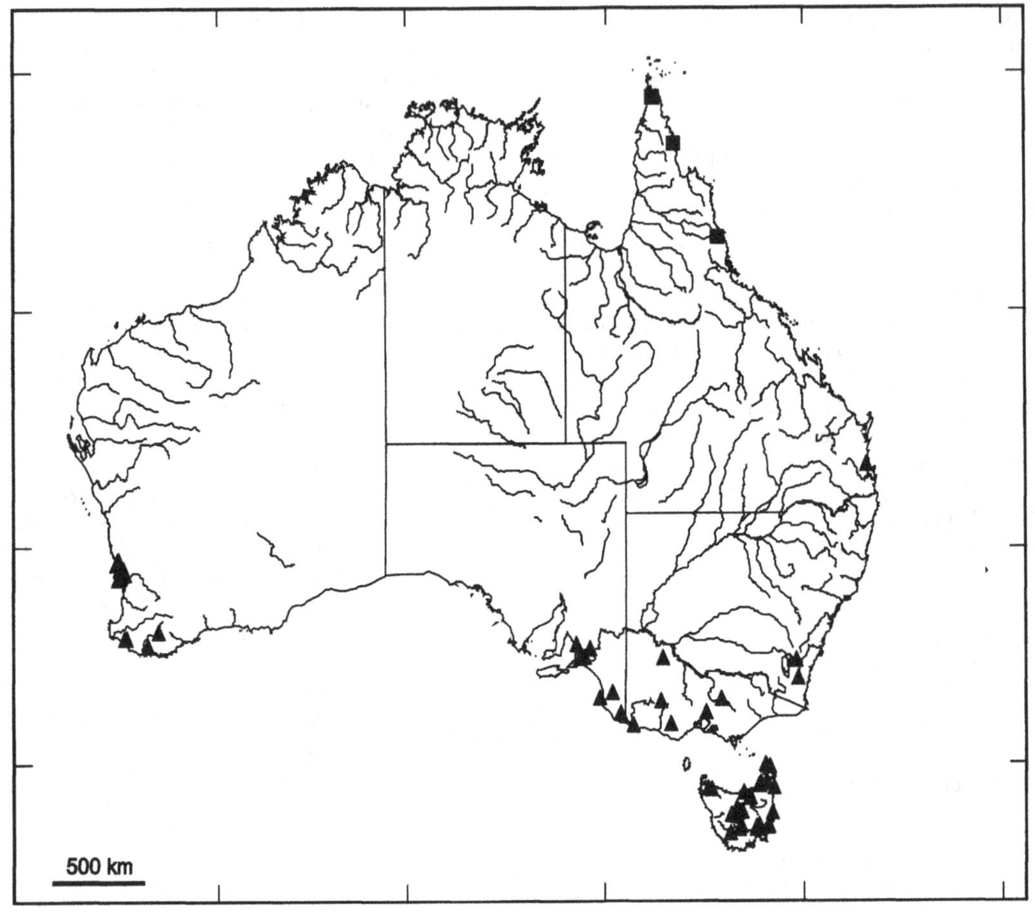

Figure 15.10. CORIXIDAE. Distribution of *Cnethocymatia nigra* (squares) and *Diaprepocoris* species (triangles) in Australia.

found in the original Lake Pedder, but also occurs in lowland localities in acid humic waters of low temperature. Whereas secretion-grooming (see above) in most corixids is performed while floating on the water surface, *Diaprepocoris* grooms on land (Kovac & Maschwitz 1991). The egg of *Diaprepocoris zealandiae* was described by Cobben (1968). The eggs are glued sideways to the substrate (Fig. 2.8I), not affixed by a stalk as in species of the two previous subfamilies.

Distribution

Diaprepocoris is confined to Australia and New Zealand and Larivière (1997) listed this genus as an example of an East Gondwanian distribution. *D. barycephalus* is the most widely distributed of the three Australian species, recorded from New South Wales, southern Queensland, South Australia, Tasmania, and Victoria (Knowles 1974), but also from southwestern Australia (Lansbury & Lake 2002). *D. pedderensis* is endemic to Tasmania (Knowles 1974), whereas *D. personatus* was originally described from Southwestern Western Australia, but also recorded from southern Queensland and Tasmania by Lansbury & Lake (2002). In total, the genus *Diaprepocoris* has a typical Bassian distribution (Fig. 15.10).

SUBFAMILY MICRONECTINAE

Identification

Posterior margin of head covering the anterior margin of the pronotum. Ocelli absent; antennae 3-segmented; rostrum broadly triangular, usually with transverse grooves. Scutellum exposed and relatively large (Fig. 15.1G; sc). Fore tarsus (pala) of ♂ one-segmented, spoon-like with long bristles (Fig. 15.12A; pa); fore tarsus and tibia fused in ♀

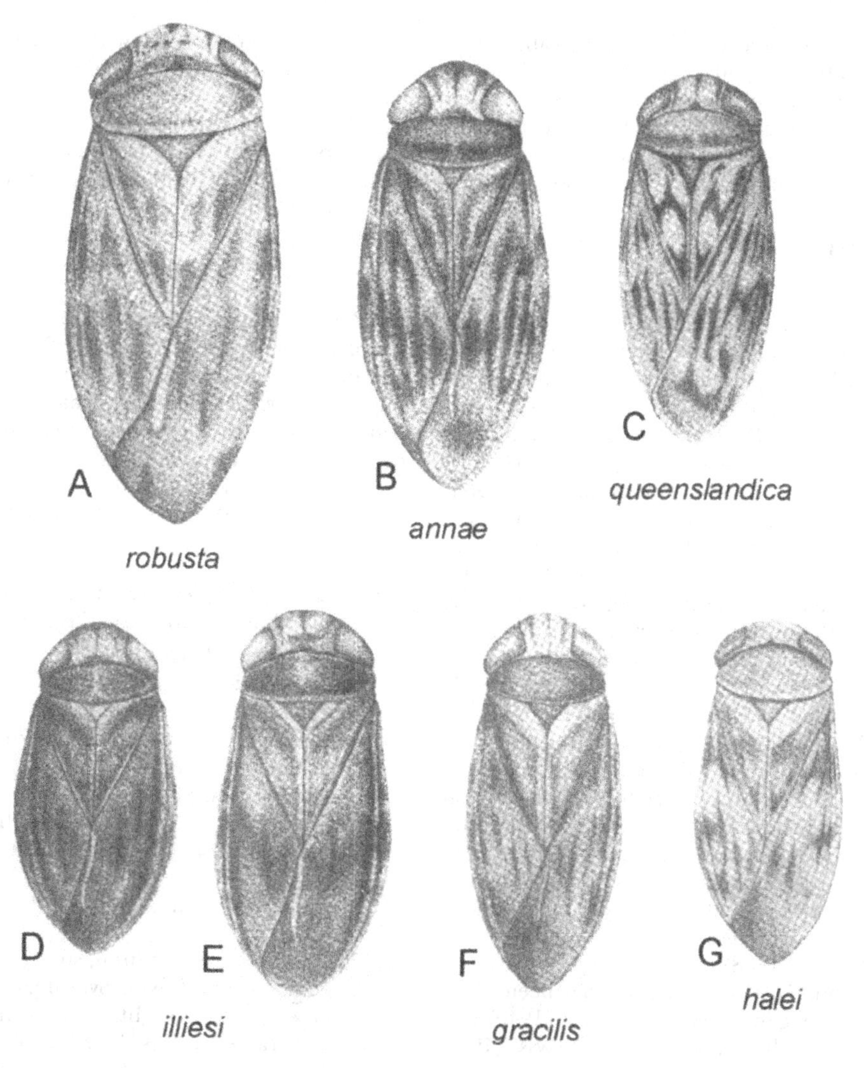

Figure 15.11. CORIXIDAE. A-G, *Micronecta* spp., dorsal view showing colour pattern; legs omitted. A, *M. robusta.* B, *M. annae.* C, *M. queenslandica.* D, *M. illiesi*, subbrachypterous form. E, *M. illiesi*, macropterous form. F, *M. gracilis.* G, *M. halei.* Reproduced from Wroblewski (1970).

(in *Synaptonecta* in both sexes). Middle tarsus one-segmented with a pair of long claws; claw of hind tarsus inserted apically. Hemelytra of uniform texture, membrane without veins (Fig. 15.1G). Distal segments of male abdomen strongly asymmetrical, usually with a strigil (Fig. 15.12C; sg).

Overview

This subfamily contains about 125 species classified in 3 genera. The genus *Tenagobia* Bergroth with 26 species is restricted to the New World. *Synaptonecta* Lundblad has two species in tropical Asia and *Micronecta* has about 100 species widely distributed in the Old World. Some authors (e.g., Chen et al. in press) treat Micronectinae as a separate family based on the results of the phylogenetic analysis by Mahner (1993) which place the Micronectinae (as subfamily) as sister group to the remaining Corixidae except the Diaprepocorinae. It is a matter of opinion, however, whether the Micronectinae are so different from the other corixids that it justifies separate taxonomic rank. We do not share this view. Parsons (1976) found that the adaptations of the thoracic region in connection with respiration are the same in *Micronecta* as in other corixids. The distribution of hydrofuge hairs also indicates that the air stores are roughly the same. The genital segments of both sexes differ considerably from those of other Corixidae. In addition, *Micronecta* transmits its sperm in a spermatophore (Larsén 1938), thus being the only nepomorphan genus besides *Aphelocheirus* (Aphelocheiridae) known to produce spermatophores (see Chapter 16).

MICRONECTA

Identification

Small or very small water boatmen, length from 0.8 to 5 mm. Pronotum and hemelytra brownish, usually with more or less distinct, transverse (pronotum) or longitudinal darker stripes (hemelytra) (Fig. 15.11, Plate 8, C). Posterior margin of pronotum straight or convex (Fig. 15.1G; pn). Adults occur in two wing forms: the macropterous form has hemelytra with a distinct membrane (Fig. 15.11E); the so-called brachypterous form has a shorter, broadly rounded membrane and relatively smaller pronotum (Fig. 15.11D). Male fore leg with tibia separated from the suboval tarsus (pala) which carries a single modified claw, variably dilated and curved (Fig. 15.12A; cl); tibia and pala fused in females. Male abdomen asymmetrical (symmetrical in females); a strigil usually present on segment 6 (Fig. 15.12C; sg); the shape of the free lobe of tergite 8 (fl) is diagnostic in some species as are the parameres; these are asymmetrical, the right paramere large, falciform, the left paramere shorter and straighter with apical part variably modified. Eggs elongate with one side flattened which is glued to the substrate (Cobben 1968).

The genus *Micronecta* Kirkaldy has been divided into six subgenera by Hutchinson (1940) and four more have been added by Wroblewski (1962, 1967). However, as some of the characters used to separate the subgenera are only visible on microscopic slides and since Wroblewski (1968, 1972) did not continue using subgenera, we refrain from using them in this book. The taxonomy of Australian *Micronecta* was confused by the almost simultaneous publication of the revisions by Chen (1965) and Wroblewski (1970), who in some cases described the same species under different names. The confusion was subsequently solved by Wroblewski (1972, 1977).

Reliable identification is in most species only possible for males. Diagnostic structures are the shape of the metaxyphus, the free lobe dorsally on the left part of the eighth segment, and the structure of both parameres. Most recently, King (1997) has demonstrated that common Australian *Micronecta* species like *M. annae* Kirkaldy consist of sibling species which are difficult to separate on museum material, but can be readily distinguished by their acoustic signals. The Australian species of *Micronecta* need revision before a reliable key for the identification of species can be constructed. The following key was adapted from Chen (1965) and Wroblewski (1970, 1972, 1977).

Key to Australian species (except *M. concordia* King, M. *dixonia* King, and *M. micra* Kirkaldy)

1. Larger species, length 3.5-4.6 mm 2
- Smaller species, length rarely exceeding 3.5 mm 3
2. Head shorter than pronotum, moderately produced in front of eyes (Fig. 15.11A). Synthlipsis very broad, 1.8x eye width. Parameres (Figs 15.12D, E). Length 3.5-4.2 mm. Australian Capital Territory, New South Wales, Queensland, South Australia, Tasmania, Victoria, Western Australia .... *robusta* Hale
- Head as long as pronotum, strongly produced in front of eyes. Synthlipsis narrower, as wide as or slightly wider than eye width. Parameres (Figs 15.12F, G). Length 4-4.6 mm. New South Wales .... *major* Chen
3. Body slender, length more than 2.2x greatest width (Figs 15.11F, G). Pronotum without a dark transverse stripe 4
- Body stout, length 1.7-2.2x greatest width (Fig. 15.11B-E). Pronotum usually with a dark transverse stripe 6

Figure 15.12. CORIXIDAE. A, *Micronecta major*, fore leg of ♂ (fe, femur; cl, claw; pa, pala (fore tarsus); ti, tibia). B, *Micronecta* sp., ventral view of ♀; head and legs except coxae of left side omitted (mx, metaxyphus; s3, s7, abdominal sterna 3 and 7). C, *Micronecta* sp., dorsal view of abdomen of ♂ (fl, free lobe of tergite 8; pf, prestrigillar lobe; s6-s8, abdominal sterna 6-8; sg, strigil). D-E, *M. robusta*, ♂: D, right paramere. E, left paramere. F-G, *M. major*, ♂: F, right paramere. G, left paramere. H-I, *M. halei*, ♂: H, right paramere. I, left paramere. J-K, *M. windi*, ♂: J, right paramere. K, left paramere. L-M, *M. gracilis*, ♂: L, right paramere. M, left paramere. N-O, *M. lansburyi*, ♂: N, right paramere. O, left paramere. P-Q, *M. australiensis*, ♂: right paramere. Q, left paramere. Reproduced with modification from: A, F-G, J-K, Chen (1965); B, Parsons (1976); C, Jansson (1986); D-E, H-I, L-M, Wroblewski (19970); N-Q, Wroblewski (1972).

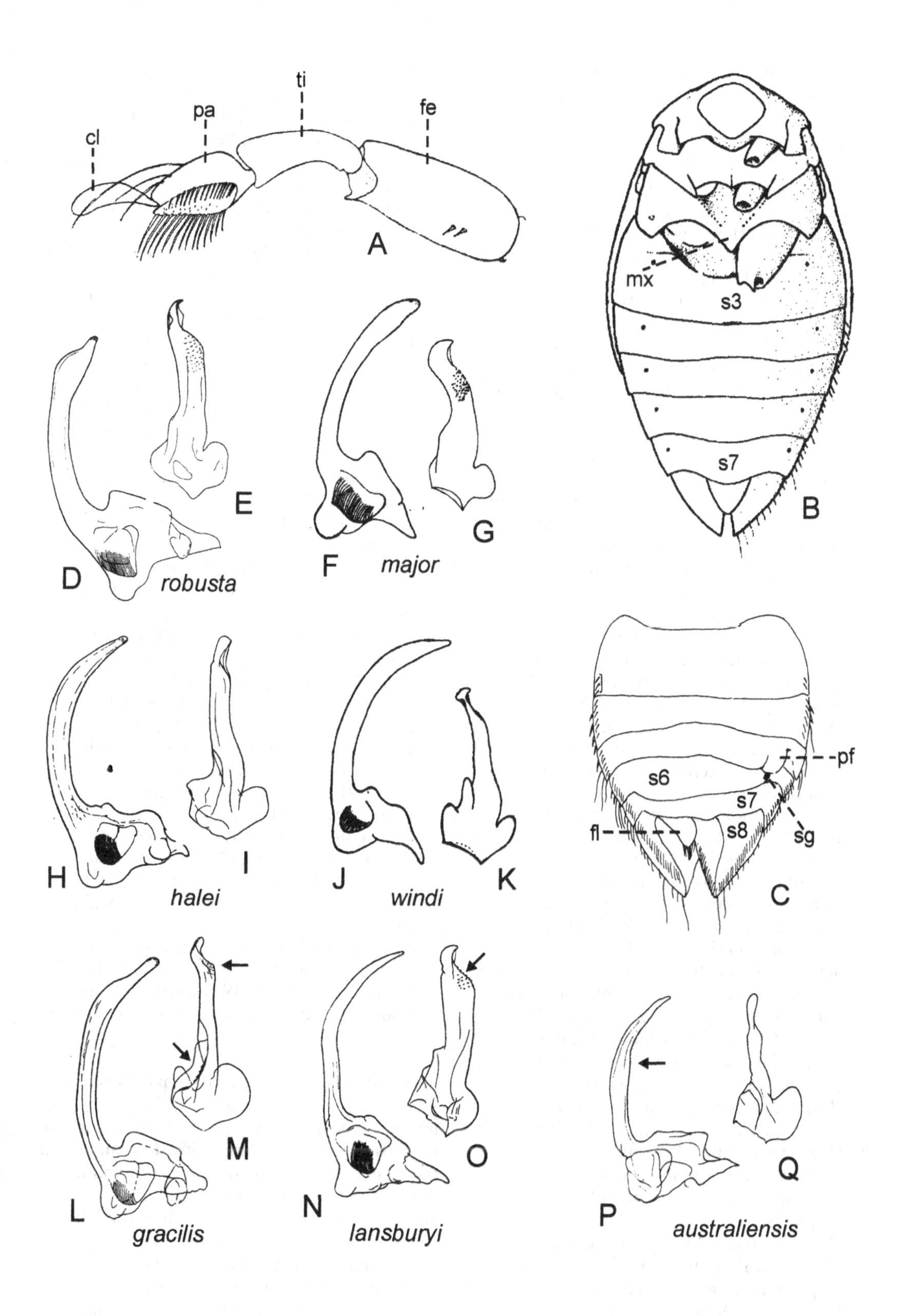

cl
pa
ti
fe
A
mx
s3
s7
B
E
G
D
F
robusta
major
pf
s6
s7
fl
s8
sg
H
I
J
K
C
halei
windi
M
O
Q
L
N
P
gracilis
lansburyi
australiensis

4. Length 2.75-3.4 mm. Hemelytra with long (50-60 µm), oppressed hairs. Left paramere with scales in distal part (Figs 15.12L-O; arrows) 5
– Length 2.5-3 mm. Hemelytra with short (20-30 µm), semierect pubescence. Left paramere without scales in distal part (Figs 15.12H-K). Northern Territory, Queensland, Western Australia .... *halei* Chen, *windi* Chen, *adelaidae* Chen
5. Length about 3.4 mm. Left paramere with barbs in basal part (Fig. 15.12M; arrow). New South Wales, Northern Territory, Queensland, South Australia, Victoria, Western Australia .... *gracilis* Hale
– Length 2.75-3.1 mm. Left paramere without barbs in basal part (Fig. 15.12O). NT. .... *ansburyi* Wroblewski
6. Synthlipsis 1.5-1.9x width of an eye. Body stout, less than 2x as long as wide 7 (5a)
– Synthlipsis narrower, 1.1-1.45x width of an eye 9
7. Large species, length 3.1-3.4 mm, always macropterous. Pronotum 2.9x as wide as long, usually with a dark transverse stripe which is interrupted in middle. Metaxyphus short, triangular, with blunt apex (Fig. 15.13A; arrow). Right paramere with hook-shaped apex (Fig. 15.13C; arrow). Queensland, Western Australia.... *virgata* Hale
– Small species, length less than 3.1 mm, usually brachypterous. Pronotum more than 3x as wide as long. Right paramere slender, apex almost straight (Figs 15.12Q, 15.13F) 8
8. Very small species, length 2.1-2.3 mm. Pronotum 4.2-4.6x as wide as long, with a scaly microsculpture. Metaxyphus broad, triangular (as Fig. 15.13M). Left paramere very slender (Fig. 15.12P). Australian Capital Territory, Victoria .... *australiensis* Chen
– Larger species, length 2.7-3.1 mm. Pronotum 3.4x as wide as long, with irregular wrinkles. Metaxyphus long, triangular (Fig. 15.13D). Left paramere stouter (Fig. 15.13E). New South Wales, Tasmania .... *illiesi* Wroblewski
9. Head very long, 1.1-1.4x as long as pronotum. Pronotum 3.5-4.3x as wide as long, distinctly narrower than head (brachypterous form) 10
– Head shorter, as long as or shorter than pronotum. Pronotum 2.6-3.3x as wide as long, almost as wide as head 12
10. Hemelytra with distinct dark colour pattern (Fig. 15.11B) 11
– Hemelytra almost uniformly pale sand-coloured, at most with faint traces of dark lines. Parameres (Figs 15.13G, H). Length 2.65-3.7 mm. Tasmania, Victoria .... *tasmanica* Wroblewski
11. Hemelytra with four central, fused broken, longitudinal dark stripes. Parameres broadly styliform (Figs 15.13I, J). Length 2.5-2.6 mm. New South Wales... *carinata* Chen
– Dark colour pattern of hemelytra not as above. Parameres slender styliform (Figs 15.13K, L). Length 2.7-3.5 mm. Australian Capital Territory, New South Wales, Queensland, South Australia, Victoria, Western Australia .... *annae* Kirkaldy
12. Larger species, length usually more than 3 mm. Head almost as long as pronotum (Fig. 15.11B). Australian Capital Territory, New South Wales, Queensland, South Australia, Victoria, Western Australia .... *annae* Kirkaldy (macropterous form)
– Smaller species, length usually about 2.7 mm, rarely exceeding 3 mm (females only). Head always distinctly shorter than pronotum 13
13. Pronotum with a distinct, dark transverse stripe, interrupted in middle (Fig. 15.11C). Metaxyphus triangular with acute apex (Fig. 15.13M). Dark pattern of corium composed by fragments of four lines. Free lobe of abdominal tergite 8 of male simple (Fig. 15.13N). Parameres (Figs 15.13O, P). Length 2.6-3.0 mm. Queensland .... *queenslandica* Chen
– Pronotum usually without traces of a dark transverse stripe. Metaxyphus triangular with blunt apex (as Fig. 15.13A). Dark lines on corium complete, though not always distinct. Free lobe of abdominal tergite 8 of male sigmoid (Fig. 15.13Q). Parameres (Figs 15.13R, S). 2.5-3.2 mm. New South Wales, Queensland, South Australia................ *qudristrigata* Breddin

Biology

Micronecta species are typically inhabitants of stagnant waters such as lakes, ponds, and stream pools. Some species prefer running waters, but then usually stay in places with slow current or take shelter behind boulders and other obstacles. Most species prefer habitats with little vegetation and stay at the bottom except when surfacing to renew their air-store. Many species come regularly and sometimes in large numbers to light. The distribution and abundance of various *Micronecta* species may be used to monitor water quality in lakes. In Lake Vasijärvi in southern Finland, *M.*

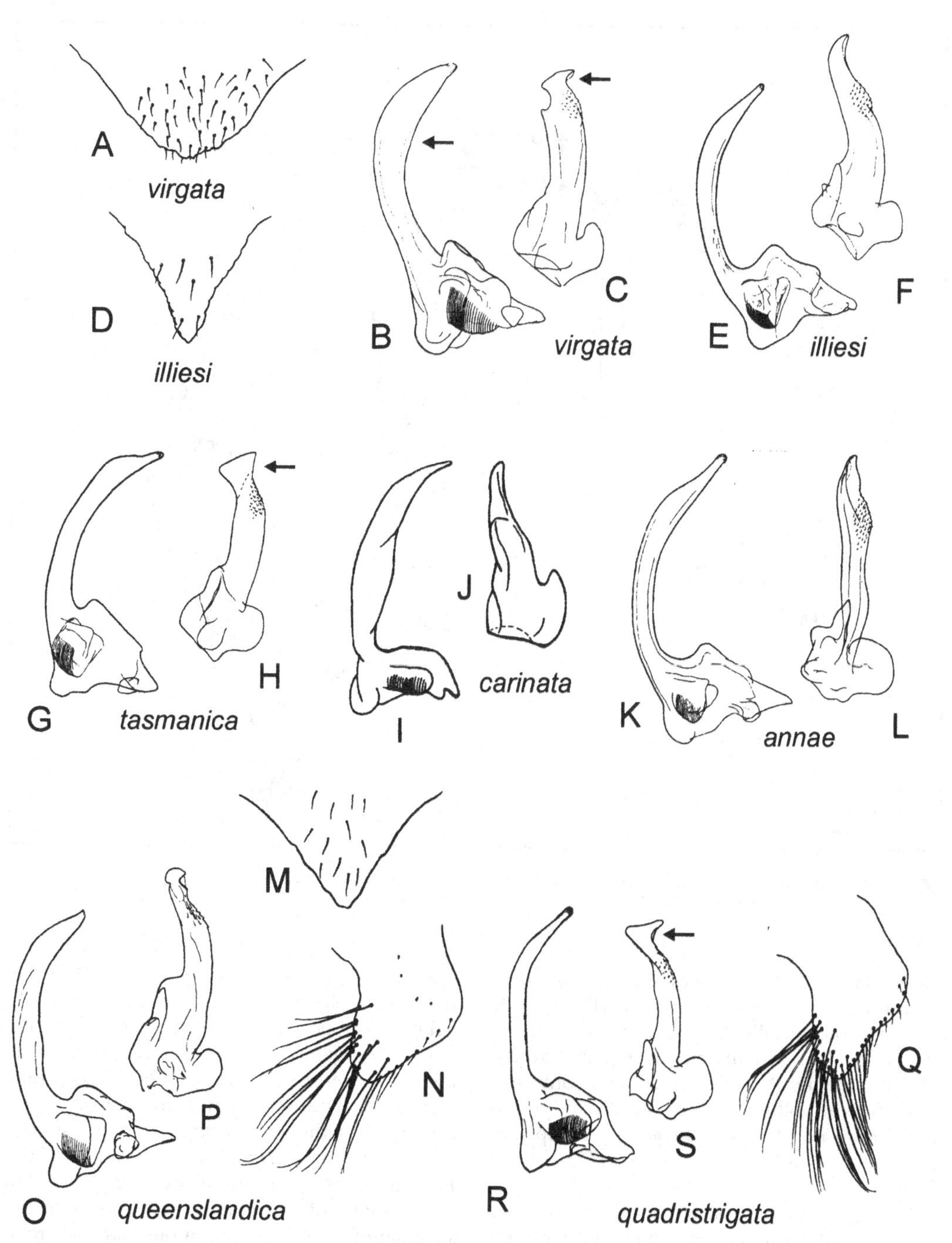

Figure 15.13. CORIXIDAE. A-C, *Micronecta virgata*, ♂: A, metaxyphus. B, right paramere. C, left paramere. D-F, *M. illiesi*, ♂: D, metaxyphus. E, right paramere. F, left paramere. G-H, *M. tasmanica*, ♂: G, right paramere. H, left paramere. I-J, *M. carinata*, ♂: right paramere. J, left paramere. K-L, *M. annae*, ♂: K, right paramere. L, left paramere. M-P, *M. queenslandica*, ♂: M, metaxyphus. N, free lobe of tergite 8. O, right paramere. P, left paramere. Q-S, *M. quadristrigata*, ♂: Q, free lobe of tergite 8. R, right paramere. S, left paramere. Reproduced with modification from: A-F, K-S, Wroblewski (1970); G, H, Wroblewski (1977); I-J, Chen (1965).

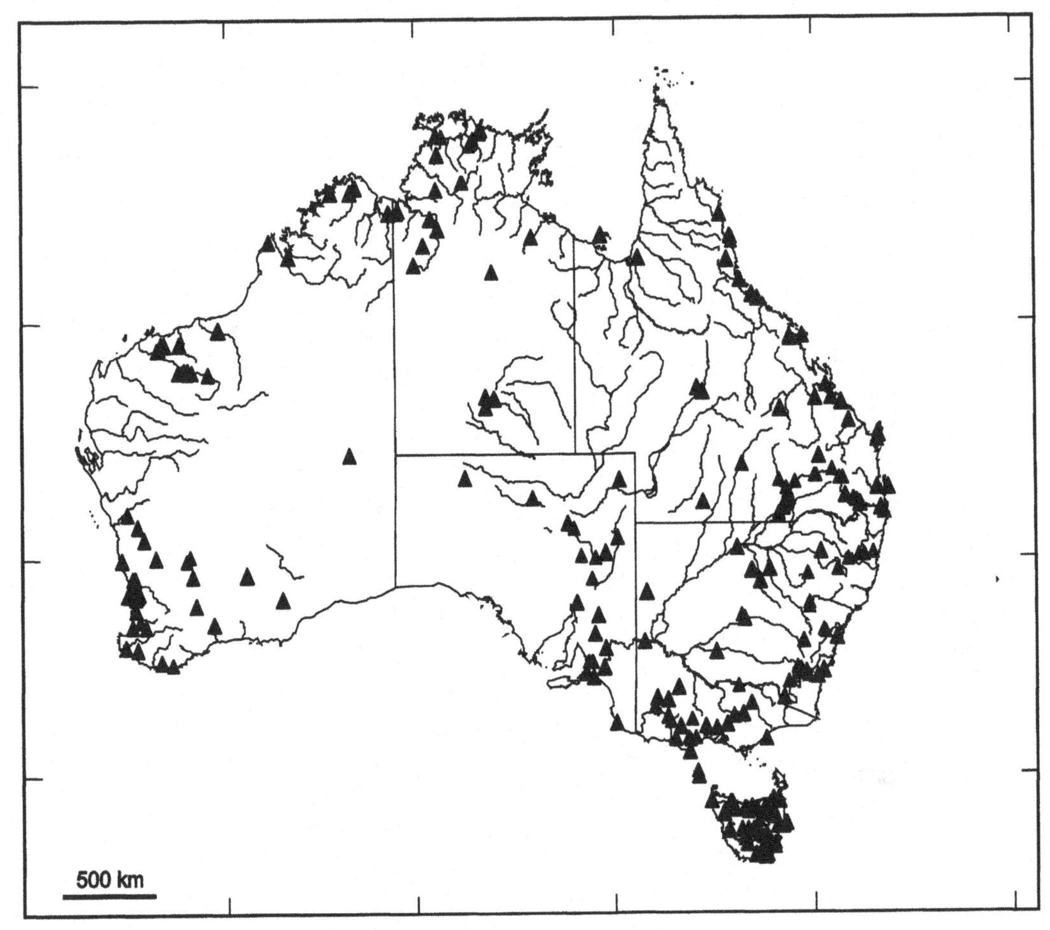

Figure 15.14. CORIXIDAE. Distribution of *Micronecta* species in Australia.

griseola Horvath was found in eutrophic or eutrophicated waters, *M. poweri* (Douglas & Scott), favoured oligotrophic waters, and *M. minutissima* (L.) was intermediate between the other two species. None of the species could survive in heavily polluted waters, but when the situation improved, *M. minutissima* was the species best tolerating pollution (Jansson 1977a, 1977b, 1987).

Most males of *Micronecta* possess a so-called "strigil" located on the right-hand side of the abdominal tergite 6. King (1976) supposed that the strigil was involved in the sound production performed only by males of the Australian species *M. batilla* (= *M. annae*). Bailey (1983) studied the same species and suggested that stridulation took place when two lobes of the abdominal sternite 8 was rubbed together. However, Jansson (1989), studied the stridulatory mechanism more thoroughly in a number of Micronectinae (including the Australian species *M. tasmanica*). He found that in *Micronecta* males, the sound is produced by rubbing a ribbed process on the base of the right paramere against one or two ridges near the median edge of the leftmost lobe of abdominal sternite 8. Thus, the first mentioned structure functions as the *stridulitrum* and the last mentioned structure as a *plectrum* (contrary to J. Polhemus 1994). Sound recordings as well as observations of stridulating males indicate that stridulation in *Micronecta* is a one way movement (unlike that of species belonging to the related South American genus *Tenagobia*).

King (1999a) studied the role of male sounds in *Micronecta concordia, M. robusta,* and *M. tasmanica* in ponds in southeastern Australia. Females were attracted to signals of conspecific males and studies of mating behaviour showed that signals were obligatory for mating. No sounds occurred during copulation. These findings strongly suggest that acoustic signalling is important in reproductive isolation in *Micronecta* (King 1999c). Furthermore, King (1999b) showed that chorusing occurred in *M. concordia* males. Pulse-trains synchronised with those of other individuals nearby were the predominant acoustic output of males throughout the stridulating season, from mid-winter to mid-summer.

The food is unknown for most *Micronecta* species which makes them difficult to keep in aquaria. Fernando & Leong (1963) reared *M. quadristrigata* Breddin on a diet of *Pleurococcus* algae although there was also bottom material from the habitat in the rearing vials. Development from egg to adult took on average 45 days at temperatures varying between 23.2-27.5°C. These authors also reported that the gut contents of various Malaysian *Micronecta* species consisted mainly of various unicellular algae, remnants of filamentous algae and some setae of freshwater annelids in addition to bacteria. The eggs of *Micronecta* differ from those of other corixids (except *Diaprepocoris*; see above) in lacking the stalk and adhesive disk They are glued to the substrate with their flattened side. The egg-shell may be covered with short wart-like processes. Eggs of *Tenagobia*, the American counterpart of *Micronecta*, however, are stalked (Cobben 1968).

Distribution

There are about 50 species in the Indo-Australian region of which 18 occur in Australia; 16 of these are as far as we know endemic to Australia. These are widely distributed in southeastern, southern, and southwestern Australia (including Tasmania), more sparsely recorded from northeastern Queensland, Northern Territory, and the northwestern Western Australia (Fig. 15.14).

16. Family APHELOCHEIRIDAE

Benthic Water Bugs

Identification

Body suboval and strongly flattened. Head produced anteriorly, inserted into pronotum; antennae 4-segmented, elongate and slender (Figs 16.1A, B). Rostrum relatively long, reaching well onto metasternum, segment 2 very short, segment 3 very long, segment 4 less than one-half of segment 3. Fore femora legs not raptorial (as compared to the Naucoridae; see below), fore femur only slightly thickened (Figs 16.1C; 16.2A); hind legs not oar-like, but fringed with swimming hairs (Fig. 16.1D). All tarsi 3-segmented, but basal segment very small; one pair of equally and well developed claws on all tarsi. Wing polymorphism common, the commonest adult form being micropterous (Figs 16.1A, E), the macropterous form has a hemelytron with well developed membrane, without veins (Fig. 16.1B). Metathoracic scent glands absent; dorsal abdominal scent glands divided, found in nymphs between mediotergites 3 and 4, but persistent in adults (Fig. 16.1E; so). Posterior, lateral angles of connexiva often spine-like produced. Venter of thorax and abdomen with a dense layer of micro-hairs (plastron; see below); abdominal spiracles 2-7 uniquely surrounded by so-called "rosettes" (Figs 16.1F, 16.3B; rs) (Larsén 1938; Thorpe & Crisp 1947a); abdominal sternum 2 with a pair of sublateral, plate-shaped sense organs (Fig. 16.1F; se).

Overview

Chen et al. (in press) gave this family the trivial name "benthic water bugs" since these bugs are able to stay submerged in water indefinitely. The Aphelocheiridae has until quite recently usually been considered a subfamily of the Naucoridae (see Chapter 17). Cladistic analyses by Rieger (1976) and Mahner (1993) give support to treating this taxon as a separate family. However, D. Polhemus and J. Polhemus (1988) in their revision of Asian *Aphelocheirus*, including a survey of the taxonomy of the family, still follow Hoberlandt & Stys (1979) in treating this taxon as a subfamily of Naucoridae.

The family contains only one genus, *Aphelocheirus* Westwood, divided into two subgenera of which *Aphelocheirus* (*Micraphelocheirus*) is restricted to the tropical part of continental Asia whereas *Aphelocheirus* (*Aphelocheirus*) is widely distributed in the Old World. The number of described species is about 65 of which 21 occur in the Indo-Australian region (Chen et al. in press). However, the number of species is probably much higher, in particular in the Afrotropical region (including Madagascar) (D. Polhemus and J. Polhemus 1988). Only one species belonging to *Aphelocheirus* (*Aphelocheirus*) occurs in Australia.

APHELOCHEIRUS

Identification

Medium sized, oval, and strongly flattened water bugs (Fig. 16.1A). Length 6.5-11.5 mm.

Inner propleural projections with an apical notch. Wing pads of micropterous form well developed, covering at least the lateral half of the metanotum (Fig. 16.1E; fw). Openings of the abdominal scent glands located closer to the midline of the abdomen than to the lateral margins (Fig. 16.1E; so). Male tergite 5 large, more or less asymmetrically developed, covering most of the distal abdominal segments. Sterna 6-8 slightly asymmetrical. Genital capsule of ♂ prominent; parameres asymmetrical and their shape is species-specific. Female abdomen symmetrical with sternum 8 forming a subgenital plate or operculum (Fig. 16.1F; sg). Only one Australian species, *Aphelocheirus* (*Aphelocheirus*) *australicus* Usinger. Length 6.5-8.75 mm. Northern Territory, Queensland.

Biology

Species of *Aphelocheirus* are well known for their ability to stay submerged in water indefinitely even if they are plastron breathers like other aquatic bugs. Physiological and biological studies were performed by Thorpe & Crisp (1947a, 1947b) on the widespread European species *Aphelocheirus aestivalis* (Fabricius). Their findings were essentially substantiated by Hinton (1976) using Scanning Electron Micrographs similar to those published by Lansbury (1991b) and reproduced in this book (Fig. 16.2). Hinton (1976) reported that body surface of *Aphelocheirus* is covered by a dense pile of micro-hairs (Figs 16.2E, F, 16.3A), each about 1.5µm long and 0.4µm wide at base; the tip of each hair is bent approximately at a right angle to its base. These micro-hairs occur at densities of about 4 million per mm^2 over most of the body surface of the bug. Using these data, Thorpe & Crisp (1947a) calculated that the hair-pile would buckle when subjected to pressures from 2.5 to 6 atmospheres. However, more refined calculations made by Hinton (1976) yield-

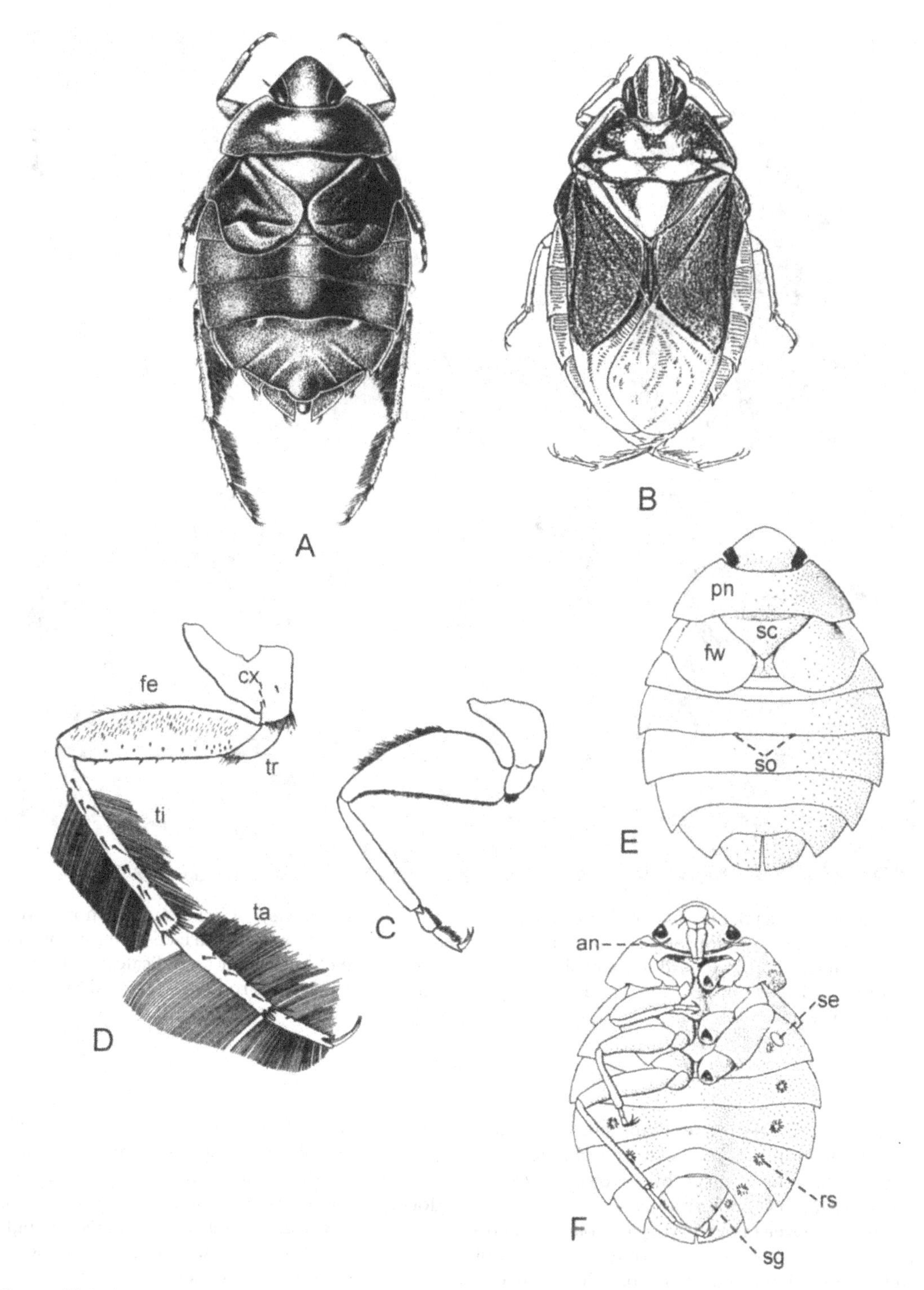

Figure 16.1. APHELOCHEIRIDAE. A-D, *Aphelocheirus australicus*: A, dorsal habitus of micropterous form. B, dorsal hbaitus of macropterous form. C, fore leg. D, hind leg (cx, coxa; fe, femur; ta, tarsus; ti, tibia; tr, trochanter). E-F, *A. aestivalis*, micropterous ♀. E, dorsal view; antennae and legs omitted (fw, fore wing rudiment; pn, pronotum; sc, scutellum; so, abdominal scent orifices). F, ventral view; legs except coxae of left side omitted (an, antenna; rs, spiracle with rosette; se, sense organ; sg, subgenital plate). Reproduced from: A, CSIRO (1991); B, Usinger (1937); C-D, Lansbury (1991b); E-F, Parsons (1969).

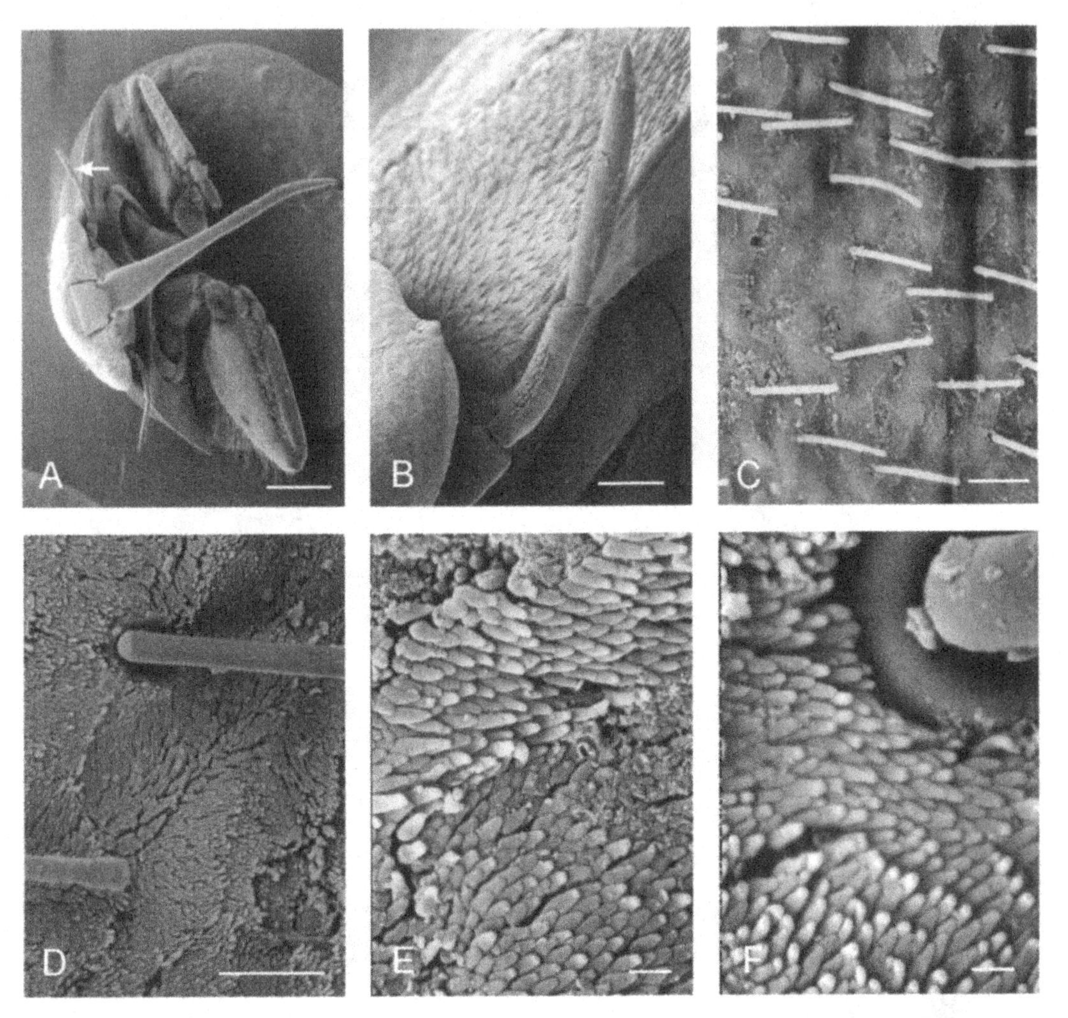

Figure 16.2. APHELOCHEIRIDAE. A-F, *Aphelocheirus australicus*: A, ventral view of head and prothorax with antenna (arrow) and fore legs. B, antenna. C, surface structure of thorax. D, surface structure of thorax with micro-hair layer and macro-hairs. E-F, detailed structure of micro-hair layer. Scales: A, 0.5 mm; B, 0.1 mm; C, 0.02 mm; D, 0.01 mm; E-F, 0.001 mm. Scanning Electron Micrographs (Oxford Museum, Oxford, U.K.).

ed a figure 16 times higher than that. This means that the hair-pile of *Aphelocheirus* is able to resist wetting and mechanical breakdown at excess pressures up to 40 atmospheres.

The thin layer of air held by the pile of micro-hairs is contiguous with the ventral spiracles of *Aphelocheirus*. However, where spiracles of normal structure would either collapse under the water pressure or become waterlogged, the spiracular openings are expanded to form a flat bag with numerous outgrowths, the so-called *rosette* (Figs 16.1F, 16.3B; rs). The rosette is perforated by minute pores on its arms through which oxygen reaches the tracheae from the thin layer of air held by the hair-pile. In addition there are many small holes (diameter 7-8m) on the abdominal dorsum and venter and the head (Messner et al. 1981). These lead to narrow channels through the cuticle ending in air sacs under the cuticula which are lined with a membrane. Tracheoles connect these air sacs with the rest of the tracheal system. The air film covering the body of *Aphelocheirus* is self-renewing since oxygen continually diffuses in and nitrogen out, the whole forming a plastron, or external physical gill. In nymphs, as with many other water bugs, respira-

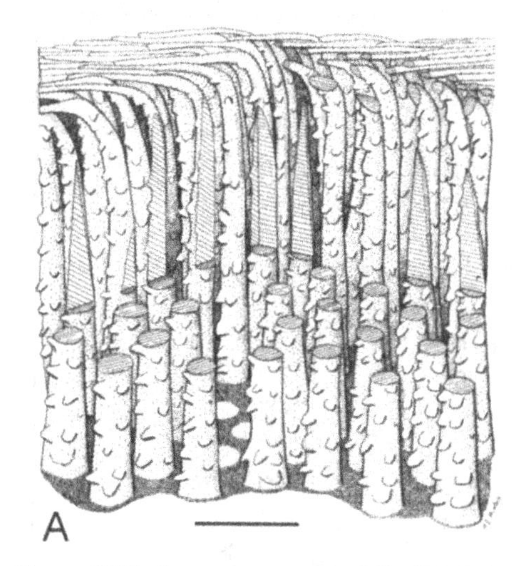

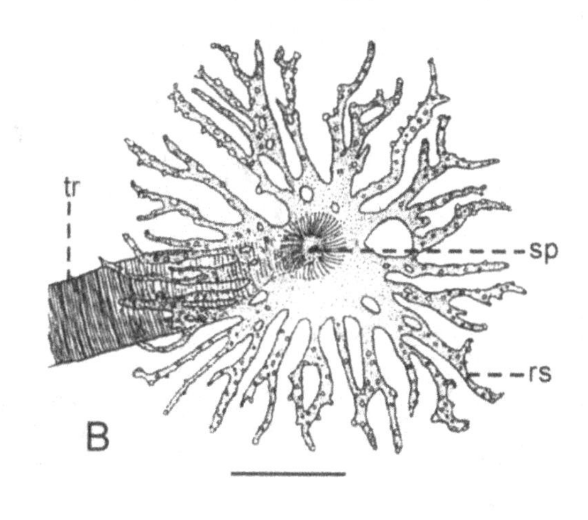

Figure 16.3. APHELOCHEIRIDAE. A-B, *Aphelocheirus aestivalis*: A, reconstruction of plastron. B, rosette-like spiracle (rs, small ducts of rosette; sp, spiracular opening; tr, trachea). Scales: A, 0.001 mm; B, 0.2 mm. Reproduced with modification from: A, Hinton (1976); B, Thorpe & Crisp (1947a).

tion takes place directly through the cuticle (cutaneous respiration).

The very efficient plastron of *Aphelocheirus* enables these bugs to complete their entire life cycle under water. The eggs are nearly cylindrical with broadly rounded ends and with a distinct hexagonal sculpture. They are glued to larger substrates like stones or mollusc shells. They may be deposited singly or in clusters or strings. In *A. aestivalis* the embryonic development takes 9-10 weeks or more and the development from first instar nymph to adult takes two years, at least in Sweden (Larsén 1927, 1932). In tropical and subtropical areas the development probably will be much faster.

Aphelocheirus species are among the few true aquatic bugs (Nepomorpha) which have a micropterous adult form, with small but distinct wing rudiments (Fig. 16.1E; fw). In most other water bugs the hemelytra (fore wings) have the important function (besides being organs of flight) of holding the subelytral air-stores no longer crucial for a permanent plastron-breather. *Aphelocheirus* does have a small air store under the wing rudiments, however, which is in contact with the so-called scolopophorous (or chordotonal) organs which are either organs of balance or pressure receptors (or perhaps both, Larsén 1957, Parsons 1969). Adult *Aphelocheirus* also have a pair of static organs on the abdominal sternum 2, situated laterally of the rosettes (Fig. 16.1F; se). Each of these consists of an oval sensory plate devoid of micro-hairs of the body surface, but covered with a large number of much larger hydrofuge hairs. These organs are assumed to be pressure receptors reacting on absolute increase or decrease of pressure in the water. When one or both of the sense organs is destroyed and the bug placed at the surface of an aquarium in the dark, the bug swims either in spirals or erratically whereas control bugs with both organs intact swim straight to the bottom (Thorpe & Crisp 1947b).

The macropterous adult form of *Aphelocheirus* (Fig. 16.1B) is generally thought of as being rare. However, *A. australicus* was described on specimens of this form (Usinger 1937) and Lansbury (1985b) only recorded macropterous specimens from Queensland, most of which were taken at light. Later, Lansbury (1991b) found micropterous specimens in one locality in Northern Territory.

Aphelocheirus species are nearly always found in well aerated streams although they may occur in areas of low current and occasionally in still side channels or pools. Lansbury (1985b) reported that the Australian species, *A. australicus*, was found crawling about beneath rock and stones, e.g., at The Boulders via Babinda near Cairns, northern Queensland. In Europe, *A. aestivalis* has been found a few times in lakes and two African species have also been reported from lakes. For demonstration or teaching purposes they can be kept alive in the laboratory for several days in dishes (Chen et al. in press). In streams, *Aphelocheirus* populations tend to have higher densities

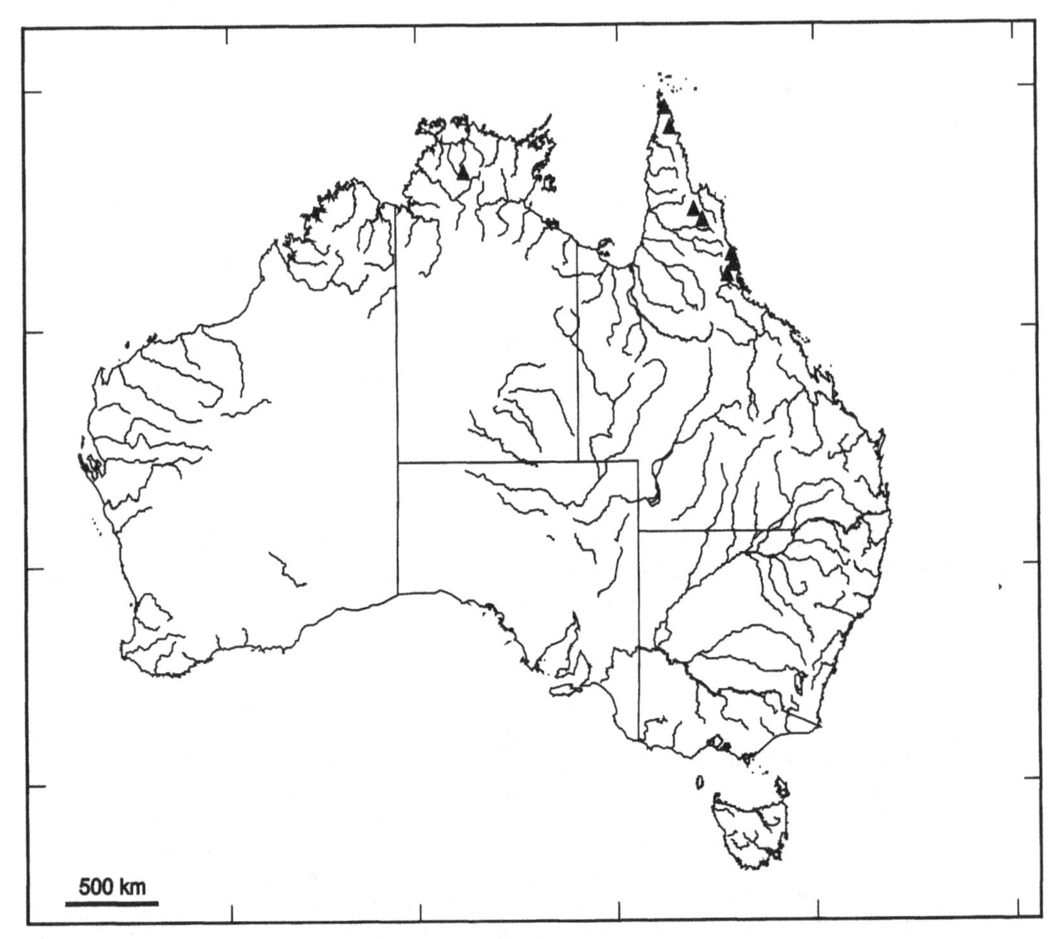

Figure 16.4. APHELOCHEIRIDAE. Distribution of *Aphelocheirus australicus* in Australia.

in areas exposed to sunshine as compared to shaded stretches. Food consists mainly of juveniles of other aquatic insects such as may-flies (Ephemeroptera) and caddis-flies (Trichoptera). *Aphelocheirus* may be useful in controlling larvae of black flies (Diptera: Simuliidae) living in much the same type of localities (Sites 2000). In daytime *Aphelocheirus* hide under pebbles or in the sand of the stream bed, at night they are more active and crawl over the bottom. This type of behaviour makes these bugs difficult to catch and until recently *Aphelocheirus* species were thought to be rare. However, in tropical Asia, they are quite common in streams, especially in the mountains (D. Polhemus & J. Polhemus 1988). The best way of collecting these bugs is to stir up the bottom substrate in front of the net facing upstream. When not handled carefully *Aphelocheirus* may occasionally "sting" or bite, which can be quite painful to man.

Distribution

Aphelocheirus australicus is an Australian endemic. Records from Queensland are from the Cairns area north to the Lockerbie area of Cape York (Lansbury 1985b; D. Polhemus & J. Polhemus 1988). Subsequently, Lansbury (1991b) found the species in South Alligator River, Northern Territory (Fig. 16.4).

17. Family NAUCORIDAE

Creeping Water Bugs, Saucer Bugs

Identification

Body suboval to slightly elongate, flattened. Eyes often overlapping antero-lateral angles of pronotum. Antennae 4-segmented, short, concealed. Rostrum very short and stout (Fig. 17.2A, B). Fore legs raptorial, fore femora conspicuously enlarged (Fig. 17.2C); fore tarsus usually fused with tibia, 1- or 2-segmented, with 1 or 2 claws or without claws. Middle and hind tarsi with 2 distinct segments (basal segment very small or absent), with 2 equally developed claws. Hind legs often modified for swimming. Membrane of forewing without venation. Metathoracic scent glands present; nymphal abdominal scent gland present between terga 3 and 4. Postero-lateral angles of abdominal connexiva sometimes produced. Some taxa with paired sublateral sense organs on abdominal sternum 2.

Overview

One of the largest families of true aquatic bugs. The most recent classification (Stys & Jansson 1988) includes five subfamilies, 40 genera, and more than 400 species. The family is distributed world-wide, but is most species-rich in tropical countries. Only one genus belonging to the subfamily Naucorinae is represented in the fauna of the Australian continent and Tasmania. Two genera of the subfamily Laccocorinae, *Heleocoris* Stål and *Laccocoris* Stål (with one species each), are recorded from Christmas Island and therefore included in the catalogue by Cassis & Gross (1995). These are not considered here.

Most creeping water bugs breathe through the cuticle as nymphs, and through the spiracles in contact with an air-store as adults. The air-store under the hemelytra is replenished by breaking the surface film with the tip of the abdomen (Fig. 2.2, Plate 8, F). This air-store is connected with layer of air which appears as a silvery sheen on the pubescent venter (Polhemus 1979). The most probable connections between the two air-stores are the first abdominal spiracular chamber and the lateral edges of abdominal segment 2 (Parsons 1970). The exposed ventral air-store, with its large surface, acts as a physical gill, which transfers oxygen in and carbon dioxide out, thus using the dissolved oxygen in the water as a secondary air source (Thorpe 1950).

NAUCORIS

Identification

Small to medium sized water bugs, length 5.5-15 mm, with a flat or very slightly convex dorsum (Fig. 17.1A). Anterior margin of head only slightly turned down and backward; inner margins of eyes converging anteriorly. Rostrum very short, originating at or near the anterior margin of the head (Figs 17.2A, B). Anterior margin of pronotum not excavated behind the head; lateral margin of pronotum without a sublateral groove. Fore tarsi fused to the tibia although the suture is still visible, 1-segmented with a minute claw (Fig. 17.2C). Middle and hind femur smooth on ventral side, without rows of bristles or spines.
Middle and hind tibia not widened, hind tarsus shorter than hind tibia. Male parameres well developed.

The genus *Naucoris* Geoffroy has about 20 species, predominantly found in the tropical and subtropical regions of the Old World. Ten of these occur in Africa and eleven in the Indo-Australian region. Six species have been recorded from Australia. The following key is adapted from Lansbury (1985b). [Note: the habitus figure in CSIRO (1991: fig. 30.54C) is probably *N. australicus*, not *N. congrex* as stated].

Key to Australian species

1. Posterior margin of pronotum not produced caudad at humeral angles (Fig. 17.A, C). Venter either appearing bare or with fine short hairs 2
- Posterior margin of pronotum produced caudad at humeral angles (Figs 17.1J, L). All abdominal sterna clothed with fine golden pubescence 5
2. Scutellum and most of embolium pale yellow (Fig. 17.1A). Length 8.0-9.6 mm. Queensland.............................. *australicus* Stål
- Scutellum unicolorous dark reddish brown or black; embolium basally variably pale yellow to yellowish brown; apically always dark brown to black (Fig. 17.1C) 3
3. Posterior angles of connexival segments 4-6 sharply produced (Fig. 17.1F). Shining area of ventral laterotergite 3 basally broader than anterior width of laterotergite 4 and broadly infuscate (Fig. 17.1G). Length 6.2-6.6 mm. Northern Territory.. .... *rhizomatus* J. Polhemus

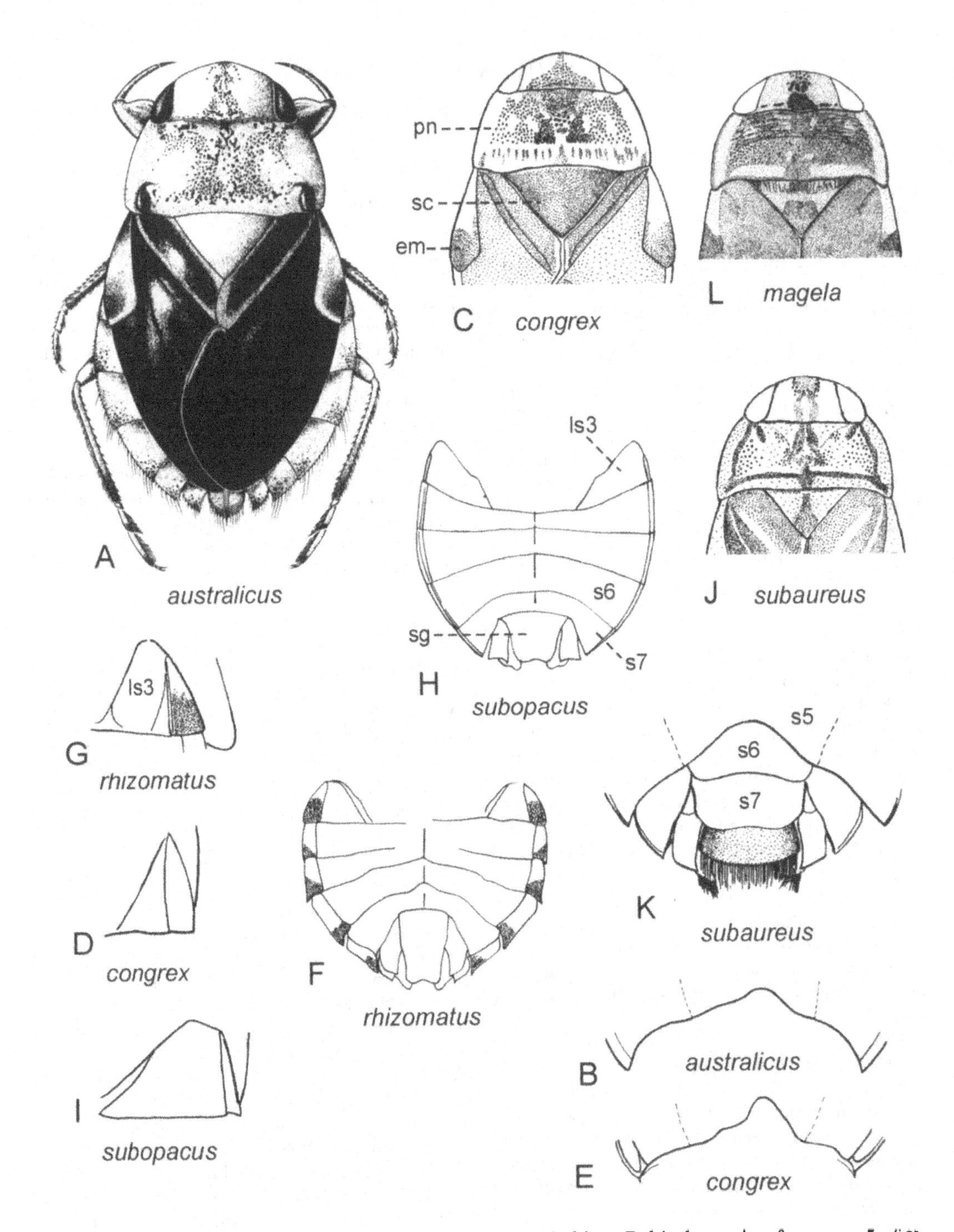

Figure 17.1. NAUCORIDAE. A-B, *Naucoris australicus*: A, dorsal habitus. B, hind margin of sternum 5 of ♂. C-E, *N. congrex*: C, dorsal view of the anterior end of ♀ (em, embolium; pn, pronotum; sc, scutellum). D, lateral part of abdominal sternum 3. E, hind margin of sternum 5 of ♂. F-G, *N. rhizomatus*: F, ventral view of abdomen of ♀. G, lateral part of abdominal sternum 3 (ls3, laterosternite 3). H-I, *N. subopacus*: H, ventral view of abdomen of ♀ (ls3, laterosternite 3; s6, s7, sterna 6 and 7; sg, subgenital plate). I, lateral part of abdominal sternum 3. J-K, *N. subaureus*: J, dorsal view of anterior end of ♂. K, ventral view of abdominal end of ♂ (s5, s6, s7, sterna 5-7). L, *N. magela*, dorsal view of anterior end of ♀. Reproduced with modification from: A, CSIRO 1991; B-K, Lansbury 1985b; L, Lansbury 1991b.

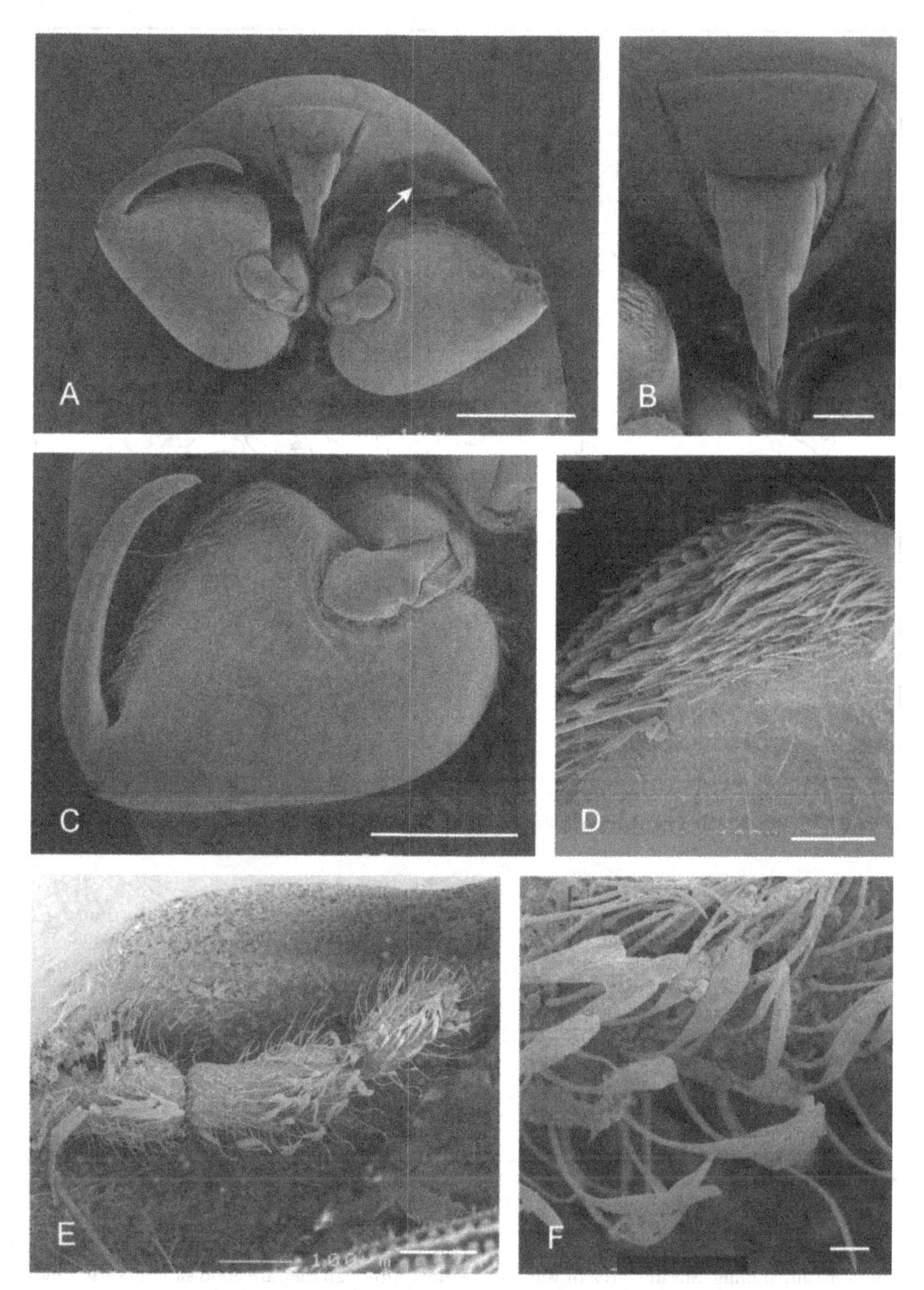

Figure 17.2. NAUCORIDAE. A-F, *Naucoris congrex*. A, Ventral view of head and prothorax with concealed antenna (arrow). B, ventral view of rostrum. C, right fore leg. D, base of femur. E, antenna. F, spatulate setae on antenna. Scales: A, 1 mm; B, 0.2 mm; C, 0.5 mm; D-E, 0.01 mm; F, 0.001 mm. Scanning Electron Micrographs (CSIRO Entomology, Canberra).

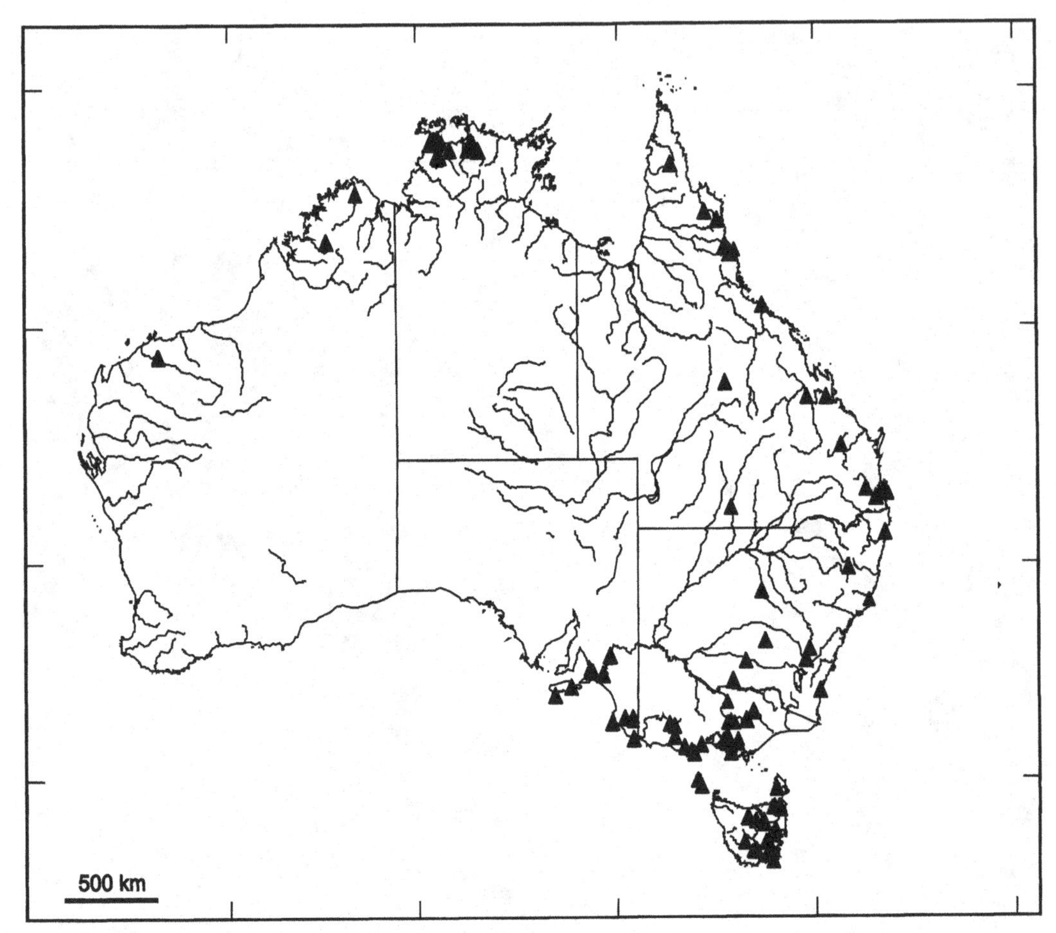

Figure 17.3. NAUCORIDAE. Distribution of *Naucoris* spp. in Australia.

– Posterior angles of connexival segments 4-6 not sharply produced (Fig. 17.1H). Shining area of ventral laterotergite 3 not basally broader than anterior width of laterotergite 4 and not broadly infuscate (Figs 17.1D, I) 4

4. Dorsum of head and pronotum rugulose and rather dull. Shining area of ventral laterotergite 3 narrow (Fig. 17.1I). Posterior margin of sternum 5 of ♂ only slightly asymmetrical (as in Fig. 17.1B). Length 6.6-6.9 mm. Northern Territory, Queensland, Western Australia .... *subopacus* Montandon

– Dorsum of head and pronotum smooth and usually shining. Shining area of ventral laterotergite 3 broad (Fig. 17.1D). Posterior margin of sternum 5 of ♂ distinctly asymmetrical (Fig. 17.1E). Length 7.7-9.0 mm. Australian Capital Territory, New South Wales, Northern Territory, Queensland, South Australia, Tasmania, Victoria .... *congrex* Stål

5. Larger species, length 7.0-7.8 mm. Corium and clavus with confused pattern of yellowish brown lines; membrane distinct. Western Australia..... *subaureus* Lansbury

– Smaller species, length 6.5-7.2 mm. Corium and clavus without such a pattern; membrane not differentiated from corium. Northern Territory...... *magela* Lansbury

Biology

Most *Naucoris* species occur in stagnant waters or in sluggish backwaters of streams with rich vegetation. Although they are good swimmers they hide between the vegetation until prey comes near rather than chase their prey. *N. rhizomatus* has, as far as is known, an aberrant habitat preference, clinging to the roots of plants at the edge of a stream (Polhemus 1984). The biology of Australian species of *Naucoris* is not well known. Much better known are the biology of some species of North American naucorids (genus *Pelocoris* Stål) and, in particular, the northern European *Ilyocoris cimicoides* (L.) (formerly placed in *Naucoris*). This species is remarkable in that it possess fully developed wings (hemelytra as well as hind wings), but such reduced musculature that no specimens yet examined have been able to fly (Larsén 1950). Southwood & Leston (1959) suggested that migration occurs by nocturnal walking. The bug can remain beneath water for a long time but surfaces mechanically under the effect of oxygen shortage.

Creeping water bugs are strong predators and will feed on water fleas, amphipods, isopods, and various aquatic insects (Plate 8, F). McCoull et al. (1998) studied the effect of temperature on the functional response and components of attack rate in *Naucoris congrex* preying on final instar mosquito larvae. The authors found that attack rate increased with temperature to a maximum at 20°C whereas handling time was negatively correlated with temperature across the entire range investigated. The frequency of reactive encounters was significantly greater at higher temperatures, but was unrelated to prey density. Creeping water bugs such as *Pelocoris* spp. and *Ilyocoris cimicoides* are notorious for inflicting a painful "bite" which Wesenberg-Lund (1943) characterised as "worse than a bee sting".

Copulation in *Ilyocoris cimicoides* was studied by Larsén (1938) who showed that the male typically is astride the back of the female with his genital segments strongly twisted and extended to engage the female. The egg is embedded in plant tissue under water whereas that of the European species *Naucoris maculatus* Fabricius is attached superficially to the substrate (Lebrun 1960). Both species have a lid or pseudoperculum at the anterior end of the egg which is penetrated by four micropyles (Cobben 1968). Both species overwinter as adults which lay eggs in early spring. The egg development takes about one month and the five nymphal instars a further two months (Southwood & Leston 1959).

Distribution

Australia has as many species of *Naucoris* as the rest of the Indo-Australian region. The most common species, *N. congrex*, is widely distributed in the Southeast (including Tasmania), other species are more sparsely distributed in northern New South Wales, Queensland, Northern Territory, and northern Western Australia (Fig. 17.3).

18. Family OCHTERIDAE

Velvety Shore Bugs

Identification

Small or medium-sized bugs, total length from 3.4 to 9.8 mm, broadly oval, moderately dorso-ventrally flattened with a velvety hair-pile on dorsum. Eyes large, with inner margins emarginated. Antennae 4-segmented, shorter than head and inserted below the eyes (Figs 18.3A; an), but usually not completely concealed in dorsal view. Ocelli present. Rostrum (ro) very long and slender, reaching the hind coxae; segment 3 much longer than other segments. All legs slender and adapted for walking (cursorial), without swimming hairs; fore legs not raptorial. Tarsal formula 2: 2: 3. Membrane of hemelytra well developed with either 7 or c. 20 cells, but without anastomosing veins. Abdominal end of ♂ almost symmetrical; sterna 7-8 partly retracted into sternum 6 (Fig. 18.5A; s6-s8). Pygophore (= segment 9; py) partly protruding. Parameres of male asymmetrically developed, left paramere reduced.

Overview

This predominantly tropical family has 3 genera with some 60 described species, nearly all belonging to the cosmopolitan genus *Ochterus* Latreille. Of the two smaller genera, the monotypic genus *Ocyochterus* Drake & Gómez-Menor is found in South America and the genus *Megochterus* Jaczewski (1934), with two species, is endemic to Australia.

Ochterids are notable for their similarity to the Saldidae (Leptopodomorpha) or true shore bugs, but they share all essential nepomorphan characters, including the relatively short, ventrally inserted antennae, asymmetrical male genitalia and abdominal segments, tubular spermatheca in the female, and the absence of cephalic trichobothria (Schuh & Slater 1995). The body surface of ochterids has peg-plates (Fig. 18.4G; called sieve-pores by Cobben 1978) of exact same structure as those found in several groups of Gerromorpha (e.g., *Mesovelia*; Fig. 7.3E). They are also of the same size in the two groups. Based upon Cobben (1978: fig. 173B), Andersen (1982: 59) erroneously stated that the peg-plates found in *Ochterus* were about ten times bigger than those of *Mesovelia*. The trumpet-shaped microtrichia (Fig. 18.4F) are remarkably similar to structures found in *Hebrus* (see Andersen 1982: plate 4, G). An extensive account of the morphology of the family is provided by Rieger (1976).

Key to the Australian genera of the family Ochteridae

1. Large species, 7.8-9.8 mm. Frontal plate strongly produced above base of rostrum (Fig. 18.1A). Antennae short and stout (Figs 18.1B, G). Membrane of hemelytra with c. 20 cells (Fig. 18.1F) .... *Megochterus* Jaczewski
- Smaller species, less than 6.5 mm. Frontal plate at most slightly produced above base of rostrum (Fig. 18.3A). Antennae, especially segments 3 and 4, elongate (Fig. 18.3A; an). Membrane of hemelytra with only 7 cells (Fig. 18.3C) .... *Ochterus* Latreille

MEGOCHTERUS

Identification

Medium-sized, broadly oval bugs, length from 7.8 mm to 9.8 mm. Frontal plate of head distinctly produced above base of rostrum (Fig. 18.1A). Antennae short and stout (Figs 18.1B, G). Membrane of hemelytra with c. 20 cells (Fig. 18.1F).

The genus *Megochterus* Jaczewski is only found in Australia with two species which are separated by the key below (adapted from Baehr 1990b).

Key to Australian species

1. Frontal plate longer than wide (Fig. 18.1C), apically rather smooth, not pilose. Antennae longer, segment 3 more than 0.33x length of segments 1 and 2 together (Fig. 18.1B). Rostrum clearly surpassing metacoxa. Middle of posterior border of pronotum, and apex of scutellum narrowly yellow (Fig. 18.1A). Apex of subgenital plate of ♂ square (Fig. 18.1D). Apex of right paramere strongly hooked and deeply notched below apical hook (Fig. 18.1E). Length 8.3-9.8 mm. New South Wales, Queensland, Tasmania .... *nasutus* (Montandon)
- Frontal plate shorter, semicircular (Fig. 18.1H), apically uneven, pilose at least in median furrow. Antennae shorter, segment 3 bare longer than segment 2 (Fig. 18.1G). Rostrum reaching middle of metacoxa. Middle of posterior border of pronotum, and apex of scutellum black. Apex of subgenital plate of ♂ oblique and somewhat pointed (Fig. 18.1I). Apex of right paramere less strongly hooked,

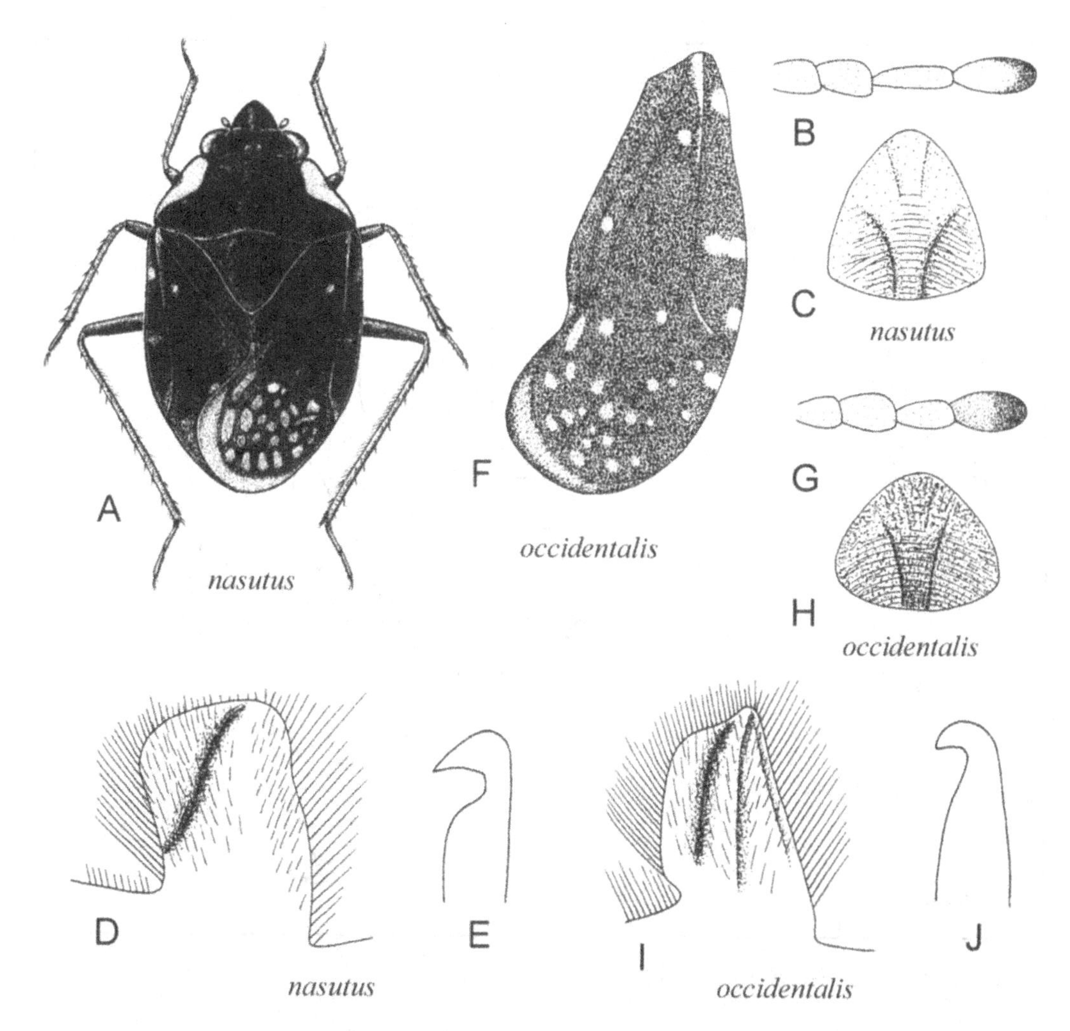

Figure 18.1. OCHTERIDAE. A-E, *Megochterus nasutus*: A, dorsal habitus. B, antenna. C, frontal plate. D, subgenital plate of ♂. E, apex of right paramere. F-J, *M. occidentalis*: F, hemelytron. G, antenna. H, frontal plate. I, subgenital plate of ♂. J, apex of right paramere. Reproduced with modification from: A, CSIRO (1991); B-J, Baehr (1990a).

not deeply notched below apical hook (Fig. 18.1J). Length 7.8-8.8 mm. Western Australia................................ *occidentalis* Baehr

Biology

Virtually unknown. Baehr (1990b) collected specimens of *Megochterus occidentalis* in bright sunlight on black, muddy soil of the banks of roadside pools in swampy heathland not far from the coast. The first specimens did not turn up before noon and they generally avoided shaded places. they ran with great agility and flew short distances when disturbed. On the black background they were extremely difficult to catch. Habitat of *M. nasutus* not known, but presumably on wet ground by lakes and pools near coast. On North Stradbroke Island found near coastal freshwater lagoon, in Tasmania found inland, perhaps in swampy sedgeland.

Distribution

This genus is endemic to Australia. *Megochterus nasutus* is sparsely recorded from southeastern Queensland, New South Wales, and Tasmania, *M. occidentalis* from southwestern Western Australia (Fig. 18.2).

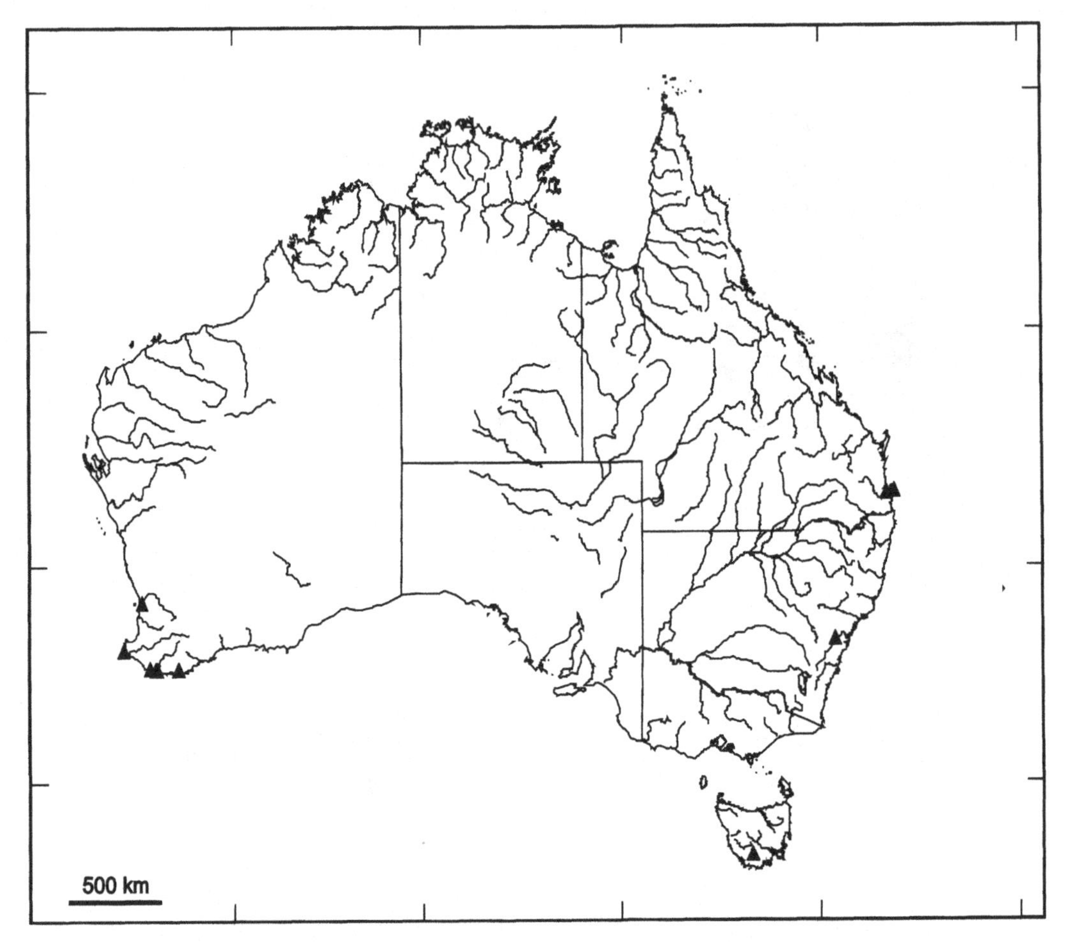

Figure 18.2. OCHTERIDAE. Distribution of *Megochterus* spp. in Australia.

OCHTERUS

Identification

Small, broadly oval bugs (Fig. 18.4A), length less than 6.5 mm. Head declivent (Fig. 18.4C), frontal plate of head not produced above base of rostrum (Fig. 18.3A). Antennae, especially segments 3 and 4, elongate (Fig. 18.4D). Membrane of hemelytra with 7 cells in two rows. Male pygophore (Fig. 18.5A; py) withdrawn into pregenital abdomen; the shape of its posterior end species-specific (Figs 18.5, C, E, G, I. Right paramere of ♂ with two appendices in Australian and other Old World species (Figs 18.5D, F, H, J, K-N).

The genus *Ochterus* Latreille is world-wide distributed with about 55 species of which about 30 occur in the Indo-Australian region. The nine Australian species were revised by Baehr (1989, 1990a). The following key is adapted from Baehr (1990a).

Key to Australian species

1. Frontal plate of head densely covered with fine hairs. Appendices of right paramere falciform, conspicuously serrate outside (Figs 18.5L, M) 2
- Frontal plate of head not covered with hairs. Appendices of right paramere not falciform nor conspicuously serrate 3
2. Appendices of right paramere longer, more curved (Fig. 18.5M). Hemelytral colour-pattern more vivid (Fig. 18.3G). Length 4.4-5.25 mm. Western Australia .... *secundus pseudosecundus* Baehr
- Appendices of right paramere shorter, less curved (Fig.18.5L). Hemelytral colour-pattern less vivid (Fig. 18.3H). Length 4.4-5.45 mm. New South Wales, Queensland, South Australia, Victoria, Western Australia.. *secundus secundus* Kormilev

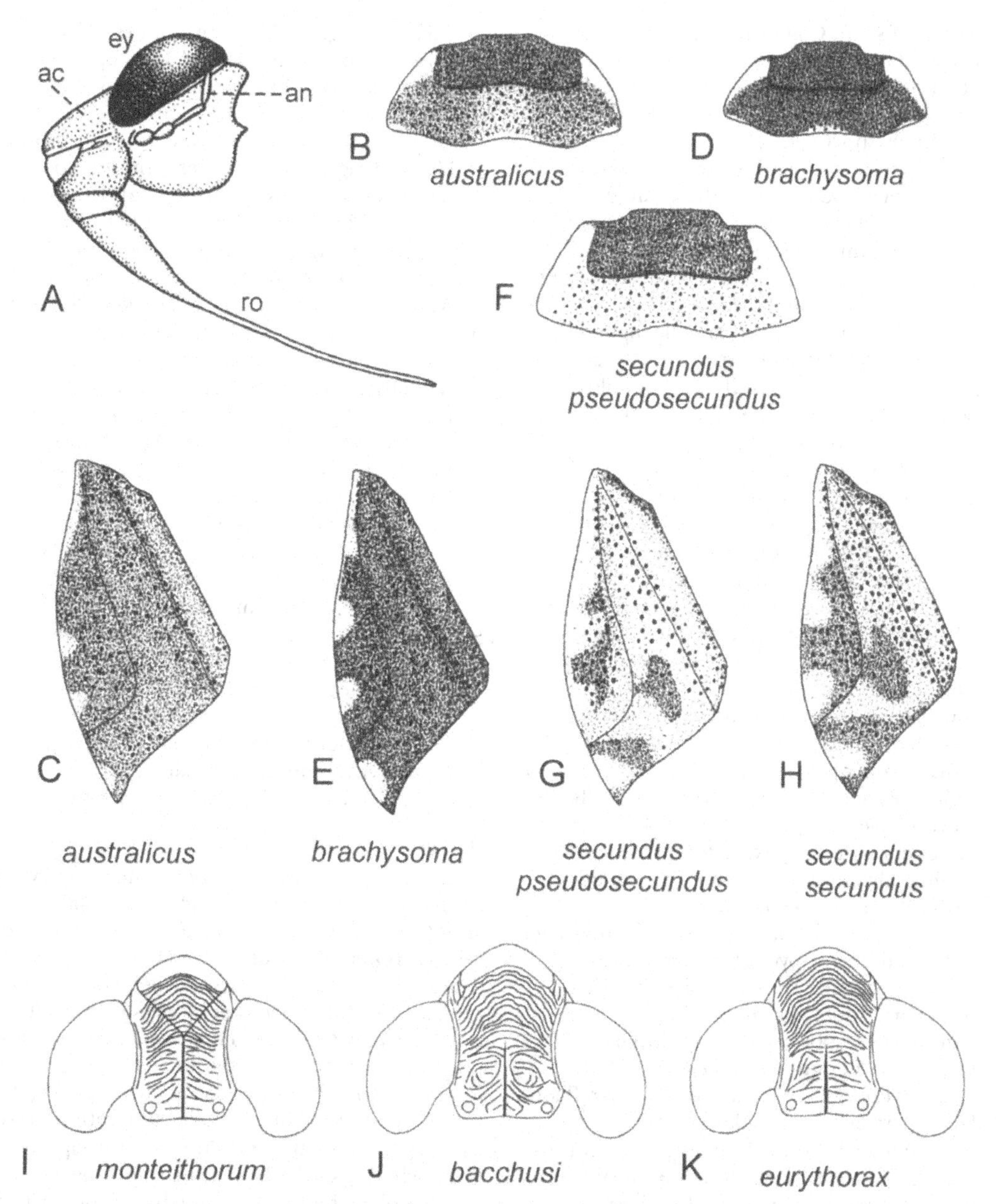

Figure 18.3. OCHTERIDAE. A, *Ochterus* sp., lateral view of head (ac, anteclypeus; an, antenna; ey, eye; ro, rostrum). B-C, *O. australicus*: B, pronotum. C, colour-pattern of hemelytron. D-E, *O. brachysoma*: D, pronotum. E, colour-pattern of hemelytron. F-G, *O. secundus pseudosecundus*: F, pronotum. G, colour-pattern of hemelytron. H, *O. secundus secundus*: colour-pattern of hemelytron. I, *O. monteithorum*, frontal area and clypeus. J, *O. bacchusi*, frontal area and clypeus. K, *O. eurythorax*, frontal area and clypeus. Reproduced with modification from: A, Popov (1971); B-H, Baehr (1989); I-K, Baehr (1990b).

3. Apical border line of male abdomen deeply incised. Appendices of right paramere very elongate, almost 0.4x length of shaft (Fig. 18.5N). Black, lateral border of corium with three very contrasting yellow spots (Fig. 18.3E). Length 3.45-3.85 mm. Queensland .... *brachysoma* Rieger

– Apical border line of male abdomen not incised. Appendices of right paramere less elongate, never longer than 0.33x length of shaft. Colour-pattern of corium not as above 4

4. Shaft of right paramere with strongly enlarged subapical lamella, appendices extremely short (Fig. 18,5H). Pygophore of ♂ excised at apex (Fig. 18.5G). Yellow border of clypeus very wide, triangularly prolonged onto frons (Fig. 18.3I). Length 3.8 mm. Queensland .... *monteithorum* Baehr
- Shaft of right paramere widened or not, though without enlarged lamella; appendices normally longer. Apex of ♂ pygophore straight or roundly acute. Yellow border of clypeus less wide, not pro longed onto frons 5

5. Head of right paramere less projecting, apex rounded; appendices short, very wide, axe-shaped, apex slightly crenulate (Fig. 18.5J). Pygophore of ♂ short and very wide, apex rather straight, lobes close to apex (Fig. 18.5I) 6
- Head of right paramere projecting, acute; appendices longer, less wide, not axe-shaped. Pygophore of ♂ longer, narrower, apex convex, sinuate or acutely rounded; lateral lobes, if present, far removed from apex 7

6. Apical border of pygophore of ♂ slightly more convex. Appendices of right paramere less wide at apex. Lateral pale spot on hemelytra wide, subcircular. Length 3.65-4.25 mm. Queensland .... *baehri baehri* Rieger
- Apical border of pygophore of ♂ rather straight (Fig. 18.5I). Appendices of right paramere wider at apex (Fig. 18.5J). Lateral pale spot on hemelytra narrow, transverse. Length 3.5-4.05 mm. Northern Territory .... *baehri riegeri* Baehr

7. Hemelytra black, only lateral border at base yellow. Posterior lobe of pronotum dark piceous or black, little contrasting from anterior lobe. Length 3.35-3.5 mm. Queensland .... *atridermis* Baehr
- Hemelytra with several distinct yellow spots at lateral border (Fig. 18.3C), or piceous with a light spot on disk. Posterior lobe of pronotum considerably paler than anterior lobe, or with contrasting paler median area (Fig. 18.3B) 8

8. Head of right paramere triangular, apex acute (Figs 18.5D, F) 9
- Head of right paramere not triangular, apex strongly rounded (Fig. 18.5K) 10

9. Head of right paramere highly convex, both appendices at apex very wide and excised (Fig. 18.5D). Pygophore of ♂ with well projecting lateral lobe on left side only (Fig. 18.5C). Length 3.7-4.55 mm. Queensland, South Australia .... *australicus* Jaczewski
- Head of right paramere less convex, appendices at apex narrower and less excised (Fig. 18.5F). Pygophore of ♂ with well projecting lateral lobe on both sides (Fig. 18.5E). Length 3.8-4.4 mm. Western Australia .... *occidentalis* Baehr

10. Oval, convex species with short hemelytra. Pronotum much narrower than hemelytra, anteriorly narrower. Anterior profile of head strongly impressed, frontal ridges few, coarse (Fig. 18.3J). Punctures of hemelytra very coarse. Head of right paramere very low (Fig. 18.5K). Length 3.75 mm. New South Wales .... *bacchusi* Baehr
- Elongate, less convex species with longer hemelytra. Pronotum barely narrower than hemelytra, anteriorly wider. Anterior profile of head less impressed, frontal ridges numerous, finer (Fig. 18.3K). Punctures of hemelytra finer. Head of right paramere more convex. Length 4-4.8 mm. Queensland .... *eurythorax* Baehr

Biology

Most of our knowledge about the biology of ochterids is based on the widespread Palearctic and Oriental species *Ochterus marginatus* (Latreille) (Boulard & Coffin 1991) and the North American species *O. banksi* Barber (Bobb 1951; Menke 1979c) *Ochterus* species are shore dwellers, not true aquatic bugs. They usually live at the margins of running freshwater and are mostly found at sandy or stony places with little vegetation (Plate 8, E). Their cryptic colouration make them nearly invisible, but they can run rapidly or jump when disturbed. At least some species are capable of flight and fly immediately away when they detect movements. Chen et al. (in press) therefore recommend: "to collect them one should look around quietly at appropriate localities. When you see one you should slap a light net over the animal to the ground, they will fly up into the net. Alternatively one can try to put with slow movement a glass tube over the animal".

Apart from being carnivores, the biology of adults is poorly known. The eggs are suboval in cross-section, one side somewhat more convex than the other (Fig. 2.8F). They are laid singly on sand grains and debris along the shore. The nymphs are sluggish and cover their dorsum with

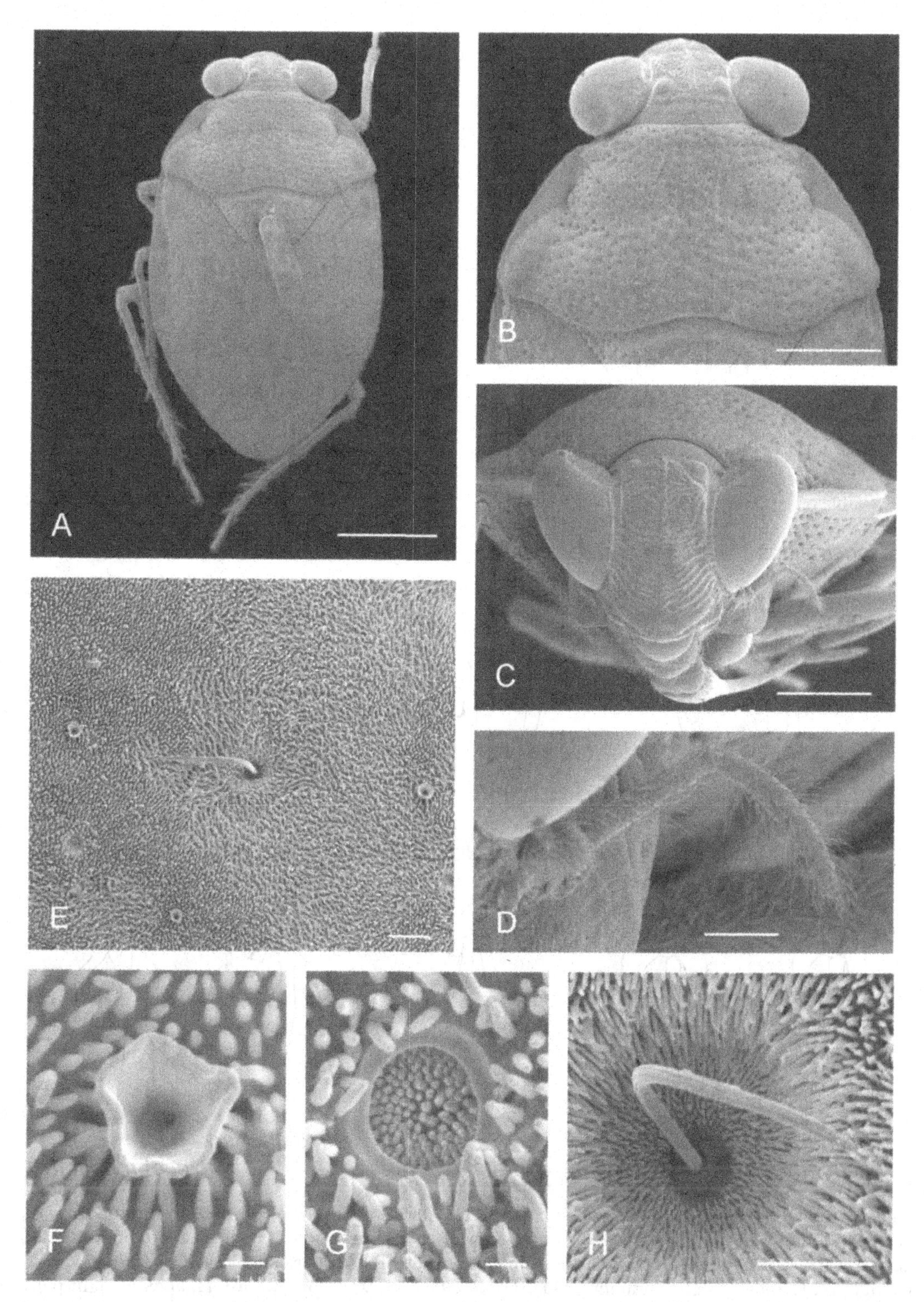

Figure 18.4. OCHTERIDAE. A-H, *Ochterus australicus*, ♀: A, dorsal view; B, dorsal view of head and pronotum. C, frontal view of head. D, antenna. E, surface of pronotum. F, trumpet-shaped micro-hair. G, cuticular peg-plate. H, seta on pronotum. Scales: A, 1 mm; B-C, 0.5 mm; D, 0.1 mm; E, 0.01 mm; F-G, 0.001 mm. Scanning Electron Micrographs (CSIRO Entomology, Canberra).

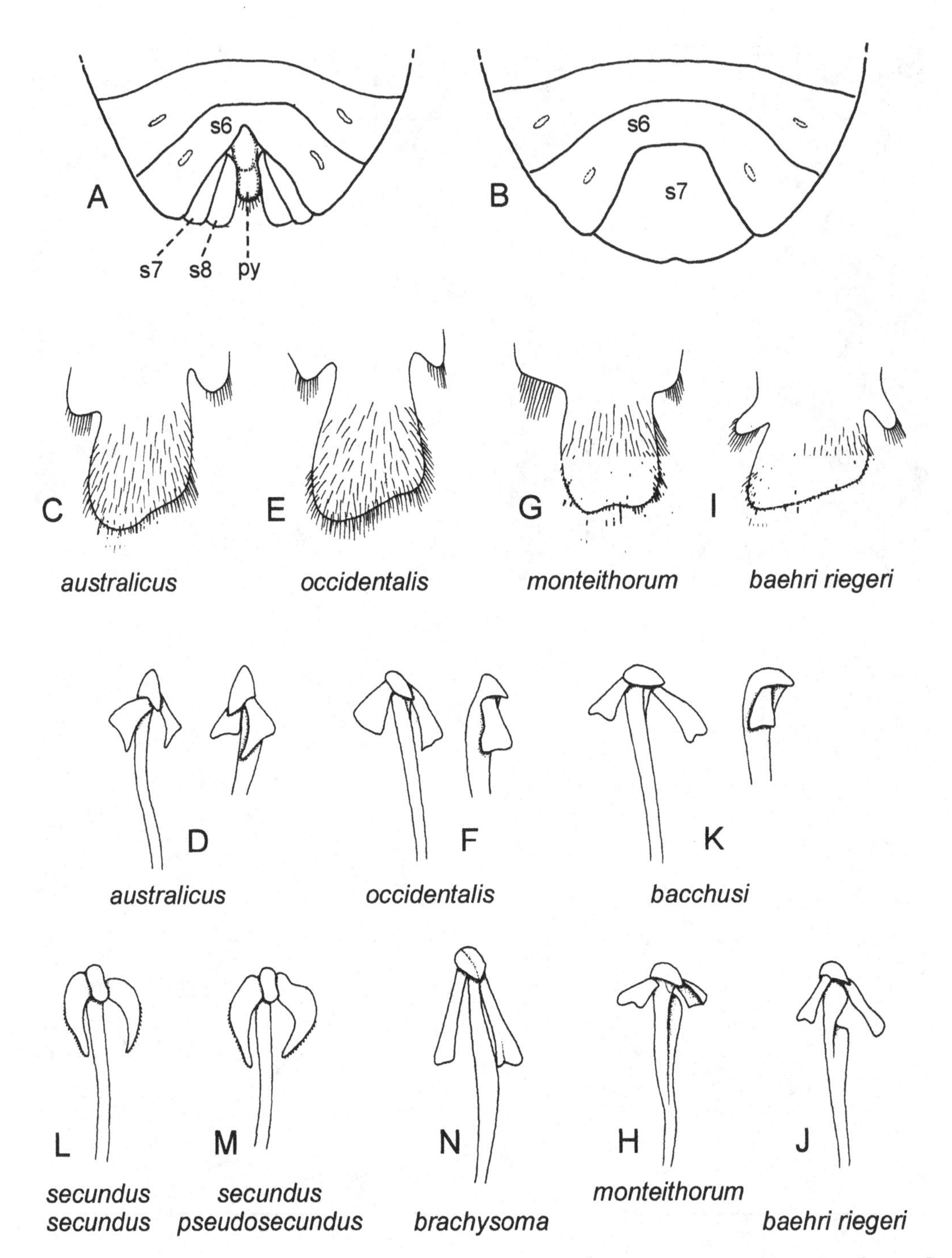

Figure 18.5. OCHTERIDAE. A-B, *Ochterus* sp.: A, ventral view of abdominal end of ♂ (py, pygophore; s6, s7, s8, abdominal sterna 6, 7, and 8. B, ventral view of abdominal end of ♀ (s6, s7, abdominal sterna 6 and 7). C-D, *O. australicus*: C, pygophore of male. D, apex of right paramere. E-F, *O. occidentalis*, ♂: E, pygophore. F, apex of right paramere. G-H, *O. monteithorum*, ♂: G, pygophore. H, apex of right paramere. I-J, *O. baehri riegeri*, ♂: I, pygophore. J, apex of right paramere. K, *O. bacchusi*, ♂: apex of right paramere. L, *O. secundus secundus*, apex of right paramere. M, *O. secundus pseudosecundus*, apex of right paramere. N, *O. brachysoma*, apex of right paramere. A-B, original; C-N, reproduced with modification from: Baehr (1990b).

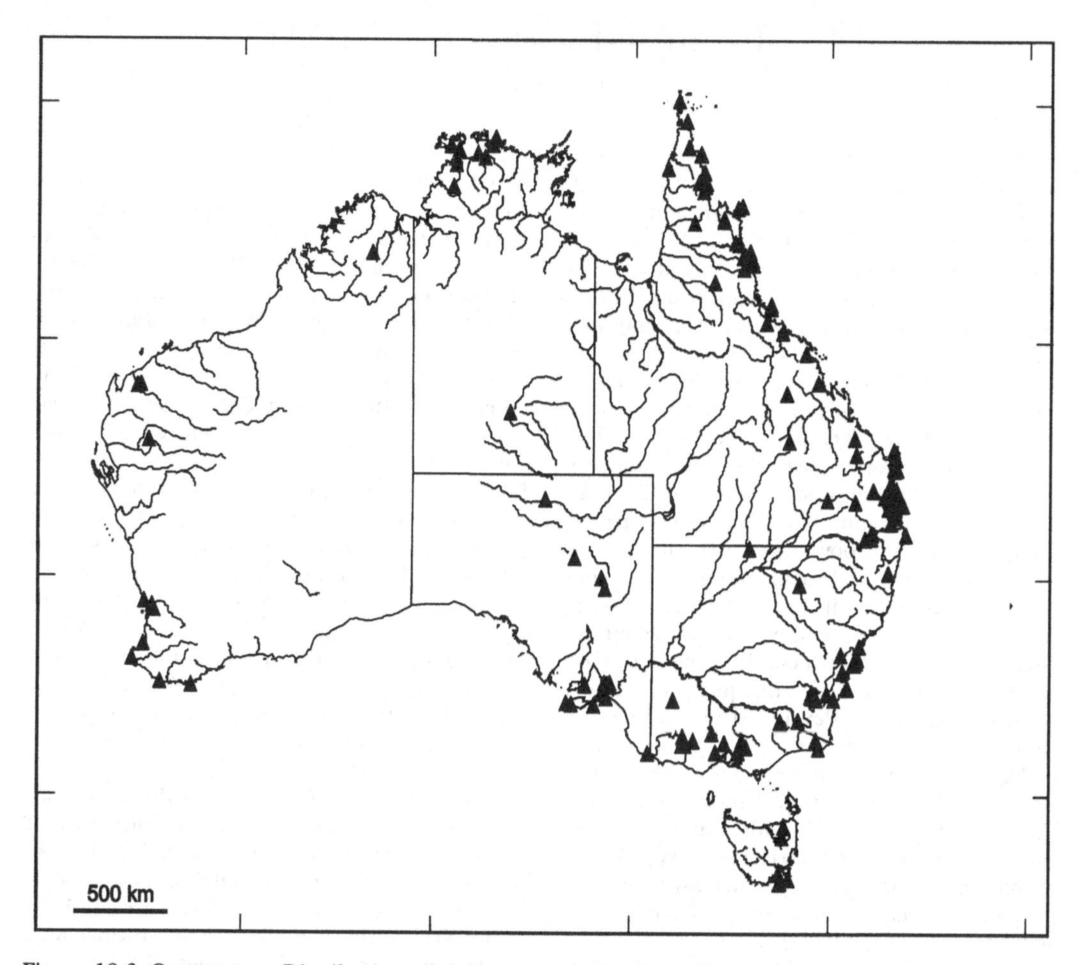

Figure 18.6. OCHTERIDAE. Distribution of *Ochterus* species in Australia.

sand and dirt. When this is removed, it is replaced actively by the nymph using a small comb anteriorly on the head, just over the labrum, to scoop up sand and dirt which is then put into place by the front and hind legs of the nymph (Boulard & Coffin 1991). At the last moult, the nymph constructs a moulting chamber where the nymph stays for two days before and after moulting. *O. marginatus formosana* (Matsumura) completes its life cycle in about one month (Takahashi 1923). *O. banksi* overwinters as fourth instar nymph in Virginia, U.S.A. (Bobb 1951).

Distribution

The genus *Ochterus* is distributed world-wide with about 30 species in the Indo-Australian region. The nine Australian species are recorded from all states (Fig. 18.6), but in particular from coastal localities in southern, eastern, and northeastern Australia, the top end of Northern Territory, and the southwestern part of Western Australia. Very few records in the interior of the continent.

19. Family GELASTOCORIDAE

Toad Bugs

Identification

Small or medium sized, broad bugs with protruding eyes. Body surface often rough and "warty" (Fig. 19.1, Plate 8, D). Head short and very broad (Figs 19.2A, B). Eyes (ey) large, with inner margins distinctly emarginated, appearing kidney-shaped. Ocelli present (oc; with few exceptions). Antennae 4-segmented (Fig. 19.2C; an), short, without finger-like processes, concealed beneath eyes. Rostrum (ro) 4-segmented, short and stout; labrum broad, flap-like. Pronotum large, subrectangular, with anterior corners more or less produced. Scutellum distinct, subtriangular. Fore legs raptorial, femur (Fig. 19.2B; fe) greatly enlarged with a groove on inner surface for reception of tibia and tarsus; middle and hind legs adapted for walking, without swimming hairs. Fore tarsus 1-segmented or sometimes fused with tibia; middle tarsus 2-segmented, hind tarsus 3-segmented; claws unequally developed on fore legs, equally developed on middle and hind legs. Hemelytra with distinct clavus (Fig. 19.2A; cl), corium (co), embolium (em), and membrane (me); membrane sometimes reduced, if fully developed, usually with numerous veins. Abdominal sterna asymmetrical in ♂ (Fig. 19.2B), symmetrical in ♀. Male genitalia asymmetrical, left paramere reduced or absent, right paramere well-developed (Fig. 19.5H).

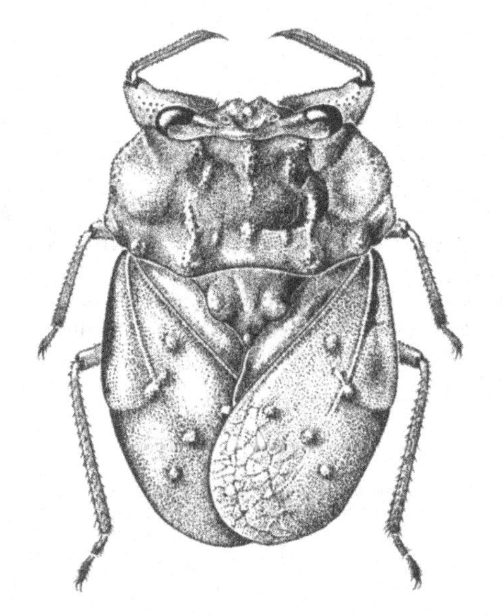

Figure 19.1. GELASTOCORIDAE. *Nerthra nudata*. Reproduced from CSIRO (1991).

Overview

Based on the number of species, this is one of the commonest groups of nepomorphans in Australia. Gelastocorids run in a staggered fashion when disturbed (Gerry Cassis, personal communication), but do not hop or jump as reported by some authors (e.g., Menke 1979d). Some species of *Nerthra* occur along the banks of streams and ponds, others are found far from water. Todd (1955) summarises reports of *Nerthra* having been found in decomposing banana trunks, rotting logs, and in leaf litter on the forest floor. Species are usually secretive, hiding under stones or other objects, and often burrowing in the soil; their mottled appearance makes them difficult to detect unless they move. Some *Nerthra* species have been seen crawling on plant stems under water, clinging to the under surface of wood floating in water, or hiding under rocks in swiftly flowing water (Menke 1979d). Perhaps associated with their burrowing habit, wing and flight muscle polymorphism is common in the Gelastocoridae (Parsons 1960).

Gelatocorids occur throughout the world, but are most abundant in the subtropical and tropical parts. There are two subfamilies and three genera with about 100 species. The subfamily Gelastocorinae with two genera and 12 species are confined to the New World. The subfamily Nerthrinae with the single genus *Nerthra* has about 90 described species distributed throughout the tropical and subtropical regions of the world except the Mediterranean. In Australia there are 24 species.

NERTHRA

Identification

Small to medium-sized bugs (Fig. 19.1), total length from 5.1 mm to 10.8 mm. Body dorsally flattened. Front of head (= clypeus according to Parsons 1959) usually provided with tooth-like tubercles (Fig. 19.2F). Rostrum 4-segmented, but first segment concealed, apparently originating on the ventral surface of the head, its apical part pointing ventrally or anteriorly (Fig. 19.2C; ro). Fore tarsus fused with the tibia, with one well-developed claw. Hemelytra variable, coriaceous (Figs 19.2A, 19.5A-G), with relatively well-defined clavus (cl), corium (co), and embolium (Fig. 19.2A; em, 19.5A; E); membrane (me, M) either well developed, reduced or absent; in the latter case, hemelytra fused together along midline.

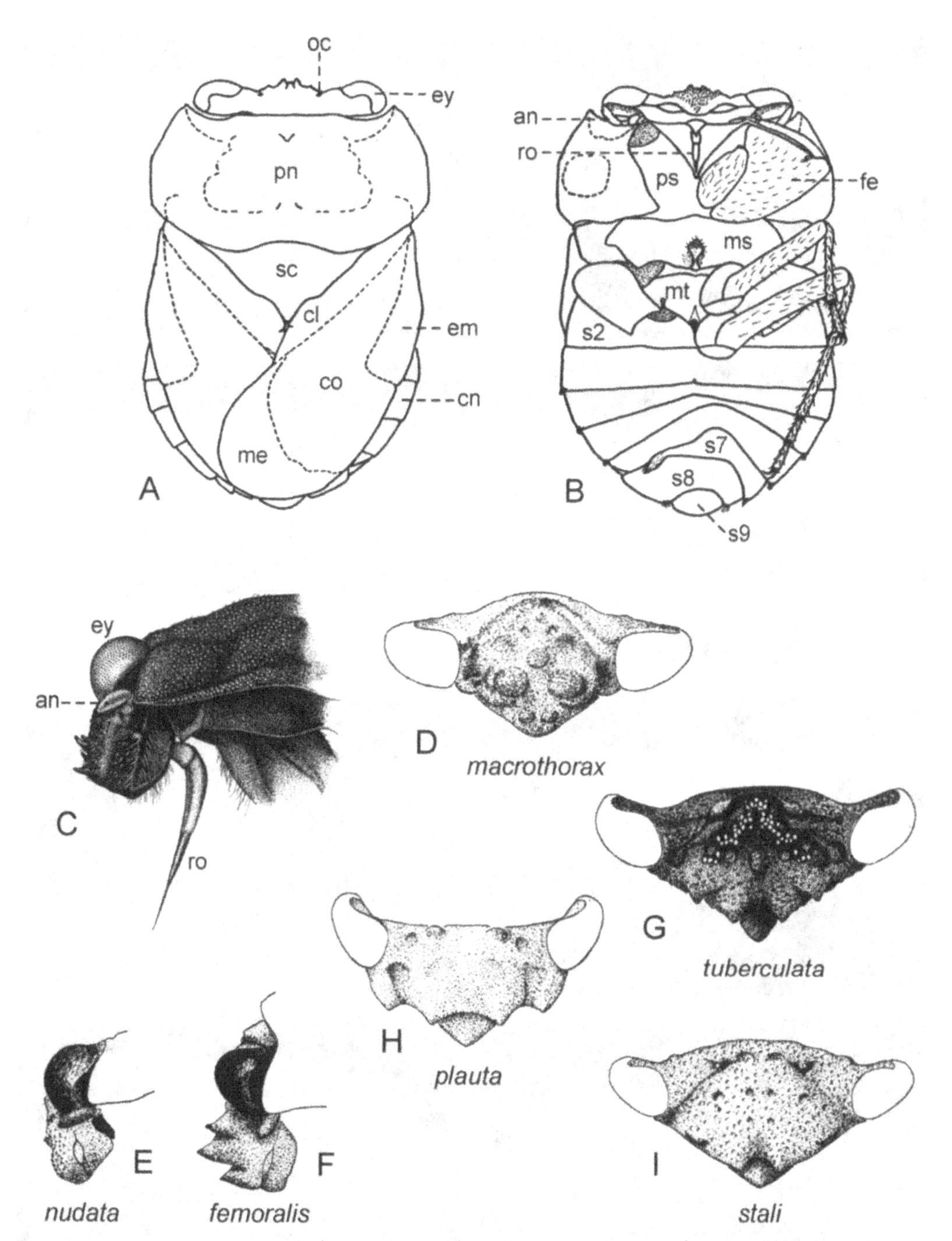

Figure 19.2. GELASTOCORIDAE. A-B, *Nerthra* sp.: A, dorsal outline of body (cl, clavus; cn, connexivum; co, corium; em, embolium; ey, eye; me, membrane; oc, ocellus; pn, pronotum; sc, scutellum). B, ventral outline of ♂; legs of right side omitted (an, antenna; fe, fore femur; ms, mesosternum; mt, metasternum; ps, prosternum; ro, rostrum; s2, s7, s8, abdominal sterna 2, 7, and 8; s9, segment 9). C, *Nerthra martini* (western U.S.A.): lateral view of head and anterior thorax (an, antenna; ey, eye; ro, rostrum). D, *N. macrothorax*: frontal view of head. E, *N. nudata*: lateral view of head. F, *Nerthra femoralis*: lateral view of head. G, *N. tuberculata*: frontal view of head. H, *N. plauta*: frontal view of head. I, *N. stali*: frontal view of head. Reproduced with modifications from: A-B, Todd (1955); C, Menke (1979d); D-I, Todd (1960).

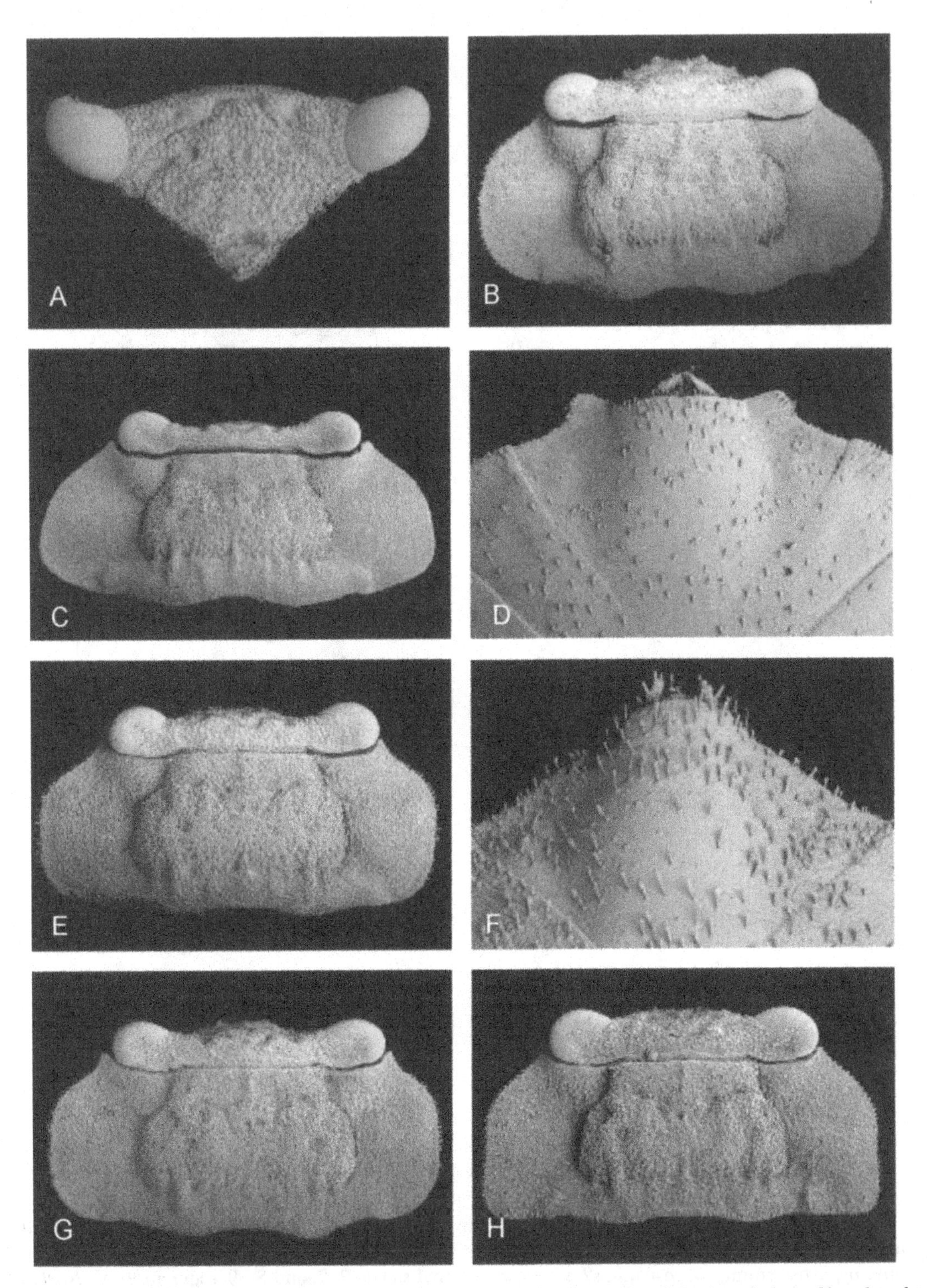

Figure 19.3. GELASTOCORIDAE. A-B, *Nerthra alaticollis*: A, frontal view of head. B, dorsal view of head and pronotum. C-D, *N. soliquetra*: C, dorsal view of head and pronotum.D, ventral view of sternite 7 of ♀. E-F, *N. hylaea*: E, dorsal view of head and pronotum. F, ventral view of sternite 7 of ♀. G, *N. appha*: dorsal view of head and pronotum. H, *N. tasmaniensis*, dorsal view of head and pronotum. Scanning Electron Micrographs. Reproduced from Cassis & Silveira (2001).

Abdominal segment 9 (= pygophore) not withdrawn into pregenital abdomen, sternum 9 visible externally (Figs 19.2B; s9; 19.5H; Py). Left paramere of male absent, right paramere strongly developed and folded longitudinally (Pa). In order to examine the paramere, dried specimens have to be softened, the paramere directed to the venter and locked against the abdominal sternites by the method described by Todd (1955: 294). Female abdominal end with large sternum 7 covering gonocoxa (Figs 19.3D, F).

Todd (1960) recognised nine species groups within *Nerthra*, four of which are represented in Australia. Cassis & Silveira (2001, 2002) characterised the Australian species groups as follows:

(1) *Nerthra elongata* group (7 species): Dorsum with dense clumps of scale-like setae. Clypeus apically rounded.

(2) *N. laticollis* group (6 species): Dorsum with dense clumps of scale-like setae. Clypeus apically excavate.

(3) *N. rugosa* group (1 species): Dorsum at most with scattered scale-like setae. Clypeus with dense clavate setae present on rounded pads

(4) *N. alaticollis* group (10 species): Dorsum at most with scattered scale-like setae. Clypeus without dense clavate setae.

The taxonomy of the genus *Nerthra* in Australia was previously known through the works by Todd (1955, 1960). Recently, Cassis & Silveira (2001, 2002) have presented excellent, well illustrated revisions of the genus which permit identifications of most Australian species. These authors have generously permitted us to use many of their excellent illustrations for this book. The following key is chiefly adapted from Cassis & Silveira (2001, 2002), but supplemented with entries from the key to Australian *Nerthra* by Todd (1960).

Key to Australian species

1. Clypeus with dense clavate setae present on rounded pads (Fig. 19.2D). Length 7.1-10.6 mm. (*rugosa* group). Northern Territory .... *macrothorax* (Montrouzier)
- Clypeus without dense clavate setae, but with some (usually 4 to 5) sharp-pointed, conical, tooth-like tubercles (Fig. 19.2F) 2
2. Distal part of hemelytral margin (from costal fracture to apex) bent up almost at a right angle to rest of hemelytron. Ocelli absent. Length 6.2 mm. (*laticollis* group). Western Australia .... *hirsuta* Todd
- Distal part of hemelytral margin not as above. Ocelli present 3
3. Clypeus ornamented with series of whitish granules in the shape of an inverted V (Fig. 19.2G). Length 8.5-9.9 mm. Western Australia .... *tuberculata* (Montandon)
- Clypeus variously coloured, but not marked as above 4
4. Clypeus apically excavate, with apical tubercle(s) (Fig. 19.2F) 5
- Clypeus rounded, without apical tubercle(s) (Fig. 19.2E) (*elongata* group) 18
5. Hemelytra variable, membrane either well developed, reduced or absent; in the latter case, hemelytra fused together. Dorsal surface with conspicuous clumps and rows of scale-like setae (*laticollis* group) 6
- Hemelytra entirely coriaceous, separate. Dorsal surface nearly glabrous, setae present extremely minute (*alaticollis* group) 10
6. Hemelytra with a well-developed membrane. Scale-like setae of scutellum and hemelytra pale yellowish to light brown in colour. Distal half of parameres broad, dorso-ventrally flattened, greatest width of visible part about 0.4x length of visible part (Fig. 19.6A). Length 7.5-9.1 mm. Northern Territory, Queensland, Western Australia .... *luteovaria* (Distant)
- Hemelytra with a membrane reduced or absent. Scale-like setae of scutellum and hemelytra black. Distal half of parameres not strongly dorso-ventrally flattened, greatest width of visible part less than 0.4x length of visible part (Fig. 19.6B) 7
7. Embolium very broad, width basally one third greater than width of eye. Broad species, more than 0.75x as wide as long 8
- Embolium narrower, width basally less than one third width of eye. More elongate species, usually less than 0.7x as wide as long 9
8. Mesosternal process of thorax pointed (caudal or frontal view). Hemelytra fused together, line of fusion usually marked by two thin, parallel rows of setae. Connexivum only slightly exposed. Scutellum with a thin, median longitudinal row of setae extending from base to apex. Length 8.9-10.2 mm. Northern Territory .... *walkeri* Todd
- Mesosternal process of thorax very broad, truncate. Hemelytra usually not fused together, but often with overlapping parts stuck together by secretion and accumulated debris. Connexivum prominent, usually slightly but sometimes distinctly crenulate. Scutellum depressed basally, the median area with

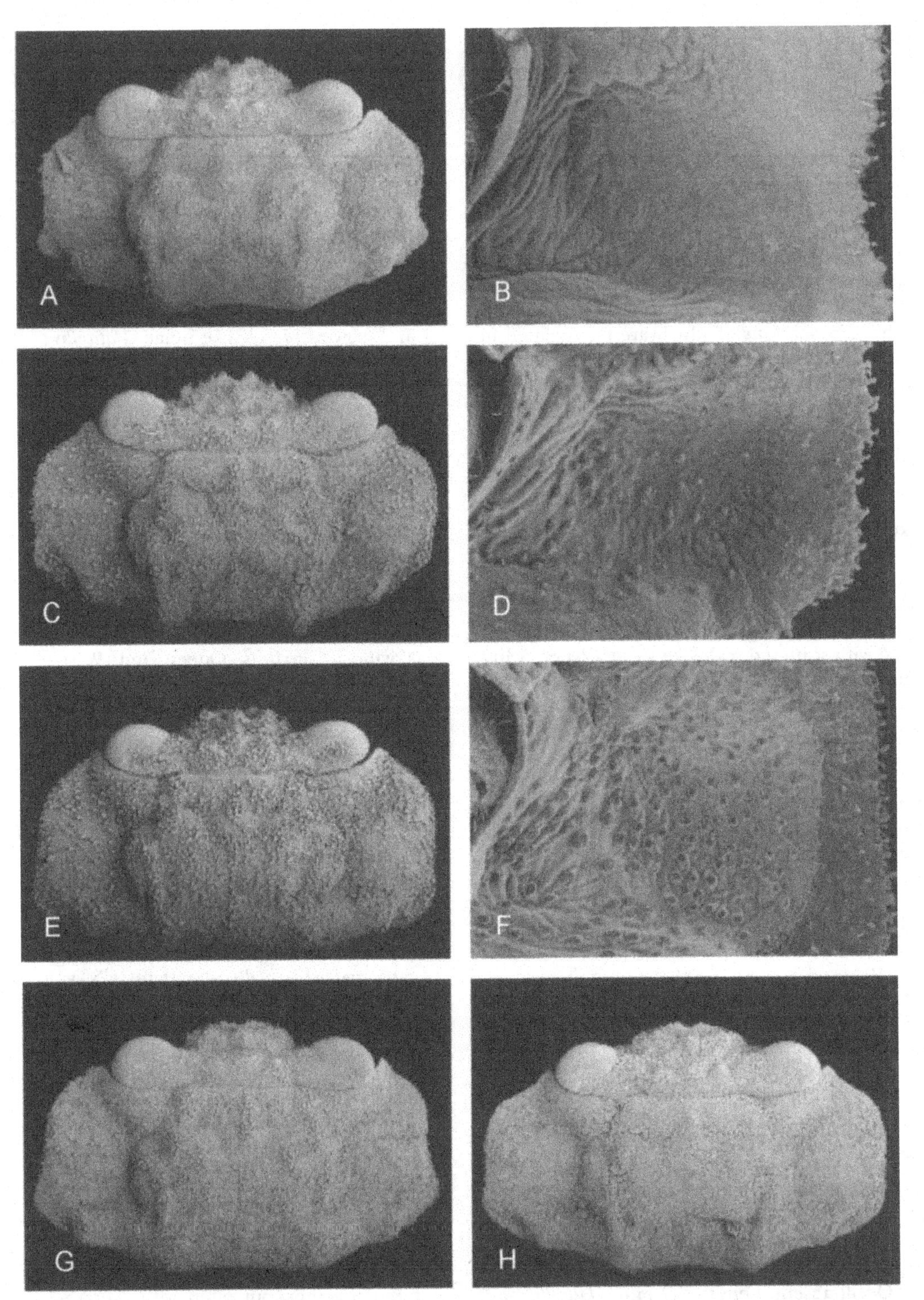

Figure 19.4. GELASTOCORIDAE. A-B, *Nerthra sinuosa*: A, dorsal view of head and pronotum. B, propleuron. C-D, *N. falcatus*: C, dorsal view of head and pronotum. D, propleuron. E-F, *N. annulipes*: E, dorsal view of head and pronotum. F, propleuron. G, *N. probolostyla*, dorsal view of head and pronotum. H, *N. monteithi*: dorsal view of head and pronotum. Scanning Electron Micrographs. Reproduced from Cassis & Silveira (2002).

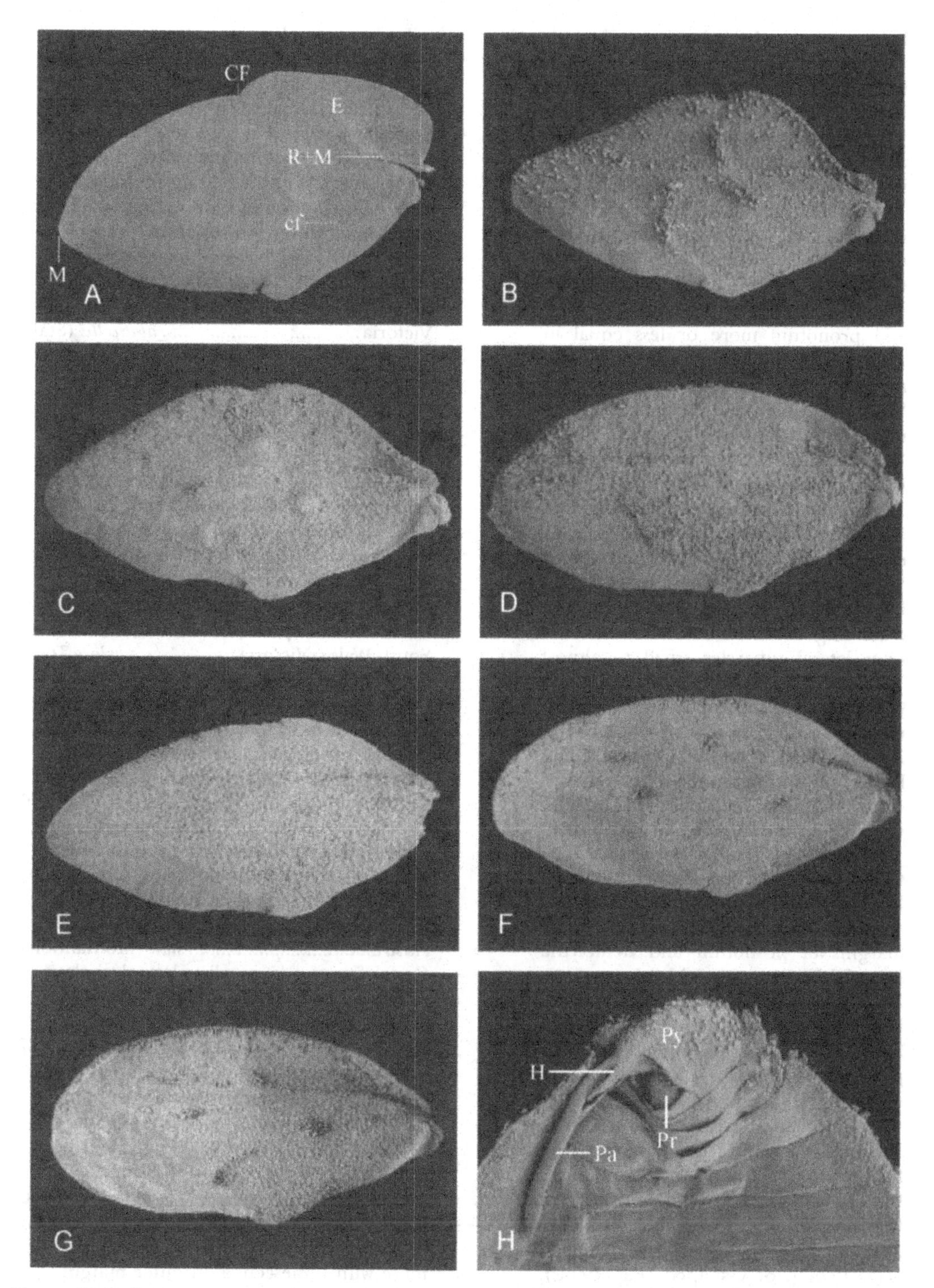

Figure 19.5. GELASTOCORIDAE. A, *Nerthra alaticollis*: left hemielytron (CF, corial fracture; cf, costal furrow; E, embolium; M, membrane; R + M, veins R + M. B, *N. falcatus*: left hemelytron. C, *N. monteithi*: left hemelytron. D, *N. sinuosa*, left hemelytron. E, *N. annulipes*: left hemielytron. F, *N. elongata*: left hemelytron. G, *N. nudata*: left hemelytron. H, *N. annulipes*: dorsal view of ♂ genitalia (H, hypandrium; Py, pygophore; Pa, right paramere; Pr, proctiger). Scanning Electron Micrographs. Reproduced from: A, Cassis & Silveira (2001); B-G, Cassis & Silveira (2002).

hairs present only in distal part. Length 8.4-10.8 mm. Victoria ... *grandis* (Montandon)

9. Legs completely black, same colour as venter. Width of pronotum usually less than 6.0 mm. Abdomen wider than pronotum. Scutellum depressed medially towards base. Length 7.7-9.1 mm. Tasmania .... *suberosa* (Erichson)

– Femora of legs yellowish brown or yellowish brown suffused with red, paler than rest of venter. Width of pronotum usually greater than 6.0 mm. Abdomen and pronotum more or less equal in width. Scutellum with a longitudinal median carina bearing a thin row of black scale-like setae. Length 7.8-10.0 mm. Western Australia .... *femoralis* (Montandon)

10. Eyes dorso-laterally oriented (Fig. 19.2H), extending above median lobe of pronotum; lateral tubercles greatly enlarged, extending well beyond dorso-lateral edge of clypeus................................. 11

– Eyes laterally oriented (Fig. 19.2I), not extending above median lobe of pronotum; lateral tubercles small to almost obsolete 12

11. Claval furrow distinct, small whitish membrane at apex of hemelytra. Dorsum mostly stramineous with median lobe of pronotum orange to fuscous and contrastingly darker. Length 5.4-6.3 mm (♂), 5.6-7.1 mm (♀). Western Australia.. .... *adspersa* (Stål) (= *membranacea* Nieser)

– Claval furrow obsolete, hemelytral membrane absent. Dorsum uniformly orange, at most with dark maculations on lateral margins of pronotum and hemelytra. Length 6.2 mm (♂), 8.1-8.4 mm (♀). South Australia .... *plauta* Todd

12. Postero-lateral margins of sternite 7 of females with triangular or subtriangular processes (Fig. 19.3D). Right paramere of male L-shaped (Fig. 19.6C), outer margin of shaft without tumescence. Mesepimeron and metepimeron uniformly brown 13

– Postero-lateral margins of sternite 7 of females without or with reduced processes (Fig. 19.3F). Right paramere S-shaped (Fig. 19.6D), or C-shaped (Fig. 19.6E), with (Fig. 19.6F) or without tumescence on outer margin of shaft. Mesepimeron and metepimeron medially stramineous.... 15

13. Embolium broad at base, more than two fifths greatest width of embolium. Lateral margins of pronotum with dense distribution of elongate clavate setae. Length 6.9-7.3 mm (♂), 8.1-8.6 mm (♀). Australian Capital Territory .... *polhemi* Cassis & Silveira

– Embolium narrow at base, less than two fifths greatest width of embolium. Lateral margins of pronotum at most with moderate distribution of short clavate setae.... 14

14. Lateral margins of pronotum arcuate (Fig. 19.3B), lacking produced humeral angles. Length 6.8-8.0 mm (♂), 7.4-8.6 mm (♀). New South Wales, Queensland, Victoria .... *alaticollis* (Stål)

– Humeral angles prominent (Fig. 19.3C), lateral margins of pronotum evenly convergent towards head. Length 6.9-7.4 mm (♂), 7.8-8.7 mm (♀).Queensland .... *soliquetra* Todd

15. Embolium narrow at base, less than two fifths greatest width of embolium. Posterior two thirds of lateral margins of pronotum divergent towards head (Fig. 19.3E). Outer margin of shaft of right paramere without tumescence. Length 6.4-7.1 mm (♂), 7.0-7.9 mm (♀). New South Wales, Victoria .... *hylaea* Todd

– Embolium broad at base, more than two fifths greatest width of embolium. Posterior two thirds of lateral margins of pronotum subparallel (Fig. 19.3G). Outer margin of shaft of right paramere with or without tumescence........................ 16

16. Costal fracture small, lateral margins of embolium and corium continuous; base of R + M strongly nodulate. Shaft of right paramere S-shaped (Fig. 19.6D). Length 7.4-8.4 mm (♂), 7.5-8.9 mm (♀).Western Australia............................. *stali* (Montandon)

– Costal fracture large, disto-lateral angle of embolium strongly arcuate; base of R + M at most weakly nodulate. Shaft of right paramere C-shaped (Fig. 19.6F) 17

17. Lateral lobes of pronotum narrow (Fig. 19.3H). Shaft of right paramere without tumescence on outer margin submedially, apex strongly deflexed (Fig. 19.6E). Length 6.5-6.6 mm (♂), 6.7-7.3 mm (♀). Tasmania .... *tasmaniensis* Todd

– Lateral lobes of pronotum moderately broad (Fig. 19.3G). Shaft of right paramere with tumescence on outer margin submedially, apex moderately deflexed (Fig. 19.6F). Length 6.5-7.3 mm (♂), 7.9-8.3 mm (♀). New South Wales .... *appha* Cassis & Silveira

18. Basal part of embolium concave (Figs 19.5B-C) 19

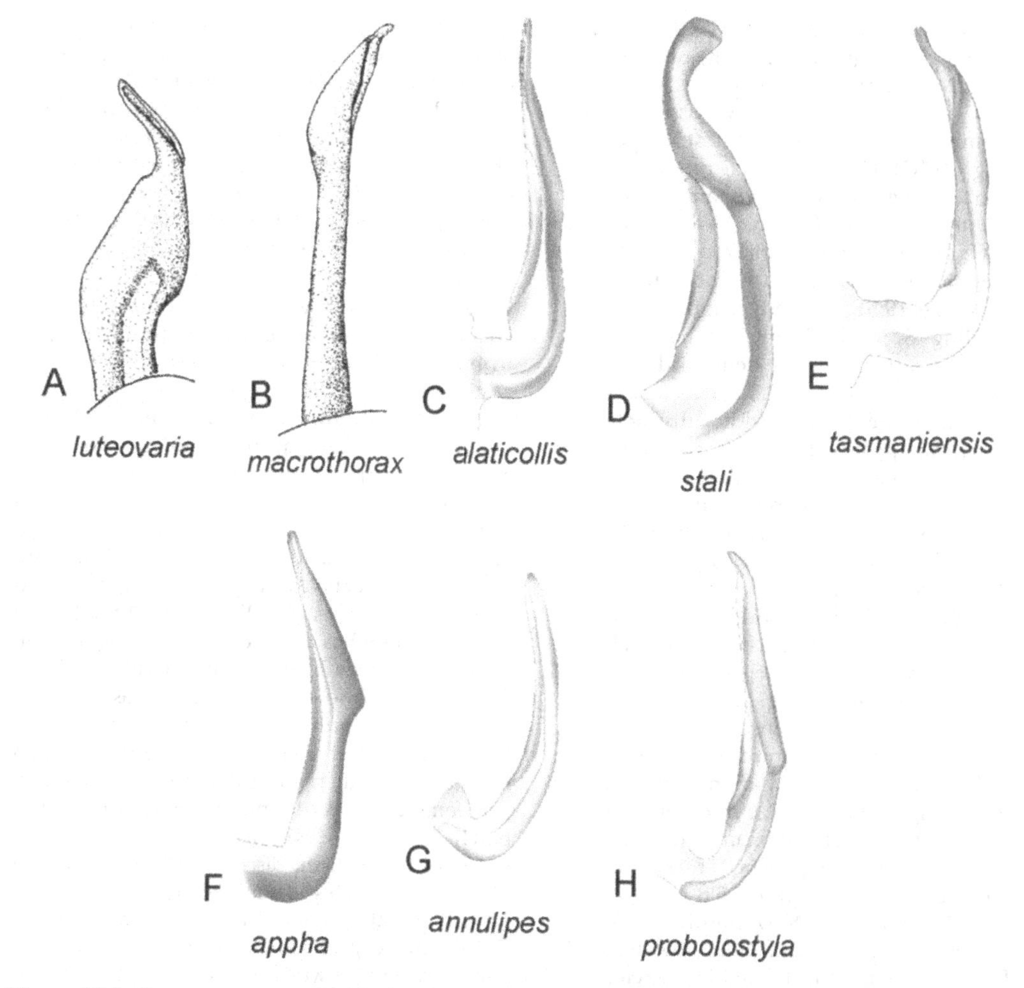

Figure 19.6. GELASTOCORIDAE. A, *Nerthra luteovaria*: right paramere. B, *N. macrothorax*: right paramere. C, *N. alaticollis*: right paramere. D, *N. stali*: right paramere. E, *N. tasmaniensis*: right paramere. F, *N. appha*: right paramere. G, *N. annulipes*: right paramere. H, *N. probolostyla*: right paramere. Reproduced with modification from: A-B, Todd (1960); C-F, Cassis & Silveira (2001); G-H, Cassis & Silveira (2002).

– Basal part of embolium straight or arcuate (Figs 19.5D-G) 21

19. Pronotum widest at humeral angles with median part of lateral margin concave (Fig. 19.4A). Nodules on propleuron absent (Fig. 19.4B). Corial fracture weakly incised. Length 6.5-7.1 mm (♂), 7.2-7.6 mm (♀).New South Wales, Queensland .... *sinuosa* Todd

– Median part of lateral margin of pronotum nearly straight (Fig. 19.4C). Nodules on propleuron moderately distributed (Fig. 19.4D) Corial fracture moderately to deeply incised 20

20. Dense distribution of short clavate setae on ventral and dorsal surfaces (Fig. 19.5B). Pronotum widest at anterior margin (Fig. 19.4C). Embolium deeply excavated. Postero-lateral margin of hemelytra moderately concave. Length not exceeding 7 mm (♂: 6.1-7.0 mm) or 8 mm (♀: 6.7-7.8 mm). New South Wales, Queensland.............. *falcatus* Cassis & Silveira

– Sparse distribution of short clavate setae on ventral and dorsal surfaces (Fig. 19.5C). Lateral margins of pronotum sub-parallel (Fig. 19.4H). Embolium weakly concave. Postero-lateral margin of hemelytra not concave. Length exceeding 7 mm (♂: 7.0-7.6 mm) or 8 mm (♀: 7.2-8.5 mm). Queensland .... *monteithi* Cassis & Silveira

21. Membrane of hemelytra well developed (Fig. 19.5G). Clumps of setae on dorsum sparse. Length exceeding 8 mm (♂: 8.1-8.9 mm, ♀: 8.7-9.8 mm). New South Wales, Queensland, Victoria....... *nudata* Todd
- Membrane of hemelytra reduced to absent (Figs 19.5E, F). Body length not exceeding 8 mm. Clumps of hairs on dorsum moderately to densely distributed... 22

22. Maximum hemelytral width distinctly wider than pronotum (Fig. 19.5F); claval furrow weakly developed; membrane of hemelytra reduced but not absent. Length 6.9-7.1 mm (♂), 7.35-7.9 mm (♀). Australian Capital Territory, New South-Wales, Victoria .... *elongata* (Montandon)
- Maximum hemelytral width as wide as pronotum (Figs 19.5E); claval furrow absent; membrane of hemelytra absent.......... 23

23. Propleura nodulate (Fig. 19.4F). Clumps of hairs on dorsum dense (Fig. 19.5E). Pronotum widest at antero-lateral angles with median anterior margin of pronotum straight (Fig. 19.4E). Apex of paramere with angulate apex and medio-lateral tumescence (Fig. 19.6G). Embolium arcuate at base. Length 6.3-6.8 mm (♂), 7.0-7.55 mm (♀). New South Wales, Queensland .... *annulipes* (Horvath)
- Propleura smooth, lacking tubercles. Clumps of hairs on dorsum less conspicuous. Pronotum widest at humeral angles with posterior part of pronotum strongly convergent towards embolium (Fig. 19.4G). Paramere with angulate apex and medio-lateral tumescence (Fig. 19.6H). Embolium mostly straight. Length 6.7-7.6 mm (♂), 7.7-8.3 mm (♀).Northern Territory, Queensland....... .... *probolostyla* Todd

Biology

In Australia, most *Nerthra* species occur on the ground and in the leaf litter in a variety of habitats, from both wet and dry sclerophyl forests to open heathland habitats. Whereas most species are terrestrial, some species also occur in riparian habitats (Cassis & Silveira 2002). *Nerthra alaticollis* was recorded from various forest ecosystems in New South Wales (Cassis & Silveira 2001). All except the rainforest sites were dominated by *Eucalyptus* and *Corymbia* species, with a range of midstory or understory elements, either with moist shrubs, *Allocasuarina*, myrtles, and peas. Grasses were consistently present, at least in the wet and dry sclerophyl sites. Adults were collected in January-June and September-December, chiefly in litter samples and pitfall traps, from sea level to almost 1,500 m. *N. adspersa* were recorded from open heathland habitats in Western Australia, particularly in disturbed situations. This maculated species exhibits cryptozoic behaviour, blending with the ground matrix. Adults were collected in all months except June and July, nymphs in March and August-November. A few adults were collected at light.

Very little is known about the biology of *Nerthra* species and then mainly based on casual observations of North American species. Many species seem to be nocturnal, hiding in daytime beneath stones, plant debris or burrowing in humid sand or mud. In captivity they have been fed with various small insects so they are probably polyphagous carnivores. The life cycle is little known and no species has been reared from egg to adult in the laboratory. The eggs are broadly oval and flattened on one side; the shell has a rough hexagonal sculpture. An American species, *N. martini* Todd, has been found to deposit its eggs in small holes in mud under stones near water. The female apparently guards the eggs until they hatch (Usinger 1956). Recently, a stridulatory mechanism has been discovered in the males of all *Nerthra* species examined (J. Polhemus & Lindskog 1994). It consists of a rastrate area on the genital capsule near the base of the right paramere against which the sclerotised posterior margin of the anal cone (proctiger) is rubbed.

Australian *Nerthra* species are usually flightless (Cassis & Silveira 2001, 2002). The hind wings are either absent or greatly reduced, and the fore wings (hemelytra) coriaceous, free, and submacropterous, that is barely attaining the length of the abdomen. The membrane is absent or reduced to a small apical stub without venation. Although macropterous specimens are common in some species there are very few records of flight of species in this genus. There seems to be a tendency to loose the ability of flight as some species have the hemelytra fused and no functional hind wings present. One of these is *N. macrothorax* which in spite of this condition is widespread in the Oriental Region, New Guinea, and the Pacific. This species has been found digging into rotten *Pandanus* logs. Todd (1959) supposed that the distribution of this species, which includes a number of small islands, had been obtained by drifting on such logs from island to island.

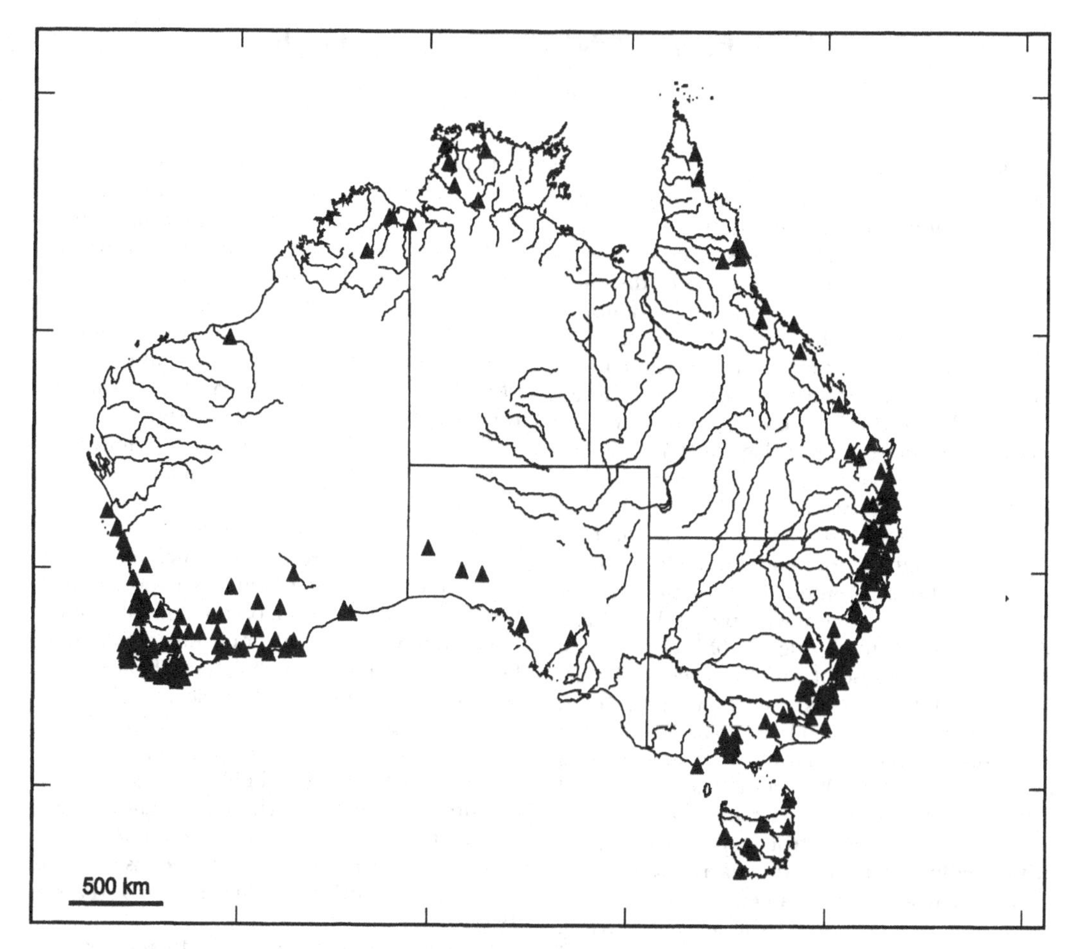

Figure 19.7. GELASTOCORIDAE. Distribution of *Nerthra* species in Australia.

Distribution

The genus *Nerthra* has 88 species worldwide revised by Todd (1955). In Australia there are 24 species distributed across Australia (Fig.19.7). Most of these occur in the coastal forests of Queensland (10 spp.), New South Wales (8 spp.), and Victoria (4 spp.), and southern Western Australia (6 species), including the semi-arid parts.

20. Family NOTONECTIDAE

Backswimmers

Identification

Backswimmers are medium-sized to large aquatic bugs and cannot be confused with other aquatic bugs on account of their wedge-shaped body with strongly convex dorsal side, hemelytra disposed in a tent-like fashion over the back, non-raptorial fore legs, and oar-like hind legs. The compound eyes are large. The antennae are 2-4 segmented, partially concealed between head and thorax (Figs 20.1D; an, 20.4A; arrow). The rostrum (ro) is relatively short and stout (Fig. 20.4C). The fore and middle legs are adapted for grasping (Figs 20.1E, 20.6A), whereas the hind legs are oar-like, with fringes of long hairs on their tibiae and tarsi (Fig. 20.6D); tarsi of fore and middle legs usually apparently two-segmented, first segment always greatly reduced, sometimes absent; hind tarsi always two-segmented (Fig. 20.6E); all pretarsi with two claws, those of hind legs reduced (Fig. 20.6E). Membrane of hemelytra without veins, longitudinally subdivided in two parts disposed in a tent-like fashion. Abdominal venter with a median longitudinal keel and heavy fringes of hairs medially and laterally forming air chambers (Fig. 20.4F); sternum 5 strongly produced anteriorly at midline; sternum 4 very narrow at midline. Male genital segments symmetrical or nearly so; parameres present, symmetrical or asymmetrical. Female ovipositor present, varying in degree of development.

Overview

Notonectids are some of the most common and widespread water bugs in Australia. They differ from other aquatic insects (except the minute Pleidae) in the habit of swimming on their backs (hence the trivial name backswimmers). They are excellent swimmers and occur in a variety of freshwater habitats although the quiet waters of pools, ponds, and lakes are preferred. All members of the family are predaceous, feeding generally on small aquatic arthropods but occasionally also on small fish and other aquatic vertebrates. Prey are grasped with the fore and middle legs. Predatory strategies differ among genera. *Enithares* and *Notonecta* species commonly rest against the surface film with the hind legs poised forward ready to dart after approaching prey or insects caught in the surface film (Plate 8, G). *Anisops* are also active predators but they hover in almost perfect equilibrium with the water, below the surface film. Although backswimmers will accept any suitable prey, their way of hunting brings them often in contact with mosquito larvae and pupae and they have been considered for mosquito control. In northern temperate countries, most notonectids overwinter as adults and lay eggs in early spring, but some species overwinter as eggs. Usually eggs are laid in or on aquatic plants, but those of *Notonecta* are often found on the surface of rocks and submerged debris. So far known all notonectids have five nymphal instars.

The primary source of oxygen in notonectids is atmospheric air which is taken in and periodically renewed by exposing the hind end of the abdomen at the water surface (Fig. 2.2). A secondary source is dissolved oxygen which enters those parts of the air stores that are exposed to the water. The seven pairs of abdominal spiracles opening into the ventral air chambers function in both inhalation and exhalation. However, the first abdominal and two thoracic spiracles, opening into separate air-filled chambers, play the most important role in respiration. These air-chambers act as "physical gills" as well as serving a hydrostatic function. The placement of the air chambers and the position of the hind legs are probably the main reasons notonectids swim with their ventral surface facing upwards (Parsons 1970).

Aquatic bugs belonging to the family Notonectidae are found world-wide, with about 350 species in two subfamilies and 11 genera (Schuh & Slater 1995). In Australia there are 42 species in 6 genera of which one are found only in Australia.

Key to the Australian genera of the family Notonectidae

1. Claval commissure of hemelytra with a prominent hair-lined pit anteriorly, close to the apex of the scutellum (Fig. 20.1A; hp). (ANISOPINAE) 2
- Claval commissure of hemelytra continuous, without a hair-lined pit anteriorly (Fig. 20.1B). (NOTONECTINAE) 4
2. Coxal plates of hind legs bare. Fore tibiae of ♂ proximally with a row of stridulatory teeth or pegs which usually are placed on a protuberance (Figs 20.1E; sr) 3
- Coxal plates of hind legs covered with long black hairs. Males without stridulatory teeth or pegs on fore tibiae..... *Paranisops*
3. Antennae 3-segmented. Rostrum of ♂ with a prominent lateral "prong" (Figs 20.1D; rp, 20.4A, C; arrow). Stridulatory

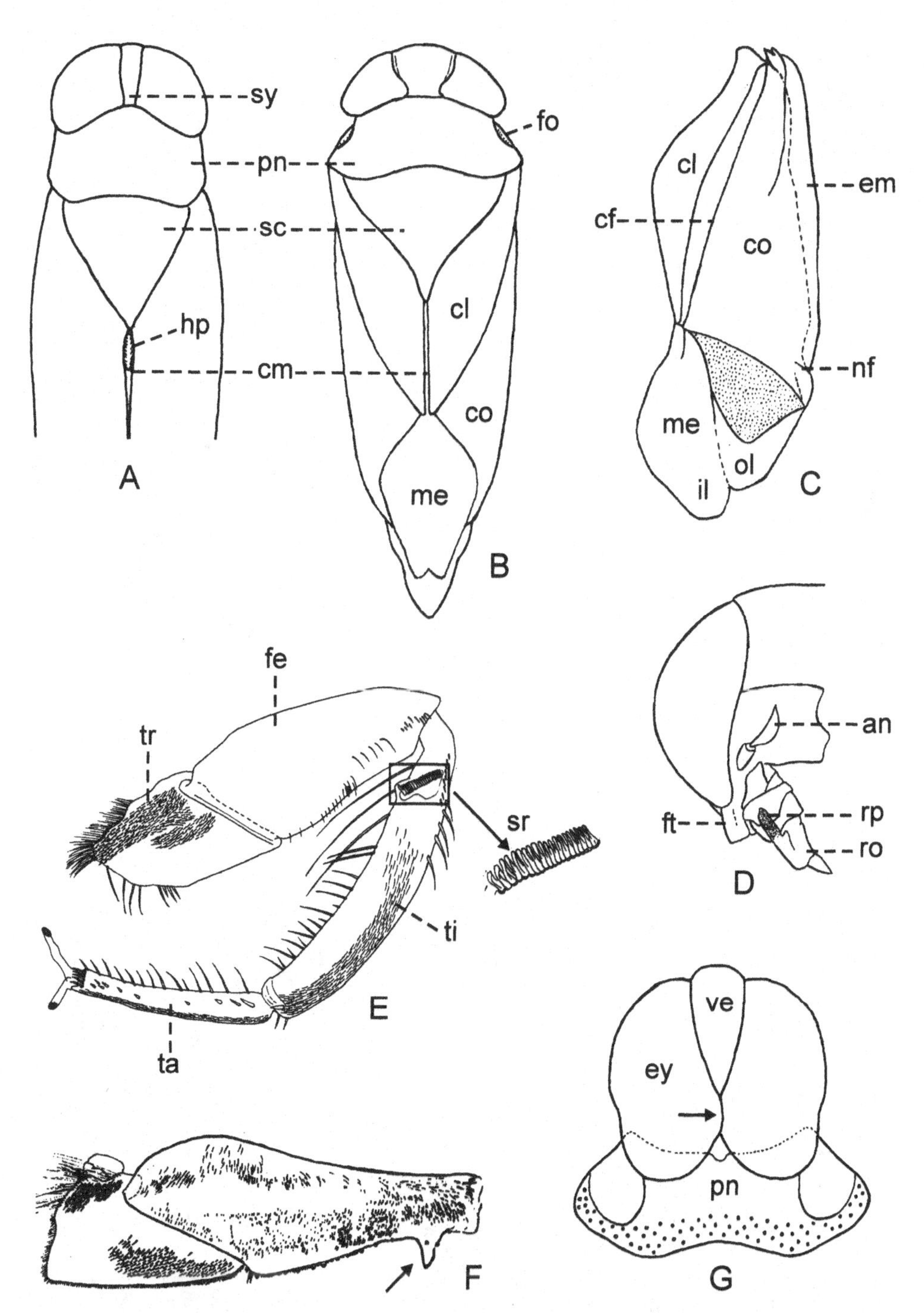

Figure 20.1. NOTONECTIDAE. A, *Anisops* sp.: dorsal view of anterior end of ♂ (cm, claval comissure; hp, hair-pit; pn, pronotum; sc, scutellum; sy, interocular space or synthlipsis). B-C, *Enithares* sp.: B, dorsal view of body (cl, clavus; co, corium; fo, pronotal fovea; me, membrane). C, right fore wing or hemelyton (cf, claval furrow; cl, clavus; co, corium; em, embolium; il, inner lobe of membrane; me, membrane; ol, outer lobe of membrane). D, *Anisops calcaratus*: lateral view of head of ♂. E, *A. tasmaniaensis*: fore leg of ♂. F, *Enithares woodwardi*: middle femur. G, *Nychia sappho*: dorsal view of head. A, original. B-G, reproduced with modification from: B-C, F, Lansbury (1968); D, Lansbury (1969); E, Lansbury & Lake (2002); G, Lansbury (1985a).

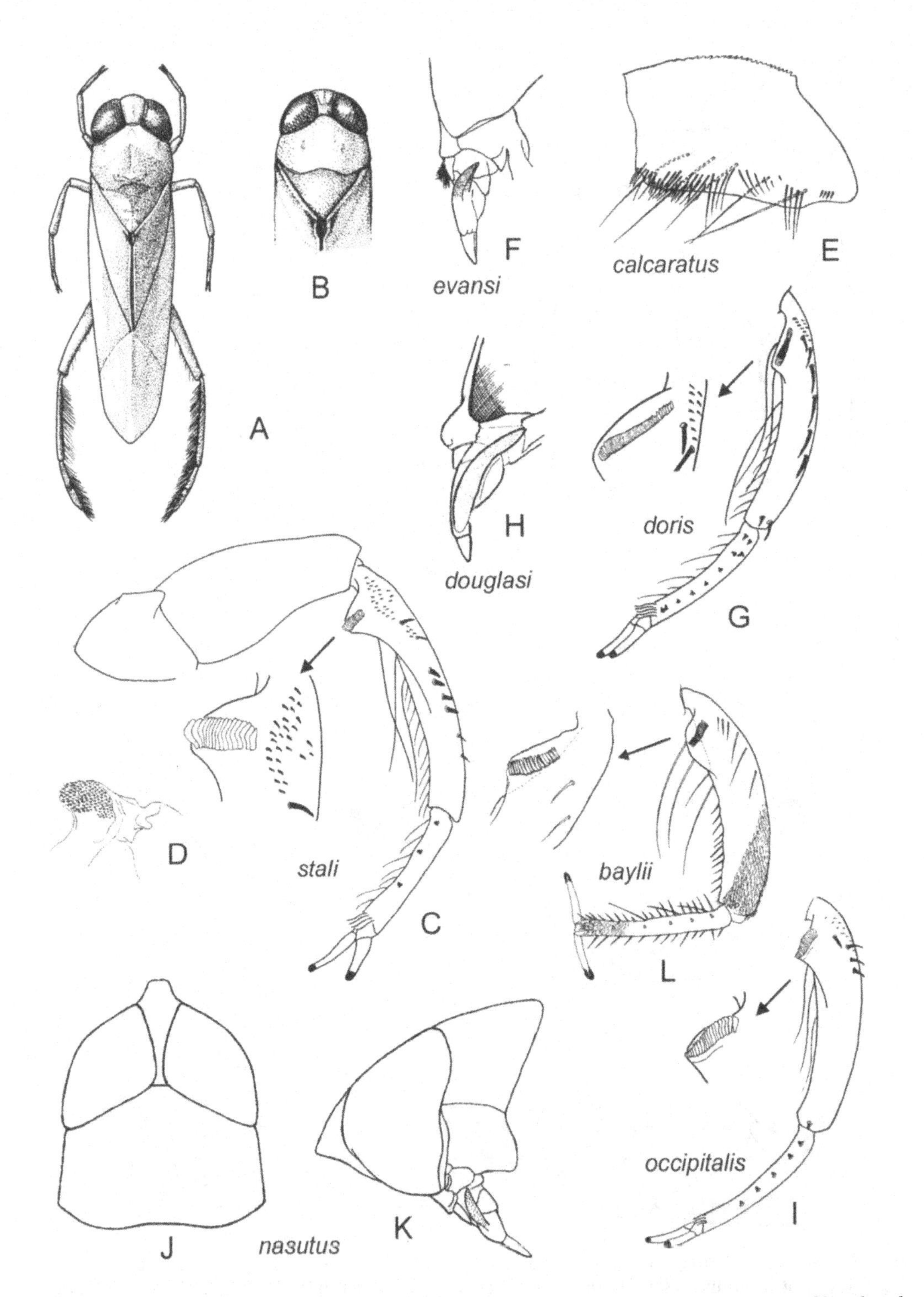

Figure 20.2. NOTONECTIDAE. A-B, *Anisops thienemanni*: A, dorsal habitus of ♂. B, dorsal view of head and thorax of ♀. C-D, *A. stali*: C, fore leg of ♂. D, base of middle tibia. E, *A. calcaratus*: fore femur of ♂. F, *A. evansi*: lateral view of rostrum of ♂. G, *A. doris*: fore leg of ♂. H, *A. douglasi*: lateral view of rostrum of ♂. I, *A. occipitalis*: fore leg of ♂. J-K, *A. nasutus*: J, dorsal view of head of ♂. K, lateral view of head of ♂. L, *A. baylii*: fore tibia of ♂. Reproduced with modification from: A-B, Lansbury & Lake (2002); C, G, I, Brooks (1951); D-F, J-K, Lansbury (1969); H, Lansbury (1984); L, Lansbury (1995b).

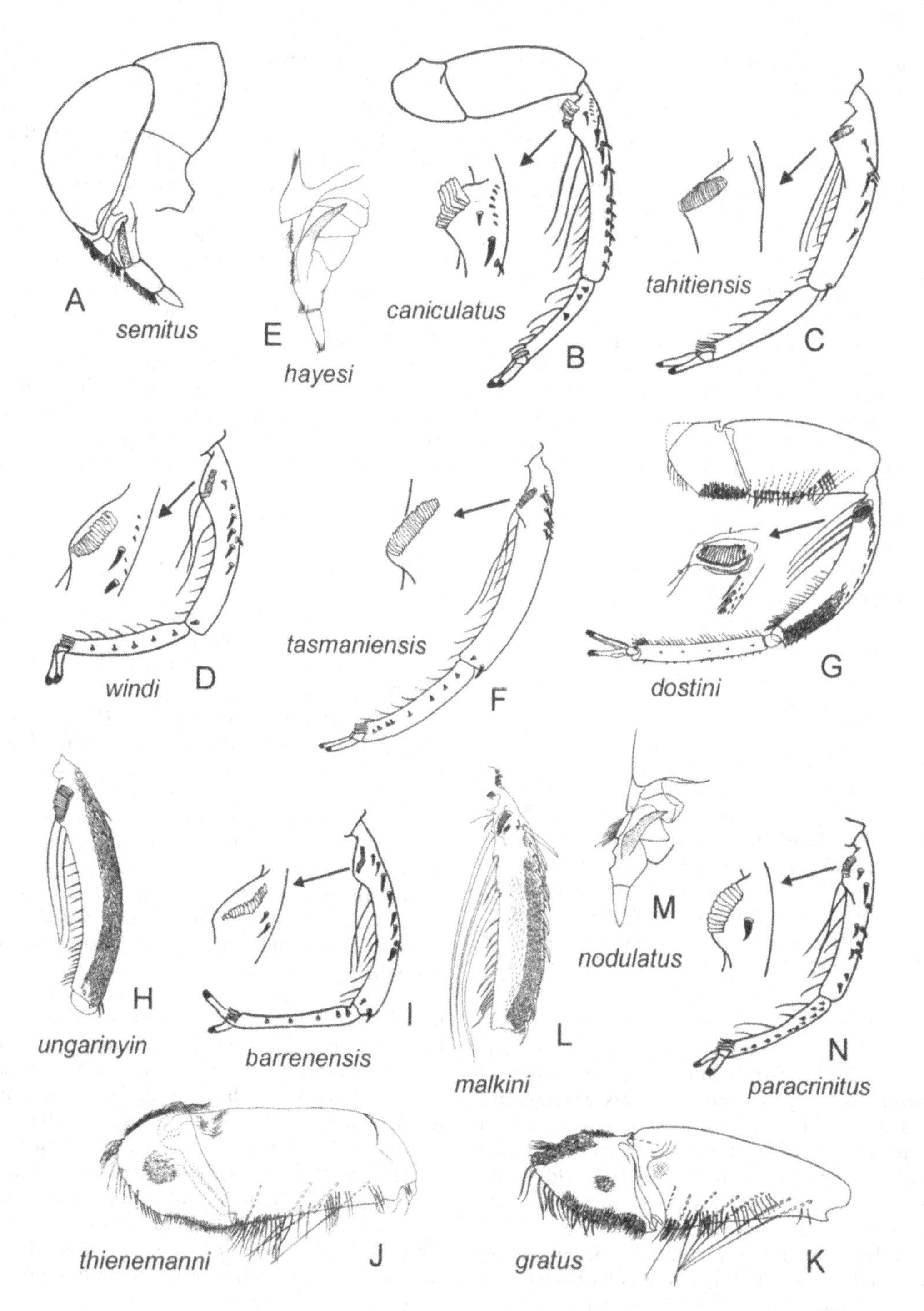

Figure 20.3. NOTONECTIDAE. A, *Anisops semitus*: lateral view of head of ♂. B, *A. caniculatus*: fore leg of ♂. C, *A. tahitiensis*: fore leg of ♂. D, *A. windi*: fore leg of ♂. E, *A. hayesi*: lateral view of rostrum of ♂. F, *A. tasmaniensis*: fore leg of ♂. G, *A. dostini*: fore leg of ♂. H, *A. ungarinyin*: fore tibia of ♂. I, *A. barrenensis*: fore leg of ♂. J, *A. thienemanni*: fore femur of ♂. K, *A. gratus*: fore femur of ♂. L, *A. malkini*: fore tibia of ♂. M, *A. nodulatus*: lateral view of rostrum of ♂. N, *A. paracrinitus*: fore leg of ♂. Reproduced with modification from: A, J, K-M, Lansbury (1969); B-D, F, I, N, Brooks (1951); E, H, Lansbury (1995a); G, Lansbury (1991b).

teeth or pegs proximally on fore tibia of ♂ packed closely together and situated on a protuberance (Fig. 20.1E; sr, 20.6A, B; arrow) ....*Anisops*
- Antennae 2-segmented (Fig. 20.10C). Rostrum of ♂ without a lateral "prong". Stridulatory pegs proximally on fore tibia of ♂ clearly separated and not situated on a protuberance (Figs 20.10D, E) .... *Walambianisops*

4. Antero-lateral margins of pronotum foveate (Fig. 20.1B, 20.12B, C; fo) 5
- Antero-lateral margins of pronotum not foveate. Middle femur with a pointed protuberance on ventral margin before apex (Fig. 20.14B) .... *Notonecta*

5. Middle femur with a pointed protuberance on ventral margin before apex (Fig. 20.1F). Eyes dorsally widely separated (Fig. 20.1B) .... *Enithares*
- Middle femur not modified as above. Eyes dorsally contiguous forming an ocular commissure, i.e. appearing to be joined or overlapping (Fig. 20.1G)....... *Nychia*

Subfamily Anisopinae

Identification

Claval commissure of hemelytra with a hair-lined pit anteriorly, near apex of scutellum (Fig. 20.1A; hp). Male fore tibiae usually (*Anisops, Walambianisops*) with a stridulatory comb (plectrum) located proximally on interior surface (Figs 20.1E; sr, 20.6A, B); rostrum with an apposing "prong" (Figs 20.1D; rp, 20.4A, C). Metathoracic scent glands absent.

Overview

These are some of the most common aquatic bugs in Australia and can be found in a variety of mostly stagnant freshwater habitats. Hungerford (1922) was the first to discover the presence of hemoglobin in the New World genus *Buenoa* Kirkaldy, which also occurs in the Old World genus *Anisops* Spinola. Hemoglobin-containing cells are concentrated in abdominal segments 3-7 and their function was studied by Miller (1964) and Wells et al. (1981). These cells contain the oxygen store which is consumed during a dive. The external gas store is relatively smaller in *Anisops* than in *Enithares* or *Notonecta*. This smaller gas store enables *Anisops* to have more or less neutral buoyancy at a certain water depth. However, the neutral buoyancy is not perfect. In the beginning of a dive they have too keep themselves submerged by actively swimming like genera of the subfamily Notonectinae. During the middle phase of the dive, they have approximately neutral buoyancy. At the end a dive the bug begins to sink gradually which again is counteracted by occasional strokes of the hind legs pushing the insect upwards. Apparently they do not have active control over the gas store and gas diffuses out of it just as in the reservoir breathers. This near neutral buoyancy makes anisopines one of only two truly planktonic insect groups (the other being larvae of the dipterous family Chaoboridae).

Species of the subfamily Anisopinae are distributed world-wide, the major genera being *Buenoa* in the New World and *Anisops* in the Old World. In Australia there are 35 species in three genera of which *Walambianisops* Lansbury is found only in Australia.

Anisops

Identification

Small to medium-sized back-swimmers, length 4.2-11.5 mm; body usually pale (Fig. 20.2A). Antennae 3-segmented (Fig. 20.4A, B). Coxal plates of hind legs bare. Ventral abdominal keel extending onto last abdominal segment (Fig. 20.4F). Usually macropterous, but some species also have a flightless, submacropterous form, where the hind wings are reduced (sometimes called "brachypterous"). Rostrum of ♂ with prominent "prong" on third labial segment (Figs. 20.1D; rp, 20.4A, C). Stridulatory teeth or pegs proximally on fore tibiae packed closely together and situated on a protuberance (Figs 20.1E; sr, 20.6A, B); fore tarsi one-segmented. Genital capsule closed behind.

The genus *Anisops* Spinola is distributed throughout the subtropical and tropical parts of the Old World with about 130 species. Australia is one of the most species-rich areas with 32 species. Most species can only be identified with certainty in the male sex. A revision of the genus with a key to species for males was presented by Brooks (1951). However, since then several species and additional records for Australia have been added by Lansbury (1964a, 1969, 1975, 1984, 1991b, 1995a, 1995b). The following key (adapted from Brooks 1951 and Lansbury 1969) applies only to males.

Key to Australian species (males only)

1. Base of middle tibia with a conspicuous process covered with densely set, stout hairs (Fig. 20.2D). Facial tubercle above labrum with a horseshoe-shaped depression. Stridulatory comb of fore tibia with

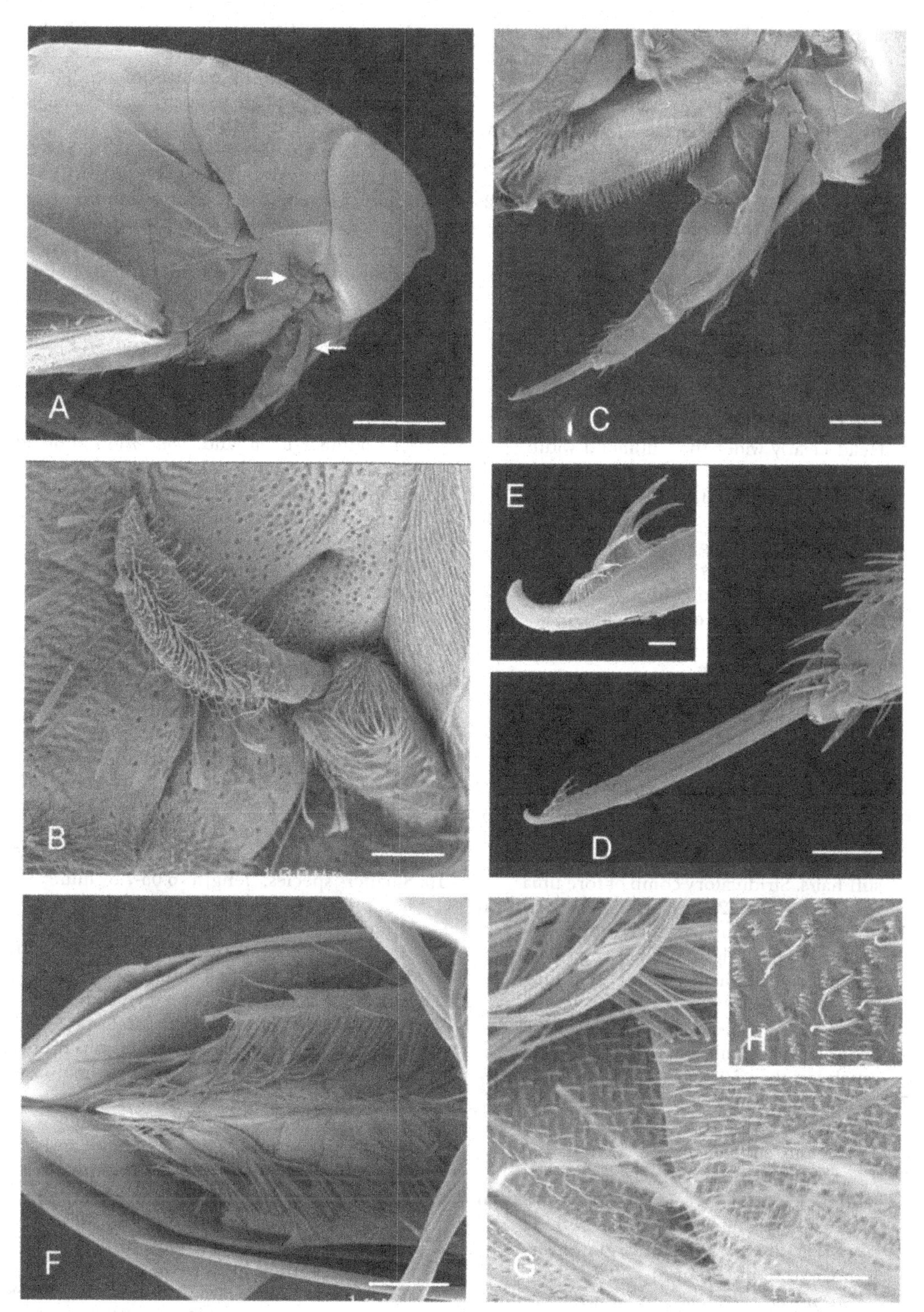

Figure 20.4. NOTONECTIDAE. A-E, *Anisops stali*, ♂: A, lateral view of head. B, antenna. C, lateral view of rostrum. D-E, apex of rostrum with stylets. F-H, *A. gratus*, ♀: F, ventral view of abdomen. G, surface of abdomen. H, detail of surface of abdomen. Scales: A, 1 mm; B, D, 0.1 mm; C, 0.2 mm; E, 0.01 mm. F, 0.5 mm; G, 0.05 mm; H, 0.01 mm. Scanning Electron Micrographs (CSIRO Entomology, Canberra).

about 25 pegs (Fig. 20.2C). Length 10.2-11.5 mm. Australian Capital Territory, New South Wales, Northern Territory, Queensland, South Australia, Western Australia .... *stali* Kirkaldy

– Base of middle tibia simple, without conspicuous process. Facial tubercle not modified as above 2

2. Facial tubercle directed downwards. Fore femur dorsally serrated (Fig. 20.2E). Stridulatory comb of fore tibia with 10 large pegs. Length 8 mm. Australian Capital Territory, New South Wales, Queensland, South Australia, Victoria ....*calcaratus* Hale

– Facial tubercle not as above. Fore femur not dorsally serrated...................................... 3

3. Head clearly wider than humeral width of pronotum... 4

– Head at most as wide as humeral width of pronotum... 8

4. Labrum with long hairs extending through entire length (Fig. 20.2F). Stridulatory comb of fore tibia with about 20 pegs. Length 7.3 mm. Tasmania......... .... *evansi* Brooks

– Labrum not as above.................................... 5

5. Rostral prong large, the tip almost reaching antenniferous tubercle (Fig. 20.2H). Facial tubercle enlarged, provided with small stiff hairs. Stridulatory comb of fore tibia with 12 pegs. Length 6.9 mm. Western Australia .... *douglasi* Lansbury

– Rostral prong smaller. Facial tubercle not enlarged, nor provided with small stiff hairs. Stridulatory comb of fore tibia with more than 12 pegs................................ 6

6. Stridulatory comb of fore tibia with about 50 pegs (Fig. 20.2G). Length 7.2-7.9 mm. Australian Capital Territory, New South Wales, Queensland, Victoria .... *doris* Kirkaldy

– Stridulatory comb of fore tibia with much less than 50 pegs (Figs 20.2I, L) 7

7. Stridulatory comb of fore tibia with 24-26 pegs (Fig. 20.2I). Dorsal margin of fore femur sinuate distally. Length 7.7-7.8 mm. Northern Territory, Queensland .... *occipitalis* Breddin

– Stridulatory comb of fore tibia with 15 pegs (Fig. 20.3L). Dorsal margin of fore femur straight. Length 7.2-7.8 mm. Western Australia.................... *baylii* Lansbury

8. Frons anteriorly produced into a cephalic horn (Figs 20.2J, K). Stridulatory comb of fore tibia with about 14 pegs. Length 6.2-7.8 mm. Northern Territory, Queensland, WA...................... *nasutus* Fieber

– Frons not produced as above 9

9. Facial tubercle with a median longitudinal groove from base to apex 10

– Facial tubercle not as above........................ 11

10. Facial tubercle with a pair of longitudinal keels fringed with hairs (Fig. 20.3A). Stridulatory comb of fore tibia with nine pegs. Length 4.2 mm. Queensland.......... .... *semitus* Brooks

– Facial tubercle not keeled or fringed with hairs. Stridulatory comb of fore tibia with seven pegs (Fig. 20.3B). Length 5.0 mm. Queensland....... *canaliculatus* Brooks

11. Facial tubercle with a median, longitudinal keel. Stridulatory comb of fore tibia with about 34 pegs (Fig. 20.3C). Length 5.1-5.5 mm. Queensland.. *tahitiensis* Lundblad

– Facial tubercle not longitudinally keeled ... 12

12. Synthlipsis (Fig. 20.1A, sy) three fourths or more anterior width of vertex 13

– Synthlipsis at most just over half anterior width of vertex 17

13. Length not exceeding 5.5 mm. Stridulatory comb of fore tibia with 16-17 pegs (Fig. 20.D). Queensland............ *windi* Brooks

– Length at least 7.1 mm 14

14. Facial tubercle distinctly raised (Fig. 20.3E), basally just above labrum with a dark brown pointed tubercle. Stridulatory comb of fore tibia with about 15 pegs. Length 7.6 mm. Northern Territory... *hayesi* Lansbury

– Facial tubercle not raised, without dark brown tubercle.. 15

15. Smaller species, length 6.05-7.3 mm. Fore tibia with 5-9 spines proximally; stridulatory comb of fore tibia with 16-18 pegs... 16

– Larger species, length 8.5 mm. Fore tibia with about 20 spines proximally; stridulatory comb of fore tibia with 22-23 pegs (Fig. 20.3F). Tasmania ..... *tasmaniensis* Brooks

16. Fore tibia with about 9 spines proximally; stridulatory comb with about 16 large pegs (Fig. 20.3G). Length 7.1-7.3 mm. Northern Territory............... *dostini* Lansbury

– Fore tibia with about 5 spines proximally; stridulatory comb with about 18 large pegs (Fig. 20.3H). Length 6.05 mm. Western Australia .... *ungarinyin* Lansbury

17. Median length of head never more than half median length of pronotum 18

– Median length of head always more than half median length of pronotum 20

18. Length not more than 5.3-5.5 mm. Pronotal disc with an elongate depression

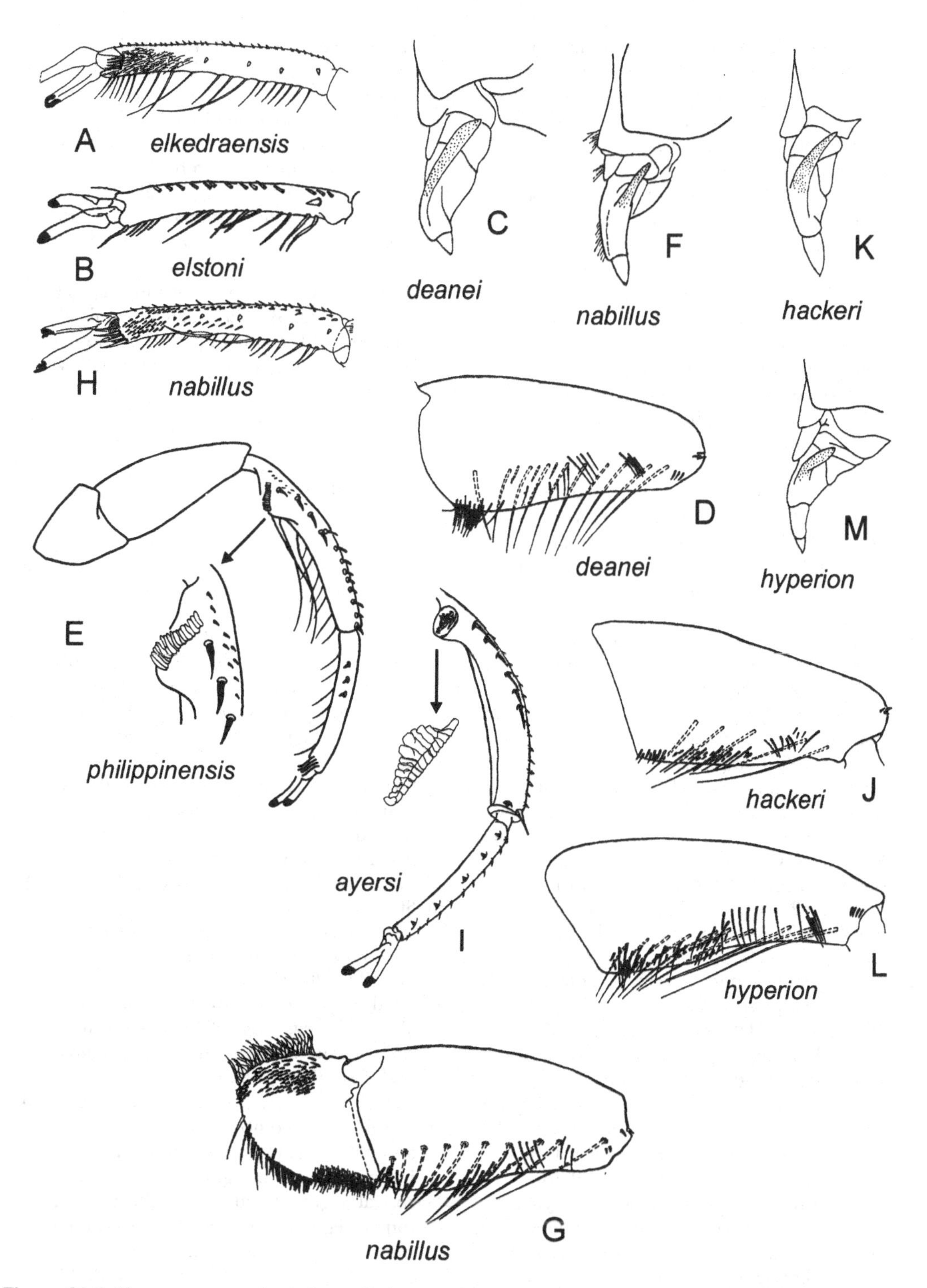

Figure 20.5. NOTONECTIDAE. A, *Anisops elkedraensis*: fore tarsus of ♂. B, *A. elstoni*: fore tarsus of ♂. C-D, *A. deanei*: C, lateral view of rostrum of ♂. D, fore femur of ♂. E, *A. philippinensis*: fore leg of ♂. F-H, *A. nabillus*: F, lateral view of rostrum of ♂. G, fore femur of ♂. H, fore tarsus of ♂. I, *A. ayersi*: fore leg of ♂. J-K, *A. hackeri*: fore femur of ♂. K, lateral view of rostrum of ♂. L-M, *A. hyperion*: fore femur of ♂. M, lateral view of rostrum of ♂. Reproduced with modification from: A, Lansbury (1995a); B-D, F-H, J-M, Lansbury (1969); E, Brooks (1951); I, Reichardt (1982).

on either side of mid-line. Stridulatory comb of fore tibia with about 18 pegs (Fig. 20.3I). Queensland.... *barrenensis* Brooks
- Length at least 6.9 mm. Pronotal disc without depressions.................................... 19

19. Fore femur parallel-sided for about four-fifths of its length, distal one-fifth of inner surface with a curved ridge from dorsal margin to almost mid-width (Fig. 20.3J). Stridulatory comb of fore tibia with 24-36 pegs. Length 6.9-7.5 mm. Australian Capital Territory, New South Wales, Northern Territory, Queensland, South Australia, Tasmania, Victoria, Western Australia......... *thienemanni* Lundblad
- Fore femur tapering in width distally, no ridge on inner surface (Fig. 20.3K). Stridulatory comb of fore tibia with 13-16 pegs (Figs 20.6A, B). Length 7.0-8.5 mm. New South Wales, Northern Territory, Queensland, South Australia, Tasmania, Victoria, Western Australia........... *gratus* Hale

20. Length at least 7.0 mm 21
- Length at most 6.75 mm............................. 23

21. Facial tubercle raised. Stridulatory comb of fore tibia with about 14 pegs. Length 8.3 mm. Queensland...... *planifascies* Lansbury
- Facial tubercles not raised 22

22. Fore tibia with at least 10 stout spines along outer margin (Fig. 20.3L). Stridulatory comb of fore tibia with about 12 pegs. Length 7.5 mm. Northern Territory, Western Australia............ *malkini* Brooks
- Fore tibia with 3-4 stout spines along outer margin. Stridulatory comb of fore tibia with 24-26 pegs. Length 7.7-7.8 mm. Northern Territory, Queensland. .... *occipitalis* Breddin

23. Third rostral segment carinate, anterior surface with two prominent teeth (Fig. 20.3M). Stridulatory comb of fore tibia with 14 large pegs. Length 5.5 mm. Northern Territory, Queensland .... *nodulatus* Brooks
- Third rostral segment not as above............ 24

24. Facial tubercle slightly depressed with a triangular patch of hairs. Stridulatory comb of fore tibia with 11 pegs. Length 4.8 mm. Northern Territory, Queensland .... *paracrinitus* Brooks
- Facial tubercle not as above........................ 25

25. Synthlipsis not more than one fourth anterior width of vertex 26
- Synthlipsis at least one third anterior width of vertex.. 27

26. Fore tarsus with a row of five stout spines along inner margin (Fig. Fig. 20.5A). Stridulatory comb of fore tibia with about 20 pegs. Length 5.4-5.8 mm. Northern Territory........ *elkedraensis* Lansbury
- Fore tarsus with a small proximal group of hairs but no stout spines along inner margin. Stridulatory comb of fore tibia with about 18 pegs. Length 5.25-5.5 mm. New South Wales, Queensland, South Australia .... *paraexigerus* Lansbury

27. Length not exceeding 5.0 mm. Fore tarsus with one stout spine on inner apical fifth (Fig. 20.5B). Stridulatory comb of fore tibia with about 14 pegs. New South Wales, Queensland, South Australia, Tasmania, Victoria, Western Australia .... *elstoni* Brooks
- Length 5.5-6.75 mm. Fore tarsus with 3-6 stout spines on inner surface....................... 28

28. Apex of third rostral segment wider than base of fourth segment 29
- Apex of third rostral segment not wider than base of fourth segment....................... 32

29. Stridulatory comb of fore tibia with 22-28 pegs. Rostral prong longer than third rostral segment (Fig. 20.5C). Apex of fore femur rounded (Fig. 20.5D). Length 5.4-6.0 mm. New South Wales, Northern Territory, Queensland, South Australia, Tasmania, Victoria, Western Australia..... .... *deanei* Brooks
- Stridulatory comb of fore tibia with less than 20 pegs. Rostral prong not longer than third rostral segment (Fig. 20.5F). Apex of fore femur truncate or acuminate, distal margin straight (Fig. 20.5G) 30

30. Stridulatory comb of fore tibia with 12-15 pegs. Fore tarsus with 4-6 more or less evenly spaced stout spines on inner surface (Fig. 20.5H).. 31
- Stridulatory comb of fore tibia with about 17 pegs (Fig. 20.5E). Fore tarsus with 3-4 stout spines on inner proximal half. Length 5.4-5.8 mm. Queensland..... .... *philippinensis* Brooks

31. Stridulatory comb of fore tibia with about 12 pegs increasing in length from inner to outer margins. Apex of third rostral segment wider than the base of fourth segment (Fig. 20.5F). Length 5.9 mm. Western Australia........ *nabillus* Lansbury
- Stridulatory comb of fore tibia with 14-15 pegs increasing in length from inner and outer margins toward the centre (Fig. 20.5I). Third and fourth rostral segments of same width at the junction. Length 5.7-5.9 mm. Northern Territory .... *ayersi* Reichart

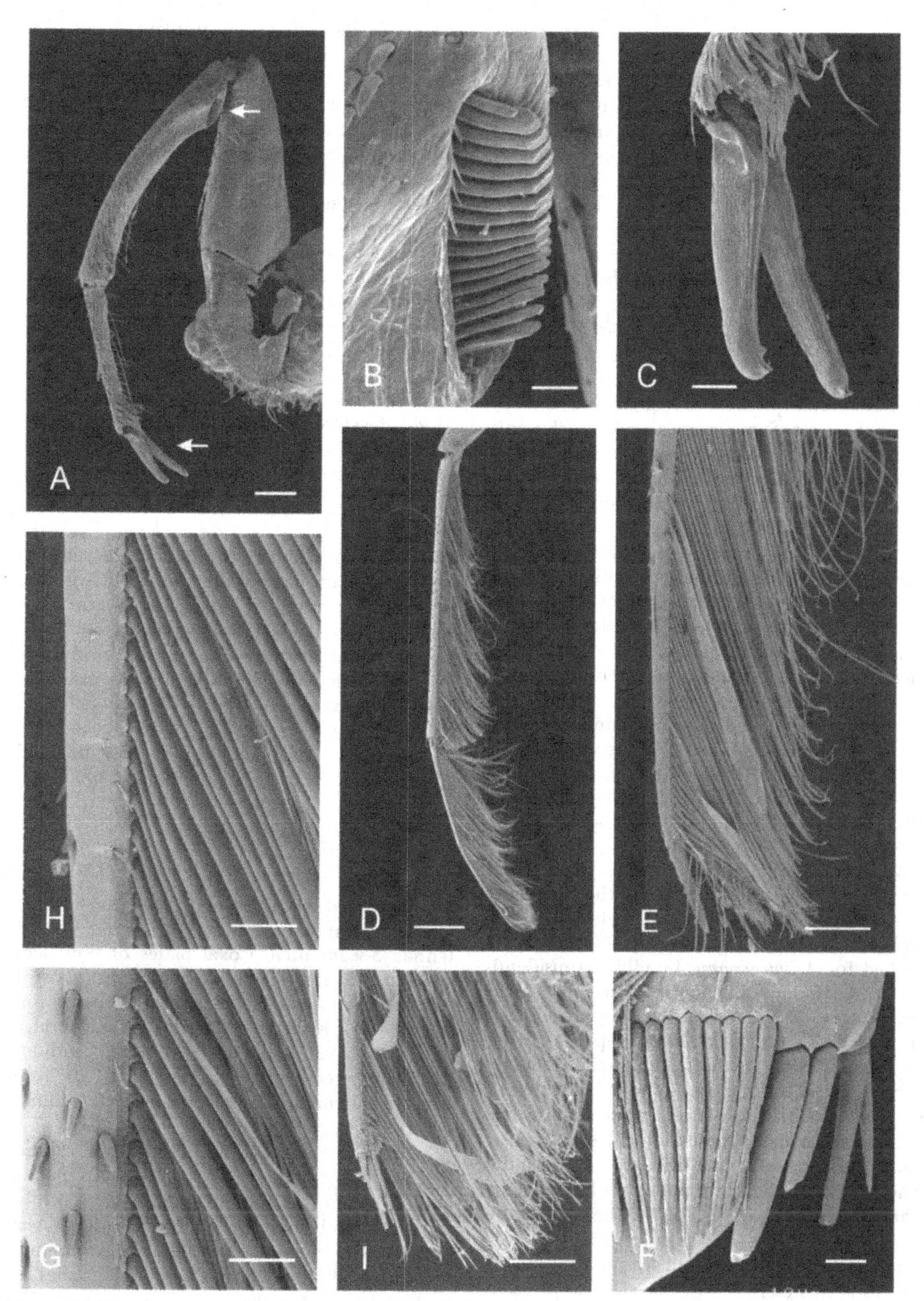

Figure 20.6. NOTONECTIDAE. A-C, *Anisops gratus*, ♂: A, fore leg. B, stridulatory pegs on fore tibia. C, claws of fore leg. D-I, *Anisops stali*, ♀: D, hind tibia and tarsus; E, swimming hairs on hind tarsus. F, junction between hind tibia and tarsus.l G, detail of hind tibia. H, detail of hind tarsus. I, apex of hind tarsus. Scales: A, 0.2 mm; B, F, 0.02 mm; C, G, H, 0.05 mm; D, 0.5 mm; E, 0.01 mm. Scanning Electron Micrographs (CSIRO Entomology, Canberra).

32. Length not exceeding 6.25 mm. Apex of fore femur truncate (Fig. 20.5J). Rostral prong evenly curved (Fig. 20.5K). Stridulatory comb of fore tibia with 23-26 pegs. New South Wales, Queensland, Tasmania, Western Australia .... *hackeri* Brooks
– Length 6.25-6.75 mm. Apex of fore femur acuminate (Fig. 20.5L). Rostral prong angular (Fig. 20.5M). Stridulatory comb of fore tibia with 21 pegs. New South Wales, Queensland, South Australia, Victoria, Western Australia................ .... *hyperion* Kirkaldy

Biology

Species of *Anisops* inhabit stagnant or virtually stagnant waters. Some species prefer rock pools or other pools on banks of streams, but several widespread species are often found in small ponds and in rice fields. They frequent a range of stagnant freshwater habitats such as pools, ponds, and lakes and some species are most abundant in coastal areas where the heavier rainfall ensures an adequate supply of habitats. Other species may have better dispersal powers and thus be better adapted to invade habitats in arid inland areas such as man-made dams and waterholes which are often isolated and many kilometres apart (Sweeney 1960). Food consists mainly of cladocerans, copepods, and ostracods as well as mosquito larvae, but any prey the bugs can subdue will be taken when they are starved (Reynolds 1981; Reynolds & Geddes 1984;Gilbert & Burns 1999).

Males of several *Anisops* species have been recorded to stridulate by rubbing the fore tibial stridulatory comb against the rostral "prong". Stridulation is used in courtship which was described for *A. thienemanni* Lundblad (misidentified as *A. hyperion* Kirkaldy) by Hale (1923) who characterises the sound as "a distant grindstone at work". A male stridulating rapidly during courtship positions itself beneath and slightly behind a female. In this position it follows every movement of the female for a while until it tries to grasp the female. Copulation takes an hour or more and the mating pair swim together with the male below and slightly to the right of the female. The finger-like claws of the fore legs help the male to embrace the female. Hale (1923) described and illustrated all nymphal instars as well as adults of *A. thienemann.* He recorded at least two annual generations in South Australia, eggs being deposited as late in the season as April, the progeny overwintering as adults and breeding the following summer. Copulation was noted at the beginning of August.

Eggs are deposited singly in plant tissues (Figs 2.8B, C). The female drills a hole in the selected plant stem and deposits an egg into it, leaving the anterior end of the egg exposed (Hale 1923). The eggs are elongate oval with a thin, transparent and smooth wall except for an area at the anterior end which shows weak hexagonal sculpturing. This area forms a lid (operculum), is positioned at the opening of the slit in the plant tissue and opens when the nymph hatches. Nymphal development of *Anisops thienemanni,* a middle-sized species (length 6-7 mm), takes about two months in South Australia (Hale 1923). In laboratory rearing of *A. breddini* Kirkaldy, a species of similar size, Leong (1962) recorded development times of 2-3 months in Singapore. He suggests that the longer development time may be due to lower temperatures with greater fluctuations in the laboratory compared to field conditions.

Distribution

The genus is distributed throughout the subtropical and tropical parts of the Old World, with the greatest concentration of species in Southeast Asia and Australia where the genus is also abundant in the temperate South. *Anisops* is probably the most widely distributed genus of Australian water bugs recorded from all states, including the interior of New South Wales, Queensland, Northern Territory, and South Australia, and large parts of Western Australia (Fig. 20.7).

PARANISOPS

Identification

Medium-sized backswimmers, length 6.5-8.75 mm, very similar to *Anisops* (Fig. 20.8A). Antennae 3-segmented. Coxal plates of hind legs covered with long black hairs. Macropterous or submacropterous ("brachypterous" sensu Lansbury 1964b). Rostrum of ♂ with a small "prong" on third labial segment. Fore tibiae without stridulatory teeth or pegs (Figs 20.8B, D); fore tarsi 2-segmented. Genital capsule cleft behind; parameres large.

The genus *Paranisops* Hale (1924a) was revised by Lansbury (1964b) who recognised two species and one variety, all from Australia. More recently, the genus has been recorded from Southeast Asia (Thailand; Nieser & Zettel 2001). The following key (adapted from Lansbury 1964b) applies to both males and females.

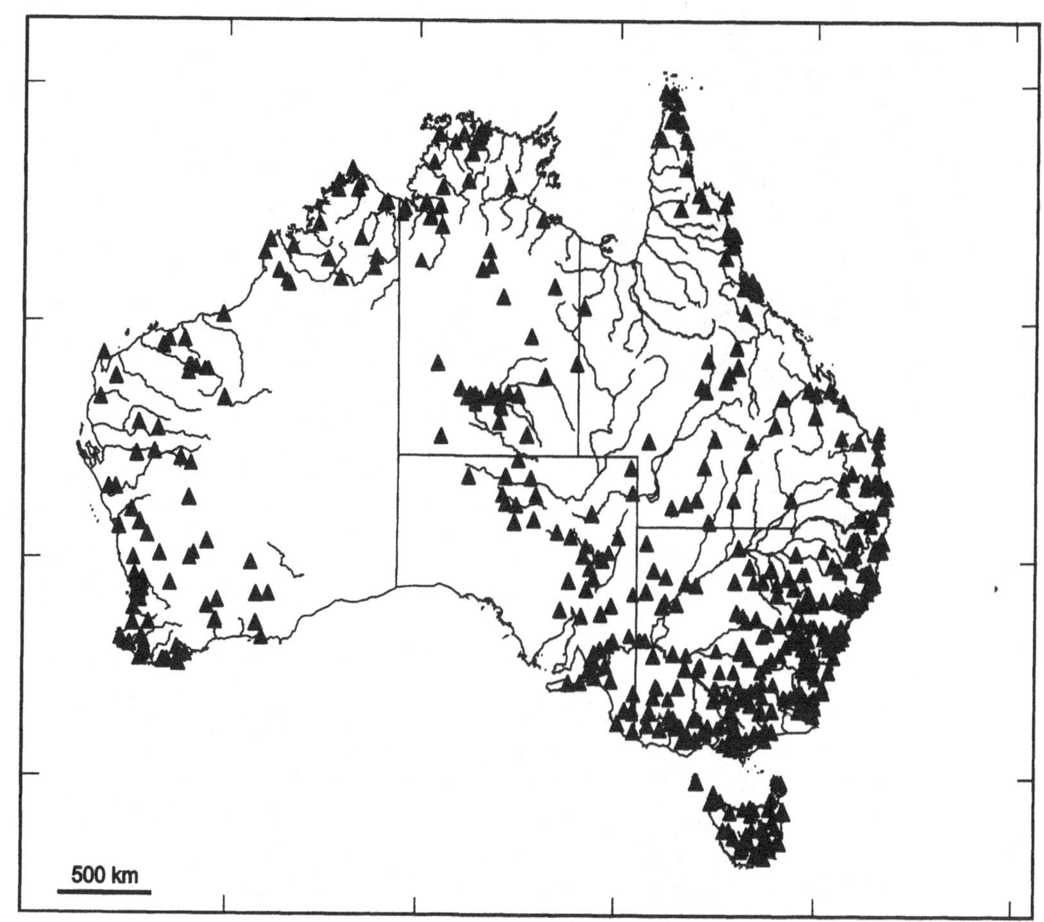

Figure 20.7. NOTONECTIDAE. Distribution of *Anisops* species in Australia.

Key to Australian species

1. Small species, length 6.5-7 mm. Male parameres asymmetrical. Inner surface of male fore tibiae with one stout spine and 6-8 short, spatulate hairs (Fig. 20.8B), that of female with two stout spines and no spatulate hairs. Female pronotum and posterior margin of head simple. New South Wales, Queensland .... *inconstans* Hale

– Large species, length 7.5-8.75 mm. Male parameres symmetrical. Inner surface of male fore tibiae with a row of stout spines (Fig. 20.8D); chaetotaxy of female fore tibiae identical to that of male. Lateral posterior margin of female head with a prominent lobe; lateral angles of female pronotum produced as a distinct nodule (Fig. 20.8C; pp). Western Australia.......... .... *endymion* (Kirkaldy)

Biology

The biology of species of *Paranisops* is virtually unknown, but probably quite similar to that of species of the related genus *Anisops* (see above). Both Australian species occur in a dark and a pale adult form (illustrated by Lansbury 1964b: fig. 1), the latter having reduced flight muscles but fully developed hind wings. For *P. inconstans*, this form was named var. *lutea* by Hale (1924a). Lansbury (1964b) described ripe ovarian eggs of *P. endymion* which are elongate oval and lack the operculum known from *Anisops*.

Distribution

The genus *Paranisops* has a disjunct distribution in Australia (Fig. 20.9). *P. inconstans* is recorded from coastal localities in southern Queensland and northern New South Wales. The record from northern Queensland represents an undescribed species. *P. endymion* is only known from southern Western Australia.

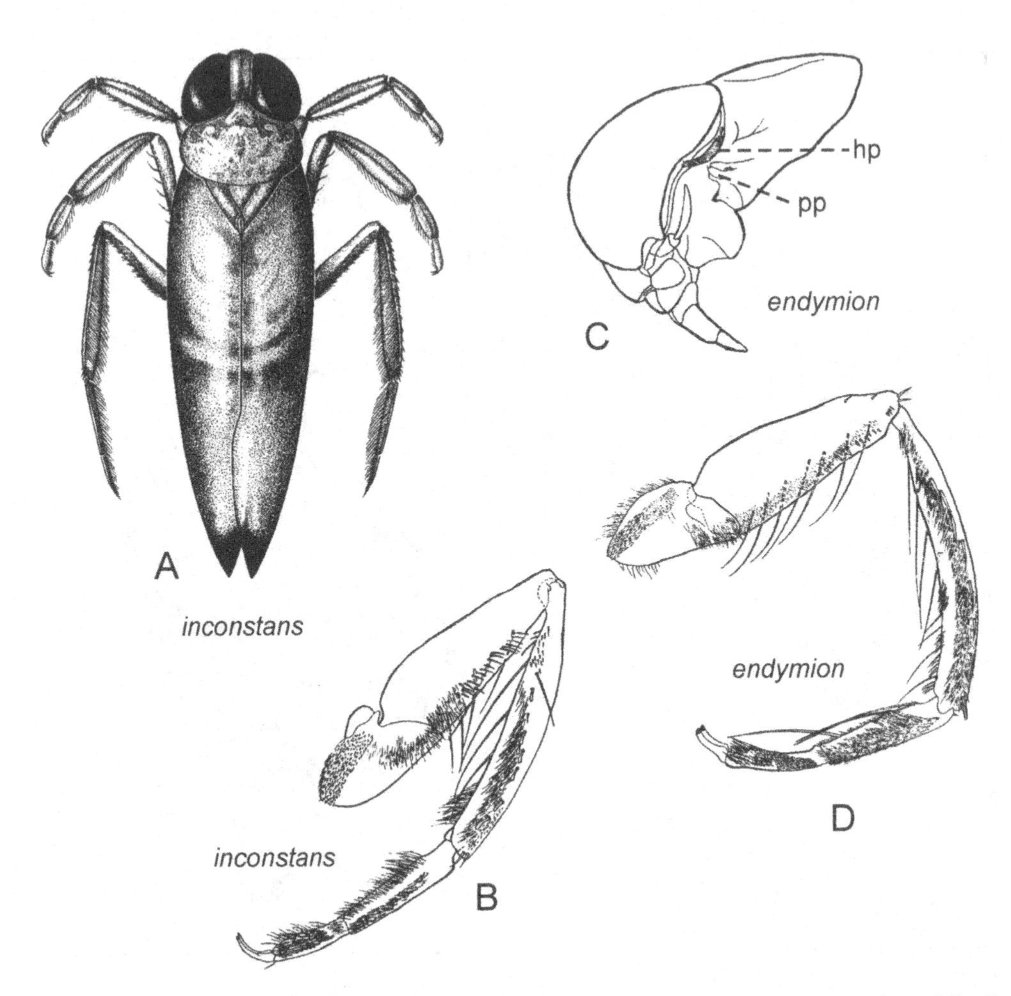

Figure 20.8. NOTONECTIDAE. A-B, *Paranisops inconstans*: A, dorsal habitus. B, fore leg of ♂. C-D, *P. endymion*: C, lateral view of head of ♀ (hp, hair-pit; pp, pronotal process). D, fore leg of ♂. Reproduced with modification from: A, CSIRO (1991); B-D, Lansbury (1964b).

WALAMBIANISOPS

Identification

Medium-sized backswimmers, length 9-9.9 mm, very similar to *Anisops*. Antennae two-segmented. Coxal plates of hind legs bare and shining, distally fringed with black hairs. Always macropterous. Male: rostrum without a lateral "prong" on third labial segment (Fig. 20.10B). Fore tibiae with stridulatory pegs clearly separated and not situated on a protuberance (Fig. 20.10D, E); fore tarsi two-segmented. Genital capsule partially cleft behind; parameres large, asymmetrical.

The genus *Walambianisops* Lansbury contains only one species, *W. wandjina* Lansbury (1984). Length 9.0-9.9 mm. Northern Territory, Western Australia.

Biology

Unknown.

Distribution

Walambianisops wandjina has been recorded from several localities in the Kimberley District, Western Australia, and a single locality in Northern Territory (Fig. 20.11).

SUBFAMILY NOTONECTINAE

Identification

Claval commissure of hemelytra without a pit anteriorly, near the apex of scutellum (Fig. 20.1B). Males lack stridulatory structures on fore legs. Metathoracic scent glands present. Obvious-

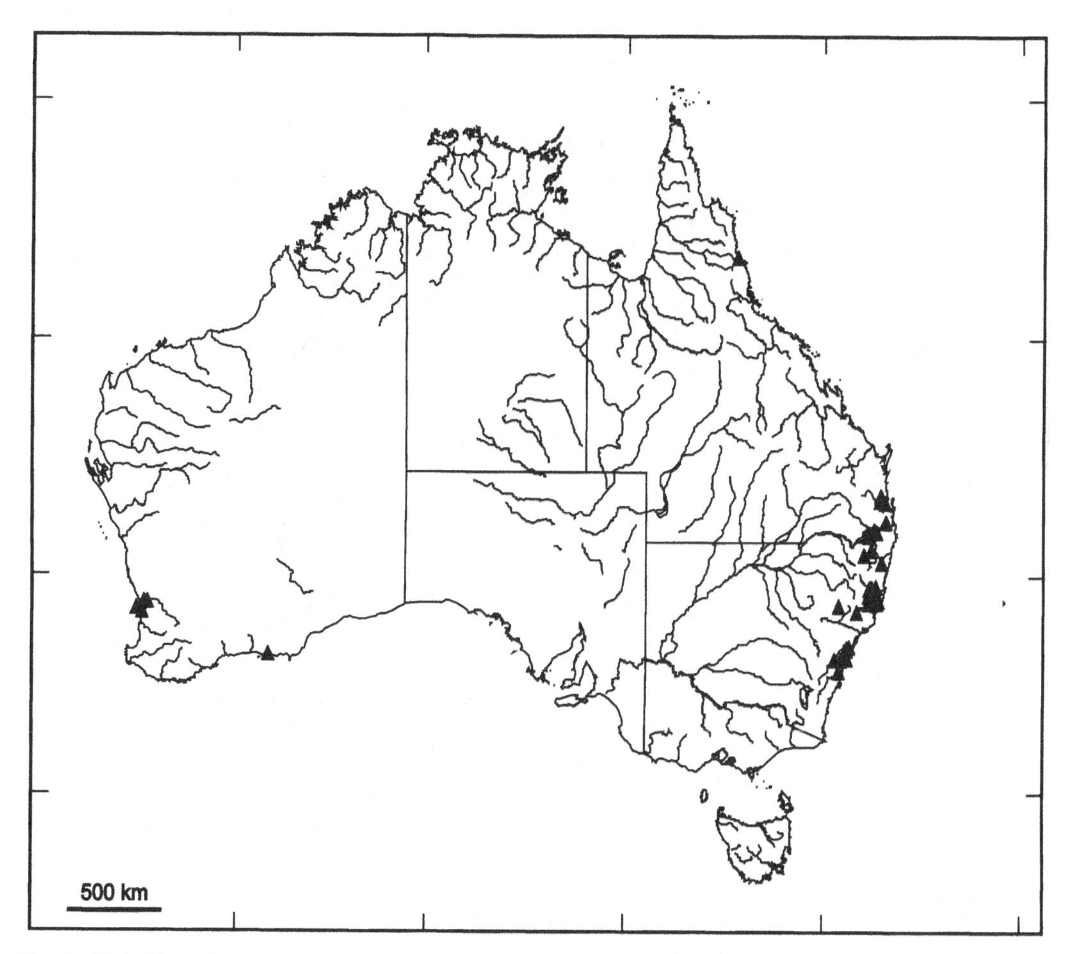

Figure 20.9. NOTONECTIDAE. Distribution of *Paranisops* species in Australia.

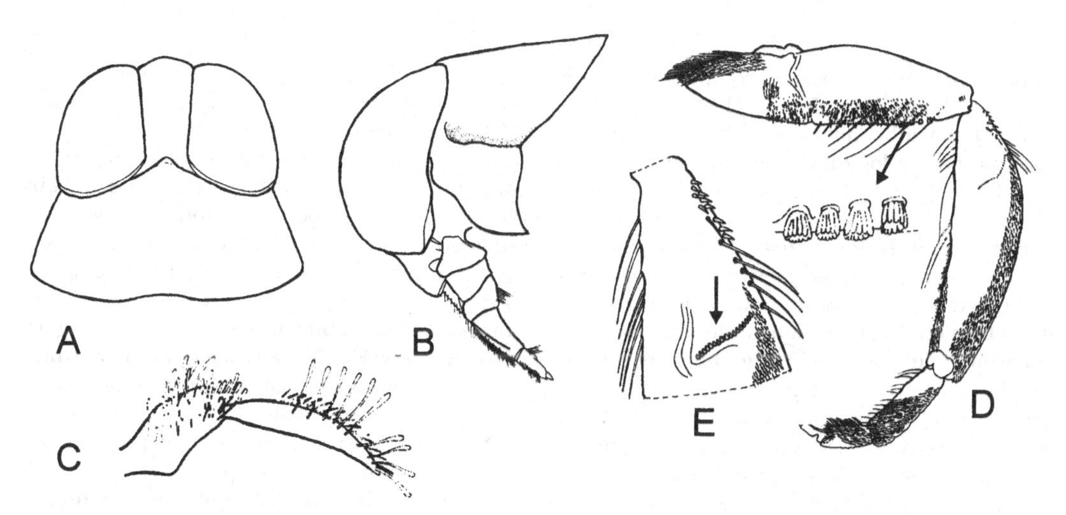

Figure 20.10. NOTONECTIDAE. A-E, *Walambianisops wandjina*: A, dorsal view of head of ♂. B, lateral view of head of ♂. C, antenna. D, fore leg of ♂. E, basal part of fore tibia of ♂. Reproduced with modification from Lansbury (1984).

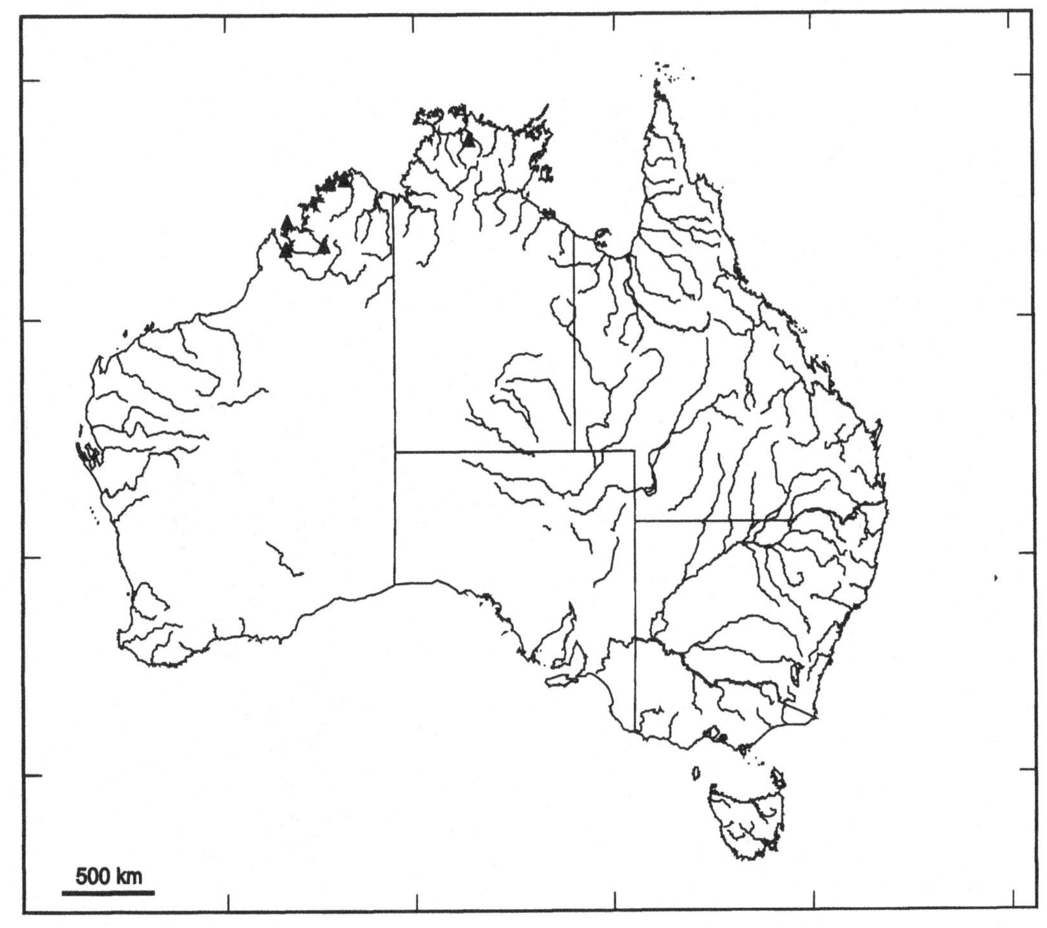

Figure 20.11. NOTONECTIDAE. Distribution of *Walambianisops wandjina* in Australia.

ly, none of these characters are unique apomorphies and as noted by Mahner (1993), the monophyly of the subfamily cannot be supported, at least on morphological characters.

Overview

The subfamily Notonectinae is far less species-rich in Australia than the previous one and only a few species of *Enithares* are common and widespread. The most thorough studies on predator behaviour and strategies in water bugs are on species belonging to the genus *Notonecta*, but the general results may also apply to species of the genus *Enithares*. These bugs detect their prey by both visual and vibratory signals. They either wait at the water surface or perch within the water body, clinging to submerged plants, until a prey comes close enough to be observed after which it is pursued and captured. Foraging behaviour is dependent on both temperature, oxygen content of the water, and prey density. Giller & McNeill (1981) compared the foraging strategies of three European *Notonecta* species, convincingly demonstrating that they are related to the type of habitat in which these species are found. *N. maculata* Fabricius forages near the water surface and is found in habitats where the vegetation is sparse and the prey density generally low. *N. glauca* L. forages at mid depths in habitats rich in vegetation and prey. Finally, *N. obliqua* Thunberg hunts in deep waters in habitats with moderate vegetation and low prey density.

Notonectines are described as polyphagous carnivores, utilizing any type of prey they can overpower. Using an electrophoretic method of gut content analysis, Giller (1986) found that the main diet of two species of *Notonecta* was *Cyclops* spp. (Copepoda) and *Cloeon dipterum* (L.) (Ephe-

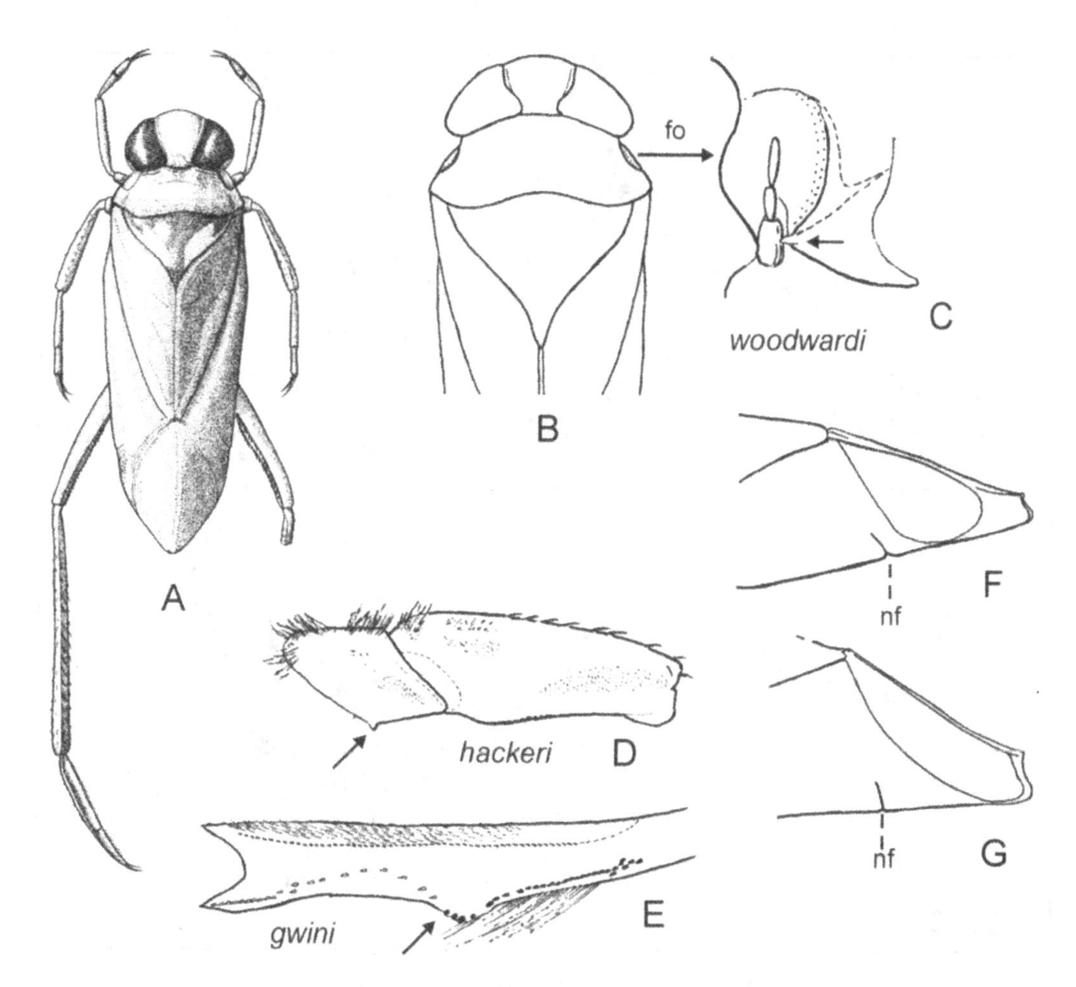

Figure 20.12. NOTONECTIDAE. A-C, *Enithares woodwardi*: A, dorsal habitus. B, dorsal view of head and thorax (fo, pronotal fovea). C, lateral view of pronotal fovea. D, *E. hackeri*: trochanter and fore femur of ♂. E, *E. gwini*: hind femur of ♂. F, *E. chinensis*: distal end of hemelytra (nf, nodal furrow). G, *E. biimpressa*: distal end of hemelytra. Reproduced with modification from: A-B, Lansbury & Lake (2002); C-D, F-G, Lansbury (1968); E, Lansbury (1984).

meroptera), the former being restricted to earlier nymphal instars and replaced by the latter in the last two instars. However, there were significant differences between species in the use of *Sigara lateralis* (Leach) (Corixidae), and larvae of Diptera Ceratopogonidae, and Coleoptera-Dytiscidae.

ENITHARES

Identification

Medium sized to large backswimmers, length 6-11 mm (Fig. 20.12A). Head narrower than pronotum (Fig. 20.12B). Eyes large, kidney-shaped; inner margins of eyes in dorsal view converging posteriorly, but widely separated from each other. Pronotum broader than long with antero-lateral margins each carrying a distinct, trough-shaped fovea (Fig. 20.12B, C; fo). Antennae four-segmented, inserted behind eyes and lying in the pronotal fovea. Corium of fore wing with a cleft known as the nodal furrow just anterior of the membrane (Fig. 20.1C; nf). Middle femur with a large, pointed process on ventral margin before apex (Fig. 20.1F). Fore and middle tarsi 3-segmented, hind tarsi 2-segmented. Male genital capsule divided by a cleft in an anterior and a posterior lobe. Parameres symmetrical, usually small, situated in the cleft. Female with short gonapophyses. Most species have pale and dark adult forms, the pale form being chiefly yellowish

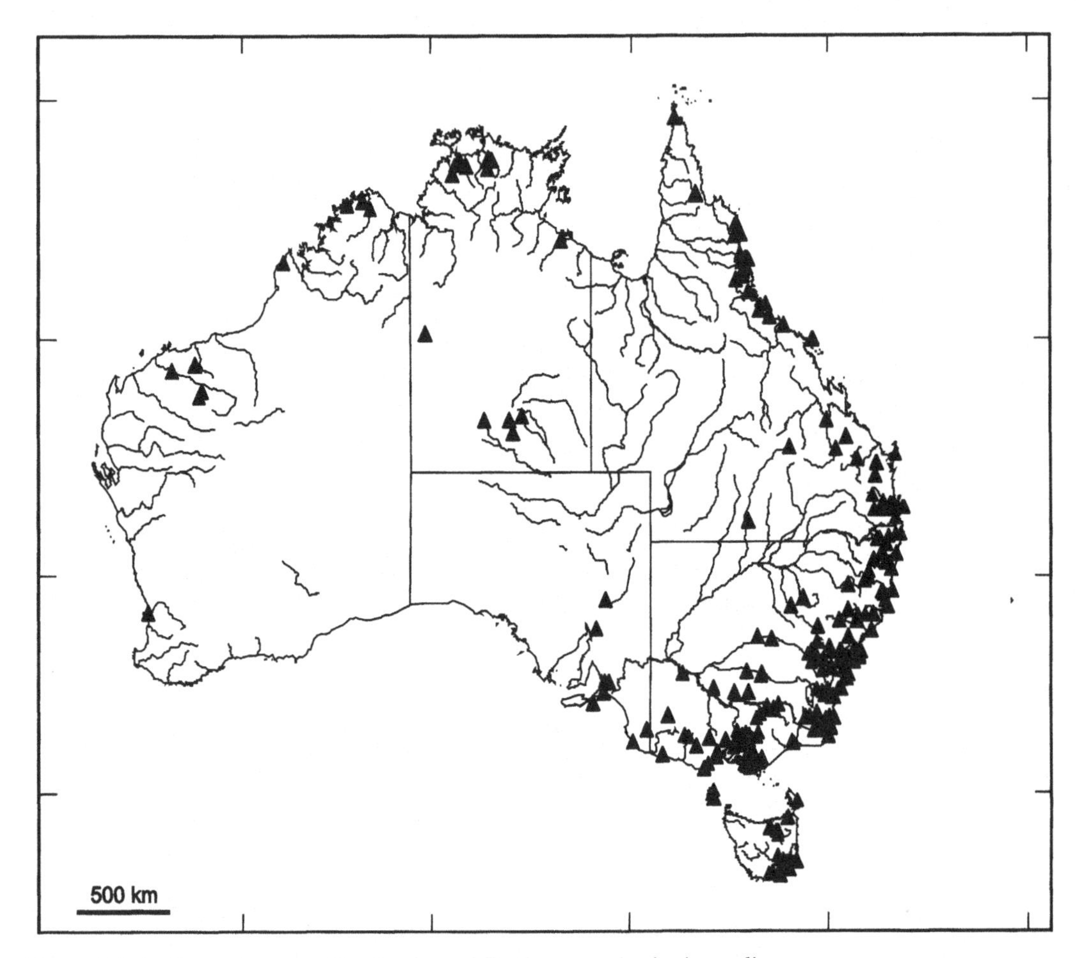

Figure 20.13. NOTONECTIDAE. Distribution of *Enithares* species in Australia.

brown, the dark form often being entirely black, but sometimes showing a variable pattern of pale marks (Plate 8, H-I).

The genus *Enithares* Spinola comprises 48 species. Five of these occur in Australia of which two are endemic to the continent. Their worldwide distribution includes tropical Asia, New Guinea, and Australia. The latest taxonomic revision is by Lansbury (1968). One additional Australian species, *E. gwini*, was added by Lansbury (1984). The following key is adapted from these works. (Note: the two sexes can be separated by the structure of the ventral abdominal end which in females show the protruding, serrated gonapophyses).

Key to Australian species

1. Males 2
– Females 6
2. Fore trochanter with a nodule on ventral surface (Fig. 20.12D). Length 9.5 mm. Australian Capital Territory, New South Wales, Queensland, Victoria .... *hackeri* Hungerford
– Fore trochanter without a nodule on ventral surface 3
3. Hind femur produced distal-ventrally (Fig. 20.12E). Length at least 11.0 mm. Western Australia .... *gwini* Lansbury
– Hind femur simple 4
4. Head as long as or longer than median pronotal length. Length 6-6.9 mm. Northern Territory, Queensland, South Australia .... *loria* Brooks
– Head clearly shorter than median pronotal length 5
5. Lateral margins of pronotal fovea produced towards eyes as a spur or tooth (Fig. 20.12C; arrow). Length 9.5-11 mm. Australian Capital Territory, New South

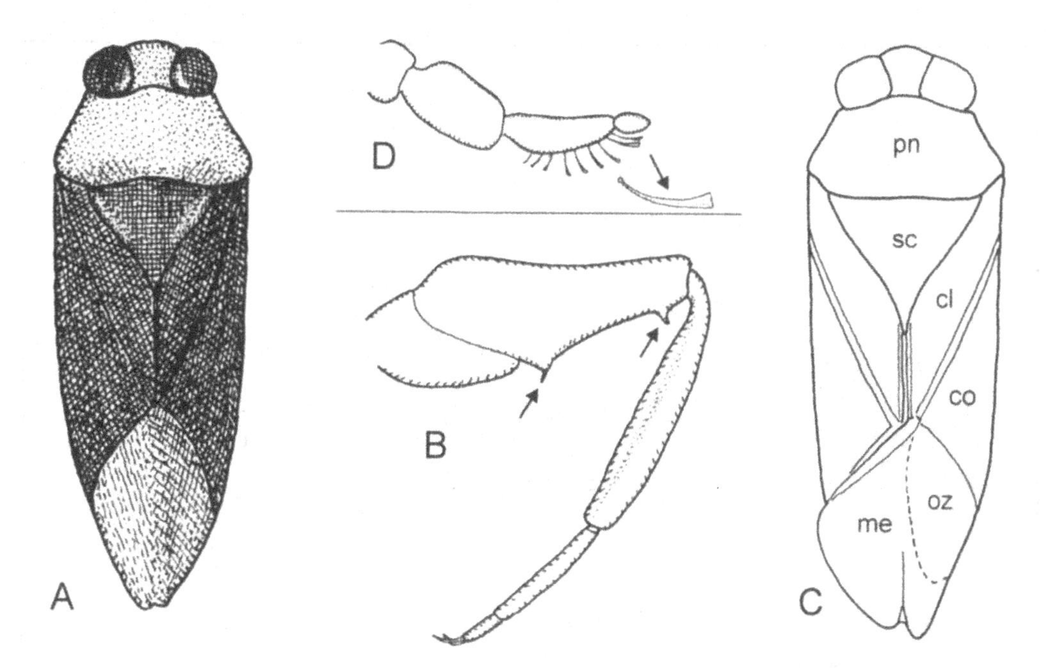

Figure 20.14. NOTONECTIDAE. A-B, *Notonecta* (*Enitharonecta*) *handlirschi*: dorsal habitus of ♂; legs omitted. B, middle leg with ventral femoral processes (arrows). C-D, *Notonecta* sp.: C, dorsal view of body; membrane folded out (cl, clavus; co, corium; me, membrane; oz, opaque zone of membrane; pn, pronotum; sc, scutellum). D, antenna with spatulate hair enlarged. Original, redrawn from: A-C, Hungerford (1933); D, Weber (1930).

Wales, Northern Territory, Queensland, South Australia, Tasmania, Victoria, Western Australia .... *woodwardi* Lansbury

– Lateral margins of pronotal fovea not as above. Length 9-10 mm. Northern Territory .... *atra* Brooks

6. Nodal furrow clearly more than its own length removed from membranal suture (as in Fig. 20.12G; nf). Length less than 7.5 mm. Northern Territory, Queensland, South Australia.................... *loria* Brooks

– Nodal furrow equal to or less than its own length removed from membranal suture (as in Fig. 20.12F; nf). Length greater than 9.0 mm 7

7. Pronotal humeral width 3x or more median length. Length 9-10 mm. Northern Territory .... *atra* Brooks

– Pronotal humeral width less than 3x median length 8

8. Lateral margin of pronotal fovea evenly rounded. Length at least 11.0 mm. Western Australia.......................... *gwini* Lansbury

– Lateral margin of pronotal fovea distinctly angulate (Fig. 20.12C). Length 9.5-11.0 mm 9

9. Head longer than pronotum. Clavus and corium of hemelytra densely punctate. Length 9.5 mm. Australian Capital Territory, New South Wales, Queensland, Victoria.................................. *hackeri* Hungerford

– Head shorter than pronotum. Clavus and corium of hemelytra more sparsely punctate. Length 9.5-11 mm. Australian Capital Territory, New South Wales, Northern Territory, Queensland, South Australia, Tasmania, Victoria, Western Australia .... *woodwardi* Lansbury

Biology

Little is known of the biology and ecology of *Enithares*. Some data on life histories are given by Hoffmann (1931) and Lansbury (1968). In these studies the eggs were found glued on submerged plant tissue. As females have relatively short gonapophyses (ovipositor), this is probably the rule in this genus. The eggs have a simple elongate shape with rounded ends (as in *Notonecta*, Fig. 2.8G). The eggs are of variable size and the number of eggs per female is also quite variable. These differences may represent different reproductive strategies, but how this would be effectuated is

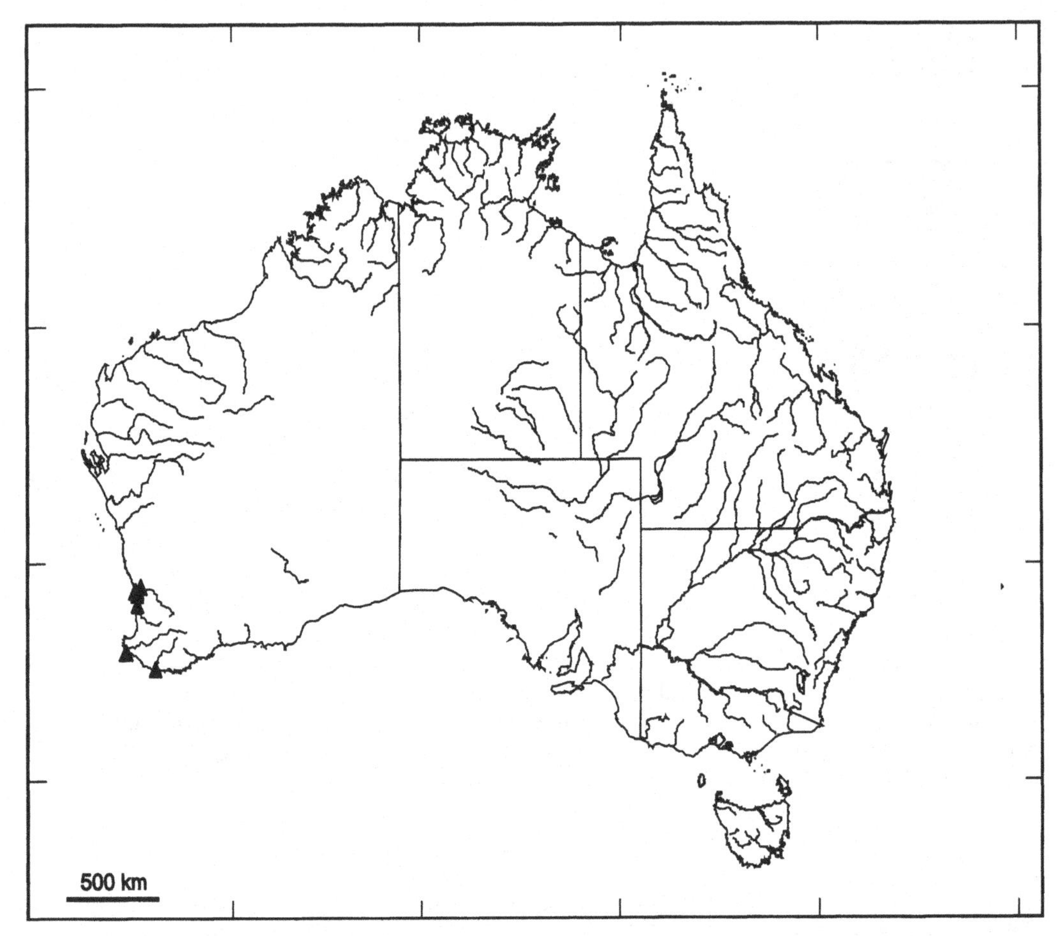

Figure 20.15. NOTONECTIDAE. Distribution of *Notonecta* (*Enitharonecta*) *handlirschi* in Australia.

unknown. The nymphs of *Enithares* are all pale with a dark transverse band (Chen et al. in press).

Based on field studies in Sulawesi, Chen et al. (in press) report two types of *Enithares* species with different general ecological strategies. Some smaller species (length 7.5-8 mm) principally live in roadside and village ponds and pools without fish. Several larger species (9.4-12.8 mm) live in virtually stagnant parts of small streams or in pools associated with forest streams. Usually one finds only a few late instar larvae or one or two adults in a pool. This led Nieser & Chen (1991) to the following hypothesis about the life history strategy of stream inhabiting species: Gravid *Enithares* females seek out suitable pools to deposit a number of eggs. The nymphs start with using other kinds of food, but in the end, when supplies become exhausted, resort to cannibalism. In this way the species stores energy to produce a few adults from a large population of nymphs.

Distribution

The genus *Enithares* is widely distributed in tropical Asia, New Guinea, and in Australia where species are recorded from all states. Most Australian records are from eastern Queensland, New South Wales, Victoria, and Tasmania (Fig. 20.13).

NOTONECTA

Identification

Large backswimmers (length of Australian species 10-10.5 mm). Head narrower than pronotum (Fig. 20.14C; pn). Eyes large, kidney-shaped; inner margins of eyes in dorsal view converging posteriorly, but widely separated from each other. Pronotum broader than long; antero-lateral margins not foveate. Antennae four-segmented (Fig. 20.14D). Ventral margin of middle femur with a pointed process before apex and sometimes also

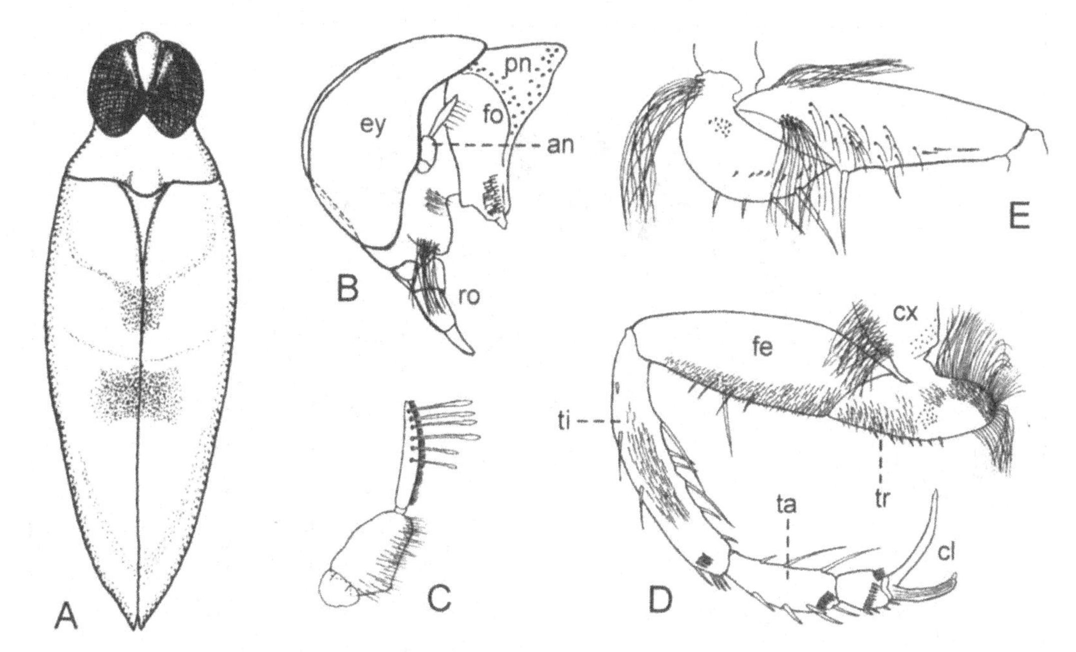

Figure 20.16. NOTONECTIDAE. A-E, *Nychia sappho*: A, dorsal habitus of submacropterous ♀; legs omitted. B, lateral view of head of ♂ (an, antenna; ey, eye; fo, pronotal fovea; pn, pronotum; ro, rostrum). C, antenna. D, fore leg of ♂ (cl, claws; cx, coxa; fe, femur; ta, tarsus; ti, tibia; tr, trochanter). E, middle femur of ♂. A, original. B-E, reproduced with modification from Lansbury (1985a).

one basally (Fig. 20.14B; arrows). Fore and middle tarsi three-segmented (basal segment very small), hind tarsi two-segmented. Male genital capsule divided by a cleft in an anterior and a posterior lobe; parameres symmetrical, situated in the cleft. Female with short gonapophyses.

The genus *Notonecta* Linnaeus is distributed world-wide with about 80 species and subspecies. Hungerford (1933) divided the genus into five subgenera. The single Australian species, *N. handlirschi* Kirkaldy (1897b) belongs to the monotypic subgenus *Enitharonecta* Hungerford. The Australian species has the following additional characters: length 10-10.5 mm; head and pronotum yellowish, hemelytra dark reddish, sometimes nearly black (Fig. 20.14A); middle femur ventrally with a basal spine (Fig. 20.14B); last two abdominal sternites of male long and slender; male genital capsule produced caudally on its ventral midline; parameres large thin plates.

Biology

Most knowledge of the biology of backswimmers pertain to northern temperate species of the genus *Notonecta* (see above). Because of the extreme scarcity of observations of the single Australian species, nothing is known about its biology and ecology.

Distribution

Of the about 80 described species and subspecies of *Notonecta*, most occur in North and South America and in the Palearctic Region (Hungerford 1933). Two species have been recorded from Australia, but reliable material exists only for *N. handlirschi* (Hungerford 1933) and all records are from southern Western Australia (Fig. 20.15). Cassis & Gross (1995) listed *N. australis* Olivier (1811) under this genus, but no type material has been examined, and it is quite likely that this species belongs to the genus *Anisops* as suggested by Hale (1923).

NYCHIA

Identification

Small, elongate oval backswimmers, length 4.0-5.0 mm. Head not as broad as pronotum (Fig. 20.1G). Eyes very large, kidney-shaped to semicircular, inner margins in dorsal view converging, touching each other in posterior third (Fig. 20.1G; arrow). Antennae 3-segmented (Fig.

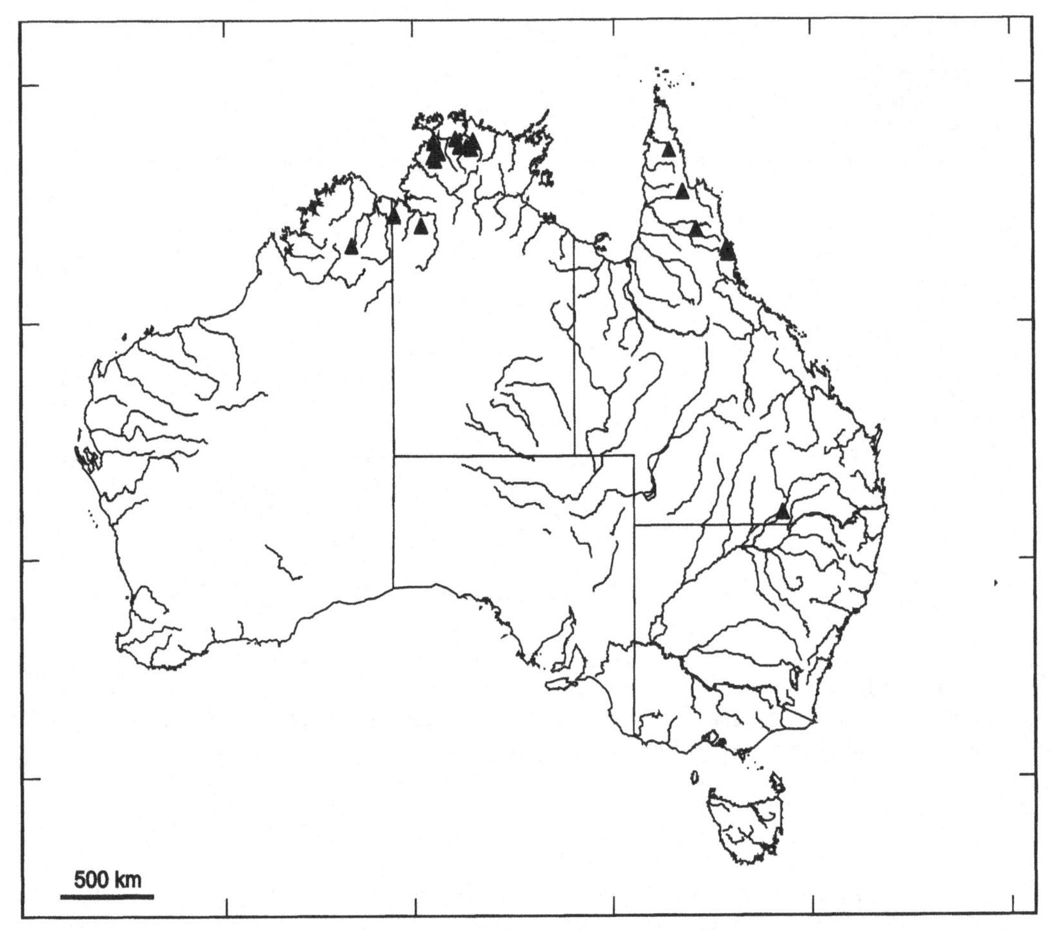

Figure 20.17. NOTONECTIDAE. Distribution of *Nychia sappho* in Australia.

20.16C). Pronotum broader than long with antero-lateral margins distinctly foveate (Figs 20.1G, 20.16B; fo). Claval commissure of hemelytra continuous, i.e. without a hair-lined pit near the apex of scutellum. Corium of fore wing without nodal furrow. Species predominantly submacropterous ("brachypterous" sensu Lansbury 1985a). Middle femur without a pointed anteapical protuberance, but with a pair of stout bristles ventrally in basal half (Fig. 20.16E). Fore and middle tarsi two-segmented in males (Fig. 20.16D), one-segmented in females. Male genital capsule not divided in an anterior and a posterior lobe; parameres asymmetrical, small, situated on the dorsal margin of the capsule. Female with short gonapophyses.

The genus *Nychia* Stål has only two confirmed species although as many as six species have been described (Lundblad 1933; Lansbury 1985a; Chen et al. in press). The single Australian species, *N. sappho* Kirkaldy, is whitish with part of the hemelytra translucent, especially in the brachypterous form. Females may show one or two dark spots on the median margin of the hemelytra (Fig. 20.16A). Length 4.0-4.7 mm. Northern Territory, Queensland, and Western Australia.

Biology

The preferred habitat of *Nychia* species is stagnant stretches of streams or, in the dry season, billabongs in stream beds. However, they have been occasionally collected in ponds not connected with a stream (Lansbury 1985a; Chen et al. in press). The biology of species belonging to *Nychia* is virtually unknown. They usually rest against the surface film or swim slowly just beneath is. When disturbed they dive to deeper water. The relatively short and non-serrate female ovipositor suggests that the eggs are deposited superficially on the substrate (Lansbury 1985a; Chen et al. in press).

Distribution

Only one species, *Nychia sappho*, occurs in Australia and has been recorded from Northern Territory, northern Queensland, and northern Western Australia (Fig. 20.17). The single record from southern Queensland is enigmatic. This species seems to be widely distributed in the Oriental Region and New Guinea (Lansbury 1985a).

21. Family PLEIDAE

Pygmy Backswimmers

Identification

Pale, greenish yellow, semi-globular and heavily punctured water bugs, which are small, body length from 1.5 to 3.3 mm. Head very broad and short, immobile relative to thorax (but not fused with thorax as in the Helotrephidae). Eyes relatively small; ocelli absent. Antennae short and hidden in a groove beneath the eyes, 3-segmented (Fig. 21.1H; an). Rostrum (ro) 4-segmented, short (Figs 21.2A, B). Scutellum relatively large (Fig. 21.1C; sc). All legs apparently cursorial although the hind legs are longer and have some swimming hairs (Figs 21.1F, 21.2E). Fore and middle tarsi 2- or 3-segmented, hind tarsi 3-segmented; all legs with 2 claws. Hemelytra coriaceous, meeting along midline, membrane not recognisable (Fig. 21.1E); hind wings sometimes reduced. Nymphal dorsal abdominal scent gland

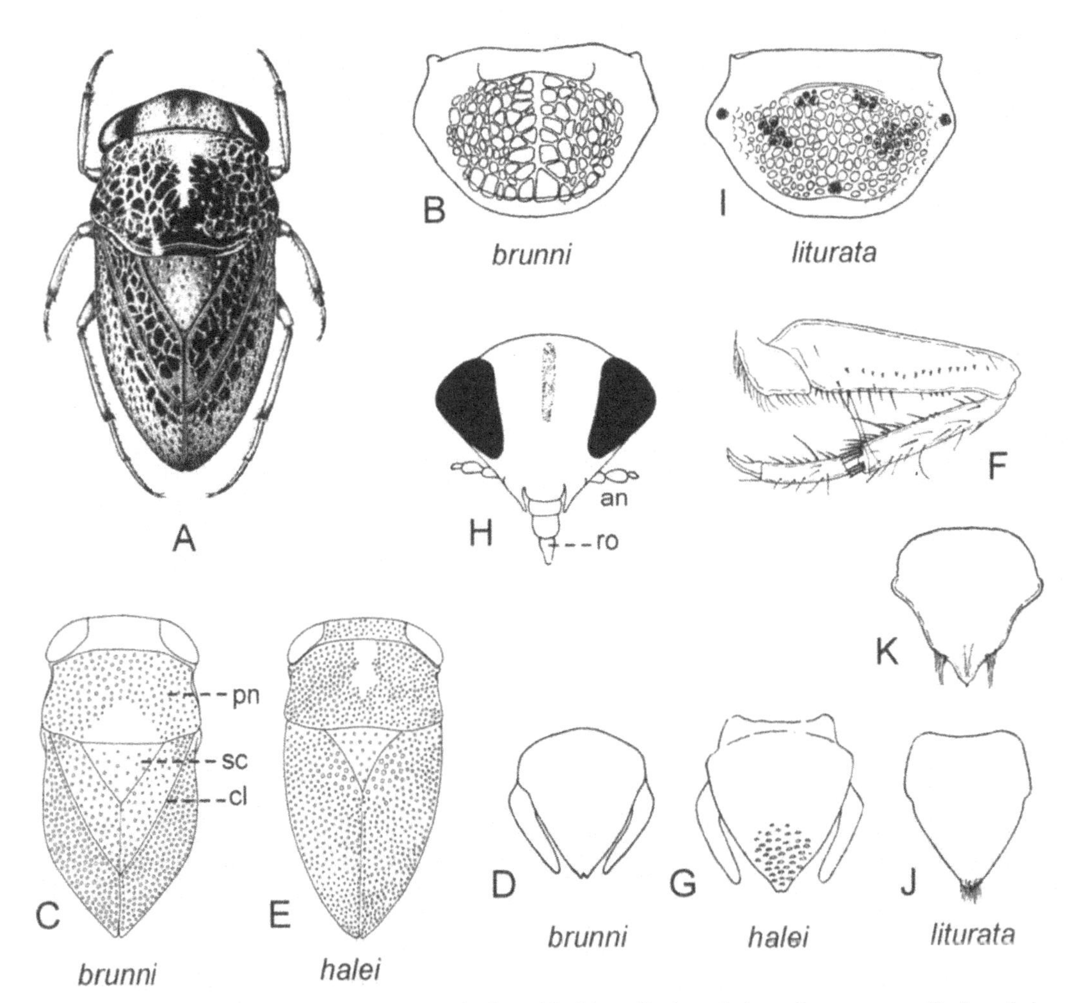

Figure 21.1. PLEIDAE. A-D, *Paraplea brunni*: A, dorsal habitus. B, dorsal view of pronotum. C, dorsal view of body (cl, claval suture; pn, pronotum; sc, scutellum). D, operculum (sternite 7) of ♂. E-G, *P. halei*: E, dorsal view of body. F, middle leg. G, operculum (sternite 7) of ♂. H-K, *P. liturata*: frontal view of head (an, antenna; ro, rostrum). I, dorsal view of pronotum. J, operculum (sternite 7) of ♂. K, operculum (sternite 7) of ♀. Reproduced with modification from: A, CSIRO (1991); B, D, F-K, Lundblad (1933); C, E, Lansbury & Lake (2002).

located between terga 3 and 4. Thoracic and abdominal venter with a laminate, longitudinal keel on segments 2-5 or 2-6 (Figs 21.2E, F). Male genitalia only weakly asymmetrical. Female ovipositor well developed with large teeth apically.

Overview

The pygmy backswimmers owe their name to their small size and habit of swimming on their back (Plate 8, J). They look like small, compact backswimmers (Notonectidae; see above) and differ from these in the relatively smaller eyes and hind legs that are neither strongly prolonged nor flattened. Despite this they are good swimmers. The abdominal venter has lateral hair fringes which hold a large ventral air bubble. In addition there are air stores under the pronotum and hemelytra. In summer temperatures pleids can stay alive under water about half a day without replenishing their air stores. In temperate climates their ability to stay under water increases significantly during autumn and winter, due to the fact that the ventral air store becomes much smaller and functions as a plastron (Kovac 1982). Pleids are typically inhabitants of stagnant waters with rich vegetation in which they hide. When not moving, they are easily overlooked due to their greenish yellow colour. Pygmy backswimmers are most easily collected by emptying the content of a pond net with plants into a tray with some water. After some time they will reveal themselves by swimming around in the tray made conspicuous by the large bubble of air which they carry on their abdominal venter. Like backswimmers they are generalised predators taking small invertebrate prey, more specifically mosquito larvae, ostracods, and cladocerans (Baum & Haddock 1978; Takahashi et al. 1979; Papacek 1985).

The morphology of pleids was studied by Wefelscheid (1912) and Papacek (1999). A unique structure in the Pleidae is a longitudinal darker stripe on the frons of the head which consists of numerous fine channels which have on top a cap with a very small bristle. Under this "pore-channel organ" lies a nerve centre and it is therefore regarded as sensory organ although its exact function has yet to be disclosed . A specialised type of behaviour, called "secretion-grooming", has been described in the European species *Plea minutissima* (Kovac & Maschwitz 1989; Kovac 1993). The bug leaves the water from time to time to apply a secretion from its metathoracic scent glands to its water-repellent ventral pubescence. This keeps the hairs free of microbial contamination and thus hydrophobe and functional. The grooming behaviour is stimulated by abiotic factors such as increase in light intensity or water temperature. The higher the temperature and light intensity, the faster the grooming act is performed and completed.

Females are usually slightly larger than males but it is otherwise very difficult to tell the sexes apart by external differences. As a rule the operculum (sternite 7) of the female has a small acute tip (Fig. 21.1K) whereas that of the male is more obtuse or has a small notch (Figs 21.1D, G, J). Alternatively, the operculum may be lifted (in alcohol material) to check whether an ovipositor (female) or a genital capsule (male) is present beneath. Copulation starts by the male grasping the dorsal side of the female, then lowers itself to the right side of the female. Next the male curves its abdominal tip against that of the female, and inserts is phallic organ. The eggs are elongate oval with a flattened ventral side (Fig. 2.8H). They are deposited by the female with its toothed ovipositor in submerged plants, preferably in leaves but also in stems. The longitudinal axis of the egg forms an angle with the plant surface leaving some of the egg exposed. When deposited the eggs are glued into their hole to prevent them from falling out. Embryonic and nymphal development take about one and one and a half month respectively in the northern temperate *Plea minutissima* Leach.

The family is distributed world-wide with about 40 species classified in three genera according to the number of segments in their fore, middle and hind tarsi (Esaki & China 1928). *Plea* Leach (1817) has a tarsal formula of 3-3-3 and includes only the palearctic species *P. minutissima*. *Neoplea* Esaki & China, with tarsal formula 3-2-3, is the predominant genus in the New World with 13 species. Finally, *Paraplea* Esaki & China, with a tarsal formula of 2-2-3, contains about 25 species distributed throughout the tropics, in particular in the Old World including Australia. The Pleidae forms the superfamily Pleioidea together with the family Helotrephidae. The latter family is much more diverse, with 16 genera and 44 species, most abundantly occurring in the Old World tropics, particularly the Oriental Region. Since the family is absent from New Guinea it can hardly be expected to occur in Australia. The Helotrephidae are easily recognisable by the unique fusion of the head with the prothorax, forming the so-called "cephalonotum" (Figs 6.2K, L) (Papacek et al. 1990). Most species are usually brachypterous and seem to be more diverse in habitat use than the pygmy backswimmers as they can be found in both running and stagnant waters.

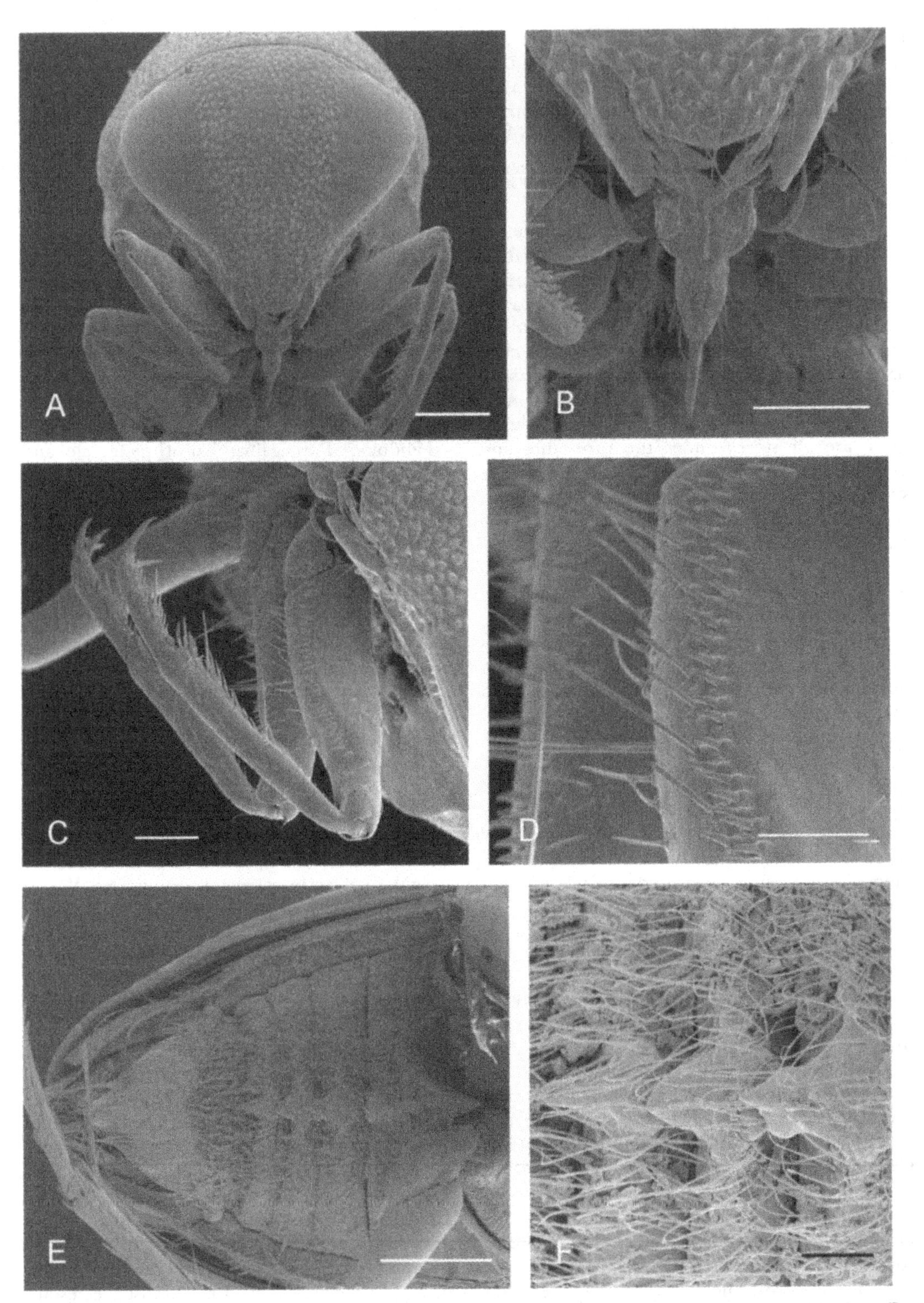

Figure 21.2. PLEIDAE. A-D, *Paraplea* sp.: A, frontal view of head. B, rostrum. C, fore and middle legs. D, detail of fore femur. E-F, *Paraplea brunni*, ♀: E, ventral view of abdomen. F, ventral median keel of abdomen. Scales: A, E, 0.2 mm; B, C, 0.1 mm; D, F, 0.05 mm. Scanning Electron Micrographs (CSIRO Entomology, Canberra).

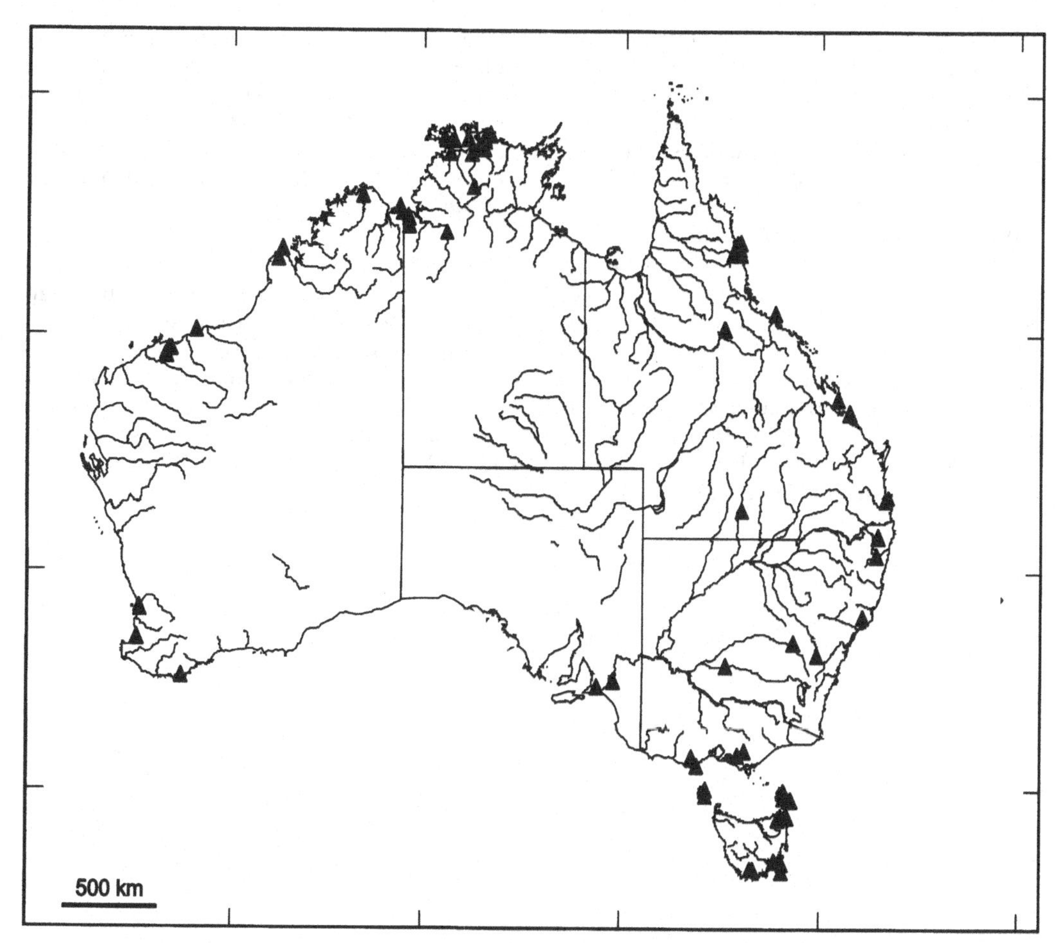

Figure 21.3. PLEIDAE. Distribution of *Paraplea* species in Australia.

PARAPLEA

Identification

Fore and middle tarsi 2-segmented, hind tarsi 3-segmented. Ventral abdominal carina extending onto sternite 6 (Figs 21.2E, F). The genus has five species in the Indo-Australian region of which two are widespread. Three species are recorded from Australia, but there are at least two undescribed species on the continent (Ivor Lansbury and Tom Weir, personal communication). Species of *Paraplea* are usually quite difficult to identify and Benzie (1989) found that the characteristics used to separate species are much more variable than was previously thought. The following key to the three described Australian species is adapted from Lundblad (1933) and Lansbury & Lake (2002).

Key to Australian species

1. Claval suture of hemelytron absent (Fig. 21.1E); hemelytra slightly reduced. Operculum (sternite 7) of ♂ indistinctly notched, with coarse punctures in distal part (Fig. 21.1G). Length 1.7-2.0 mm. South Australia, Tasmania, Victoria .... *halei* (Lundblad)
– Claval suture of hemelytron present (Fig. 21.1A, C); hemelytra not reduced. Operculum (sternite 7) of ♂ of different configuration (Fig. 21.1D, J) 2
2. Pronotum with numerous punctures throughout, normally with 5 distinct dark spots (Fig. 21.1I). Operculum (sternite 7) of ♂ simple, pointed, with a tuft of hairs in distal part (Fig. 21.1J). Hemelytra with a pale median band. Length 1.8-2.0 mm. Northern Territory, Western Australia .... *liturata* (Fieber)

– Pronotal disc largely devoid of punctures in the middle, at least posteriorly, without distinct dark spots (Figs 21.1A, B, C). Operculum (sternite 7) of ♂ distinctly bifurcate, without punctures (Fig. 21.1D). Hemelytra without pale medial band. Length 1.9-2.1 mm. New South Wales, Northern Territory, Queensland, South Australia, Tasmania, Western Australia................................ *brunni* (Kirkaldy)

Biology

Almost nothing is known about the biology of Australian species. In Tasmania, *Paraplea* species are found in small to medium-sized, coastal stagnant localities rich in macrophytes (Lansbury & Lake 2002). In South and Southeast Asia, *Paraplea* species usually occur in small stagnant waters, notably village ponds, with dense vegetation often covered with floating plants such as Lemnacea, *Azolla* or *Salvinia* (Chen et. al. in press). The nymphal instars were described by Benzie (1989) for a Sri Lankan population of *P. frontalis* (Lundblad).

Distribution

The genus *Paraplea* has five species in the Indo-Australian region of which two are widespread. The three Australian species are sparsely recorded from northern and southwestern Western Australia, South Australia, Tasmania, eastern Australia, more frequently recorded from the top end of Northern Territory (Fig. 21.3).

References

Andersen, N. M. (1967). A contribution to the knowledge of Philippine semiaquatic Hemiptera-Heteroptera. *Entomologiske Meddelelser* 35: 260-282.

Andersen, N. M. (1969). A new *Microvelia* from Australia with a check-list of Australian species (Hemiptera, Veliidae). *Entomologiske Meddelelser* 37: 253-261.

Andersen, N. M. (1971). Zoogeography and evolution of Pacific freshwater Gerridae (Hemiptera-Heteroptera). *Proceedings of XIII International Congress of Entomology, Moscow* 1968, 1: 469-470.

Andersen, N. M. (1973). Seasonal polymorphism and developmental changes in organs of flight and reproduction in bivoltine pondskaters (Hem. Gerridae). *Entomologica Scandinavica* 4: 1-20.

Andersen, N. M. (1975). The *Limnogonus* and *Neogerris* of the Old World with character analysis and a reclassification of the Gerrinae (Hemiptera: Gerridae). *Entomologica Scandinavica Supplement* 7: 1-96.

Andersen, N. M. (1976). A comparative study of locomotion on the water surface in semiaquatic bugs (Insecta, Hemiptera, Gerromorpha). *Videnskabelige Meddelelser fra Dansk Naturhistorisk Forening* 139: 337-396.

Andersen, N. M. (1977). Fine structure of the body hair layers and morphology of the spiracles of semiaquatic bugs (Insecta, Hemiptera, Gerromorpha) in relation to life on the water surface. *Videnskabelige Meddelelser fra Dansk Naturhistorisk Forening* 140: 7-37.

Andersen, N. M. (1978). A new family of semiaquatic bugs for *Paraphrynovelia* Poisson with a cladistic analysis of relationships (Insecta, Hemiptera, Gerromorpha). *Steenstrupia* 4: 211-225.

Andersen, N. M. (1979). Phylogenetic inference as applied to the study of evolutionary diversification of semiaquatic bugs (Hemiptera: Gerromorpha). *Systematic Zoology* 28: 554-578.

Andersen, N. M. (1982). The Semiaquatic Bugs (Hemiptera, Gerromorpha). Phylogeny, adaptations, biogeography, and classification. *Entomonograph* 3: 1-455.

Andersen, N. M. (1989a). The Old World Microveliinae (Hemiptera: Veliidae). II. Three new species of *Baptista* Distant and a new genus from the Oriental region. *Entomologica Scandinavica* 19 [1988]: 363-380.

Andersen, N. M. (1989b). The coral bugs, genus *Halovelia* Bergroth (Hemiptera, Veliidae). I. History, classification, and taxonomy of species except the *H. malaya*-group. *Entomologica. Scandinavica* 20: 75-120.

Andersen, N. M. (1989c). The coral bugs, genus *Halovelia* Bergroth (Hemiptera, Veliidae). II. Taxonomy of the *H. malaya*-group, cladistics, ecology, biology, and biogeography. *Entomologica Scandinavica* 20: 179-227.

Andersen, N. M. (1990). Phylogeny and taxonomy of water striders, genus *Aquarius* Schellenberg (Insecta, Hemiptera, Gerridae), with a new species from Australia. *Steenstrupia* 16: 37-81.

Andersen, N. M. (1991a). Cladistic biogeography of marine water striders (Hemiptera, Gerromorpha) in the Indo-Pacific. *Australian Systematic Botany* 4: 151-163.

Andersen, N. M. (1991b). Marine insects: genital morphology, phylogeny and evolution of sea skaters, genus *Halobates* (Hemiptera, Gerridae). *Zoological Journal of the Linnean Society* 103: 21-60.

Andersen, N. M. (1993a). The evolution of wing polymorphism in water striders (Gerridae): a phylogenetic approach. *Oikos* 67: 433-443.

Andersen, N. M. (1993b). Classification, phylogeny, and zoogeography of the pond skater genus *Gerris* Fabricius (Hemiptera, Gerridae). *Canadian Journal of Zoology* 71: 2473-2508.

Andersen, N. M. (1994). The evolution of sexual size dimorphism and mating systems in water striders (Hemiptera, Gerridae): a phylogenetic approach. *Écoscience* 1: 208-214.

Andersen, N. M. (1995a). Infraorder Gerromorpha Popov, 1971 - semiaquatic bugs. Pp 77-114 *in* Aukema, B. & Rieger, C. (eds): *Catalogue of the Heteroptera of the Palaearctic Region, Vol. 1.* xxvi + 222 pp. Netherlands Entomological Society, Amsterdam.

Andersen, N. M. (1995b). Cladistics, historical biogeography, and a check list of gerrine water striders (Hemiptera, Gerridae) of the World. *Steenstrupia* 21: 93-123.

Andersen, N. M. (1995c). Cladistic inference and evolutionary scenarios: locomotory structure, function, and performance in water striders. *Cladistics* 11: 279-295.

Andersen, N. M. (1995d). Phylogeny and classification of aquatic bugs (Heteroptera, Nepomorpha). An essay review of Mahner's "Systema Cryptoceratum Phylogeneticum". *Entomologica Scandinavica* 26: 159-166.

Andersen, N. M. (1996). Heteroptera Gerromorpha, semiaquatic bugs. Pp. 77-90 *in* Nilsson, A.

(Ed.) *Aquatic Insects of Northwest Europe.* 274 pp. Apollo Books, Stenstrup, Denmark.

Andersen, N. M. (1997). A phylogenetic analysis of the evolution of sexual dimorphism and mating systems in water striders (Hemiptera, Gerridae). *Biological Journal of the Linnean Society* 61: 345-368.

Andersen, N. M. (1998a). Marine water striders (Hemiptera, Gerromorpha) of the Indo-Pacific: cladistic biogeography and Cenozoic palaeogeography. Pp. 341-354 *in* R. Hall & J.D. Holloway (eds): *Biogeography and Geological Evolution of SE Asia.* ii + 417 pp. Backhuys Publishers, Leiden.

Andersen, N. M. (1998b). Water striders from the Paleogene of Denmark with a review of the fossil record and evolution of semiaquatic bugs (Hemiptera: Gerromorpha). *Det Kongelige Danske Videnskabernes Selskab, Biologiske Skrifter* 50: 1-152.

Andersen, N. M. (2000a). A new species of *Tetraripis* from Thailand, with a critical assessment of the generic classification of the subfamily Rhagoveliinae (Hemiptera, Veliidae). *Tijdschrift voor Entomologie* 142 [1999]: 185-194.

Andersen, N. M. (2000b). The evolution of dispersal dimorphism and other life history traits in water striders (Hemiptera: Gerridae). Entomological Science (Japan) 3: 187-199.

Andersen, N. M. (2000c). Fossil water striders in the Eocene Baltic amber (Hemiptera, Gerromorpha). *Insect Systematics & Evolution* 31: 257-284.

Andersen, N. M. (2001a). Fossil water striders in the Oligocene/Miocene Dominican amber (Hemiptera, Gerromorpha). *Insect Systematics & Evolution* 31 [2000]: 411-431.

Andersen, N. M. (2001b). The impact of W. Hennig's 'phylogenetic systematics on contemporary entomology. *European Journal of Entomology* 98: 133-150.

Andersen, N. M. & Cheng, L. (in press). The marine insect *Halobates* (Heteroptera: Gerridae) - Biology, adaptations, distribution and phylogeny. *Oceanography and Marine Biology. An annual review.*

Andersen, N. M., Farma, A., Minelli, A. & Piccoli, G. (1994). A fossil *Halobates* from the Mediterranean and the origin of sea skaters (Hemiptera, Gerridae). *Zoological Journal of the Linnean Society* 112: 479-489.

Andersen, N. M. & Grimaldi, D. (2001). A fossil water measurer (Hemiptera: Gerromorpha: Hydrometridae) from the mid-Cretaceous Burmese amber. *Insect Systematics & Evolution* 32: 381-392.

Andersen, N. M. & Poinar, Jr., G. O. (1992). Phylogeny and classification of an extinct water strider genus (Hemiptera, Gerridae) from Dominican amber, with evidence of mate guarding in a fossil insect. *Zeitschrift für Zoologisches Systematik und Evolutionsforschung* 30: 256-267.

Andersen, N. M. & Poinar, Jr. G. O. (1998). A marine water strider (Hemiptera, Veliidae) from Dominican amber. *Entomologica Scandinavica* 29: 1-9.

Andersen, N. M. & Polhemus, J. T. (1976). Water-striders (Hemiptera: Gerridae, Veliidae, etc.). Pp 187-224 *in* Cheng, L. (Ed.): *Marine Insects.* 581 pp. North-Holland Publishing Company, Amsterdam.

Andersen, N. M. & Spence, J. R. (1992). Classification and phylogeny of the Holarctic water strider genus *Limnoporus* Stål (Hemiptera, Gerridae). *Canadian Journal of Zoology* 70: 753-785.

Andersen, N. M. & Weir, T. A. (1994a). *Austrobates rivularis* gen. et sp.nov., a freshwater relative of *Halobates* (Hemiptera, Gerridae) with a new perspective on the evolution of sea skaters. *Invertebrate Taxonomy* 8: 1-15.

Andersen, N. M. & Weir, T. A. (1994b). The sea skaters, genus *Halobates* Eschscholtz (Hemiptera, Gerridae), of Australia: taxonomy, phylogeny, and zoogeography. *Invertebrate Taxonomy* 8: 861-909.

Andersen, N. M. & Weir, T. A. (1997). The gerrine water striders of Australia (Hemiptera: Gerridae): taxonomy, distribution, and ecology. *Invertebrate Taxonomy* 11: 203-299.

Andersen, N. M. & Weir, T. A. (1998). Australian water striders belonging to the subfamilies Rhagadotarsinae and Trepobatinae (Hemiptera, Gerridae). *Invertebrate Taxonomy* 12: 509-544.

Andersen, N. M. & Weir, T. A. (1999). The marine Haloveliinae (Hemiptera: Veliidae) of Australia, New Caledonia, and southern New Guinea. *Invertebrate Taxonomy* 13: 309-350.

Andersen, N. M. & Weir, T. A. (2000). The coral treaders, *Hermatobates* Carpenter (Hemiptera, Hermatobatidae), of Australia and New Caledonia, with notes on biology and ecology. *Invertebrate Taxonomy* 14: 327-345.

Andersen, N. M. & Weir, T. A. (2001). New genera of Veliidae (Hemiptera-Heteroptera) from Australia, with notes on the generic classification of the Microveliinae. *Invertebrate Taxonomy* 15: 217-258.

Andersen, N. M. & Weir, T. A. (2003a). A new species of sea skaters, *Halobates* Eschscholtz (Hemiptera: Gerridae) from Robinson River, Western Australia. *Aquatic Insects* 25: 9-18.

Andersen, N. M. & Weir, T. A. (2003b). The genus *Microvelia* Westwood in Australia (Hemiptera: Heteroptera: Veliidae). *Invertebrate Systematics* 17: 261-348.

Andersen, N. M. & Weir, T. A. (in press). The families Mesoveliidae, Hebridae, and Hydrometridae of Australia (Hemiptera, Heteroptera, Gerromorpha), with a reanalysis of phylogenetic relationships between families. *Invertebrate Systematics.*

Arnqvist, G. (1997). The evolution of water strider mating systems: causes and consequences of sexual conflicts. Pp 146-163 *in* Choe, J. C. & Crespi, B. J. (eds): *Evolution of Mating Systems in Insects and Arachnids.* ix + 387 pp. Cambridge University Press, Cambridge.

Arnqvist, G., Jones, T. M. & Elgar, M. A. (2003). Reversal of sex roles in nuptial feeding. *Nature* 424: 387.

Arnqvist, G. & Mäki, M. (1990). Infection rates and pathogenicity of trypanosomatid gut parasites in the water strider *Gerris odontogaster* (Zett.) (Heteroptera: Gerridae). *Oekologia* 84: 194-198.

Aukema, B. & Rieger, C. (eds.) (1995). *Catalogue of the Heteroptera of the Palaearctic Region. Volume 1. Enicocephalomorpha, Dipsocoromorpha, Nepomorpha, Gerromorpha and Leptopodomorpha.* xxv + 222 pp. The Netherlands Entomological Society, Amsterdam.

Bacon, J. A. (1956). A taxonomic study of the genus *Rhagovelia* of the Western Hemisphere *Kansas University Science Bulletin* 38: 695-913.

Baehr, M. (1989). Review of the Australian Ochteridae (Insecta, Heteroptera). *Spixiana* 11: 111-126.

Baehr, M. (1990a). Revision of the genus *Ochterus* Latreille in the Australian region (Heteroptera: Ochteridae). *Entomologica Scandinavica* 20 [1989]: 449-477.

Baehr; M. (1990b). Revision of the genus *Megochterus* Jaczewski (Insecta: Heteroptera: Ochteridae). *Invertebrate Taxononomy* 4: 197-203.

Bailey, P. C. E. (1986a). The feeding behavior of a sit-and-wait predator, *Ranatra dispar,* (Heteroptera: Nepidae): Description of behavioral components of prey capture and the effect of food deprivation on predator arousal and capture dynamics. *Behaviour* 97: 66-93.

Bailey, P. C. E. (1986b). The effect of predation risk on the predatory behavior of a sit-and-wait predator, *Ranatra dispar* (Heteroptera: Nepidae), the water stick insect. *Journal of Ethology* 4: 17-26.

Bailey, P. C. E. (1987). The effect of water depth on the predatory behavior of the water stick insect, *Ranatra dispar* (Heteroptera: Nepidae). *Australian Journal of Zoology* 35: 443-450.

Bailey, P. C. E. (1989). The effect of water temperature on the functional response of the water stick insect *Ranatra dispar* (Heteroptera: Nepidae). *Australian Journal of Ecology* 14: 381-386.

Bailey, W. J. (1983). Sound production in *Micronecta batilla* Hale (Hemiptera: Corixidae) - an alternative structure. *Journal of the Australian Entomological Society* 22: 35-38.

Bakonyi, G. (1978). Contribution to the knowledge of the feeding habits of some water boatmen: *Sigara* spp. (Heteroptera: Corixidae). *Folia Entomologica Hungarici* 31: 19-24.

Baum, R. T. & Haddock, J. D. (1978). Studies on the predation of mosquito larvae by pleid bugs. *Proceedings of the Indiana Academy of Science* 87 [1978]: 243.

Baunacke, W. (1912). Statische Sinnesorganen bei den Nepiden. *Zoologisches Jahrbücher Abteilung Anatomie* 34: 179-346.

Bendell, B. E. & McNicol, D. K. (1987). Fish predation, lake acidity and the composition of aquatic insect assemblages. *Hydrobiologica* 150: 193-202.

Benjamin, R. K. (1986). Laboulbeniales on semiaquatic Hemiptera: V. Triceromyces: with a description of monoecious-dioecous dimorphism in the genus. *Aliso* 11: 245-278.

Benzie, J. A. H. (1989). The immature stages of *Plea frontalis* (Fieber, 1844) (Hemiptera, Pleidae), with a redescription of the adult. *Hydrobiologia* 179: 157-171.

Bergroth, E. (1893). On two halophilous Hemiptera. *Entomologist's Monthly Magazine* 29: 277-279.

Bergroth, E. (1916). Heteropterous Hemiptera collected by Professor W. Baldwin Spenser during the Horn Expedition into Central Australia. *Proceedings of the Royal Society of Victoria* 29: 19-39.

Birch, M., Cheng, L. & Treherne, J. E. (1979). Distribution and environmental synchronization of the marine insect, *Halobates robustus,* in the Galapagos Islands. *Proceedings of the Royal Society of London (B)* 206: 33-52.

Bobb, M. L. (1951). Life history of *Ochterus banksi* Barber. *Bulletin of Brooklyn Entomological Society* 46: 92-100.

Boulard, M. & Coffin, J. (1991). Sur la biologie juvénile d'*Ochterus marginatus* (Latreille, 1804) camouflage et construction (Hemiptera: Ochteridae). *EPHE, Travaux Laboratoire Biologique d'Evolution des Insectes* 4: 57-68.

Breddin, G. (1905a). Rhynchota Heteroptera aus Java, gesammlt von Prof. K. Kraepelin 1904.

Mitteilungen aus dem Naturhistorischen Museum in Hamburg 22: 109-159.
Breddin, G. (1905b). Übersicht der javanischen *Micronecta*-Arten. *Societas Entomologica* 20: 57.
Brooks, G. T. (1951). A revision of the genus *Anisops* (Notonectidae: Hemiptera). *University of Kansas Science Bulletin* 34: 301-519.
Brönmark, C., Malmquist, B. & Otto, C. (1984). Anti-predator adaptations in a neustonic insect (*Velia caprai*). *Oecologia* 61: 189-191.
Carpenter, F. M. (1992). *Arthropoda. Superclass Hexapoda. Treatise on Invertebrate Palaeontology, Part R, Volume 3.* 277 pp. Geological Society of America & The University of Kansas.
Carpenter, G. H. (1892). Rhynchota from Murray Island and Mabuiag. Reports on the Zoological collections made in Torres Straits by Professor A.C. Haddon, 1888-1889. *Scientific Proceedings of the Royal Dublin Society, N.S.* 7: 137-146.
Cassis, G. & Gross, G. F. (1995). Hemiptera: Heteroptera (Coleorrhyncha to Cimicomorpha). Pp 1-506 *in* Houston, W. W. K. & Maynard, G. V. (eds): *Zoological Catalogue of Australia. Vol. 27.3A.* CSIRO Australia, Melbourne.
Cassis, G. & Silveira, R. (2001). A revision and phylogenetic analysis of the *Nerthra alaticollis* species-group (Heteroptera: Gelastocoridae: Nerthrinae). *Journal of the New York Entomological Society* 109: 1-46.
Cassis, G. & Silveira, R. (2002). A revision and phylogenetic analysis of the *Nerthra elongata* species-group (Heteroptera: Gelastocoridae: Nerthrinae). *Journal of the New York Entomological Society* 110: 143-181.
Chen, L. (1965). A revision of *Micronecta* of Australia and Melanesia (Heteroptera: Corixidae). *University of Kansas Science Bulletin* 46: 147-165.
Chen; P., Nieser, N. & Zettel, H. (in press). The aquatic and semiaquatic bugs of Malesia and adjacent areas (Heteroptera: Nepomorpha & Gerromorpha). *Fauna Malesiana.*
Cheng, L. (1977). The elusive sea bug *Hermatobates. Pan-Pacific Entomologist* 53: 87-97.
Cheng, L. (1985). Biology of *Halobates* (Heteroptera: Gerridae). *Annual Review of Entomology* 30: 111-135.
Cheng, L. (1989). Factors limiting the distribution of *Halobates* species. Pp. 357-362 *in* Ryland, J. S. & Tyler, P. A. (eds): *Reproduction, genetics and distributions of marine organisms. 23rd European Marine Biology Symposium.* Olsen & Olsen, Fredensborg, Denmark.
Cheng, L. & Fernando, C. H. (1971). Life history and biology of the riffle bug *Rhagovelia obesa* Uhler in southern Ontario. *Canadian Journal of Zoology* 49: 435-442.
Cheng, L. & Leis, E.W. (1980). Notes on the seabug, *Hermatobates hawaiiensis* China (Heteroptera: Hermatobatidae). *Proceedings of the Hawaiian Entomological Society* 23: 193-197.
Cheng, L. & Schmitt, P. D. (1982). Marine insects of the genera *Halobates* and *Hermatobates* (Heteroptera) from neuston tows around Lizard Island, Great Barrier Reef. *Australian Journal of Marine and Fresh Water Research* 33: 1109-1112.
China, W. E. (1933). A new family of Hemiptera-Heteroptera withnotes on the phylogeny of the suborder. *Annals and Magazine of Natural History* (10) 12: 180-196.
China, W. E. (1955). The evolution of the water bugs. *Symposium on Organic Evolution Bulletin of the National Institute of Science, India* 7: 91-103.
China, W.E. (1957). The marine Hemiptera of the Monte Bello Islands, with descriptions of some allied species. *Journal of the Linnean Society of London, Zoology* 40: 342-357.
China, W. E.& Usinger, R. L. (1949). Classification of the Veliidae (Hemiptera) with a new genus from South Africa. *Annals and Magazine of Natural History* (12) 2: 343-354.
Cloarec, A. (1972). Activité respiratoire de *Ranatra linearis* (Insecte, Hétéroptère aquatique). *Bulletin de la Société Zoologigique de France* 97: 729-736.
Cloarec, A. (1976a). Variations of behavioral patterns during development in the water stick insect. *Journal of Ethology* 14: 15-20.
Cloarec, A. (1976b). Interactions between different receptors involved in prey capture in *Ranatra linearis. Biology of Behavior* 1: 251-266.
Cloarec, A. (1989). Variations of foraging tactics in a water bug, *Diplonychus indicus. Journal of Ethology* 7: 27-34.
Cloarec, A. (1990a). Variations of predatory tactics of a water bug during development. *Ethology* 86: 33-46.
Cloarec, A. (1990b). Factors influencing the choice of predatory tactics in a water bug, *Diplonychus indicus* Venk. & Rao (Heteroptera: Belostomatidae). *Animal Behaviour* 40: 262-271.
Cloarec, A. (1991). Predatory versatility in the water-bug *Diplonychus indicus. Behavioural Processes* 23: 231-242.
Cloarec, A. (1992). The influence of feeding on predatory tactics in a water bug. *Physiological Entomology* 17: 25-32.
Cloarec, A. & Joly, D. (1988). Choice of perch by the water stick insect. *Behavioural Processes* 17: 131-144.
Cobben, R. H. (1965). Egg-life and symbiont

transmission in a predatory bug, *Mesovelia furcata* Ms. & Rey (Heteroptera, Mesoveliidae). *Proceedings of the 12th International Congress of Entomology, London* 1964: 166-168.

Cobben, R. H. (1968). *Evolutionary trends in Heteroptera. Part I. Eggs, architecture of the shell, gross embryology and eclosion.* 475 pp. Centre for Agricultural Publishing and Documentation, Wageningen.

Cobben, R. H. (1978). Evolutionary trends in Heteroptera. Part. II. Mouthpart-structures and feeding strategies. *Mededelingen Landbouwhogeschool Wageningen* 78-5: 1-407.

Coutière, H., and Martin, J. (1901a). Sur un nouvelle sous-famille d'Hémiptères marins, les Hermatobatidae. *Comptes Rendus de l'Académie des Sciences* 132: 1066-1068.

Coutière, H., and Martin, J. (1901b). Sur un nouvel Hémiptère halophile, *Hermatobatodes marchei* n.gen., n.sp. *Bulletin de Muséum d'Histoire Naturelle, Paris* 5: 214-226.

Cranston, P. S. & Naumann, I. D. (1991). 7. Biogeography. Pp 180-197 *in* CSIRO (eds.): *The Insects of Australia. A textbook for students and research workers, 2nd edition, Vol. 1.* xvi + 542 pp. Melbourne University Press, Melbourne.

CSIRO (eds.) (1991). *The Insects of Australia. A textbook for students and research workers, 2nd edition.* Vols 1-2. xvi + 1137 pp. Melbourne University Press, Melbourne.

Cullen, M. J. (1969). The biology of giant water bugs in Trinidad. *Proceedings of the Royal Entomological Society of London (A)* 44: 123-136.

Damgaard, J., Andersen, N. M., Cheng, L. & Sperling, F. A. H. (2000). Phylogeny of sea skaters, *Halobates* (Hemiptera, Gerridae), based on mtDNA sequence and morphology. *Zoological Journal of the Linnean Society* 130: 511-526.

Damgaard, J. & Andersen, N. M. (1996). Distribution, phenology, and conservation status of the larger water striders in Denmark. *Entomologiske Meddelelser* 64: 289-306.

Darwin, C. (1859). *On the Origin of Species by Means of Natural Selection.* 502 pp. John Murray, London.

Davis, J. & Cristidis, F. (1997). *A Guide to Wetland Invertebrates of Southwestern Australia.* vi + 177 pp. Western Australia Museum, Perth.

Decu, V., Gruia, M., Keffer, S. L. & Sarbu, S. M. (1994). Stygobiotic waterscorpion, *Nepa anophthalma,* n. sp. (Heteroptera: Nepidae), from a sulfurous cave in Romania. *Annals of the Entomological Society of America* 87: 755-761.

Deshefy, G. S. (1980). Anti-predator behaviour in swarms of *Rhagovelia obesa* (Hemiptera: Veliidae). *Pan-Pacific Entomologist* 56: 111-112.

Distant, W. L. (1903). *The Fauna of British India, including Ceylon and Burma. Rhynchota 2(1).* i-x, 1-242 pp. Taylor & Francis: London.

Distant, W. L. (1904). Rhynchotal notes. XXIV. *Annals and Magazine of Natural History* (7) 14: 61-66.

Distant, W. L. (1909). Oriental Rhynchota Heteroptera. *Annals and Magazine of Natural History* (8) 3: 491-507.

Don, A. W. (1967). Aspects of the biology of *Microvelia macgregori* Kirkaldy. *Proceedings of the Royal Entomological Society of London (A)* 42: 171-179.

Drake, C. J. (1917). A survey of the North American species of *Merragata. Ohio Journal of Science* 17: 101-105.

Dudgeon, D. (1990). Feeding by the aquatic heteropteran, *Diplonychus rusticum* (Belostomatidae): An effect of prey density on meal size. *Hydrobiologia* 190: 93-96.

Dufour, L. (1833). Recherches anatomiques et physiologiques sur les Hémiptères, accompagnées de considerations relatives a l'histoire naturelle et a la classification de ces insectes. *Mem. Pres. Acad. Sci. Paris* 4: 129-462, pls. I-XIX.

Dufour, L. (1863). Essai monographique sur les Bélostomides. *Annales de la Societé Entomologique Francaíse* (4) 3: 373-400.

Dunn, C.E. (1976). The archaic nature of the genus *Diaprepocoris* Kirkaldy, as indicated by the male and female genitalia (Heteroptera, Corixidae). *Journal of Georgia Entomological Society* 11: 373-375.

Ekblom, T. (1926). Morphological and biological studies of the Swedish families of Hemiptera-Heteroptera. Part I. The families Saldidae, Nabidae, Lygaeidae, Hydrometridae, Veliidae and Gerridae. *Zoologiska Bidrag, Uppsala* 10: 29-179.

Ekblom, T. (1930). Morphological and biological studies of the Swedish families of Hemiptera-Heteroptera. Part II. The families Mesoveliidae, Coreidae and Corixidae. *Zoologiska Bidrag, Uppsala* 12: 113-150.

Erichson, W. F. (1842). Beitrag zur Insekten-Fauna von Vandiemensland, mit besonderer Berüchsichtigung der geographischen Verbreitung der Insekten. *Archiv für Naturgeschicte* 8: 83-287.

Erlandsson, A, & Giller, P. S. (1992). Distribution and feeding behaviour of field populations of the water cricket *Velia caprai* (Hemiptera). *Freshwater Biology* 28: 231-236.

Esaki, T. (1924). On the curious halophilous water strider, *Halovelia maritima* Bergroth (Hemiptera: Gerridae). *Bulletin of Brooklyn Entomological Society* 14: 29-34.

Esaki, T. (1926). The water-striders of the subfamily Halobatinae in the Hungarian National Museum. *Annales Historico-Naturales Musei Nationalis Hungarici* 23: 117-164.

Esaki, T. (1927). An interesting new genus and species of Hydrometridae (Hem.) from South America. *Entomologist* 60: 181-184.

Esaki, T. (1928). Aquatic and semiaquatic Heteroptera. *Insects of Samoa and other Samoan terrestrial arthropods* 2 (2): 67-80.

Esaki, T. (1930). New or little-known Gerridae from the Malay Peninsula. *Journal of the Federated Malay States Museum* 16: 13-24.

Esaki, T. & China, W. R. (1928). A monograph of the Helotrephidae, subfamily Helotrephinae (Hem. Heteroptera). A new family of aquatic Heteroptera. *Transactions of the Entomological Society of London* 1927: 279-295.

Eschscholtz, J. F. (1822). *Entomographien, vol. 1 (1ste lieferung)*. Berlin.

Fabricius, I. C. (1775). *Systema Entomologiae, sistens insectorum classes, ordines, genera, species, adiectis synonymis, locis, descriptionibus, observationibus.* xxviii + 832 pp. Flensburg & Leipzig.

Fabricius, I. C. (1781). *Species insectorum exhibentes eorum differentias specificas, synonyma auctorum, loca natalia, metamorphosin adjectis observationibus, describtionibus 2.* 517 pp. Bonn, Hamburg & Kiel.

Fernando, C. H. (1961). Notes on aquatic insects caught at light in Malaya, with a discussion of to their distribution and dispersal. *Bulletin of the National Museum, State of Singapore* 30: 19-31.

Fernando, C. H. (1964). Notes on aquatic insects colonizing an isolated pond in Mawai, Johore. *Bulletin of the National Museum, State of Singapore* 32: 80-89.

Fernando, C. H. & Leong, C. Y. (1963). Miscellaneous notes on the biology of Malayan Corixidae (Hemiptera: Heteroptera) and a study of the life histories of two species, *Micronecta quadristrigata* Bredd. and *Agraptacorixa hyalinipennis* (F.). *Annals and Magazine of Natural History* (13) 6: 545-558.

Fieber, F. X. (1844). *Entomologische Monographien.* 138 pp. Verlag von Johann Ambrosius Barth, Leipzig.

Fieber, F. X. (1851a). Genera Hydrocoridum secundum ordinem naturalem in familias deposita. *Abhandlungen Königlichen Böhmischen Gesellschaft für Wissenschaften, Pragae* (5) 7: 181-212.

Fieber, F. X. (1851b). Rhynchotographien, drei monographischen Abhandlungen. *Abhandlungen Königlichen Böhmischen Gesellschaft für Wissenschaften, Pragae* (5) 7: 425-486.

Foster, W. A. (1989). Zonation, behaviour and morphology of the intertidal coral-treader *Hermatobates* (Hemiptera: Hermatobatidae) in the south-west Pacific. *Zoological Journal of the Linnean Society* 96: 87-105.

Foster, W. A. & Treherne, J. E. (1980). Feeding, predation and aggregation behaviour in a marine insect, *Halobates robustus* Barber (Hemiptera: Gerridae), in the Galapagos Islands. *Proceedings of the Royal Society of London (B)* 209: 539-553.

Foster, W. A. & Treherne, J. E. (1982). Reproductive behaviour of the ocean skater *Halobates robustus* (Hemiptera: Gerridae) in the Galapagos Islands. *Oecologia* 55: 202-207.

Foster, W. A. & Treherne, J. E. (1986). The ecology and behaviour of a marine insect, *Halobates fijiensis* (Hemiptera: Gerridae). *Zoological Journal of the Linnean Society* 86: 391-412.

Fraser, N. C., Grimaldi, D. A., Olsen, P. E. & Axsmith, B. (1996). The Triassic Lagerstatte from eastern North America. *Nature* 380: 615-619.

Galbreath, J. E. (1973). Diapause in *Mesovelia mulsanti* (Hemiptera: Mesoveliidae). *Journal of Kansas Entomological Society* 46: 224-233.

Galbreath, J. E. (1976). The effect of age of the female on diapause in *Mesovelia mulsanti* (Hemiptera: Mesoveliidae). *Journal of Kansas Entomological Society* 49: 27-31.

Gilbert, J. J. & Burns, C. W. (1999). Some observations on the diet of the backswimmer, *Anisops wakefieldi* (Hemiptera: Notonectidae). *Hydrobiologia* 412: 111-118.

Giller, P. S. (1986). The natural diet of the Notonectidae: field trials using electrophoresis. *Ecological Entomology* 11: 163-172.

Giller, P. S. & McNeill, S. (1981). Predation strategies, resource partitioning and habitat selection in *Notonecta* (Hemiptera/Heteroptera). *Journal of Animal Ecology* 50: 789-808.

Gooderham, J. & Tsyrlin, E. (2002). *The Waterbug Book.* 232 pp. CSIRO Publishing, Melbourne.

Groffith, J. (1945). The environment, life history and structure of the water boatman, *Ramphocorixa acuminata* (Uhler). *University of Kansas Science Bulletin* 20: 5-61.

Hale, H. M. (1922). Studies in Australian aquatic Hemiptera. No. I. *Records of South Australian Museum* 2: 309-330.

Hale, H. M. (1923). Studies in Australian aquatic Hemiptera. No. II. *Records of South Australian Museum* 2: 397-424.

Hale, H. M. (1924a). Studies in Australian aquatic Hemiptera. No. III. *Records of South Australian Museum* 2: 503-520.

Hale, H.M. (1924b). Studies in Australian aquatic Hemiptera. No. IV. *Transactions of the Royal Society of South Australia* 48: 7-9.

Hale, H. M. (1925). The aquatic and semi-aquatic Hemiptera. Results of Dr. E. Mjöberg's Swedish Scientific Expeditions to Australia 1910-1913. *Arkiv för Zoologi* 17A(20): 1-19.

Hale, H. M. (1926). Studies in Australian aquatic Hemiptera, No. VII. *Records of South Australian Museum* 3: 195-217.

Harris, H. M. & Drake, C. J. (1941). Note on the family Mesoveliidae (Hemiptera) with descriptions of two new species. *Iowa State College Journal of Science* 15: 275-276.

Hawking, J. H. & Smith, F. J. (1997). *Colour Guide to Invertebrates of Australian Inland Waters.* Cooperative Research Centre for Freshwater Ecology. Identification and Ecology Guide No. 8, 213 pp.

Heming-Van Battum, K. & Heming B. (1986). Structure, function and evolution of the reproductive system in females of *Hebrus pusillus* and *H. ruficeps* (Hemiptera, Gerromorpha, Hebridae). *Journal of Morphology* 190: 121-167.

Heming-Van Battum, K. & Heming, B. (1989). Structure, function, and evolutionary significance of the reproductive system in males of *Hebrus pusillus* and *H. ruficeps* (Heteroptera, Gerromorpha, Hebridae). *Journal of Morphology* 202: 281-323.

Hennig W. (1966). *Phylogenetic Systematics.* 280 pp. University of Illinois Press, Urbana (translated by D. D. Davis & R. Zangerl).

Henrikson, L. & Oscarson, H. (1978). Fish predation limiting abundance and distribution of *Glaenocorisa p. propinqua. Oikos* 31: 102-105.

Henrikson, L. & Oscarson, H. (1981). Corixids (Hemiptera-Heteroptera), the new top predators in acidified lakes. *Verhandlungen der internationalen Vereinigung für theoretische und angewandte Limnologie* 21: 1616-1620.

Henrikson, L. & Oscarson, H. (1985). Water bugs (Corixidae, Hemiptera-Heteroptera) in acidified lakes: Habitat selection and adaptations. *Ecological Bulletins* 37: 232-238.

Henry, T. J. & Froeschner, R. C. (eds.) (1988). *Catalog of the Heteroptera, or True Bugs, of Canada and the Continental United States.* xix + 958 pp. E. J. Brill, Leiden, etc.

Herring, J. L. (1961). The genus *Halobates* (Hemiptera: Gerridae). *Pacific Insects* 3: 223-305.

Hinton, H. E. (1961). The structure and function of the egg-shell in the Nepidae (Hemiptera). *Journal Insect Physiology* 7: 224-257.

Hinton, H. E. (1976). Platron respiration in bugs and beetles. *Journal of Insect Physiology* 22: 1529-1550.

Hoberlandt, L. & Stys, P. (1979). *Tampocoris asiaticus* gen. and sp. n. - A new aphelocheirine from Vietnam and further studies on Naucoridae (Heteroptera). *Acta Musei Nationalis Pragae* 33(B) [1977]: 1-20.

Hoffmann, W. E. (1927). Biological notes on *Laccotrephes* (Hemiptera, Nepidae). *Lingnan Agricultural Review* 4: 77-93.

Hoffmann, W. E. (1931). Life history notes on *Enithares sinica* Stål (Hemiptera Notonectidae). *Lingnan Science Journal* 9: 432-433.

Hoffmann, W. E. (1933). The life history of a second species of *Laccotrephes* (Hemiptera, Nepidae). *Lingnan Science Journal* 12: 245-257.

Hoffmann, W. E. (1936a). The life history of *Limnogonus fossarum* (Fabr.) in Canton (Hemiptera: Gerridae). *Lingnan Science Journal* 15: 289-300.

Hoffmann, W.E. (1936b). Life history notes on *Rhagadotarsus kraepelini* Breddin (Hemiptera: Gerridae) in Canton. *Lingnan Science Journal* 15: 477-482.

Horváth, G. (1895). Hémiptères nouveaux d'Europe et des pays limitrophes. *Revue d'Entomologie* 14: 152-165.

Horváth, G. (1902). Descriptions of new Hemiptera from New South Wales. *Természetr. Füzetek* 25: 593-600.

Horváth, G. (1918). De Hydrocorisis nonnullis extraeuropaeis. *Annales Historico-Naturales Musei Nationalis Hungarici* 16: 140-146.

Hu, D. L., Chan, B. & Bush, J. W. M. (2003). The hydrodynamics of water strider locomotion. *Nature* 424: 663-666.

Hungerford, H. B. (1920). The biology and ecology of aquatic and semiaquatic Hemiptera. *Kansas University Sciience Bulletin* 11 [1919]: 1-328.

Hungerford, H. B. (1922). Oxyhaemoglobin present in backswimmer, *Buenoa margaritacea* Bueno. *Canadian Entomologist* 54: 262-263.

Hungerford, H. B. (1933). The genus *Notonecta* of the world. *University of Kansas Science Bulletin* 21: 5-195.

Hungerford, H.B. (1934). Concerning some aquatic and semiaquatic Hemiptera from Australia. *Bulletin of the Brooklyn Entomological Society* 29: 68-73.

Hungerford, H. B. (1938). A new *Hydrometra* from New Caledonia and Australia. *Pan-Pacific Entomologist* 14: 81-83.

Hungerford, H. B. (1940). A new *Enithares* for Australia (Notonectidae-Hemiptera). *Journal of the Kansas Entomological Society* 13: 130-131.

Hungerford, H. B. (1948). The Corixidae of the Western Hemisphere (Hemiptera). *Kansas University Science Bulletin* 32: 1-827.

Hungerford, H. B. (1953). A new *Agraptocorixa* from Australia. *Journal of Kansas Entomological Society* 26: 42-44.

Hungerford, H. B. & Evans, N. E. (1934). The Hydrometridae of the Hungarian National

Museum and other studies in the family (Hemiptera). *Annales Historico-Naturales Musei Nationalis Hungarici* 28: 31-112.

Hungerford, H. B. & Matsuda, R. (1958). The *Tenagogonus-Limnometra* complex of the Gerridae. *Kansas University Science Bulletin* 39: 371-457.

Hungerford, H. B. & Matsuda, R. (1961). A new species of *Limnogonus* from Australia (Hemiptera: Gerridae). *Bulletin of the Brooklyn Entomological Society* 56: 117-120.

Hutchinson, G. E. (1940). A revision of the Corixidae of India and adjacent regions. *Transactions of the Connecticut Academy of Arts and Sciences* 33: 339-476.

Hynes, H. B. N. (1955). Biological notes on some East African aquatic Heteroptera. *Proceedings of the Royal Entomological Society of London (A)* 30: 43-54.

Ichikawa, N. (1989). Breeding strategy of the male brooding water bug, *Diplonychus major* Esaki (Heteroptera: Belostomatidae): Is male back space limiting? *Journal of Ethology* 7: 133-140.

Ichikawa, N. (1990). Egg mass destroying behavior of the female giant water bug *Lethocerus deyrollei* Vuillefroy (Heteroptera: Belostomatidae). *Journal of Ethology* 8: 5-12.

Ichikawa, N. (1991a). Egg mass destroying and guarding behavior of the giant water bug, *Lethocerus deyrollei* Vuillefroy (Heteroptera: Belostomatidae). *Journal of Ethology* 9: 25-30.

Ichikawa, N. (1991b). Additional benefit of egg mass destruction by giant water bug *Lethocerus deyrollei* Vuillefroy (Heteroptera: Belostomatidae) females. *Journal of Ethology* 9: 34-36.

International Commission on Zoological Nomenclature (1996). Opinion 1850. *Nepa rustica* Fabricius, 1781 and *Zaitha stollii* Amyot & Serville, 1843 (currently *Diplonychus rusticus* and *Belostoma stollii*; Insecta, Heteroptera): specific names conserved. *Bulletin of Zoological Nomenclature* 53: 213-214.

Jackson, R. R. & Walls, E. I. (1998). Predatory and scavenging behaviour of *Microvelia macgregori* (Hemiptera: Veliidae). *New Zealand Journal of Zoology* 25: 23-28.

Jaczewski, T. (1934). Notes on the Old World species of Ochteridae (Heteroptera). *Annals and Magazine of Natural History* (10) 11: 597-613.

Jansson, A. (1972). Mechanisms of sound production and morphology of the stridulatory apparatus in the genus *Cenocorixa*. *Annales Zoologici Fennici* 9: 120-129.

Jansson, A. (1973). Stridulation and its significance in the genus *Cenocorixa*. *Behaviour* 46: 1-36.

Jansson, A. (1974). Annual periodicity of male stridulation in the genus *Cenocorixa*. *Freshwater Biology* 4: 93-98.

Jansson, A. (1976). Audiospectrographic analysis of stridulatory signals in North American Corixidae. *Annales Zoologici Fennici* 13: 48-62.

Jansson, A. (1977a). Distribution of Micronectae (Heteroptera, Corixidae) in Lake Päijänne, central Finland: Correlation with eutrophication and pollution. *Annales Zoologici Fennici* 14: 105-117.

Jansson, A. (1977b). Micronectae as indicators of water quality in two lakes in southern Finland. *Annales Zoologici Fennici* 14: 118-124.

Jansson, A. (1982). Notes on some Corixidae (Heteroptera) from New Guinea and New Caledonia. *Pacific Insects* 24: 95-103.

Jansson, A. (1984). The stridulatory apparatus of Micronectinae (Heteroptera, Corixidae). *International Congress of Entomology Proceedings* 17: 80 (abstract).

Jansson, A. (1986). The Corixidae (Heteroptera) of Europe and some adjacent regions. *Acta Entomologica Fennica* 47: 1-94.

Jansson, A. (1987). Micronectinae (Heteroptera, Corixidae) as indicators of water quality in Lake Vesijärvi, southern Finland, during the period 1976-1986. *Biological Research reports of the University of Jyväskylä* 10: 119-128.

Jansson, A. (1989). Stridulation of Micronectinae (Heteroptera, Corixidae). *Annales Entomologici Fennici* 55: 161-175.

Jansson, A. (1996). Heteroptera Nepomorpha, aquatic bugs. Pp. 91-104 *in* Nilsson, A. (Ed.) *Aquatic Insects of Northwest Europe.* 274 pp. Apollo Books, Svendstrup, Denmark.

Jansson, A. & Scudder, G. (1972). Corixidae as predators: rearing on frozen brine shrimp. *Journal of the Entomological Society of British Columbia* 69: 44-45.

Jansson, A. & Scudder, G. (1974). The life cycle and sexual development of *Cenocorixa* species (Hemiptera, Corixidae) in the Pacific Northwest of North America. *Freshwater Biology* 4: 73-92.

Jell, P. A. & Duncan, P. M. (1986). Invertebrates, mainly insects, from the freshwater, Lower Createceous, Koonwarra Fossil Bed (Korumburra Group), South Gippsland, Victoria. *Memoirs of the Association of Australasian Palaeontologists* 3: 111-205.

Jordan, K. H. C. (1952). Wasserläufer. *Die Neue Brehm Bücherei* 52: 1-32.

Keffer, S. L. (1996). Systematics of the new World waterscorpion genus *Curicta* Stål (Heteroptera: Nepidae). *Journal New York Entomological Society* 104: 117-215.

Kellen, W. R. (1959). Notes on the biology of *Halovelia marianarum* Usinger in Samoa (Veliidae: Heteroptera). *Annals of the Entomological Society of America* 52: 53-62.

King, I. M. (1976). Underwater sound production in *Micronecta batilla* Hale (Heteroptera: Corixidae). *Journal of the Australian Entomological Society* 15: 35-43.

King, I. M. (1997). Two new species of *Micronecta* Kirkaldy (Heteroptera: Corixidae) from Victoria, Australia. *Australian Entomologist* 24: 145-152.

King, I. M. (1999a). Species-specific sounds in water bugs of the genus *Micronecta.* Part 1. Sound analysis. *Bioacoustics* 9: 297-323.

King, I. M. (1999b). Species-specific sounds in water bugs of the genus *Micronecta.* Part 2, chorusing. *Bioacoustics* 10: 19-29.

King, I. M. (1999c). Acoustic communication and mating behaviour in water bugs of the genus *Micronecta. Bioacoustics* 10: 115-130.

Kirkaldy, G. W. (1897a). Aquatic Rhynchota: descriptions and notes. No. 1. *Annals and Magazine of Natural History* (6) 20: 52-60.

Kirkaldy, G. W. (1897b). Revision of the Notonectidae. Part 1. Introduction, and systematics revision of the genus *Notonecta. Transactions of the Entomological Society of London* 1897 (4): 393-426.

Kirkaldy, G. W. (1898). Neue und seltene Notonectiden-Arten. *Wiener Entomologische Zeitung* 17: 141-142.

Kirkaldy, G. W. (1899a) Sur quelques Hemipteres aquatiques nouveax ou peu connues. *Revue d'Entomologie, Caen* 18: 85-96.

Kirkaldy, G. W. (1899b). Aquatic Rhynchota in the collection of the Royal Museum of Belgium. *Annales de la Societé Entomologique de Belgique* 43: 505-510.

Kirkaldy, G. W. (1901). On some Rhynchota, principally from New Guinea (Amphibicorisae and Notonectidae). *Annali del Museo Civico di Storia Naturale Giacomo Doria* 20: 804-810.

Kirkaldy, G. W. (1902a). Miscellanea Rhynchotalia No. 3. *Entomologist* 35: 136-138.

Kirkaldy, G. W. (1902b). Miscellanea Rhynchotalia No. 5. Entomologist 35: 280-284.

Kirkaldy, G. W. (1903). Miscellanea Rhynchotalia. No. 7. *Entomologist* 36: 179-181.

Kirkaldy, G. W. (1904). Über Notonectiden (Hemiptera). *Wiener Entomologische Zeitung* 23: 93-135.

Kirkaldy, G. W. (1905). Five new species of *Micronecta* Kirkaldy. *Entomological News* 16: 260-263.

Kirkaldy, G. W. (1908). Memoir on a few Heteropterous Hemiptera from Eastern Australia. *Proceedings of the Linnean Society of New South Wales* 32: 768-788.

Knowles, J. N. (1974). A revision of Australian species of *Agraptocorixa* Kirkaldy and *Diaprepocoris* Kirkaldy (Heteroptera: Corixidae). *Australian Journal of Marine and Freshwater Research* 25: 173-191.

Knowles, J. N. & Williams, W. D. (1973). Salinity range and osmoregularity ability of Corixids (Hemiptera: Heteroptera) in south-east Australian inland waters. *Australian Journal of Marine and Freshwater Research* 24: 297-302.

Kormilev, N. A. (1971). Ochteridae from the Oriental and Australian Regions. *Pacific Insects* 13: 429-444.

Kovac, D. (1982). Zur Uberwinterung der Wasserwanze *Plea minutissima* Leach (Heteroptera, Pleidae): Diapause mit Hilfe der Plastronatmung. *Nachrichten des Entomologischen Vereins Apollo* 3: 59-76.

Kovac, D. (1993). A quantitative analysis of secretion-grooming behaviour in the water bug *Plea minutissima* Leach (Heteroptera, Pleidae): control by abiotic factors. *Ethology* 93: 41-61.

Kovac, D. & Maschwitz, U. (1989). Secretion-grooming in the water bug *Plea minutissima*: a chemical defence against microorganisms interfering with the hydrofuge properties of the respiratory region. *Ecological Entomology* 14: 403-411.

Kovac, D. & Maschwitz, U. (1991). The function of the metathoracic scent gland in corixid bugs (Hemiptera, Corixidae): secretion grooming on the water surface. *Journal of Natural History* 25: 331-340.

Kovac, D. & Yang, C. M. (2000). Revision of the Oriental bamboo-inhabiting semiaquatic bugs genus *Lathriovelia* Andersen, 1989 (Heteroptera: Veliidae) with description of *L. rickmersi*, new species, and notes on the genus *Baptista* Distant, 1903. *Raffles Bulletin of Zoology* 48: 153-165.

Laird, M. (1956). Studies of mosquitoes and freshwater ecology in the South Pacific. *Bulletin of the Royal Society of New Zealand* 6: 1-213.

Lansbury, I. (1964a). The genus *Anisops* in Australia (Hemiptera: Notonectidae). Part I. *Journal of the Entomological Society of Queensland* 3: 52-65.

Lansbury, I. (1964b). A revision of the genus *Paranisops* Hale (Hemiptera: Notonectidae). *Proceedings of the Royal Entomological Society of London* 33: 181-188.

Lansbury, I. (1967). Comparative morphology of the male Australian Nepidae (Hemiptera: Heteroptera). *Australian Journal of Zoology* 15: 641-649.

Lansbury, I. (1968). The genus *Enithares* (Hemiptera-Heteroptera: Notonectidae). *Pacific Insects* 10: 353-442.

Lansbury, I. (1969). The genus *Anisops* in Australia (Hemiptera-Heteroptera: Notonectidae). *Journal of Natural History* 3: 433-458.

Lansbury, I. (1970). Revision of the Australian *Sigara* (Hemiptera-Heteroptera: Corixidae). Journal of Natural History 4: 39-54.

Lansbury, I. (1972). A review of the Oriental species of *Ranatra* Fabricius (Hemiptera-Heteroptera: Nepidae). *Transactions Royal Entomological Society of London* 124: 287-341.

Lansbury, I. (1973). A review of the genus *Cercotmetus* Amyot & Serville, 1843 (Hemiptera: Nepidae). *Tijdschrift voor Entomologie* 116: 83-106.

Lansbury, I. (1974a). A new genus of Nepidae from Australia with a revised classification of the family (Hemiptera: Heteroptera). *Journal of the Australian Entomological Society* 13: 219-227.

Lansbury, I. (1974b). Notes on the genus *Enithares* Spinola (Hem., Notonectidae). *Entomologist's Monthly Magazine* 109: 226-231.

Lansbury, I. (1975). Notes on additions, changes and the distribution of the Australia water-bug fauna (Hemiptera-Heteroptera). *Memoirs of the National Museum of Victoria* 36: 17-23.

Lansbury, I. (1978). A review of *Goondnomdanepa* Lansbury (Heteroptera: Nepidae). *Australian Journal of Marine and Freshwater Research* 29: 117-126.

Lansbury, I. (1983). Notes on the Australasian species of *Cymatia* Flor s.l. (Insecta, Heteroptera: Corixidae). *Transactions Royal Society of South Australia* 107: 51-57.

Lansbury, I. (1984). Some Nepomorpha (Corixidae, Notonectidae and Nepidae) (Hemiptera-Heteroptera) of north-west Australia. *Transactions Royal Society of South Australia* 108: 35-49.

Lansbury, I. (1985a). Notes on the identity of *Nychia* Stål (Hemiptera-Heteroptera: Notonectidae) in Australia. *Occasional Papers, Northern Territory Museum of Arts and Science* 2: 1-9.

Lansbury, I. (1985b). The Australian Naucoridae (Insecta, Hemiptera-Heteroptera) with description of a new species. *Transactions of the Royal Society of South Australia* 109: 109-119.

Lansbury, I. (1989). Notes on the Haloveliinae of Australia and the Solomon Islands (Insecta, Hemiptera, Heteroptera: Veliidae). *Reichenbachia* 26: 93-109.

Lansbury, I. (1990). Notes on the Hebridae (Insecta: Hemiptera-Heteroptera) of Australia with descriptions of three new species. *Trans actions of the Royal Society of South Australia* 114: 55-66.

Lansbury, I. (1991a). Cuticular blades and other structures of *Diaprepocoris* Kirkaldy and *Stenocorixa* Horvath (Heteroptera: Corixidae). *Tijdschrift voor Entomologie* 134: 35-36.

Lansbury, I. (1991b). Naucoridae and Notonectidae (Hemiptera-Heteroptera) of the Northern Territory, Australia. *Beagle, Records of the Northern Territory Museum of Arts and Science* 8: 103-114.

Lansbury, I. (1993). *Rhagovelia* of Papua New Guinea, Solomon Islands and Australia (Hemiptera-Veliidae). *Tijdschrift voor Entomologie* 136: 23-54.

Lansbury, I. (1995a). Notes on the genus *Anisops* Spinola (Hemiptera-Heteroptera, Notonectidae) of the Northern Territory and Western Australia. *Beagle, Records of the Northern Territory Museum of Arts and Science* 12: 65-74.

Lansbury, I. (1995b). Notes on the Corixidae and Notonectidae (Hemiptera-Heteroptera) of southern Western Australia. *Records of the Western Australian Museum* 17: 181-189.

Lansbury, I. (1996). Notes on the marine veliid genera *Haloveloides, Halovelia* and *Xenobates* (Hemiptera-Heteroptera) of Papua New Guinea. *Tijdschrift voor Entomologie* 139: 17-28.

Lansbury, I. & Lake, P. S. (2002). *Tasmanian Aquatic & Semi-Aquatic Hemipterans.* Cooperative Research Centre for Freshwater Ecology. Identification and Ecology Guide No. 40, 64 pp.

Lariviere, M. C. (1997). Composition and affinities of the New Zealand heteropteran fauna (including Coleorrhyncha). *New Zealand Entomologist* 20: 37-44.

Larsén, O. (1927). Über die Entwicklung und biologie von *Aphelocheirus aestivalis* Fabr. *Entomologisk Tidskrift* 48: 181-206.

Larsén, O. (1932). Beiträge zur Ökologie und Biologie von *Aphelocheirus aestivalis* Fabr. *Int. Rev. Gesellschaft für Hydrobiologie und Hydrographie* 26: 1-19.

Larsén, O. (1938). Untersuchungen über den Geschlechtsapparat der aquatilen Wanzen. *Opuscula Entomologica Supplementum* 1: 1-388.

Larsén, O. (1949). Die Ortsbewegungen von *Ranatra linearis* L. *Lunds Universitets Årsskrift N. F. Avd. 2* 15(6): 1-82.

Larsen, O. (1950). Die Veränderungen der Heteropteren bei der Reduktion des Flugapparates. *Opuscula Entomologica* 15: 17-51.

Larsén, O. (1957). Truncale scolopalorgane in den pterothorakalen und bei den beiden ersten abdominalen Segmenten der aquatilen Heteropteren. *Acta Universitatis Lundensis (N.S.)* 53(1): 1-67.

Larsson, S. G. (1978). Baltic Amber - a Palaeobiological Study. *Entomonograph* 1: 1-192.

Lauck, D. R. (1979). Family Corixidae. Pp 87-123.*in* Menke, A. S. (Ed.): *The Semiaquatic and Aquatic Hemiptera of California (Heteroptera: Hemiptera).* ix + 166 pp. Bulletin of the California Insect Survey 21.

Lauck, D. R. & Menke, A. S. (1961). The higher classification of the Belostomatidae (Hemiptera). *Annals of the Entomologica Society of America* 54: 644-657.

Lebrun, D. (1960). Recherches sur la biologie et l'éthologie de quelques Héteroptères aquatiques. *Annales de la Societé d'Entomologique Francaise* 129: 179-199.

Leong, C. Y. (1962). The life-history of *Anisops breddini* Kirkaldy (Hemiptera, Notonectidae). *Annals and Magazine of Natural History* (13) 5: 377-383.

Leston, D., Pendergrast, J. G. & Southwood, T. R. E. (1954). Classification of the terrestrial Heteroptera (Geocorisae). *Nature* 174: 91-92.

Lundbeck, W. (1914). Some remarks on the eggs and egg-deposition of *Halobates. Mindeskrift for Japetus Steenstrup* 2: 1-13.

Lundblad, O. M. (1928). Die Australischen Arten der Gattung *Agraptocorixa. Arkiv för Zoologi* 20A: 1-19.

Lundblad, O. M. (1933). Zur Kenntniss der aquatilen und semiakvatilen Hemipteren von Sumatra, Java und Bali. *Archiv für Hydrobiologie Supplementum* 12: 1-195, 263-489.

Lundblad, O. M. (1935). Aquatic and semiaquatic Heteroptera of Tahiti. *Bulletin of Bernice P. Bishop Museum* 113: 121-126.

Macan, T. T. (1938). Evolution of aquatic habitats with special reference to the distribution of Corixidae. *Journal of Animal Ecology* 7: 1-19.

Macan, T. T. (1976). A twenty-one year study of the water bugs in a moorland fish pond. *Journal of Animal Ecology* 45: 913-922.

Mahner, M. (1993). Systema Cryptoceratorum Phylogeneticum (Insecta, Heteroptera). *Zoologica* 143: i-ix, 1-302.

Malipatil, M. B. (1980). Review of Australian *Microvelia* Westwood (Hemiptera: Veliidae) with the description of two new species from eastern Australia. *Australian Journal of Marine and Freshwater Research* 31 85-108.

Malipatil, M. B. (1988). Two new species of *Halobates* Eschscholtz (Hemiptera: Gerridae) from Australia. *Journal of the Australian Entomological Society* 27: 157-160.

Malipatil, M. B. & Monteith, G. B. (1983). One new genus and four new species of terrestrial Mesoveliidae (Hemiptera: Gerromorpha) from Australia and New Caledonia. *Australian Journal of Zoology* 31: 943-955.

Markl, H. & Wiese, K. (1969). The sensitivity of *Notonecta glauca* L. to ripples of water surface. *Zeitschrift für Vergleichende Physiologie* 62: 413-420.

Martin, N. A. (1969). *The food, feeding mechanism and ecology of the Corixidae (Hemiptera-Heteroptera), with special reference to Leicestershire.* Dissertation, University of Leicester.

Matsuda, R. (1960). Morphology, evolution and a classification of the Gerridae (Hemiptera-Heteroptera). *Kansas University Science Bulletin* 41: 25-632.

McCafferty, W. P. (1998). *Aquatic Entomology. The Fisherman's and Ecologist's Illustrated Guide to Insects and Their Relatives.* 480 pp. Jones & Bartlett, Boston.

McCoull, C. J., Swain, R. & Barnes, R.W. (1998). Effect of temperature on the functional response and components of attack rate in *Naucoris congrex* Stal (Hemiptera: Naucoridae). *Australian Journal of Entomology 37*: 323-327.

Menke, A. S. (1960). A review of the genus *Lethocerus* (Hemiptera: Belostomatidae) in the Eastern Hemisphere with the description of a new species from Australia. *Australian Journal of Zoology* 8: 285-288.

Menke, A. S. (1979a). Family Nepidae. Pp 70-75 *in* Menke, A. S. (Ed.): *The Semiaquatic and Aquatic Hemiptera of California (Heteroptera: Hemiptera).* ix + 166 pp. Bulletin of the Californian Insect Survey 21.

Menke, A. S. (1979b). Family Belostomatidae, pp. 76-86 *in* Menke, A. S. (Ed.): *The Semiaquatic and Aquatic Hemiptera of California (Heteroptera: Hemiptera).* ix + 166 pp. Bulletin of the Californian Insect Survey 21.

Menke, A. S. (1979c.) Family Ochteridae. Pp 124-125 *in* Menke, A. S. (Ed.): *The Semiaquatic and Aquatic Hemiptera of California (Heteroptera: Hemiptera).* ix + 166 pp. Bulletin of the Californian Insect Survey 21.

Menke, A. S. (1979d). Family Gelastocoridae. Pp 126-130 *in* Menke, A. S. (Ed.): *The Semiaquatic and Aquatic Hemiptera of California (Heteroptera: Hemiptera).* ix + 166 pp. Bulletin of the Californian Insect Survey 21.

Menke, A. S. & Stange, L. A. (1964). A new genus of Nepidae from Australia with notes on the higher classification of the family. *Proceedings of the Royal Society of Queensland* 75: 67-72.

Merritt, R. W. & Cummins, K. W. (eds) (1996). *An Introduction to the Aquatic Insects of North America. 3rd edition.* xii + 441 pp. Kendall/Hunt, Dubuque.

Messner, B., Lunk, A., Groth, I., Subklew, H.-J. & Taschenberger, D. (1981). Neue Befunde zum Atmungssystem der Grundwanze *Aphelocheirus aestivalis* Fab. (Heteroptera, Hydrocorisae). I. Imagines. *Zoologische Jahrbücher für Anatomie*

105: 474-496.

Miller, P. L. (1964). The possible rôle of haemoglobin in *Anisops* and *Buenoa* (Hemiptera: Notonectidae). *Proceedings of the Royal Entomologicl Society of London (A)* 39: 166-175.

Miyamoto, S. (1964). Veliidae of the Ryukyus (Hemiptera, Heteroptera). *Kontyu* 32: 137-150.

Montandon, A. L. (1898). Hémiptèra Cryptocerata. Notes et descriptions d'espèces nouvelles. *Bulletin de la Societé Bucarest* 7: 430-432.

Montandon, A. L. (1899a). Hémiptèra Cryptocerata s. fam. Mononychinae. Notes et descriptions d'espèces nouvelle. 1ère partie. *Bulletin Societé Bucarest* 8: 392-407.

Montandon, A. L. (1899b). Hémiptèra Cryptocerata s. fam. Mononychinae. Notes et descriptions d'espèces nouvelle. 2ème partie. *Bulletin Societé Bucarest* 8: 774-788.

Montandon, A. L. (1903). Hémiptères aquatiques. Notes synonymiques et geographiques, descriptions d'espèces nouvelles. *Bulletin de la Societé Bucarest* 12: 97-121.

Montandon, A. L. (1907). Quelques espèces du genre *Ranatra* des collections du Muséum de Paris. *Annales de la Societé Entomologique Francaise* 76: 49-66.

Montandon, A. L. (1913). Hémiptères aquatiques. Notes et descriptions de deux espèces nouvelles. Bulletin de la Section Scientifique de l'Academie Roumaine 1: 219-224.

Montrouzier, P. (1855). Essai sur la faune de l'île de Woodlark ou Mouiou. *Annales de la Societé Agricoles de Lyon* (2) 7: 1-114.

Montrouzier, P. (1864). Essai sur la faune entomologique de Kanala (Nouvelle Calédonie) et description de quelques espèces nouvelles ou peu connue. *Annales de la Societé Linneenne de Lyon* 11: 46-257.

Muraji, M., Miura, T. & Nakasuji, F. (1989a). Change in photoperiodic sensitivity during hibernation in a semi-aquatic bug, *Microvelia douglasi* (Heteroptera: Veliidae). *Applied Entomology & Zoology* 24: 450-457.

Muraji, M., Miura, T. & Nakasuji, F. (1989b). Phenological studies on the wing dimorphism of a semi-aquatic bug, *Microvelia douglasi* (Heteroptera: Veliidae). *Research in Population Ecology* 31: 129-138.

Muraji, M. & Nakasuji, F. (1988). Comparative studies on life history traits of three wing dimorphic water bugs, *Microvelia* spp. Westwood (Heteroptera: Veliidae). *Research in Population Ecology* 30: 315-327.

Muraji, M. & Nakasuji, F. (1990). Effect of photoperiodic shifts on egg production in a semiaquatic bug, *Microvelia douglasi* (Heteroptera, Veliidae). *Applied Entomology & Zoology* 25: 405-407.

Muraji, M. & Tachikawa, S. (2000). Phylogenetic analysis of water striders (Hemiptera: Gerroidea) based on partial sequences of mitochondrial and nuclear ribosomal RNA genes. *Entomological Science* 3: 616-626.

Nakasuji, F. & Dyck, V. A. (1984). Evaluation of the role of *Microvelia douglasi atrolineata* (Bergroth) as predator of the brown planthopper *Nilaparvata lugens* Stal. *Research in Population Ecology* 26: 134-149.

Nel, A. & Paicheler, J. (1992-1993). Les Heteroptera aquatiques fossiles, état actuel des connaissances (Heteroptera: Nepomorpha et Gerromorpha). Entomologica Gallica 3-4: 159-182, 15-21, 79-89.

Nieser, N. (1977). Gelastocoridae in the Zoologisches Museum der Humboldt-Universität zu Berlin (Heteroptera). *Deutsches Entomologisches Zeitschrift (NS)* 24: 293-303.

Nieser, N. & Chen, P. (1991). Notes on Malesian aquatic bugs (Heteroptera) I. Naucoridae, Nepidae and Notonectidae, mainly from Sulawesi and Pulau Buton (Indonesia). *Tijdschrift voor Entomologie* 134: 47-67.

Nieser, N. & Chen, P. (1992). Revision of *Limnometra* Mayr (Gerridae) in the Malay Archipelago. Notes on Malesian aquatic and semiaquatic bugs (Heteroptera), II. *Tidjschrift voor Entomologie* 135: 1126.

Nieser, N. & Wasscher, M. (1986). The status of the larger waterstriders in the Netherlands (Heteroptera: Gerridae). *Entomologische Berichten (Amsterdam)* 46: 68-76.

Nieser, N. & Zettel, Herbert (2001). First record of *Paranisops* Hale, 1924 (Insecta: Heteroptera: Notonectidae) from Southeast Asia, with description of *P. leucopardalos* sp.n. *Annales Naturhistorisches Museum Wien* 103B: 243-247.

Nilsson, A. (Ed.) (1996). *Aquatic Insects of Northwest Europe, Vol I.* 274 pp. Apollo Books, Stenstrup, Denmark.

Nummelin, M. (1988). Waterstriders (Het: Gerridae) as predators of hatching mosquitoes. *Bicovas Proceedings* 1: 121-125.

Oscarson, H. G. (1987). Habitat segregation in a water boatman (Corixidae) assemblage - the role of predation. *Oikos* 49: 133-140.

Pajunen, V. I. & Ukkonen, M. (1987). Intra- and interspecific predation in rock-pool corixids (Hemiptera, Corixidae). *Annales Zoologici Fennici* 24: 295-304.

Papacek, M. (1985). Der Lebenszyklus und die Entwicklung des Zwergruckenschwimmers *Plea leachi* McGregor & Kirkaldy 1899 (Heteroptera, Pleidae) im becken von Ceske Budejovice. *Sbornik Jihoceskeho Muzea v Ceskych Budejovicich*

Prirodni Vedy 25: 73-85.
Papacek, M. (1999). Does the architecture of lateral pterothoracic region and thoracico-abdominal junction in the Notonectidae, Pleidae, and Helotrephidae (Heteroptera: Nepomorpha) reflect phylogenetic relationships? *Acta Societatis Zoologicae Bohemicae* 63: 165-178.
Papacek, M., Stys, P. & Tonner, M. (1988). A new subfamily of Helotrephidae (Heteroptera, Nepomorpha) from Southeast Asia. *Acta Entomologica Bohemoslovaca* 85: 120-152.
Papacek, M., Stys, P. & Tonner, M. (1990). Formation of cephalothorax in the Pleoidea (Heteroptera, Nepomorpha). *Acta Entomologica Bohemoslovaca* 87: 326-331.
Parsons, M. (1959). Skeleton and musculature of the head of *Gelastocoris oculatus* (Fabricius). *Bulletin of the Museum of Comparative Zoology* 122: 1-53.
Parsons, M. (1960). Skeleton and musculature of the thorax of *Gelastocoris oculatus* (Fabricius). *Bulletin of the Museum of Comparative Zoology* 122: 299-357.
Parsons, M. C. (1966). Modification of the food pumps of Hydrocorisae. *Canadian Journal of Zoology* 44: 585-620.
Parsons, M. C. (1968). The cephalic and prothoracic skeletomusculature and nervous system in *Lethocerus* (Heteroptera, Belostomatidae). *Journal of the Linnean Society, Zoology* 47: 349-406.
Parsons, M. C. (1969). Skeletomusculature of the pterothorax and first abdominal segment in micropterous *Aphelocheirus aestivalis* F. *Transactions of the Royal Entomological Society of London* 121: 1-39.
Parsons, M. C. (1970). Respiratory significance of the thoracic and abdominal morphology of the three aquatic bugs *Ambrysus, Notonecta* and *Hesperocorixa* (Insecta, Heteroptera). *Zeitschrift für Morphologie das Tiere* 66: 242-298.
Parsons, M. C. (1972a). Respiratory significance of the thoracic and abdominal morphology of *Belostoma* and *Ranatra* (Insecta, Het.). *Zeitschrift für Morphologie das Tiere* 73: 163-194.
Parsons, M. C. (1972b). Morphology of the three anterior pairs of spiracles of *Belostoma* and *Ranatra* (Aquatic Het., Belostomatidae, Nepidae). *Canadian Journal of Zoology* 50: 865-876.
Parsons, M. C. (1973). Morphology of the eight abdominal spiracles of *Belostoma* and *Ranatra* (Aquatic Heteroptera: Belostomatidae, Nepidae). *Journal of Natural History* 7: 255-265.
Parsons, M. C. (1976). Respiratory significance of the thoracic and abdominal morphology of three Corixidae, *Diaprepocoris, Micronecta* and *Hesperocorixa* (Hemiptera: Heteroptera: Corixidae). *Psyche* 83: 132-179.
Peters, W. & Spurgeon, J. (1971). Biology of the water boatman *Krizousacorixa femorata* Guérin. *Journal of Morphology* 100: 141-156.
Poinar, G. O., Jr. (1992). *Life in Amber.* xiii + 350 pp. Stanford University Press, Stanford, California.
Poisson, R. A. (1924). Contributions á l'étude des Hémiptères aquatiques. *Bulletin Biologique de la France et Belgique* 58: 49-305.
Poisson, R. A. (1957). Hémiptères aquatiques. *Faune de France* 61: 1-263.
Poisson, R. A. (1965a). Catalogue des Hétéroptères Hydrocorises africano-malgaches de la famille des *Nepidae* (Latreille) 1802. *Bulletin Institut Français Afrique Noire, Serie A* 27: 229-269.
Poisson, R. A. (1965b). Catalogue des Insectes Hétéroptères Gerridae Leach, 1807, africano-malgaches. *Bulletin Institute Francais Afrique Noire, Serie A* 27: 1466-1503.
Polhemus, D. A. (1990). Heteroptera of Aldabra Atoll and nearby islands, western Indian Ocean, Part 1. Marine Heteroptera (Insecta); Gerridae, Veliidae, Hermatobatidae, Saldidae and Omaniidae, with notes on ecology and insular zoogeography. *Atoll Research Bulletin* 345: 1-16.
Polhemus, D. A. (1993). Conservation of aquatic insects: worldwide crisis or localized threats? *American Zoologist* 33: 588-598.
Polhemus, D. A. (1997). *Systematics, phylogeny, and zoogeography of the genus* Rhagovelia *(Heteroptera: Veliidae) in the western hemisphere (exclusive of the* angustipes *complex).* 410 pp. Thomas Say Monographs, Entomological Society of America.
Polhemus, D. A. & Polhemus, J. T. (1988). The Aphelocheirinae of tropical Asia (Heteroptera: Naucoridae). *Raffles Bulletin of Zoology* 36: 167-300.
Polhemus, D. A. & Polhemus, J. T. (1997). A review of the genus *Limnometra* Mayr in New Guinea, with the description of a very large new species (Heteroptera: Gerridae). *Journal of the New York Entomological Society* 105: 24-39.
Polhemus, D. A. & Polhemus, J. T. (2000). Additional new genera and species of Microveliinae (Heteroptera: Veliidae) from New Guinea. *Tijdschrift voor Entomologie* 143: 91-123.
Polhemus, J. T. (1979). Family Naucoridae. Pp. 131-138 *in* Menke, A. S. (Ed.): *The Semiaquatic and Aquatic Hemiptera of California (Heteroptera: Hemiptera).* ix + 166 pp. Bulletin of the Californian Insect Survey 21.
Polhemus, J. T. (1982). Marine Hemiptera of the Northern Territory, including the first freshwater species of *Halobates* Eschscholtz (Gerridae, Veliidae, Hermatobatidae and Corix-

idae). *Journal of Australian Entomological Society* 21: 5-11.

Polhemus, J. T. (1984). A review of the Naucoridae of Australia (Heteroptera: Naucoridae). *Journal of the Australian Entomological Society* 23: 157-60.

Polhemus, J. T. (1994a). Stridulatory mechanisms in aquatic and semiaquatic Heteroptera. *Journal of the New York Entomological Society* 102: 270-274.

Polhemus, J. T. (1994b). The identity and synonymy of the Belostomatidae (Heteroptera) of Johan Christian Fabricius 1775-1803. *Proceedings of the Entomological Society of Washington* 96: 687-695.

Polhemus, J. T. (1995). A new genus of Hebridae from Chiapas amber (Heteroptera). *Pan-Pacific Entomologist* 71: 78-81.

Polhemus, J. T. & Chapman, H. C. (1979a). Family Mesoveliidae. Pp 39-42 *in* Menke, A. S. (Ed.): *The Semiaquatic and Aquatic Hemiptera of California (Heteroptera: Hemiptera)*. ix + 166 pp. Bulletin of the Californian Insect Survey 21.

Polhemus, J.T. & Chapman, H. C. (1979b). Family Hebridae. Pp 34-38 *in* Menke, A. S. (Ed.): *The Semiaquatic and Aquatic Hemiptera of California (Heteroptera: Hemiptera)*. ix + 166 pp. Bulletin of the Californian Insect Survey 21.

Polhemus, J. T. & Chapman, H. C. (1979c). Family Hydrometridae, pp. 43-45 *in* Menke, A. S. (Ed.): *The Semiaquatic and Aquatic Hemiptera of California (Heteroptera: Hemiptera)*. ix + 166 pp. Bulletin of the Californian Insect Survey 21.

Polhemus, J. T. & Chapman, H. C. (1979d). Family Veliidae, pp. 49-57 *in* Menke, A. S. (Ed.): *The Semiaquatic and Aquatic Hemiptera of California (Heteroptera: Hemiptera)*. ix + 166 pp. Bulletin of the Californian Insect Survey 21.

Polhemus, J. T. & Chapman, H. C. (1979e). Family Gerridae, pp. 58-69 *in* Menke, A. S. (Ed.): *The Semiaquatic and Aquatic Hemiptera of California (Heteroptera: Hemiptera)*. ix + 166 pp. Bulletin of the Californian Insect Survey 21.

Polhemus, J.T. & Cheng, L. (1982). Notes on marine water-striders with descriptions of new species. Part I. Gerridae (Hemiptera). *Pacific Insects* 24: 219-227.

Polhemus, J. T. & Copeland, R. S. (1996). A new genus of Microveliinae from treeholes in Kenya (Heteroptera: Veliidae). *Tijdschrift voor Entomologie* 139: 73-77.

Polhemus, J. T. & Karunaratne, P. B. (1993). A review of the genus *Rhagadotarsus* with descriptions of three new species (Heteroptera: Gerridae). *Raffles Bulletin of Zoology* 41: 95-112.

Polhemus, J. T. & Keffer, S. L. (1999). Notes on the genus *Laccotrephes* Stål (Heteroptera: Nepidae) in the Malay Archipelago, with the description of two new species. *Journal of the New York Entomological Society* 107: 1-13.

Polhemus, J. T. & Lansbury, I. (1997). Revision of the genus *Hydrometra* Latreille in Australia, Melanesia, and the Southwest Pacific (Heteroptera: Hydrometridae). *Bishop Museum Occasional Papers* 47: 1-67.

Polhemus, J. T. & Lindskog, P. (1994). The stridulatory mechanism of *Nerthra* Say, a new species, and synonymy (Heteroptera: Gelastocoridae). *Journal of the New York Entomological Society* 102: 242-248.

Polhemus, J.T. & Polhemus, D. A. (1988). Zoogeography, ecology, and systematics of the genus *Rhagovelia* Mayr (Heteroptera: Veliidae) in Borneo, Celebes, and the Moluccas. *Insecta Mundi* 2: 161-230.

Polhemus, J. T. & Polhemus, D. A. (1991). Three new species of marine water-striders from the Australasian region, with notes on other species (Gerridae: Halobatinae, Trepobatinae). *Raffles Bulletin of Zoology* 39: 1-13.

Polhemus, J. T. & Polhemus, D. A. (1993). The Trepobatinae (Heteroptera: Gerridae) of New Guinea and surrounding regions, with a review of the world fauna. Part 1. Tribe Metrobatini. *Entomologica Scandinavica* 24: 241-284.

Polhemus, J. T. & Polhemus, D. A. (1994a). Four new genera of Microveliinae (Heteroptera) from New Guinea. *Tijdschrift voor Entomologie* 137: 57-74.

Polhemus, J. T. & Polhemus, D. A. (1994b). The Trepobatinae (Heteroptera: Gerridae) of New Guinea and surrounding regions, with a review of the world fauna. Part 2. Tribe Naboandelini. *Entomologica Scandinavica* 25: 333-359.

Polhemus, J. T. & Polhemus, D. A. (1995a). Revision of the genus *Hydrometra* Latreille in Indochina and the western Malay Archipelago (Heteroptera: Hydrometridae). *Bishop Museum Occasional Papers* 43: 9-72.

Polhemus, J. T. & Polhemus, D. A. (1995b). The Trepobatinae (Heteroptera: Gerridae) of New Guinea and surrounding regions, with a review of the world fauna. Part 3. Tribe Trepobatini. *Entomologica Scandinavica* 26: 97-118.

Polhemus, J. T. & Polhemus, D. A. (1996). The Trepobatinae (Heteroptera: Gerridae) of New Guinea and surrounding regions, with a review of the world fauna. Part 4. Tribe Stenobatini. *Entomologica Scandinavica* 27: 279-346.

Polhemus, J. T. & Polhemus, D. A. (2000a). The Trepobatinae (Heteroptera: Gerridae) of New Guinea and surrounding regions, with a review of the world fauna. Part 5. Taxonomic and distributional addenda. *Insect Systematics &*

Evolution 31: 291-316.

Polhemus, J. T. & Polhemus, D. A. (2000b). The genus *Mesovelia* Mulsant & Rey in New Guinea (Heteroptera: Mesoveliidae). *Journal of the New York Entomological Society* 108: 205-230.

Polhemus, J. T. & Polhemus, D. A. (2002). The Trepobatinae (Heteroptera: Gerridae) of New Guinea and surrounding regions, with a review of the world fauna. Part 6. Phylogeny, biogeography, world checklist, bibliography, and final taxonomic addenda. *Insect Systematics & Evolution* 33: 253-290.

Popham, E. J. (1942). The variation in the colour of certain species of *Arctocorisa* and its significance. *Proceedings of the Zoological Society of London (A)* 111: 135-172.

Popham, E. J. (1960). On the respiration of aquatic Hemiptera Heteroptera with special reference to the Corixidae. *Proceedings of the Zoological Society of London (A)* 135: 209-242.

Popham, E. J. (1966). An ecological study of the predatory action of the Three Spined Stickleback (*Gasterosteus aculeatus* L.). *Archiv für Hydrobiologie* 62: 70-81.

Popham, E. J., Bryant, M. T. & Savage, A. A. (1984). The function of the abdominal strigil in male corixid bugs. *Journal of Natural History* 18: 441-444.

Popham, E. J. & Lansbury, I. (1960). The uses and limitations of light traps in the study of the ecology of Corixidae. *Entomologist* 93: 162-169.

Popov, Y. (1971a). Historical development of Hemiptera infraorder Nepomorpha (Heteroptera). *Trud. Paleont. Inst. Acad. Sci. SSSR* 129: 1-228. [In Russian; informal English translation by H. Vaitaitis & M. Parsons].

Popov, Y. (1971b). Origin and main evolutionary trends of Nepomorpha bugs. *Proceedings 13th International Congress of Entomology, Moscow 1968* 1: 282-283.

Popov, Y. A. (1996). The first record of a fossil water bug from the Lower Jurassic of Poland (Heteroptera: Nepomorpha: Belostomatidae). *Polskie Pismo Entomologiczne* 65: 101-105.

Prager, J. (1973). Die Hörschwelle des mesothorakalen Tympanalorgan von *Corixa punctata* Ill. (Heteroptera, Corixidae). *Journal of Comparative Physiology* 86: 55-58.

Rasnitsyn, A. P. & Quicke, D. L. J. (2002). *History of Insects.* xii + 517 pp. Kluwer Academic Publishers, Dordrecht, Boston, London.

Reichart, C. V. (1982). An addition to the genus *Anisops* of Australia (Hemiptera: Notonectidae). *Proceedings of the Entomological Society of Washington* 84: 366-368.

Rensing, L. (1962). Beiträge zur vergleichenden Morphologie, Physiologie und Ethologie der Wasserläufer. *Zoologisher Beiträge N.F.* 7: 447-485.

Reynolds, J. (1981). Size selective predation of *Daphnia* by the backswimmer *Anisops deanei. Newsletter Australian Society of Limnology* 19: 14.

Reynolds, J. G. & Geddes, M. C. (1984). Functional response analysis of size-selective predation by the notonectid predator *Anisops deanei* (Brooks) on *Daphnia thomsoni* (Sars). *Australian Journal of Marine and Freshwater Research* 35: 725-733.

Rieger, C. (1976). Skelett und Muskulatur des Kopfes und Prothorax von *Ochterus marginatus* Latreille. *Zoomorphologie* 83: 109-191.

Rieger, C. (1977). Neue Ochteriden aus der Alten Welt (Heteroptera). *Deutsches Entomologisches Zeitschrift (N.F.)* 24: 213-217.

Ruhoff, F. A. (1964). The proposal of a new name (Hem.-Hydrom.). *Proceedings of the Entomological Society of Washington* 66: 32.

Sailer, R. I. & Lienk, I. (1967). Insect predators of mosquito larvae and pupae in Alaska. *Mosquito News* 14: 14-16.

Savage, A. A. (1982). Use of water boatmen (Corixidae) in the classification of lakes. *Biological Conservation* 23: 55-70.

Savage, A. A. (1989). Adults of the British Aquatic Hemiptera Heteroptera. A key with ecological notes. *Freshwater Biological Association, Scientific Publication* 50: 1-173.

Schaefer, C. & Panizzi, A. R. (eds) (2000). *Heteroptera of Economic Importance.* ??? pp. CRC Press, Boca Raton, Fla.

Schuh, R. T. (1979). Review of: *Evolutionary trends in Heteroptera. Part. II. Mouthpart-structures and feeding strategies* by R. H. Cobben. *Systematic Zoology* 28: 653-656.

Schuh, R. T. & Slater, J. A. (1995). *True Bugs of the World (Hemiptera: Heteroptera).* xii + 336 pp. Cornell University Press, Ithaca and London.

Scudder, G. G. E. (1976). Water-boatmen of saline waters (Hemiptera: Corixidae). Pp 263-289 *in* Cheng, L. (Ed.): *Marine Insects.* 581 pp. North-Holland Publishing Company, Amsterdam.

Scudder, G. G. E. (1983). A review of factors governing the distribution of two closely related corixids in the saline lakes of British Columbia. *Hydrobiologia* 105: 143-154.

Scudder, G. G. E. & Meredith, J. (1972). Temperature induced development in the indirect flight muscle of adult *Cenocorixa. Developmental Biology* 29: 330-336.

Selvanayagam, M. & Rao, T. K. R. (1988). Biological aspects of two species of gerrids, *Limnogonus fossarum fossarum* and *L. nitidus* Mayr. *Bombay Natural History Society Journal* 85: 474484.

Sites, R. W. (2000). Creeping Water Bugs (Naucoridae). Pp 571-576 *in:* Schaefer, C. W. & Panizzi, A. R. (eds): *Heteroptera of Economic Importance.* 828 pp. CRC Press, Boca Raton.

Skuse, F. A. A. (1891). Description of a new pelagic Hemipteron from Port Jackson. *Records of the Australia Museum* 1: 174-177.

Skuse, F. A. A. (1893). Notes on Australian aquatic Hemiptera No. 1. *Records of the Australian Museum* 2: 42-45.

Smith, B. P. (1989). Impact of parasitism by larval *Limnochares aquatica* (Acari: Hydrachnidia; Limnocharidae) on juvenile *Gerris comatus, Gerris alacris,* and *Gerris buenoi* (Insecta: Hemiptera; Gerridae). *Canadian Journal of Zoology* 67: 2238-2243.

Smith, R. L. (1997). Evolution of paternal care in the giant water bugs (Heteroptera: Belostomatidae). Pp 116-149 *in* Choe, J. C. & Crespi, B. J. (eds): *The evolution of social behavior in insects and arachnids.* xiii + 541 pp. Cambridge University Press, Cambridge, New York & Oakleigh.

Smithers, C. N. (1981). *Handbook of Insect Collecting (Collection, Preparation, Preservation and Storage).* 120 pp. Reed, Sydney.

Southwood, T. R. E. & Leston, D. (1959). *Land and Water Bugs of the British Isles.* 436 pp. Frederick Warne & Co. Ltd, London & New York.

Spence, J. R. (1986). Interactions between the scelionid egg parasitoid *Tiphodytes gerriphagus* (Hymenoptera) and its gerrid hosts (Heteroptera). *Canadian Journal of Zoology* 64: 2728-2738.

Spence, J. R. & Andersen, N. M. (1994). Biology of water striders: interactions between systematics and ecology. *Annual Review of Entomology* 39: 101-128.

Spence, J. R. & Andersen, N. M. (2000). Semiaquatic Bugs (Gerromorpha). Pp 601-606 *in* C. Schaefer & A. R. Panizzi (eds): *Heteroptera of Economic Importance.* 828 pp. CRC Press, Boca Raton, Fla.

Spence, J. R. & Wilcox, R. S. (1986). The mating system of two hybridizing species of water striders (Gerridae). II. Alternative tactics of males and females. *Behavioral Ecology and Sociobiology* 19: 87-95.

Sprague, I. B. (1956). The biology and morphology of *Hydrometra martini* Kirkaldy. *Kansas University Science Bulletin* 38: 579-693.

Stål, C. (1854). Nya Hemiptera. *Öfversigt Kongliga Svenska Vetenskabs-Akademiens Förhandlinger, Stockholm* 11: 231-255.

Stål, C. (1863). Verzeichniss der Mononychiden. *Berliner Entomologische Zeitung* 7: 405-408.

Stål, C. (1876). Enumeratio Hemipterorum. Bidrag till en förteckning öfver aller hittills kända Hemiptera, jämte systematiska meddelanden. 5. *Kongliga Svenska Vetenskabs-Akademiens Handlinger* 14(4): 1-162.

Staddon, B. W. (1971). Metasternal scent glands in Belostomatidae. *Journal of Entomology (A)* 46: 69-71.

Stys, P. (1985). The present state of beta-taxonomy in Heteroptera. *Prace Slov. ent. spol.* 4 [1984]: 205-235. [In Czech with English summary].

Stys, P. & Jansson, A. (1988). Check-list of recent family-group and genus-group names of nepomorpha (Heteroptera) of the world. *Acta. Entomologica Fennica* 50: 1-44.

Stys, P. & Kerzhner, I. (1975). The rank and nomenclature of the higher taxa in Heteroptera. *Acta Entomologica Bohemoslovaca* 72: 65-79.

Sulzer, J. H. (1776). *Abgekürzte Geschichte der Insekten nach dem Linnéischen System.* Pt. 1: 27 + 274 pp., Pt. 2: 71 pp. H. Steiner & Co., Winterthur.

Sweeney, A.W. (1960). The distribution of the Notonectidae (Hemiptera) in south-eastern Australia. *Proceedings of the Linnean Society of New South Wales* 90: 87-94.

Takahashi, R.(1923). Observations on the Ochteridae. *Bulletin of the Brooklyn Entomological Society* 18: 67-68.

Takahashi, R. M., Stewart, R. J., Schaeffer, C. H. & Sjogren, R. D. (1979). An assessment of *Plea striola* (Hemiptera: Pleidae) as a mosquito control agent in California. *Mosquito News* 39: 514-519.

Thorpe, W. H. (1950). Plastron respiration in aquatic insects. *Biological Reviews* 25: 344-390.

Thorpe, W. H. & Crisp, D. J. (1947a). Studies on plastron respiration I. The biology of *Aphelocheirus* (Hemiptera, Aphelocheiridae (Naucoridae)) and the mechanism of plastron retention. *Journal Experimental Biology* 24: 227-269.

Thorpe, W. H. & Crisp, D. J. (1947b). Studies on plastron respiration. II. The respiratory efficeincy of the plastron in *Aphelocheirus. Journal Experimental Biology* 24: 270-303.

Todd, E. L. (1955). A taxonomic revision of the family Gelastocoridae (Hemiptera). *University of Kansas Science Bulletin* 37: 277-475.

Todd, E. L. (1959). The Gelastocoridae of Melanesia. *Nova Guinea, N.S.* 10: 61-94.

Todd, E. L. (1960). The Gelastocoridae of Australia (Hemiptera). *Pacific Insects* 2: 171-194.

Todd, E. L. (1961). A checklist of the Gelastocoridae (Hemiptera). *Proceedings of the Hawaiian Entomological Society* 17: 461-476.

Torre-Bueno, J. R. de la (1926). *Limnometra skusei:*

a new name. *Bulletin of the Brooklyn Entomological Society* 21: 129.
Towns, D. R. (1978). Some little-known benthic insect taxa from a northern New Zealand river and its tributaries. *New Zealand Entomologist* 6: 409-419.
Upton, M. S. (1991). *Methods for collecting, preserving and studying insects and allied forms.* Miscellaneous Publications Australian Entomological Society 3.
Usinger, R. L. (1937). A new species of *Aphelocheirus* from Australia (Hemiptera, Naucoridae). *Australian Zoologist* 8: 341-342.
Usinger, R. L. (1956). Aquatic Hemiptera. Pp 182-228 *in* Usinger, R. L. (Ed.): *Aquatic Insects of California.* x + 508 pp. University of California Press, Berkeley and Los Angeles.
Weber, H. (1930). *Biologie der Hemipteren. Eine Naturgeschichte der Schnabelkerfe.* J. Springer, Berlin.
Wefelscheid, H. (1912). Über die Biologie und Anatomie von *Plea minutissima* Leach. *Zoologische Jahrbücher, Systematik* 32: 389-474.
Weitschat, W. & Wichard, W. (1998). *Atlas der Pflanzen und Tiere im Baltischen Bernstein.* 256 pp. Vorlag Dr. Friedrich Pfeil, München.
Wells, R. M. G, Hudson, M. J. & Brittain, T. (1981). Function of the hemoglobin and the gas bubble in the backswimmer *Anisops assimilis* (Hemiptera: Notonectidae). *Journal of Comparative Physiology* 142: 515-522.
Wesenberg-Lund, C. (1943). *Biologie der Süsswasserinsekten.* viii + 682 pp. Gyldendal, Copenhagen.
Wheeler, W. C., Schuh, R. T. & Bang, R. (1993). Cladistic relationships among higher groups of Heteroptera: congruence between morphological and molecular data sets. *Entomologica Scandinavica* 24: 121-137.
White, F. B. (1883). Report on the pelagic Hemiptera. *Voyage of Challenger, Reports, Zoology* 7: 1-82.
Wichard, W., Arens, W. & Eisenbeis, G. (2002). *Biological Atlas of Aquatic Insects.* 330 pp. Apollo Books, Stenstrup, Denmark.
Wichard, W. & Weitschat, W. (1996). Wasserinsekten im Bernstein. Eine paläobiologische Studie. *Entomologische Mitteilungen aus dem Löbbecke-Museum + Aquazoo* 4: 1-122.
Wilcox, R. S. (1972). Communication by surface waves. Mating behaviour of a water strider (Gerridae). *Journal of Comparative Physiology* 80: 255-266.
Wilcox, R. S. (1979). Sex discrimination in *Gerris remigis*: Role of a surface wave signal. *Science* 206: 1325-1327.
Wilcox, R. S. (1995). Ripple communication in aquatic and semiaquatic insects. *Écoscience* 2: 109-115.
Wilcox, R. S. & Spence, J. R. (1986). The mating system of two hybridizing species of water striders (Gerridae). I. Ripple signal functions. *Behavioral Ecology and Sociobiology* 19: 79-85.
Williams, D. D. & Feltmate, B. W. (1992). *Aquatic Insects.* xii + 358 pp. C.A.B. International, Wellingford, U.K.
Williams, W. D. (1980). *Australian Freshwater Life. 2nd edition.* 321 pp. Macmillan, Melbourne.
Williamson, K. B. (1949). Chapter 62. Naturalistic measures of Anopheline Control. Pp 1360-1384 *in* Boyd, M. F. (Ed.) *Malariology II.* Saunders, Philadelphia.
Wise, K. A. J. (1965). An annotated list of the aquatic and semiaquatic insects of New Zealand. *Pacific Insects* 7: 191-216.
Wise, K. A. J. (1990). Lacewings and aquatic insects of New Zealand. *Records of the Auckland Institute & Museum* 27: 195-198.
Wroblewski, A. (1962). Notes on the Micronectinae from Viet-Nam (Heteroptera, Corixidae). *Bulletin de l'Académie polonaise des Sciences* 10: 175-180.
Wroblewski, A. (1967). Further notes on Micronectinae from Viet-Nam (Heteroptera, Corixidae). *Polskie Pismo Entomologiczne* 37: 229-251.
Wroblewski, A. (1968). Notes on Oriental Micronectinae (Heteroptera, Corixidae). *Polskie Pismo Entomologiczne* 38: 753-779.
Wroblewski, A. (1970). Notes on Australian Micronectinae (Heteroptera, Corixidae). *Polskie Pismo Entomologiczne* 40: 681-703.
Wroblewski, A. (1972). Supplementary notes on Australian Micronectinae (Heteroptera, Corixidae). *Polskie Pismo Entomologiczne* 42: 517-526.
Wroblewski, A. (1977). Further notes on Australian Micronectinae (Heteroptera, Corixidae). *Polskie Pismo Entomologiczne* 47: 683-690.
Young, E. C. (1962). The Corixidae and Notonectidae (Hemiptera-Heteroptera) of New Zealand. *Records of the Canterbury Museum New Zealand* 7: 327-374.
Young, E. C. (1965a). Teneral development in British Corixidae. *Proceedings of the Royal Entomological Society of London (A)* 40: 159-168.
Young, E. C. (1965b). Flight muscle polymorphism in British Corixidae: ecological observations. *Journal of Animal Ecology* 34: 353-390.
Young, E. C. (1966). Observations on migration in Corixidae in Southern England. *Entomologist's Monthly Magazine* 101: 217-229.
Young, E. C. (1978). Seasonal cycles of ovarian development in Corixidae and Notonectidae, aquatic Hemiptera-Heteroptera. *The New*

Zealand Entomologist 6: 361-362.
Zborowski, P. & Storey, R. (1995). *A Field Guide to Insects in Australia.* 207 pp. Reed Books, Chatswood, N.S.W.
Zettel, H. & Sehnal, C. (1995). Description of male genitalia of the Australian species *Hebrus nourlangiei* Lansbury, 1990 (Heteroptera: Hebridae). *Linzer Biologische Beiträge* 27: 1085-1087.
Zimmermann, G. (1984). Heteroptera aus dem Nepal-Himalaya. *Geovelia* n.gen., eine Gattung terrestrischer Microveliinae (Insecta: Gerromorpha: Veliidae). *Senckenbergiana Biologica* 65: 65-74.
Zimmermann, G. (1986). *Zur Phylogenie der Corixidae Leach, 1815 (Hemiptera, Heteroptera, Nepomorpha).* Dissertation, Universität Marburg.
Zimmermann, M. (1984). Population structure, life cycle and habitat of the pondweed bug *Mesovelia furcata* (Hemiptera, Mesoveliidae). *Revue Suisse Zoologie* 91: 1017-1035.

APPENDIX 1:
Checklist and Distribution of Australian Water Bugs

With information on type(s), type locality, type repository (list of abbreviations below), synonyms (only those relevant for the Australian region), and distribution.

Infraorder GERROMORPHA Popov, 1971

Superfamily MESOVELOIDEA Douglas & Scott, 1867

Family MESOVELIIDAE Douglas & Scott, 1867

Subfamily MESOVELIINAE Douglas & Scott, 1867

Austrovelia queenslandica Malipatil & Monteith, 1983: 949. Holotype ♂, Mt Sorrow summit, Cape Tribulation, Queensland (QMB).
Distribution: QLD.

Mesovelia ebbenielseni Andersen & Weir, in press: 000. Holotype ♂, Burster Creek, Queensland (ANIC).
Distribution: NT, QLD.

Mesovelia hackeri Harris & Drake, 1941: 277. Holotype ♂, "Asharove" [Ashgrove], Queensland (USNM).
Distribution: NSW, Norfolk Island, QLD. - EL: New Zealand.

Mesovelia horvathi Lundblad, 1933: 190. Syntypes, Sumatra and Java, Indonesia (NHRS).
(Syn. *Mesovelia japonica* Miyamoto, 1964: 199).
Distribution: NT, QLD. WA. - EL: Southeast Asia, Malay Archipelago, Japan, Philippines, and New Guinea.

Mesovelia hungerfordi Hale, 1926: 198. Lectotype ♂ (Cassis & Gross 1995: 128), Adelaide, South Australia (SAMA).
Distribution: NSW, NT, QLD, SA, TAS, VIC, WA.

Mesovelia stysi J. Polhemus & D. Polhemus, 2000: 215. Holotype ♂, Bosett's Lagoon, Western Prov., Papua New Guinea (JTPC).
Distribution: NT, QLD, WA. - EL: Papua New Guinea.

Mesovelia vittigera Horváth, 1895: 160. Syntypes, Cairo and Abukir, Egypt (HNHM).
(Syn. *Mesovelia orientalis* Kirkaldy, 1901: 808).
Distribution: NT, QLD, WA. - EL: South, East, and Southeast Asia, the Malay Archipelago, New Guinea, islands of western Pacific, Africa incl. Madagascar, southern and southeastern Europe, and the Middle East.

Superfamily HEBROIDEA Amyot & Serville, 1843

Family HEBRIDAE Amyot & Serville, 1843

Subfamily HEBRINAE Amyot & Serville, 1843

Austrohebrus apterus Andersen & Weir, in press: 000. Holotype ♀, Broome, Western Australia (WAM).
Distribution: WA.

Hebrus axillaris Horváth, 1902: 606. Lectotype ♀ (Lansbury 1990: 56), Tweed River, New South Wales (HNHM).
(Syn. *Naeogeus latensis* Hale, 1926: 196; *Hebrus woodwardi* Lansbury, 1990: 59).
Distribution: NSW, NT, QLD, SA, TAS, VIC, WA.

Hebrus monteithi Lansbury 1990: 61. Holotype ♂, Wallaman Falls, via Ingham, Queensland (QMB).
Distribution: QLD.

Hebrus nourlangiei Lansbury, 1990: 62. Holotype ♀, Nourlangie Rock area, Kakadu National Park, Northern Territory (NTMD).
Distribution: NSW, NT, QLD, WA.

Hebrus pilosus Andersen & Weir, in press: 000. Holotype ♀, Blackmore River at Southport, Norhern Territory (ANIC).
Distribution: NT.

Merragata hackeri Hungerford, 1934: 70. Holotype ♂, Brisbane, Queensland (SEMC).
Distribution: NSW, NT, QLD, VIC, WA.

Superfamily HYDROMETROIDEA Billberg, 1820

Family HYDROMETRIDAE Billberg, 1820

Subfamily HYDROMETRINAE Billberg, 1820

Hydrometra claudie J. Polhemus & Lansbury, 1997: 8. Holotype ♂, Claudie River, Iron Range, Queensland (ANIC).
Distribution: QLD.

Hydrometra darwiniana J. Polhemus & Lansbury, 1997: 9. Holotype ♂, 9 km N by E of Mudginberri Homestead, Northern Territory (ANIC).
Distribution: NT, QLD, WA.

Hydrometra feta Hale, 1925: 4. Holotype ♀, Bellenden Ker, Queensland (NHRS).
(Syn. *Hydrometra halei* Hungerford & Evans, 1934: 61).
Distribution: NSW, NT, QLD, WA.
Hydrometra illingworthi Hungerford & Evans, 1934: 59. Holotype ♂, Cairns, Queensland (USNM).
Distribution: QLD.
Hydrometra jourama J. Polhemus & Lansbury, 1997: 17. Holotype ♂, Jourama Falls National Park, Queensland (ANIC).
Distribution: QLD.
Hydrometra novaehollandiae J. Polhemus & Lansbury, 1997: 22. Holotype ♂, Royal National Park, Upper Causeway, New South Wales (ANIC).
Distribution: NSW, QLD.
Hydrometra orientalis Lundblad, 1933: 430. Holotype ♂, Sumatra (NHRS).
(Syn. *H. insularis* Hungerford & Evans, 1934: 76; *H. sumatrana* Ruhoff, 1964: 32)
Distribution: NT, QLD. - EL: Philippines (Luzon, Lubang, Miandanao), Indonesia (Sulawesi, Java, Sumatra, Sumba), Papua New Guinea, Malaysia (W), Thailand, Burma, Vietnam.
Hydrometra papuana Kirkaldy, 1901: 807. Neotype ♂ (J. Polhemus & Lansbury 1997: 26), Fly River area, Papua New Guinea (BPBM).
(Syn. *Hydrometra hoplogastra* Hale, 1925: 2).
Distribution: NSW, NT, QLD, WA. - EL: Malaysia (Pahang), Indonesia (Irian Jaya, Kalimantan), Papua New Guinea.
Hydrometra strigosa (Skuse, 1893: 43). Lectotype ♂ (J. Polhemus & Lansbury 1997: 29), Botany Swamps, New South Wales (AMS).
(Syn. *Hydrometra risbeci* Hungerford, 1938: 81).
Distribution: ACT, Norfolk Island, NSW, NT, QLD, SA, TAS, VIC, WA. - EL: New Caledonia, Vanuatu, New Zealand, and Society Islands (Tahiti).

Superfamily GERROIDEA Leach, 1815

Family HERMATOBATIDAE Coutière & Martin, 1901

[***Hermatobates armatus*** Andersen & Weir, 2000: 340. Holotype ♂, Bennett & Long I., Chesterfield Islands [New Caledonia] (ZMUC).
Distribution: - EL: Chesterfield Islands, New Caledonia.]
Hermatobates haddoni Carpenter, 1892: 143, pl. 12. Holotype ♂, Mabuiag, Torres Strait, Queensland (BMNH).
(Syn. *Hermatobates walkeri* China, 1957: 346).
Distribution: NT
Hermatobates marchei (Coutiere & Martin, 1901b: 215). Syntypes, "ile Paragua, baie de Honda", Philippines (MNHN?).
(Syn. *Hermatobates weddi* China, 1957: 346).
Distribution: NT, QLD, WA. - EL: New Caledonia, Indonesia (Java), Fiji, Tonga, Philippines, Ryukyu Islands,

Family VELIIDAE Brullé, 1836

Subfamily HALOVELIINAE ESAKI, 1930

Halovelia corallia Andersen, 1989a: 92. Holotype ♂, Malupore Is., Central Prov., Papua New Guinea (BPBM).
Distribution: QLD. - EL: Papua New Guinea.
Halovelia heron Andersen, 1989a: 102. Holotype ♂, Heron Island, Queensland (UQIC).
Distribution: QLD.
Halovelia hilli China, 1957: 352.Holotype ♂, South Hermite, Western Australia (BMNH).
Distribution: NT, QLD, WA. - EL: New Caledonia, Papua New Guinea, East Timor.
Halovelia maritima Bergroth, 1893: 277. Lectotype ♂ (China 1957: 2), Cartier Island, Timor Sea, Northern Territory (BMNH).
Distribution: NT.
Halovelia polhemi Andersen, 1989a: 109. Holotype ♂, East Point, Darwin, Northern Territory (ANIC).
Distribution: NT.
Xenobates angulanus (J. Polhemus, 1982: 7). Holotype ♂, Frances Bay, Darwin, Northern Territory (ANIC).
Distribution: NT.
Xenobates chinai Andersen & Weir, 1999: 322. Holotype ♂, 3 ml. N mouth of Finniss River, Northern Territory (ANIC).
Distribution: NT.
Xenobates lansburyi Andersen & Weir, 1999: 321. Holotype ♂, Darwin Harbour, Northern Territory (NTMD).
Distribution: NT.
Xenobates major Andersen & Weir, 1999: 323. Holotype ♂, Somerset, Queensland (ANIC).
Distribution: QLD.
Xenobates mangrove Andersen & Weir, 1999: 313. Holotype ♂, 3-mile Creek, Townsville, Queensland (ANIC).
Distribution: QLD.
Xenobates myorensis (Lansbury, 1989: 97). Holotype ♂, Myora Swamp, Stradbroke Island, Queensland (OXUM).
Distribution: QLD.

Xenobates ovatus Andersen & Weir, 1999: 319. Holotype ♂, Bizant River, Lakefield National Park, Queensland (ANIC).
Distribution: QLD.
Xenobates spinoides Andersen & Weir, 1999: 328. Holotype ♂, Portland Roads, Queensland (ANIC).
Distribution: QLD.

Subfamily MICROVELIINAE
China & Usinger, 1949 [1861]

Drepanovelia biceros Andersen & Weir, 2001: 229. Holotype ♂, Minnamurra River u/s Museum, New South Wales (AMS).
Distribution: NSW.
Drepanovelia dubia (Hale, 1926: 214). Lectotype ♂ (Andersen & Weir 2001: 223), Devonport, Tasmania (SAMA).
Distribution: NSW, TAS, VIC.
Drepanovelia millennium Andersen & Weir, 2001: 226. Holotype ♂, Mt. Tamborine, Queensland (ANIC).
Distribution: NSW, QLD.
Drepanovelia nielseni Andersen & Weir, 2001: 230. Holotype ♂, Emerald Creek, Queensland (ANIC).
Distribution: QLD.
Lacertovelia hirsuta Andersen & Weir, 2001: 232. Holotype ♂, Dilgry River, Barrington Tops S. F., New South Wales (ANIC).
Distribution: NSW.
Microvelia (Austromicrovelia) alisonae Andersen & Weir, 2003b: 303. Holotype ♂, Mary River, 37 ml. E of Pine Creek, Northern Territory (ANIC).
Distribution: NT.
Microvelia (Austromicrovelia) angelesi Andersen & Weir, 2003b: 292. Holotype ♂, 19 km NE by E of Mt. Cahill, Northern Territory (ANIC).
Distribution: NT.
Microvelia (Austromicrovelia) annemarieae Andersen & Weir, 2003b: 289. Holotype ♂, Roaring Meg Valley, Queensland (QMB).
Distribution: QLD.
Microvelia (Austromicrovelia) apunctata Andersen & Weir, 2003b: 314. Holotype ♂, Campbell Spring, 38 km E by SE of Mt. Panton, Northern Territory (ANIC).
Distribution: NT, QLD.
Microvelia (Austromicrovelia) australiensis Andersen & Weir, 2003b: 316. Holotype ♂, Mistake Creek, Queensland (ANIC).
Distribution: NT, QLD, WA.
Microvelia (Austromicrovelia) carnavon Andersen & Weir, 2003b: 287. Holotype ♂, Ward's Canyon, Carnarvon National Park, Queensland (AMS).
Distribution: QLD.
Microvelia (Austromicrovelia) childi Andersen, 1969: 253. Holotype ♂, Kuringai Chase, 20 miles N of Sydney, New South Wales (ANIC)
Distribution: NSW.
Microvelia (Austromicrovelia) distincta Malipatil, 1980: 96. Holotype ♂, Thomson River, Thomson-Jordan Divide Road, New South Wales (MVM).
Distribution: NSW, VIC.
Microvelia (Austromicrovelia) eborensis Andersen & Weir, 2003b: 309. Holotype ♂, Ebor, New South Wales (ANIC).
Distribution: NSW.
Microvelia (Austromicrovelia) fluvialis Malipatil, 1980: 101. Holotype ♂, Thomson River, Thomson-Jordan Divide Road, Victoria (MVM).
(Syn. *Microvelia fluvialis weiri* Malipatil, 1980: 105).
Distribution: NSW, QLD, VIC.
Microvelia (Austromicrovelia) herberti Andersen & Weir, 2003b: 300. Holotype ♂, Bertie creek, Queensland (ANIC).
Distribution: NT, QLD, WA.
Microvelia (Austromicrovelia) hypipamee Andersen & Weir, 2003b: 274. Holotype ♂, Mt Hypipamee National Park, Queensland (ANIC).
Distribution: QLD.
Microvelia (Austromicrovelia) malipatili Andersen & Weir, 2003b: 303. Holotype ♂, Limestone Gorge, junction East Baines River and Limestone Creek, Gregory National Park, Northern Territory (ANIC).
Distribution: NT, QLD, WA.
Microvelia (Austromicrovelia) margaretae Andersen & Weir, 2003b: 275. Holotype ♂, Bald Mt. Area, Queensland (ANIC).
Distribution: NSW, QLD, VIC.
Microvelia (Austromicrovelia) milleri Andersen & Weir, 2003b: 322. Holotype ♂, creek flowing to Yowaka River, New South Wales (ANIC).
Distribution: NSW.
Microvelia (Austromicrovelia) mjobergi Hale, 1925: 6. Holotype ♂, Herberton, Queensland (NHRS).
Distribution: QLD.
Microvelia (Austromicrovelia) monteithi Andersen & Weir, 2003b: 277. Holotype ♂, Mt. Spec. National Park via Paluma, Queensland (ANIC).
Distribution: QLD.
Microvelia (Austromicrovelia) mossman Andersen & Weir, 2003b: 290. Holotype ♂, Mossman Gorge via Mossman, Queensland (ANIC).

Distribution: QLD.

Microvelia (Austromicrovelia) myorensis Andersen & Weir, 2003b: 283. Holotype ♂, Myora Creek, North Stradbroke Islands, Queensland (ANIC).

Distribution: QLD.

Microvelia (Austromicrovelia) odontogaster Andersen & Weir, 2003b: 319. Holotype ♂, Muirella Park, Nourlangie Creek, Kakadu National Park, Northern Territory (ANIC).

Distribution: NT.

Microvelia (Austromicrovelia) pennicilla Andersen & Weir, 2003b: 293. Holotype ♂, Florence Falls, Litchfield National Park, Northern Territory (ANIC).

Distribution: NT, WA.

Microvelia (Austromicrovelia) peramoena Hale, 1925:8. Holotype ♂, Myponga, South Australia (SAMA).

Distribution: ACT, NSW, NT, QLD, SA, TAS, VIC, WA.

Microvelia (Austromicrovelia) queenslandiae Andersen & Weir, 2003b: 280. Holotype ♂, 7 km SE Rossville, Queensland (ANIC).

Distribution: NSW, QLD.

Microvelia (Austromicrovelia) spurgeon Andersen & Weir, 2003b: 272. Holotype ♂, Stewart Creek, 4 km NNE Mt Spurgeon, Queensland (QMB).

Distribution: QLD.

Microvelia (Austromicrovelia) torresiana Andersen & Weir, 2003b: 311. Holotype ♂, Roper River at 4 Mile, Elsey National Park, Northern Territory (ANIC).

Distribution: NT, QLD, WA.

Microvelia (Austromicrovelia) tuberculata Andersen & Weir, 2003b: 279. Holotype ♂, Mt. Tamborine, Queensland (ANIC).

Distribution: NSW, QLD.

Microvelia (Austromicrovelia) ventrospinosa Andersen & Weir, 2003b: 321. Holotype ♂, Escarpment Creek near Twin Falls [Canyon Creek], Warrie National Park, Queensland (ANIC).

Distribution: NSW, QLD.

Microvelia (Austromicrovelia) woodwardi Andersen & Weir, 2003b: 284. Holotype ♂, Carnarvon Gorge, Queensland (ANIC).

Distribution: QLD.

Microvelia (Barbivelia) barbifer Andersen & Weir, 2003b: 324. Holotype ♂, Holroyd River, Queensland (ANIC).

Distribution: QLD.

Microvelia (Barbivelia) falcifer Andersen & Weir, 2003b: 325. Holotype ♂, Georgetown Billabong nr. Jabiru, Northern Territory (NTMD).

Distribution: NT.

Microvelia (Pacificovelia) kakadu Andersen & Weir, 2003b: 337. Holotype ♂, Jarrnarm Walks, Keep River National Park, Northern Territory (ANIC).

Distribution: NT, QLD, WA.

Microvelia (Pacificovelia) lilliput Andersen & Weir, 2003b: 334. Holotype ♂, 27 km NW by W of Rokeby, Queensland (ANIC).

Distribution: NT, QLD, WA.

Microvelia (Pacificovelia) macgregori (Kirkaldy, 1899a: 91). Syntypes, New Zealand (Museum of Perth, Scotland).

Distribution: Norfolk Island. - EL: New Zealand.

Microvelia (Pacificovelia) oceanica Distant, 1914: 383. Syntypes, Outbache, New Caledonia (BMNH).
(Syn. *Microvelia australica* Bergroth, 1916: 38; *Microvelia halei* Esaki, 1928: 69).

Distribution: ACT, Norfolk Island, NSW (incl. Lord Howe Island), NT, QLD, SA, TAS, VIC, WA. - EL: New Caledonia, Loyalty Islands, Vanuatu, Fiji.

Microvelia (Pacificovelia) tasmaniensis Andersen & Weir, 2003b: 331. Holotype ♂, Cape Portland lagoon, Tasmania (ANIC).

Distribution: TAS.

Microvelia (Picaultia) cassisi Andersen & Weir, 2003b: 345. Holotype ♂, Cowarra Ck, Houston Mitchell Drive, New South Wales (AMS).

Distribution: NSW.

Microvelia (Picaultia) douglasi Scott, 1874: 448. Syntypes, Japan (type repository unknown).
(Syn. *Microvelia repentina* Distant, 1903; *Microvelia singalensis* Kirkaldy, 1903: 180; *Microvelia kumaonensis* Distant, 1909: 500; *Microvelia samoana* Esaki, 1928: 67).

Distribution: NT, QLD, WA. - EL: India, Indonesia (Irian Jaya, Bali, Java, Sumatra), Japan, Marianna Islands (Guam), Philippines, Samoa, Sri Lanka, Taiwan, West Malaysia.

Microvelia (Picaultia) justi Andersen & Weir, 2003b: 342. Holotype ♂, 4 km S by W of Mining Camp, Mitchell Plateau, Kimberley District, Western Australia (ANIC).

Distribution: NT, QLD, WA.

Microvelia (Picaultia) paramega Andersen & Weir, 2003b: 343. Holotype ♂, Jarrnarm Walks, Keep River National Park, Northern Territory (ANIC).

Distribution: NT, QLD, WA.

Microvelopsis exuberans Andersen & Weir, 2001: 246. Holotype ♂, Mt Spurgeon, Queensland (QMB).

Distribution: QLD.

Microvelopsis melancholica (Hale, 1925: 5). Holotype ♂, Herberton, Queensland (NHRS).

Distribution: QLD.

Microvelopsis minor Andersen & Weir, 2001: 248. Holotype ♂, 5 km E Rossville, Queensland (ANIC).
Distribution: QLD.

Nesidovelia howensis (Hale, 1926: 211). Lectotype ♂ (Andersen & Weir 2001: 242), Mt Gower, Lord Howe Island (SAMA).
Distribution: NSW (Lord Howe Island).

Petrovelia agilis Andersen & Weir, 2001: 235. Holotype ♂, Pioneer River at Balnagowan Crossing, Qld. (ANIC).
Distribution: QLD.

Petrovelia katherinae Andersen & Weir, 2001: 238. Holotype ♂, Katherine, N.T. (ANIC).
Distribution: NT.

Phoreticovelia disparata D. Polhemus & J. Polhemus, 2000: 100. Holotype ♀, Upper Mulgrave River at Goldsborough Road Bridge, Queensland (ANIC).
Distribution: QLD.

Phoreticovelia rotunda D. Polhemus & J. Polhemus, 2000: 97. Holotype ♀, Aramia River near Balimo, Papua New Guinea (BPBM).
Distribution: NSW, NT, QLD, VIC. - EL: Papua New Guinea.

Tarsoveloides brevitarsus Andersen & Weir, 2001: 249. Holotype ♂, Leo Creek Road, McIlwraith Range, 20 km N.E. of Coen, Queensland (QMB).
Distribution: QLD.

Subfamily RHAGOVELIINAE China & Usinger, 1949

Rhagovelia australica Kirkaldy, 1908: 783. Holotype ♀, Kuranda, Queensland (BPBM).
Distribution: QLD.

Family GERRIDAE Leach, 1815

Subfamily GERRINAE Leach, 1815

Aquarius antigone (Kirkaldy, 1899b: 507). Lectotype ♀ (Andersen & Weir 1997: 279), Port Denison, Queensland (SEMC).
Distribution: ACT, NSW, QLD, SA, VIC.

Aquarius fabricii Andersen, 1990: 46. Holotype ♂, Morgan Falls, Drysdale River system, Western Australia (ANIC).
Distribution: NT, WA.

Limnogonus (Limnogonus) fossarum gilguy Andersen & Weir, 1997: 255. Holotype ♂, Burster Creek, Queensland (ANIC).
(Syn. *Limnogonus pacificus* Esaki, 1928: 71; *L. (Limnogonus) fossarum skusei* [Torre-Bueno, 1926] Andersen, 1975: 36).
Distribution: NT, QLD, SA, WA. - EL: Indonesia (Java, Lesser Sunda Islands, Kalimantan, Sulawesi, Irian Jaya), Malaysia (Sarawak), Papua New Guinea, Solomon Islands, New Caledonia, Vanuatu, Fiji, Samoa, Cook Islands.

Limnogonus (Limnogonus) hungerfordi Andersen, 1975: 46. Holotype ♂, Port Moresby Dist., Papua New Guinea (BMNH).
Distribution: NT, QLD, Christmas Island. - EL: Southeast Asia, Taiwan, Philippines, Guam, Palau Islands, Indo-Malayan Archipelago, Papua New Guinea.

Limnogonus (Limnogonus) luctuosus (Montrousier, 1864: 242). Holotype ♀, Kanala, New Caledonia (MRAC).
(Syn. *Limnometra lineata* Carpenter, 1892: 141; *Limnogonus australis* [Skuse, 1893: 42]; *L. skusei* [Torre-Bueno, 1926: 129]; *L. pacificus* Esaki, 1928: 71).
Distribution: NT, QLD, WA. - EL: Solomon Islands, New Caledonia, Vanuatu, Fiji, Samoa, Society Islands.

Limnogonus (Limnogonus) windi Hungerford & Matsuda, 1961: 117. Holotype ♂, Barron River, Queensland (SEMC).
Distribution: NT, QLD, WA.

Limnometra ciliodes Andersen & Weir, 1997: 242. Holotype ♂, Lockerbie, Cape York, Queensland (ANIC).
Distribution: QLD. - EL: Indonesia (Irian Jaya), Papua New Guinea.

Limnometra cursitans (Fabricius, 1775: 729). Lectotype ♂ (Andersen & Weir 1997: 211), "Nova Hollandia" [Australia] (ZMUC).
(Syn. *Limnometra poliakanthina* Nieser & Chen, 1992: 23).
Distribution: NT, QLD.

Limnometra lipovskyi Hungerford & Matsuda, 1958: 399. Holotype ♂, Guadalcanal, Solomon Islands (SEMC).
Distribution: NT, QLD. - EL: Solomon Islands, Papua New Guinea, Indonesia (Irian Jaya, Halmahera).

Tenagogerris euphrosyne (Kirkaldy, 1902a: 138). Lectotype ♀ (Andersen & Weir 1997: 211), Alexandra, Victoria (SEMC).
Distribution: ACT, NSW, QLD, SA, VIC.

Tenagogerris femoratus Andersen & Weir, 1997: 224. Holotype ♂, 19 km E by N of Mt. Cahill, Northern Territory (ANIC).
Distribution: NT, WA.

Tenagogerris pallidus Andersen & Weir, 1997: 220. Holotype ♂, Cooper Creek, 19 km E by S Mt. Borradaile, Northern Territory (ANIC).
Distribution: NT, WA.

Tenagogonus australiensis Andersen & Weir, 1997: 229. Holotype ♂, Mt Spec Nat. Park via

Paluma, Queensland (ANIC).
Distribution: QLD.

Subfamily HALOBATINAE Bianchi, 1896

Austrobates rivularis Andersen & Weir, 1994a: 6. Holotype ♂, Lydia Creek, Queensland (ANIC).
Distribution: QLD.

Halobates (Halobates) acherontis J. Polhemus, 1982: 6. Holotype ♂, Daly River, Northern Territory (ANIC).
Distribution: NT.

Halobates (Halobates) darwini Herring, 1961: 278. Holotype ♂, Port Darwin, Northern Territory (BMNH).
Distribution: NT, QLD. - EL: Papua New Guinea.

Halobates (Halobates) germanus White, 1883: 50. Holotype ♀, North Pacific Ocean (BMNH).
Distribution: NT, QLD, WA. - EL: Indian Ocean, West and Central Pacific Ocean, Aldabra & Cosmoledo Atolls, China, India (Bay of Bengal), Iran, Japan, Madagascar, Malaysia (Johore), Maldives, Red Sea (Israel, Oman, Saudi Arabia), Singapore, Somalia.

Halobates (Halobates) hayanus White, 1883: 52. Holotype ♂, Red Sea near Aden (BMNH). (Syn. *Halobates australiensis* Malipatil, 1988: 157).
Distribution: NT, QLD. - EL: China, Indonesia (Bali, Irian Jaya, Java, Moluccas), Malaysia (Johore), Nicobar Is., Papua New Guinea, Philippines (Luzon, Mindanao, Palawan), Red Sea (Djibouti, Egypt, Saudi Arabia, Sudan, Yemen), Singapore, Thailand (Phuket), Vietnam.

Halobates (Halobates) herringi J. Polhemus & Cheng, 1982: 224. Holotype ♂, Gladstone, Auckland Creek, Queensland (ANIC).
Distribution: NT, QLD. - EL: Indonesia (Irian Jaya).

Halobates (Halobates) micans Eschscholtz, 1822: 107. Syntypes, southern Pacific Ocean and southern Atlantic Ocean (IZBE).
Distribution: NSW, NT, QLD, WA. - EL: Atlantic Ocean between 40°N and 30-40°S, Caribbean Sea, Indian Ocean between 20°N and 10-40°S, western Pacific Ocean between 30°N and 5°S, central and eastern Pacific Ocean between 20°N and 20°S; Aldabra & Cosmoledo Atolls, Cape Verde Is., China, Japan, Sri Lanka, Taiwan, U.S.A. (Texas).

Halobates (Halobates) princeps White, 1883: 44, 80. Holotype ♀, Celebes Sea (BMNH). (Syn. *Halobates ashmorensis* Malipatil, 1988: 158).
Distribution: NT, QLD. - EL: Caroline Is. (Palau Is.), Malaysia (Penang?, Sabah), Indonesia (Halmahera, Java, Madura, S Moluccas, Sulawesi, Sumbawa), Solomon Is.

Halobates (Halobates) regalis Carpenter, 1892: 144. Holotype ♂, Torres Strait [Australia] (BMNH).
Distribution: QLD, WA.

Halobates (Halobates) sericeus Eschscholtz, 1822: 108, pl. 2, fig. 4. Syntypes, N Pacific (IZBE).
Distribution: NSW ((Lord Howe I.), QLD. - EL: Pacific Ocean between 40ºN and 5ºN and between 5ºS and 40ºS; Australia, China, Hawaiian Is., Japan (Honshu), Vietnam.

Halobates (Halobates) whiteleggei Skuse, 1891: 174. Holotype ♂, Port Jackson, New South Wales (AMS).
Distribution: NSW, QLD.

Halobates (Halobates) zephyrus Herring, 1961: 276. Holotype ♂, Bribie Island, Queensland (USNM).
Distribution: NSW, QLD.

Halobates (Hilliella) lannae Andersen & Weir, 1994b: 874. Holotype ♂, Frances Bay, Darwin, Northern Territory (ANIC).
Distribution: NT, WA.

Halobates (Hilliella) mjobergi Hale, 1925: 12. Holotype ♂, Broome, Western Australia (NHRS).
Distribution: NT, QLD, WA. - EL: Papua New Guinea.

Halobates (Hilliella) robinsoni Andersen & Weir, 2003a: 10. Holotype ♂, Robinson River, Western Australia (ANIC).
Distribution: WA.

Subfamily RHAGADOTARSINAE Lundblad, 1933

Rhagadotarsus (Rhagadotarsus) anomalus J. Polhemus & Karunaratne, 1993: 103. Holotype ♂, Jourama Falls National Park, Queensland (ANIC).
Distribution: NSW, NT, QLD, WA. - EL: Papua New Guinea.

Subfamily TREPOBATINAE Matsuda, 1960

Calyptobates jourama J. Polhemus & D. Polhemus, 1994: 339. Holotype ♂, Jourama Falls National Park, Queensland (ANIC).
Distribution: QLD.

Calyptobates minimus J. Polhemus & D. Polhemus, 1994: 340. Holotype ♂, Robin Falls, Northern Territory (ANIC).
Distribution: NT.

Calyptobates rubidus J. Polhemus & D. Polhemus, 1994: 342. Holotype ♂, Wenlock River at Iron Range Road, Queensland (ANIC).
Distribution: QLD.

Rheumatometra dimorpha Andersen & Weir, 1998: 519. Holotype ♂, Birthday Creek via Paluma, Queensland (ANIC).
Distribution: NSW, QLD, VIC.
Rheumatometra philarete Kirkaldy, 1902b: 281. Lectotype ♂ (Andersen & Weir 1998: 517), Alexandria [Alexandra], Victoria (IRSN).
Distribution: ACT, NSW, QLD, TAS, VIC.
Rheumatometroides carpentaria (J. Polhemus & D. Polhemus, 1991: 5). Holotype ♂, Groote Eylandt, Northern Territory (USNM).
Distribution: NT.
Stenobates australicus J. Polhemus & D. Polhemus, 1991: 3. Holotype ♂, Deeral Landing, Lower Mulgrave River, Queensland (ANIC).
Distribution: QLD. - EL: Papua New Guinea.

Infraorder NEPOMORPHA Popov, 1968

Superfamily NEPOIDEA Latreille, 1802

Family NEPIDAE Latreille, 1802

Subfamily NEPINAE Latreille, 1802

Laccotrephes (Laccotrephes) tristis (Stål, 1854: 241). Holotype (probable), "New Holland" [Australia] (NHRS).
Distribution: ACT, NSW, NT, QLD, SA, VIC, WA. - EL: Oriental Region, Papua New Guinea.

Subfamily RANATRINAE Douglas & Scott, 1865

Austronepa angusta (Hale, 1924: 508). Lectotype ♂ (Menke & Stange 1964: 67), Groote Eylandt, Northern Territory (SAMA).
Distribution: NT, QLD, WA.
Cercotmetus brevipes australis Lansbury, 1975: 17. Holotype ♂, Birraduk Creek, 16 km W by SW of Nimbuwah Rock, Northern Territory (NTDPIF).
Distribution: NT, QLD.
Goondnomdanepa brittoni Lansbury, 1978: 121. Holotype ♂, Baroalba Creek Springs, 19 km NE by E of Mt Cahill, Northern Territory (ANIC).
Distribution: NT.
Goondnomdanepa prominens Lansbury, 1978: 118. Holotype ♂, 15 km E of Mt Cahill, Northern Territory (ANIC).
Distribution: NT.
Goondnomdanepa weiri Lansbury, 1974: 220. Holotype ♂, 15 km E by N of Mt Cahill, Northern Territory (ANIC).
Distribution: NT, WA.
Ranatra diminuta Montandon, 1907: 57. Lectotype ♂ (Lansbury 1972: 323), "E. Australia" (NHRS).
(Syn. *Ranatra longipes* Stål, 1861 *sensu* Hale, 1924: 518).
Distribution: NT, QLD, SA, VIC, WA.
Ranatra dispar Montandon, 1903: 104. Lectotype ♂ (Lansbury 1972: 323), New South Wales (NHRS).
(Syn. *Ranatra australiensis* Hale, 1924: 510).
Distribution: ACT, NSW, QLD, SA, TAS, VIC, WA.
Ranatra occidentalis Lansbury, 1972: 326. Holotype ♂, Millstream, Western Australia (WAM).
Distribution: WA.

Family BELOSTOMATIDAE Leach, 1815

Subfamily BELOSTOMATINAE Leach, 1815

Diplonychus eques (Dufour, 1863: 394). Holotype (probable), "Australia" (NHRS).
Distribution: NSW, NT, QLD, TAS, VIC, WA
Diplonychus rusticus (Fabricius, 1781: 333). Syntypes, "aus Amerika" [in error; see Polhemus 1994: 692] (ZMUC). [ICZN 1996, Opinion 1850].
(Syn. *Diplonychus planus* [Sulzer, 1776: 92]).
Distribution: NT, QLD, SA, WA.

Subfamily LETHOCERINAE Lauck & Menke, 1961

Lethocerus (Lethocerus) distinctifemur Menke, 1960: 285. Holotype ♂, Brisbane, Queensland (QMB).
Distribution: NT, QLD, SA, WA.
Lethocerus (Lethocerus) insulanus (Montandon, 1898: 430). Syntypes (probable), New Caledonia (MNHN).
Distribution: NSW, NT, QLD. - EL: New Caledonia, Papua New Guinea, Oriental Region (Philippines).

Superfamily CORIXOIDEA Leach, 1815

Family CORIXIDAE Leach, 1815

Subfamily CORIXINAE Leach, 1815

Agraptocorixa eurynome (Kirkaldy, 1897a: 54). Holotype ♂, Adelaide, South Australia (BMNH).
Distribution: NSW, QLD, SA, TAS, VIC, WA. - EL: Indonesia (Irian Jaya).
Agraptocorixa gambrei Lansbury, 1984: 35. Holotype ♂, Port Warrender, Western Australia (WAM).

Distribution: WA.

Agraptocorixa halei Hungerford, 1953: 43. Holotype ♂, Barron River nr. Barron Waters, Queensland (SEMC).

Distribution: NSW, NT, QLD, WA.

Agraptocorixa hirtifrons (Hale, 1922: 321). Lectotype ♂ (Knowles 1974: 179), Cooper Creek, South Australia (SAMA).

Distribution: NSW, QLD, SA, VIC.

Agraptocorixa macrops Hungerford, 1953: 38. Holotype ♂, Lake Paniai Meer, Central New Guinea (SEMC).

Distribution: WA. - EL: Papua New Guinea.

Agraptocorixa parvipunctata (Hale, 1922: 320). Lectotype ♂ (Knowles 1974: 177), Adelaide, South Australia (SAMA).

Distribution: NSW, QLD, SA, TAS, VIC, WA

Sigara (Tropocorixa) australis (Fieber, 1851: 20). Holotype ♂ [lost], Port Phillip, Victoria (ZMHB).

Distribution: NSW, SA, TAS, VIC.

Sigara (Tropocorixa) mullaka Lansbury, 1970: 42. Holotype ♂, Mt Yokine, Western Australia (MVM).

Distribution: WA.

Sigara (Tropocorixa) neboissi Lansbury, 1970: 52. Holotype ♂, Shannon Lagoon, Tasmania (MVM).

Distribution: TAS.

Sigara (Tropocorixa) sublaevifrons (Hale, 1922: 316). Lectotype ♂ (Lansbury 1970: 47), Coromby, Victoria (SAMA).

Distribution: NSW, SA, VIC.

Sigara (Tropocorixa) tadeuszi Lundblad, 1933: 81. Holotype ♂, Manly, Sydney, New South Wales (?ZMUH).
(Syn. *Sigara halei* Hungerford, 1934: 69).

Distribution: NSW, QLD, SA. - EL: New Caledonia; Indonesia (Irian Jaya).

Sigara (Tropocorixa) tasmaniae Jaczewski, 1939: 298. Holotype ♂, National Park, Tasmania (TDA).

Distribution: TAS

Sigara (Tropocorixa) truncatipala (Hale, 1922: 314). Lectotype ♂ (Lansbury 1970: 51), Adelaide, South Australia (SAMA).

Distribution: NSW, QLD, SA, TAS, VIC. - EL: Papua New Guinea, New Caledonia.

Subfamily Cymatiainae Walton, 1940

Cnethocymatia nigra (Hungerford, 1947: 154). Holotype ♀, Prince of Wales Island, Torres Strait Islands, Queensland (SEMC).

Distribution: QLD. – EL: Papua New Guinea; Indonesia (Irian Jaya).

Subfamily Diaprepocorinae Lundblad. 1928

Diaprepocoris barycephalus Kirkaldy, 1897: 53. Syntype(s) ♀, Launceston, Tasmania and Melbourne, Victoria (BMNH).

Distribution: NSW, QLD, SA, TAS, VIC, WA.

Diaprepocoris pedderensis Knowles, 1974: 189. Holotype ♂, E shore of Lake Pedder, Tasmania (MVM).

Distribution: TAS.

Diaprepocoris personatus Hale, 1924: 8. Holotype ♂, Swan River, Western Australia (SAMA).

Distribution: QLD, TAS, WA.

Subfamily Micronectinae Jaczewski, 1924

Micronecta adelaidae Chen, 1965: 155. Holotype ♂, Adelaide River, 70 miles S of Darwin, Northern Territory (USNM).

Distribution: NSW, NT.

Micronecta annae Kirkaldy, 1905: 262. Holotype (probable), Victoria (BPBM).
(Syn. *Micronecta erato* Kirkaldy, 1905: 263; *Micronecta annae pallida* Kirkaldy, 1908: 788; *Micronecta batilla* Hale, 1922: 323)

Distribution: ACT, NSW, QLD, SA, VIC, WA.

Micronecta australiensis Chen, 1965: 156. Holotype ♂, near Canberra, Australian Capital Territory (SEMC).

Distribution: ACT, VIC.

Micronecta carinata Chen, 1965: 157. Holotype ♂, Dorrigo, New South Wales (MCZC).

Distribution: NSW.

Micronecta concordia King, 1997: 149. Holotype ♂, Dixon's Creek, Victoria (MVM).

Distribution: VIC.

Micronecta dixonia King, 1997: 146. Holotype ♂, Dixon's Creek, Victoria (MVM).

Distribution: VIC.

Micronecta gracilis Hale, 1922: 326. Holotype (probable), Quorn, South Australia (SAMA).

Distribution: NSW, NT, QLD, SA, VIC, WA.

Micronecta halei Chen, 1965: 157. Holotype ♂, De Grey at Yarrie Station, Western Australia (SEMC).
(Syn. *Micronecta halei* Wroblewski, 1970: 695 [nec Chen, 1965])

Distribution: QLD, WA.

Micronecta illiesi Wroblewski, 1970: 687 (as ssp. of *M. annae* Kirkaldy). Syntypes, several localities, New South Wales (repository unknown; King, 1997).

Distribution: NSW, TAS.

Micronecta lansburyi Wroblewski, 1972: 521. Holotype ♂, Katherine, Northern Territory (IZW).

Distribution: NT, QLD.

Micronecta major Chen, 1965: 160. Holotype ♂, Coolabah, New South Wales (MNH).
Distribution: NSW.
Micronecta micra Kirkaldy, 1908: 788. Holotype (probable), Kuranda, Queensland (BPBM).
Distribution: NT, QLD.
Micronecta quadristrigata Breddin, 1905: 57. Holotype (probable), Java (?ZMHB).
Distribution: NSW, QLD, SA. - EL: Oriental Region, New Guinea.
Micronecta queenslandica Chen, 1965: 159. Holotype ♂, N Queensland (SEMC). (Syn. *Micronecta annae kirkaldyi* Wroblewski, 1970: 688).
Distribution: QLD.
Micronecta robusta Hale, 1922: 325. Holotype (probable) ♂, Adelaide, South Australia (SAMA).
Distribution: ACT, NSW, QLD, SA, TAS, VIC, WA.
Micronecta tasmanica Wroblewski, 1977: 685 (as ssp. of *M. annae* Kirkaldy). Syntypes, Tasmania & King Is. (OXUM).
Distribution: TAS, VIC.
Micronecta virgata Hale, 1922: 325. Holotype (probable) ♂, Townsville, Queensland (SAMA).
Distribution: QLD, WA. - EL: Solomon Is.
Micronecta windi Chen, 1965: 161. Holotype ♂, Kuranda, Queensland (SEMC).
Distribution: QLD.

Superfamily NAUCOROIDEA Leach, 1815

Family APHELOCHEIRIDAE Fieber, 1851

Aphelocheirus (Aphelocheirus) australicus Usinger, 1937: 341. Holotype ♀, Cairns, Qld. (CAS).
Distribution: NT, QLD.

Family NAUCORIDAE Leach, 1815

Subfamily NAUCORINAE Leach, 1815

Naucoris australicus Stål, 1876: 145. Lectotype ♀ (Lansbury 1985: 111), "N. Australia" (NHRS).
Distribution: QLD.
Naucoris congrex Stål, 1876: 145. Holotype ♀, Moreton Bay, Qld. (NHRS).
Distribution: ACT, NSW, NT, QLD, TAS, SA, VIC.
Naucoris magela Lansbury, 1991: 107. Holotype ♂, Bowerbird billabong, N.T. (NTMD).
Distribution: NT.
Naucoris rhizomatus J. Polhemus, 1984: 157. Holotype ♂, Coomalie Creek nr Darwin, N.T. (ANIC).
Distribution: NT. - EL: Malaysia (Johore).
Naucoris subaureus Lansbury, 1985: 116. Holotype ♂, Millstream, W.A. (WAM).
Distribution: WA.
Naucoris subopacus Montandon, 1913: 223. Lectotype ♀ (Lansbury 1985: 115), Adelaide River, N.T. (BMNH).
Distribution: NT, QLD, WA.

Superfamily OCHTEROIDEA Kirkaldy 1906 [1815]

Family OCHTERIDAE Kirkaldy 1906 [1815]

Megochterus nasutus (Montandon, 1898: 72). Lectotype ♀ (Baehr, 1990a: 198), "Australia" (MNHN).
Distribution: QLD, SA, TAS.
Megochterus occidentalis Baehr, 1990a: 200. Holotype ♂, Broke Inlet, Walpole, Western Australia (WAM).
Distribution: WA.
Ochterus atridermis Baehr, 1989: 122. Holotype ♂, Beerway [Beerwah], Queensland (QMB).
Distribution: NT, QLD.
Ochterus australicus Jaczewski, 1934: 607. Lectotype ♂ (Baehr 1990: 452), New Caledonia (BMNH).
Distribution: NSW, QLD, SA, TAS, VIC, WA. - EL: New Caledonia, New Hebrides, Solomon Is.
Ochterus bacchusi Baehr, 1990: 465. Holotype ♂, Wentworth Falls, Blue Mountains, New South Wales (BMNH).
Distribution: NSW.
Ochterus baehri baehri Rieger, 1977: 215. Holotype ♂, Stewart River, Cape York Peninsula, Queensland (ZSMC).
Distribution: QLD.
Ochterus baehri riegeri Baehr, 1989: 119. Holotype ♂, Humpty Doo, Northern Territory (QMB).
Distribution: NT.
Ochterus brachysoma Rieger, 1977: 214. Holotype ♂, Stewart River, Cape York Peninsula, Queensland (Rieger Collection).
Distribution: QLD.
Ochterus eurythorax Baehr, 1989: 116. Holotype ♂, Dunwich, Stradbroke Is., Queensland (QMB).
Distribution: ACT, NSW, QLD, SA, TAS, VIC.
Ochterus monteithorum Baehr, 1990: 468. Holotype ♂, Leo Creek Road, McIlwraith Range, Coen, Queensland (QMB).
Distribution: QLD.
Ochterus occidentalis Baehr, 1990: 460. Holotype

♂, Margaret River, Western Australia (BMNH).
Distribution: WA.
Ochterus secundus secundus Kormilev, 1971: 441. Holotype ♂, Cronulla, New South Wales (AMS).
Distribution: ACT, NSW, QLD, SA, VIC, WA.
Ochterus secundus pseudosecundus Baehr, 1989: 124. Holotype ♂, Gascoyne River, N of Carnarvon, Western Australia (QMB).
Distribution: WA.

Family GELASTOCORIDAE Kirkaldy 1897

Nerthra adspersa (Stål, 1863: 407). Holotype ♀, "Austral[ia] occidentalis" (NHRS).
(Syn. *Nerthra membranacea* Nieser, 1977: 301).
Distribution: WA.
Nerthra alaticollis (Stål, 1854: 239). Holotype ♂, Sydney, New South Wales (NHRS).
Distribution: NSW, QLD, VIC.
Nerthra annulipes (Horváth, 1902: 611). Holotype ♀, Clarence River, New South Wales (MNH).
Distribution: NSW, QLD.
Nerthra appha Cassis & Silveira, 2001: 19. Holotype ♂, Marengo State Forest, New South Wales (AMS).
Distribution: NSW.
Nerthra elongata (Montandon, 1899b: 778). Holotype ♀, "Australia" (NHMW).
Distribution: ACT, NSW, VIC.
Nerthra falcatus Cassis & Silveira, 2002: 164. Holotype ♂, Wonga Walk, about 600m N of Tristiana Falls, Dorrigo National Park, New South Wales (AMS).
Distribution: NSW, QLD.
Nerthra femoralis (Montandon, 1899a: 402). Holotype (probable), Champion Bay, Western Australia (BMNH).
Distribution: WA.
Nerthra grandis (Montandon, 1899b: 777). Holotype (probable), "Australie" (MNHN).
Distribution: VIC.
Nerthra hirsuta Todd, 1955: 417. Holotype ♂, Augusta, Western Australia (MCZC).
Distribution: WA.
Nerthra hylaea Todd, 1960: 184. Holotype ♀, Beech Forest, Victoria (MVM).
Distribution: NSW, VIC.
Nerthra luteovaria (Distant, 1904: 63). Holotype ♂, Townsville, Queensland (BMNH).
Distribution: NT, QLD, WA.
Nerthra macrothorax (Montrouzier, 1855: 110). Holotype (probable), Woodlark Is. (MNHN).
Distribution: NT. - EL: Oriental Reg., New Guinea, Micronesia, Solomon Is., Tonga Is.
Nerthra monteithi Cassis & Silveira, 2002: 167. Holotype ♂, Bunya Mountains, Queensland (QMB).
Distribution: QLD.
Nerthra nudata Todd, 1955: 425. Holotype ♂, Brisbane, Queensland (AMNH).
Distribution: NSW, QLD, VIC.
Nerthra plauta Todd, 1960: 186. Holotype ♀, Ooldea, South Australia (SAMA).
Distribution: SA.
Nerthra polhemi Cassis & Silveira, 2001: 27. Holotype ♂, Picadilly Circus, Australian Capital Territory (ANIC).
Distribution: ACT.
Nerthra probolostyla Todd, 1960: 193. Holotype ♂, Tolga, Queensland (NHRS).
Distribution: NT, QLD.
Nerthra sinuosa Todd, 1955: 440. Holotype ♀, Dorrigo, New South Wales (AMNH).
Distribution: NSW, QLD.
Nerthra soliquetra Todd, 1960: 183. Holotype ♂, Stanthorpe, Queensland (USNM).
Distribution: QLD.
Nerthra stali (Montandon, 1899b: 776). Holotype ♀, "Australia occidentalis" (NHRS).
Distribution: WA.
Nerthra suberosa (Erichson, 1842: 285). Lectotype ♂ (Nieser 1977: 302), "Vandiemensland" [Tasmania] (ZMHB).
Distribution: TAS.
Nerthra tasmaniensis Todd, 1955: 437. Holotype ♂, Lake St Clair Res[erve], Tasmania (BMNH).
Distribution: TAS.
Nerthra tuberculata (Montandon, 1899a: 403). Holotype (probable), Port Denison, Western Australia (ISNB).
Distribution: WA.
Nerthra walkeri Todd, 1955: 439. Holotype ♂, Adelaide River, Northern Territory (BMNH).
Distribution: NT.

Superfamily NOTONECTOIDEA Latreille, 1802

Family NOTONECTIDAE Latreille, 1802

Subfamily ANISOPINAE Hutchinson, 1929

Anisops ayersi Reichardt, 1982: 366. Holotype ♂, Initi waterhole, Ayers Rock, Northern Territory (ANIC).
Distribution: NT.
Anisops barrenensis Brooks, 1951: 402. Holotype ♂, Barron River [as Barren River], Queensland (SEMC).
Distribution: QLD.
Anisops baylii Lansbury, 1995b: 186. Holotype ♂,

Lake Mt Brown, Western Australia (WAM).
Distribution: WA

Anisops calcaratus Hale, 1923: 416. Lectotype ♂ (Brooks, 1951: 461), Bordertown, South Australia (SAMA).
Distribution: ACT, NSW, QLD, SA, VIC.

Anisops canaliculatus Brooks, 1951: 367. Holotype ♂, Barron River [as Barren River], Queensland (SEMC).
Distribution: QLD.

Anisops deanei Brooks, 1951: 381. Holotype ♂, Bogan River [as Began River], New South Wales (SEMC).
Distribution: NSW, NT, QLD, SA, TAS, VIC, WA.

Anisops doris Kirkaldy, 1904: 112. Syntypes ♂♀, Alexandra [as Alexandria], Victoria (probably SEMC).
Distribution: ACT, NSW, QLD, SA, TAS, VIC, (WA).

Anisops dostini Lansbury, 1991: 110. Holotype ♂, Gulungul billabong, Northern Territory (NTMD).
Distribution: NT

Anisops douglasi Lansbury, 1984: 37. Holotype ♂, Port Warrender, Kimberley Region, Western Australia (WAM).
Distribution: WA.

Anisops elkedraensis Lansbury, 1995a: 70. Holotype ♂, Elkedra River, Northern Territory (NTMD).
Distribution: NT.

Anisops elstoni Brooks, 1951: 326. Holotype ♂, Myponga, South Australia (SEMC).
Distribution: NSW, QLD, SA, TAS, VIC, WA. - EL: Palearctic and Oriental Region, Papua New Guinea, Melanesia.

Anisops evansi Brooks, 1951: 350. Holotype ♂, Lake Dukerton, Tasmania (SEMC).
Distribution: TAS.

Anisops gratus Hale, 1923: 413. Holotype ♂, Broken Hill, New South Wales (SAMA).
Distribution: NSW, NT, QLD, SA, TAS, VIC, WA.

Anisops hackeri Brooks, 1951: 331. Holotype ♂, Brisbane, Queensland (SEMC).
Distribution: NSW, QLD, TAS, WA.

Anisops hayesi Lansbury, 1995a: 72. Holotype ♂, Alice Springs, Northern Territory (OXUM).
Distribution: NT.

Anisops hyperion Kirkaldy, 1898: 141. Syntypes ♂, Rockhampton, Queensland (ZMUH).
Distribution: NSW, QLD, SA, VIC, WA. - EL: Oriental Region, Micronesia, Melanesia.

Anisops malkini Brooks, 1951: 349. Holotype ♂, Darwin, Northern Territory (USNM).
Distribution: NT, WA.

Anisops nabillus Lansbury, 1969: 451. Holotype ♂, 14 miles W of Marillana [as Marilana], Western Australia (WAM).
Distribution: WA.

Anisops nasutus Fieber, 1851b: 484. Holotype ♂, type locality unknown (ZMHB).
Distribution: NT, QLD, WA. - EL: Oriental Region, Micronesia, Papua New Guinea.

Anisops nodulatus Brooks, 1951: 336. Holotype ♂, Pangasinan, Philippines (SEMC).
Distribution: NT, QLD. - EL: Philippines, Papua New Guinea.

Anisops occipitalis Breddin, 1905a: 152. Lectotype ♀ (Brooks 1951: 345), Buitenzorg, Java (DEIC).
(Syn. *Anisops ocularis* Hale, 1923: 412).
Distribution: NT, QLD, WA. - EL: Oriental Region, Papua New Guinea, New Caledonia.

Anisops paracrinitus Brooks, 1951: 329. Holotype ♂, Coen, Queensland (MCZC).
Distribution: NT, QLD. - EL: Indonesia.

Anisops paraexigerus Lansbury, 1964: 55. Holotype ♂, Longreach, Queensland (SAMA).
Distribution: NSW, QLD, SA.

Anisops philippinensis Brooks, 1951: 383. Holotype ♂, Lake Linau, Mt Apo, Davao Prov., Mindanao, Philippines (FMNH).
Distribution: QLD. - EL: Philippines.

Anisops planifascies Lansbury, 1975: 19. Holotype ♂, Hahn River, 112 km S of Coen, Queensland (MVM).
Distribution: QLD.

Anisops semitus Brooks, 1951: 366. Holotype ♂, Barron River [as Barren River], Queensland (SEMC).
Distribution: QLD, WA.

Anisops stali Kirkaldy, 1904: 132. Syntypes, "Australia" (MNHN).
Distribution: ACT, NSW, NT, QLD, SA, WA. - EL: –Palearctic and Oriental Region, Melanesia.

Anisops tahitiensis Lundblad, 1935: 121. Holotype (probable), Tahiti (BMNH).
Distribution: QLD. - EL: Oriental and Palearctic Region, Micronesia, Papua New Guinea, Melanesia, Polynesia.

Anisops tasmaniensis Brooks, 1951: 378. Holotype ♂, Lake Leake, Tasmania (SEMC).
Distribution: TAS.

Anisops thienemanni Lundblad, 1933: 167. Holotype (probable), Dieng plateau, Java (NHRS).
Distribution: ACT, NSW, NT, QLD, SA, TAS, VIC, WA. - EL: Indonesia (Java).

Anisops ungarinyin Lansbury, 1995a: 72. Holotype ♂, Kimberley, Western Australia (OXUM).
Distribution: WA.

Anisops windi Brooks, 1951: 392. Holotype ♂, Barren [Barron] River, Queensland (SEMC).
Distribution: QLD.

Paranisops endymion (Kirkaldy, 1904: 133). Holotype ♂, Swan River, Western Australia (BMNH).
Distribution: WA.
Paranisops inconstans Hale, 1924: 463. Holotype ♂, Berrowa Creek, New South Wales (SAMA).
Distribution: NSW, QLD.
Walambianisops wandjina Lansbury, 1984: 42. Holotype ♂, Port Warrender, Kimberley, Western Australia (WAM).
Distribution: WA.

Subfamily NOTONECTINAE Latreille, 1802

Enithares atra Brooks, 1948: 48. Holotype ♂, Rigo, New Guinea (USNM).
Distribution: NT. - EL: Papua New Guinea.
Enithares gwini Lansbury, 1984: 40. Holotype ♂, Port Warrender, Kimberley, Western Australia (WAM).
Distribution: WA.
Enithares hackeri Hungerford, 1940: 130. Holotype ♂, Brisbane, Queensland (SEMC).
Distribution: ACT, NSW, QLD, VIC. - EL: Papua New Guinea.
Enithares loria Brooks, 1948: 45. Holotype ♂, Rigo, New Guinea (SEMC).
Distribution: NT, QLD, SA. - EL: Papua New Guinea, Solomon Is.
Enithares woodwardi Lansbury, 1968: 398. Holotype ♂, Brisbane, Queensland (QMB).
Distribution: ACT, NSW, NT, QLD, SA, TAS, VIC, WA. - EL: New Guinea, Solomon Is.
Notonecta (Enitharonecta) handlirschi Kirkaldy, 1897b: 408. Holotype ♂, "Australia" (NHMW).
Distribution: WA.
Nychia sappho Kirkaldy, 1901: 809. Holotype (probable), New Guinea (BPBM?).
(Syn. *Nychia marshalli* Kirkaldy var. *atavia* Hale, 1925: 17).
Distribution: NT, QLD, WA. - EL: Oriental Region, New Guinea.

Superfamily PLEOIDEA Fieber, 1851

Family PLEIDAE Fieber, 1851

Paraplea brunni (Kirkaldy, 1898: 141). Holotype (probable), Rockhampton, Qld. (MNH).
(Syn. *Plea australis* Horvath, 1918: 145).
Distribution: NSW, NT, QLD, SA, TAS, WA.
Paraplea halei (Lundblad, 1933: 144). Holotype ♂, Myponga, SA (NHRS).
Distribution: SA, TAS, VIC.
Paraplea liturata (Fieber, 1844: 19). Holotype (probable), India [as Ostindien] (NHMW).
(Syn. see Lundblad 1933: 129)
Distribution: NT, WA. - EL: Oriental Region, New Caledonia.

Abbreviations

Acronyms for repositories for type material

AMNH	American Museum of Natural History, New York, U. S. A.
AMS	Australian Museum, Sydney, NSW.
ANIC	Australian National Insect Collection, CSIRO Entomology, Canberra.
BMNH	The Natural History Museum (formerly British Museum, Natural History), London, U. K.
BPBM	Bernice P. Bishop Museum, Honolulu, U. S. A.
CAS	Californian Academy of Science, San Francisco, U. S. A.
DEIC	Deutsches Entomologisches Institut, Eberswalde, Germany.
FMNH	Field Museum of Natural History, Chicago, Illinois, U. S. A.
HNHM	Hungarian Natural History Museum, Budapest, Hungary.
ISNB	Instutute Royal des Sciences Naturelle de Belgique, Brussels, Belgium.
IZBE	Institute of Zoology and Botany, Tartu, Estonia.
IZW	Zoological Museum, Polish Academy of Sciences, Warsaw, Poland.
JTPC	John T. Polhemus collection, Englewood, Colorado, U.S.A.
MCZC	Museum of Comparative Zoology, Harvard University, Cambridge, Massachusetts, U. S. A.
MNHN	Muséum National d'Histoire Naturelle, Paris, France.
MRAC	Musée Royal de l'Afrique Centrale, Tervuren, Belgium.
MVM	Museum of Victoria, Melbourne, VIC.
NHMW	Natural History Museum, Wien, Austria.
NHRS	Naturhistoriska Riksmuseet (Natural History Museum), Stockholm, Sweden.
NTDPIF	Northern Territory Department of Primary Industry and Fisheries, Darwin, NT.
NTMD	Northern Territory Museum, Darwin, NT.
OXUM	Oxford University Museum, Hope Entomological Collections, Oxford, U. K.
QMB	Queensland Museum, Brisbane, QLD.
SAMA	South Australian Museum, Adelaide, SA.
SEMC	Snow Entomological Museum, University of Kansas, Lawrence, Kansas, U. S. A.
TDA	Department of Agriculture, Hobart, TAS.
UQIC	University of Queensland Insect Collection, Brisbane, QLD.
USNM	National Museum of Natural History, Smithsonian Institution, Washington, D. C., U. S. A.
WAM	Western Australian Museum, Perth, WA.
ZMHB	Zoologisches Museum, Humboldt Universität, Berlin, Germany.
ZMUC	Zoological Museum, University of Copenhagen, Denmark.
ZMUH	Zoologisches Museum, Universität Hamburg, Germany.
ZSMC	Zoologisches Staatssammlung, Münich, Germany.

Geographical locations

ACT	Australian Capital Territory.
EL	Extra-limital (outside Australia).
NSW	New South Wales.
NT	Northern Territory.
QLD	Queensland.
SA	South Australia.
TAS	Tasmania.
VIC	Victoria.
WA	Western Australia.

Index

The index contains names of subgenera, genera, and higher taxa as well as morphological terms and subjects treated in this book. Principal entries of taxa and subjects are indicated by **bold face** page numbers.